Designer Biochar Assisted Bioremediation of Industrial Effluents

This book provides useful information and applications of biochar produced from agricultural waste for removal of contaminants from industrial effluent and reutilization of waste sludge in the production of biofuel/bioenergy. It describes how designer or modified biochar or combined application (biochar + microbes) can be applied successfully for reuse of wastewater and contaminated soil for ecorestoration, environment protection, and sustainable development. It also deals with the unique features, advantages, and disadvantages of techniques for biochar production and analyses. It underlines a road map in development of future strategy for pollution abatement and sustainable development.

Features:

- Provides exhaustive coverage of biochar and its production and properties.
- Highlights use of biochar in pollution control and environment protection.
- Covers use of agricultural waste/waste biomass for dye decolorization and degradation.
- Explores synergistic approaches for contaminants removal for better insights into basic and advanced biotechnological applications.
- Describes how biochar treatment can be successfully applied for reuse of wastewater and contaminated soil ecorestoration and environment protection.

This book is aimed at graduate students and researchers in chemical/biochemical engineering, biotechnology, environmental sciences/engineering, and agriculture engineering.

Designer Biochar Assisted Bioremediation of Industrial Effluents

A Low-Cost Sustainable Green Technology

Edited by
Riti Thapar Kapoor and Maulin P. Shah

CRC Press
Taylor & Francis Group
Boca Raton London New York

CRC Press is an imprint of the
Taylor & Francis Group, an **informa** business

First edition published 2023
by CRC Press
6000 Broken Sound Parkway NW, Suite 300, Boca Raton, FL 33487–2742

and by CRC Press
4 Park Square, Milton Park, Abingdon, Oxon, OX14 4RN

CRC Press is an imprint of Taylor & Francis Group, LLC

ISBN: 978-1-032-06694-3 (hbk)
ISBN: 978-1-032-06696-7 (pbk)
ISBN: 978-1-003-20343-8 (ebk)

DOI: 10.1201/9781003203438

Typeset in Times
by Apex CoVantage, LLC

Contents

Editor Biographies

Riti Thapar Kapoor is Associate Professor in Amity Institute of Biotechnology, Amity University Uttar Pradesh, Noida, India. Dr. Kapoor received her PhD from the University of Allahabad and worked as postdoctoral fellow at Banaras Hindu University, Varanasi, India. Dr. Kapoor has 14 years of teaching and research experience, and her area of specialization is environmental biotechnology, bioremediation, and abiotic stresses. Dr. Kapoor is author of three books and editor of five books of international repute (Elsevier, Springer, DeGruyter, etc.). She has published over 80 research papers in various journals of national and international repute. Dr. Kapoor has visited eight countries for participation in different academic programs. Dr. Kapoor has received a prestigious travel award from Bill & Melinda Gates Research Foundation (CGIAR project) for participation in an international training program held at International Rice Research Institute (IRRI), Manila, Philippines in 2010. She is also recipient of a DST travel grant for participation in an International Conference held at Sri Lanka in 2013. Dr. Kapoor has been awarded a Teacher's Research Fellowship from the Indian Academy of Sciences, Bengaluru, in 2019.

Maulin P. Shah is an active researcher and scientific writer in his field for over 20 years. He received a BSc degree (1999) in Microbiology from Gujarat University, Godhra (Gujarat), India. He also earned his PhD degree (2005) in Environmental Microbiology from Sardar Patel University, Vallabh Vidyanagar (Gujarat), India. His research interests include Biological Wastewater Treatment, Environmental Microbiology, Biodegradation, Bioremediation, and Phytoremediation of Environmental Pollutants from Industrial Wastewaters. He has published more than 250 research papers in national and international journals of repute on various aspects of microbial biodegradation and bioremediation of environmental pollutants. He is the editor of 90 books of international repute (RSC, Elsevier, Springer, DeGruyter, Wiley, and CRC Press).

Contributors

Abhishek, Kumar
Centre for Oceans, Rivers, Atmosphere and
    Land Sciences
Indian Institute of Technology
    Kharagpur
Kharagpur, West Bengal, India

Bhattacharjee, Abhik
Department of Chemical Engineering
Indian Institute of Technology Guwahati
Guwahati, Assam, India

Chakraborty, Arun
Centre for Oceans, Rivers, Atmosphere
    and Land Sciences
Indian Institute of Technology
    Kharagpur
Kharagpur, West Bengal, India

Das, Pranjal P.
Department of Chemical Engineering
Indian Institute of Technology Guwahati
Guwahati, Assam, India

Dash, Subhakanta
Department of Chemistry
Synergy Institute of Engineering and
    Technology
Dhenkanal, Odisha, India

Dhara, Simons
Department of Chemical Engineering
Indian Institute of Technology
    Guwahati
Guwahati, Assam, India

Duarah, Prangan
Centre for the Environment
Indian Institute of Technology Guwahati
Guwahati, Assam, India

Galyan, Veerupaksh
Department of Life Sciences
School of Basic Sciences and Research
Sharda University
Greater Noida, Uttar Pradesh, India

Ghosh, Dipita
Department of Environmental Science and
    Engineering
Indian Institute of Technology (Indian School
    of Mines)
Dhanbad, Jharkhand, India

Gupta, Namrata
Department of Chemistry
RBS Engineering Technical Campus
Bichpuri, Agra, Uttar Pradesh, India

Gupta, Piyush
Department of Chemistry
Faculty of Engineering and Technology
SRM Institute of Science and Technology
Modinagar, Ghaziabad, Uttar Pradesh,
    India

Jain, Ankur
Centre for Renewable Energy & Storage
Suresh Gyan Vihar University
Mahal Road, Jagatpura, Jaipur, India

Joshi, Sidhika
Department of Life Sciences
School of Basic Sciences and Research
Sharda University
Greater Noida, Uttar Pradesh, India

Kanupriya
Department of Life sciences
School of Basic Sciences and Research
Sharda University
Greater Noida, Uttar Pradesh, India

Kapoor, Riti Thapar
Amity Institute of Biotechnology
Amity University
Uttar Pradesh, Noida, India

Kumar, Ankit
Department of Life Sciences
School of Basic Sciences and Research
Sharda University
Greater Noida, Uttar Pradesh, India

Kumar, Sanjay
Department of Life Sciences
School of Basic Sciences and Research
Sharda University
Greater Noida, Uttar Pradesh, India

Kuttippurath, Jayanarayanan
Centre for Oceans, Rivers, Atmosphere and
    Land Sciences
Indian Institute of Technology Kharagpur
Kharagpur, West Bengal, India

Maiti, Subodh Kumar
Department of Environmental Science and
    Engineering
Indian Institute of Technology (Indian School
    of Mines)
Dhanbad, Jharkhand, India

Mishra, Bikram
Indian Institute of Technology
    Bhubaneswar
Bhubaneswar, Argul
Odisha, India

Mohanty, Swati Sambita
Department of Chemical Engineering
National Institute of Technology Rourkela
Rourkela, Odisha, India

Mondal, Piyal
Department of Chemical Engineering
Indian Institute of Technology Guwahati
Guwahati, Assam, India

Pandit, Soumya
Department of Life Sciences
School of Basic Sciences and Research
Sharda University
Greater Noida, Uttar Pradesh, India

Prasad, Kamal
Absolute Foods
Plant Science Research, Division of Microbiology
Gurugram, Haryana, India

Purkait, Mihir K.
Department of Chemical Engineering
Indian Institute of Technology Guwahati
Guwahati, Assam, India

Raghudhas, Rahul
Centre for Oceans, Rivers, Atmosphere and
    Land Sciences
Indian Institute of Technology Kharagpur
Kharagpur, West Bengal, India

Rana, Neha
Department of Pharmacy
SRM Modinagar College of Pharmacy
SRM Institute of Science and Technology
Modinagar, Ghaziabad, Uttar Pradesh, India

Remya, Neelancherry
Indian Institute of Technology Bhubaneswar
Bhubaneswar, Argul, Odisha, India

Sahni, Mohit
Department of Physics
School of Basic Sciences and Research
Sharda University
Greater Noida, Uttar Pradesh, India

Sahu, Sai Shankar
Department of Environmental Science and
    Engineering
Indian Institute of Technology (Indian School
    of Mines)
Dhanbad, Jharkhand, India

Salar, Sapna
Department of Pharmaceutical Sciences
Apex University
Jaipur, Rajasthan, India

Samanta, Niladri Shekhar
Centre for the Environment
Indian Institute of Technology Guwahati
Guwahati, Assam, India

Shah, Maulin P.
Industrial Waste Water Research Laboratory
Division of Applied and Environmental
    Microbiology
Enviro Technology Limited
Ankleshwar, Gujarat, India

Shankara Narayanan, S.
Department of Physics
School of Basic Sciences and Research
Sharda University
Greater Noida, Uttar Pradesh, India

Sharma, Mukesh
Centre for Oceans, Rivers, Atmosphere and
  Land Sciences
Indian Institute of Technology Kharagpur
Kharagpur, West Bengal, India

Sharma, V. P.
CSIR–Indian Institute of Toxicology Research
Lucknow, Uttar Pradesh, India

Shrivastava, Kriti
School of Applied Sciences
Suresh Gyan Vihar University
Jagatpura, Jaipur, India

Singh, Ajay Kumar
A G Bio Systems Private Limited
Cherlapally, Hyderabad, Telangana, India

Singh, Vandana
Department of Allied Health Science
Sharda University
Greater Noida, Uttar Pradesh, India

Sontakke, Ankush D.
Department of Chemical Engineering
Indian Institute of Technology Guwahati
Guwahati, Assam, India

Tiwari, Shreya
Department of Chemical Engineering
Indian Institute of Technology Guwahati
Guwahati, Assam, India

Preface

Water is a natural treasure, and the availability of safe and clean water is essential for human health, the ecosystem, and sustainable development. Water resources are getting scarcer due to the exponential increase in population, agriculture, urbanization, and industrialization. Industries consume large quantities of water that are returned to the environment as wastewater. The continuous decline in the groundwater table and deterioration of water quality are matters of serious concern. Dyes are used in various industries like the textile, paper, leather, rubber, plastic, cosmetics, food, pharmaceutical, and petroleum industries, among others. During the dyeing process, a significant amount of dyes remain unbound and are released with wastewater as they are resistant to fading due to their complex structure. The lack of proper treatment facilities has proliferated the discharge of effluents enriched with toxic pollutants such as dyes, heavy metals, organic compounds, and other hazardous chemicals in the environment. Various physicochemical strategies have been devised to remove contaminants from industrial effluents, but due to the high cost, low efficiency, and sludge generation, these methods are not feasible at a large scale. However, biotechnological approaches have attracted worldwide attention for their relative cost-effectiveness and environmentally friendly nature. Most biotechnological approaches rely on the use of microbes that have the potential to enzymatically degrade contaminants present in the industrial effluents. However, there are still many challenges in scaling up microbial and enzymatic technologies for the decontamination of wastewater containing metals/metalloids, salts, and other toxic compounds.

The burning of agro-waste is another major problem that produces thick smoke in the air with charred carbon particles, inflicting severe respiratory and cardiovascular problems. Agro-waste burning releases harmful gases with hazardous chemicals such as polycyclic aromatic hydrocarbons and polychlorinated dibenzofurans into the atmosphere that have toxicological effects. The burning of crop residues raises topsoil temperature, which destroys all the potential microbes present in the soil. Therefore, the development of economical and efficient control measures against pollution is imperative to safeguard our ecosystem and natural resources. Agro-wastes can be converted into biochar via a pyrolysis process as they are available at zero cost and can be used in wastewater treatment, removal of heavy metals, pesticides, pharmaceutical compounds, and toxic substances present in industrial effluent. The adsorption of contaminants by the application of synergistic approaches, i.e., the use of biochar prepared from agricultural biomass and microbes has received considerable attention because of the flexibility, high efficiency, renewability, faster contaminant removal rate, and ability to treat concentrated effluent. The main objective of the proposed book is to provide useful information on the application of biochar, alone or in combination with microbes, to academicians, agronomists, scientists, and environmentalist working in the field of environment protection, bioremediation, effluent treatment, and waste management, and related areas.

1 Recent Advances in Biochar Production and Its Applications toward Textile Industry Effluent Treatment

Ankush D. Sontakke, Shreya Tiwari, Abhik Bhattacharjee, Piyal Mondal, and Mihir K. Purkait

CONTENTS

1.1   INTRODUCTION

Water is essential to human and animal existence, serving as a foundation for climate protocols and natural ecosystems. Water shortage has jeopardized human health and the quality of life. The fast expansion of the human population and environmental degradation caused by rapid industrialization

DOI: 10.1201/9781003203438-1

are major contributors to the water crisis. For example, for textile industries, it has been stated that almost 7×10^5 tons of dye-related products are produced worldwide, and for each ton of the product, approximately 200–350 m^3 of contaminated water are inevitably generated (Sontakke et al. 2021). As a result, it has become necessary to manage freshwater resources and develop wastewater treatment technologies in order to maintain adequate quantity and quality of water. However, the agricultural sector uses most of the fresh water, followed by the industrial and household sectors. It has been noted that the shortage of fresh water combined with the rapid growth in its use may result in severe water stress in most European nations. Presently, humankind is facing a water crisis throughout the world (Pendolino and Armata 2017).

Meanwhile, the scarcity of water resources necessitates the development of effective wastewater decontamination technologies and saltwater desalination technologies. Wastewater treatment and water purification include conventional and nonconventional methods. The conventional methods comprise disinfection by oxidation/ozonation (Das et al. 2021), carbon adsorption, chemical precipitation (Deepti et al. 2020), membrane filtration (Yaranal, Subbiah, and Mohanty 2020; Sontakke and Purkait 2020; Sontakke and Purkait 2021), and chemical or biological degradation (Baláž 2014; Smith and Rodrigues 2015). The well-known adsorption technology is based on the chemical and physical properties of the adsorbent. The choice of adsorbent material is based on its characteristics, accessibility, cost, and toxicity to human health and the environment. Recent advancements in adsorption materials have resulted in a greater capacity to adsorb developing pollutants. Additionally, newly developed nanomaterials with higher surface area and characteristics have presented novel water treatment approaches. The combination of removal techniques with novel materials such as biochar is a potential methodology for enhancing water quality and reuse (Upadhyay, Soin, and Roy 2014; Pendolino and Armata 2017).

Biochar is regarded as an eco-friendly and low-cost material, mainly produced from organic wastes like forest residues, biomass, and agricultural and municipal waste. Biochar with their material characteristic and improved applicability have garnered great attention in various environmental remediation applications. Organic waste can be transformed into biochar via numerous techniques such as pyrolysis, gasification, liquefaction, hydrothermal carbonization (HTC), and torrefaction. However, pyrolysis is considered the conventional carbonization method, whereas gasification HTC and torrefaction methods do not meet the definition of biochar provided by the guidelines of the European Biochar Certificate (EBC). Biochar, along with its activated derivatives, has superior surface characteristics such as higher surface area, carbon content, stable structure, and good ion exchange capacity. These properties make biochar a prominent material for removing several water contaminants, including organic dyes, inorganic heavy metals, and pathogens (Mohan et al. 2014; Enaime et al. 2020; Dey, Singh, and Purkait 2014). Previously, most of the studies are mainly focused over the application of biochar in soil modifications for adsorption of inorganic nutrients and enhancement in the soil quality. Various research works have shown a great interest for biochar in enhancing soil properties and crop yields, which eventually contributes to the reduction in greenhouse gases via soil carbon sequestration. In recent years, several efforts have been made for development in the production of biochar in order to expand its applicability in multidisciplinary areas like wastewater treatments. Biochar research has been conducted in various nations for its diverse and broad applications, which are based on feedstock availability, local environment and economy, production and functionalization methods (X. Wang et al. 2020).

Meanwhile, the studies related to wastewater decontamination via biochar are related chiefly to batch experiments. However, the information related to the optimization and design of biochar-based processes for water purification is limited. The absorption capacity of biochar is a decisive parameter for its applicability in the removal of organic and inorganic pollutants from wastewater. Although, the adsorption capacity of biochar is dependent on the physicochemical properties such as surface area, functional groups, elemental composition, and pore structure and size. However, these physicochemical properties vary with the nature and quality of feedstock and methods of preparation. The functionalization or modification of biochar may assist in improving these properties and provides additional benefits by providing greater efficiency for the removal of water contaminants. Functionalization of biochar can be done via chemical processes such as oxidative functionalization, acid and alkaline treatment, as well as via physical methods like gas purging (Enaime et al. 2020). Moreover, the inherent characteristics of biochar make them an ideal candidate for the adsorptive removal of organic dyes and heavy metals from the water. Biochar is an established adsorbent with viable nontoxicity and environmental friendliness for the adsorption of various types of water pollutants. However, it was discovered that, depending on the type of feedstock and the manufacturing procedures used, biochar might include numerous heavy metals and other contaminants that could be released upon its application in aqueous media. Despite the fact that biochar has demonstrated vast usage visions for environmental remediation due to advancements in manufacturing methods and functionalization or modification possibilities, it is also necessary to assess its negative repercussions (Enaime et al. 2020). Therefore, biochar stability and its correlations with experimental conditions during production need to be investigated.

Considering the significance and utilization of biochar for various applications such as environmental remediation and wastewater treatment, this chapter discusses in detail the feedstocks and latitudes of biochar production processes. The present chapter also explores the factors affecting the synthesis of biochar, along with various functionalization or modification approaches for improving the effectiveness of biochar. Moreover, the recent advancements and application of biochar toward textile industry effluent treatments to remove dyes and heavy metals are also discussed in detail. Further, the scope of enhancements for improving the overall efficiency of biochar as an adsorbent has been highlighted.

1.2 ADVANTAGES AND PROPERTIES OF BIOCHAR FOR ENVIRONMENTAL REMEDIATION

Biochar is an eco-friendly and low-cost adsorbent material synthesized from organic wastes and has attracted significant attention for its application in environmental remediation. It provides the same benefits as those of organic materials but with a more perpetual alternative. Biochar has the ability to attract and retain water and nutrients like phosphorous and nitrogen, which leads to an increase in the fertility of the soil. Besides, biochar can sequester carbon and absorb greenhouse gases (GHG). It also generates gas and oil as by-products and can be used as a sustainable energy source. Biochar makeup's properties of higher porosity and surface area affect its sorption capacity. During the pyrolysis process of biomass, dehydration occurs and leads to the formation of microvoids within the biochar. The biochar exists in different pore sizes such as micro (< 2 nm), macro (> 50 nm), and nano (< 0.9 nm). The temperature of the pyrolysis process significantly affects the porosity and surface area of the biochar. It was observed that increased temperature also increases the surface area and porous structure of the biochar. The structure of the biochar is changing upon heating to an aromatic structure, which is beneficial for microbial decomposition. Biochar is mainly comprised of moisture, labile and fixed carbon, ash, and volatile matter (Oni, Oziegbe, and Olawole 2019). As stated, biochar, along with its activated derivatives, has superior surface characteristics such as higher surface area, carbon content, stable structure, and good ion exchange capacity. The cation exchange property of biochar showed a better capability for adsorption of cations such as Ca^{2+}, NH_4, Pb^{2+}, Cr^{4+}. As textile industry effluent consists of cationic dyes and heavy metals, biochar can be a suitable candidate for the adsorption of these pollutants. The ion exchange capacity of biochar is dependent on its structure, surface area, and functional groups, which provide the surface charge to the biochar. To determine the number of exchangeable ions, the material must be added to the solution, which is later affected by the pH of the solution and the type of solvent. Higher-pH solutions are subject to higher cation exchange capacity. It was observed that the cation exchange capacity of biochar is higher for lower pyrolysis temperatures and that the surface area is higher than that of feedstock for the identical temperature. The retardation in the cation exchange capacity of biochar at higher temperatures is subject to the aromatization of biochar along with the deterioration in the functional group sites (Rahman et al. 2018; Oni, Oziegbe, and Olawole 2019).

Moreover, environmental stability is a significant characteristic of biochar for its long-term amendment in ecological remediation application. The stability of biochar depends on the type of feedstock and pyrolysis temperature. It was observed that the higher temperature of the pyrolysis process resulted in enhanced biochar stability (Mašek, Buss, and Sohi 2018). Biochar has shown its utilization in a wide range of applications related to wastewater treatment due to its unique characteristics. The major properties of biochar include adsorption capacity, higher surface area, ion exchange capacity, and microporosity. The adsorption mechanism of biochar is directed by pyrolysis temperature, type of feedstocks, and its interaction with different pollutants. These aspects are discussed in subsequent sections.

1.3   FEEDSTOCKS AND CHALLENGES FOR BIOCHAR PRODUCTION

Biochar, with its unique characteristics, has shown a superior advancement in the field of biomass utilization and application in soil enrichment and environmental remediations. These characteristics of biochar are primarily affected by the type of feedstock and the pyrolysis conditions. However, the rate of heat transfer and temperature during pyrolysis and the process residence time are essential factors in producing biochar with desirable characteristics such as increased surface area, porosity, adsorption capacity, and ion exchange capacity. It was discovered that various feedstocks resulted in varying magnitudes of pores, functional sites, and surface area in the biochar, which substantially influence its adsorption capacity. The feedstocks used for biochar production comprise agricultural waste such as rice husk, wheat straw, bagasse, empty fruit bunches, sugar beet tailing, poultry manure, and other organic debris. Alongside forest biomass, plant residues, dairy manure, woodchips, pinewood, human manure, municipal waste, and sewage sludge were also being utilized for the production of biochar (Oni, Oziegbe, and Olawole 2019; Duarah, Haldar, and Purkait 2020). However, the feedstocks from industrial waste and sewage sludge must be evaluated before being used for biochar synthesis since they should not contain heavy metals, as these end up in the final product biochar. The adverse effects of enhancement in competition for the use of agricultural land and rising food prices may cause a threat to biodiversity. Therefore, it is essential to adopt all the possible feedstock supply channels for biochar production to maintain sustainability and diversity. In developing nations, biomass is frequently used as a heating fuel for cooking. The biomass and municipal sludge, crop residues, and agroindustrial by-products are considered sustainable feedstocks available globally to achieve the next-level production of biochar. Prior to pyrolysis, the feedstock must be dehydrated, which may be accomplished by utilizing the alternative energy obtained from the previous batch of the pyrolysis process. Depending on the choice of feedstock and the storage and dehydration process, a series of advancements in energy utilization, labor, and dehydrators must be made (Ghodake et al. 2021).

The biochar obtained from animal waste persists less specific pore areas as compared to plants at identical pyrolysis conditions. The reason could be higher ash and inorganic content in animal manure. The quality of feedstock depends on its composition, which includes fixed carbon, nitrogen, calcium, potassium, ash content, moisture, and volatile matter, which can be determined via elemental and proximate analysis of feedstocks. Although the pyrolytic conditions also greatly affect the nutrient characteristics, biochar must be tested under different operating conditions to determine the specific properties of the biochar. The properties of biochar obtained from various feedstocks at different operating conditions are listed in Table 1.1 (Oni, Oziegbe, and Olawole 2019).

To accomplish the aims of efficient biochar synthesis, supply variables such as feedstock type, storage, collection, handling, and analysis of feedstocks must be carefully examined. The goals of biochar production include fine-tuning biochar characteristics, decreasing environmental emissions

TABLE 1.1
Properties of Biochar Obtained from Various Feedstocks

Feedstock	Heating rate (°C/min)	Temp. (°C)	Atm.	Residence time (h)	pH	Ash content (%)	Surface area (m²/g)	% Yield
Rice husk	10	400	Limited oxygen	5	10.1	23	55.9	33.5
Wood bark	5	500	N_2	4	11.9	27.6	67.5	27.8
Empty fruit bunches	7	600	N_2	6	9.2	24.2	102.9	37.4
Sugar beet tailing	10	400	N_2	4	10.7	11.1	40.1	26.2
Dairy manure	5	500	N_2	4	8.8	14.7	44.4	33.5
Wood chips	7	400	N_2	6	12.9	22.9	39.4	29.2
Plant residue	7	400	N_2	5	10.5	15.5	84.0	31.8
Organic waste	4	500	N_2	5	9.4	13.4	55.4	22.1
Poultry manure	500	500	N_2	4	9.6	22.7	51.0	23.7

during pyrolysis, improving process economics, and improving coproducts. The significant challenges related to biochar feedstocks are their availability, cost-effective transport, the uncertainty of cost and quality, and a requirement for pretreatment processes. These challenges can be subsided via blending, comminution, pelleting, a combination of chemical and physical pretreatments, and a supply of more reliable, higher-volume, and cheaper biomass (Ghodake et al. 2021). Besides, several other factors must be considered for the sustainability of biochar production via pyrolysis, such as biomass choice, yield, and quality. The details related to various production processes, factors affecting, and the functionalization methodologies for biochar are discussed in subsequent sections.

1.4 METHODS FOR THE SYNTHESIS OF BIOCHAR

Biochar can be synthesized through biochemical or thermochemical routes. However, thermochemical methods are more commonly employed, and biochar is primarily synthesized using either pyrolysis, carbonization, gasification, or torrefaction method. Each of these techniques has its advantages, and moreover the yield of the product depends on parameters such as quality of biomass and reaction conditions. Some commonly employed synthesis techniques and the parameters that influence the production of biochar are discussed next.

1.4.1 PYROLYSIS

Pyrolysis is the most widely used method for converting biomass into biochar via a thermochemical breakdown process in an oxygen-depleted environment at high temperatures (Chatterjee et al. 2020). When any organic material undergoes pyrolysis, its lignocellulosic components such as cellulose, hemicellulose, and lignin go through depolymerization, fragmentation, and cross-linking at specified temperatures, which results in char or bio-oil (Yaashikaa et al. 2020). The properties of biochar, like its composition, surface area, surface functional groups, and pore structures, depend mainly on the temperature conditions (Chatterjee et al. 2020). Based on the temperature and heating rate, pyrolysis is subdivided into fast pyrolysis and slow pyrolysis. Fast pyrolysis is characterized by a pyrolysis temperature of 500°C and a high heating rate up to 1000°C/min for short residence

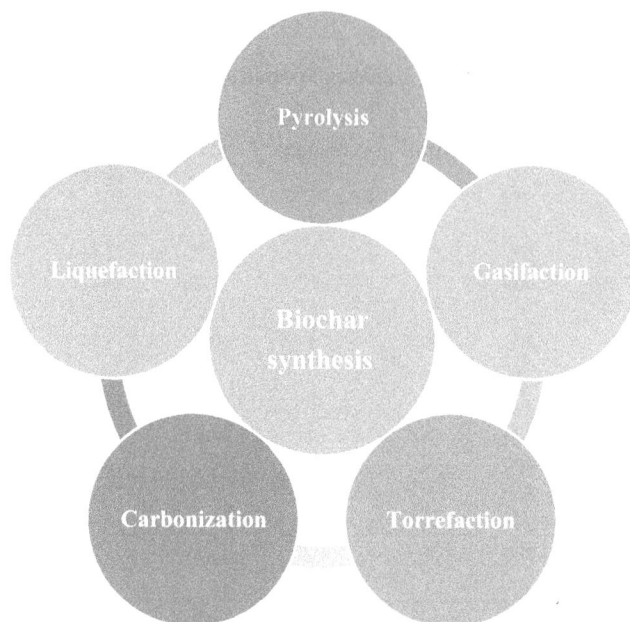

FIGURE 1.1   Biochar production methods.

time (<2 s) (D. Wang et al. 2020). The reactors employed for fast pyrolysis are bubbling and circulating fluidized bed, rotating cone, and ablative reactor (Garcia-Nunez et al. 2017). Slow pyrolysis is a batch process carried out at atmospheric pressure, a long residence time of more than 1 hour, and under lower temperature (350–500°C) and heating rate (1–30°C/min) than fast pyrolysis (Kim, Oh, and Park 2020; Yaashikaa et al. 2020; D. Wang et al. 2020). Agitated drum sand rotating kilns, wagon reactors, and paddle pyrolysis kilns are examples of reactors employed in slow pyrolysis (Garcia-Nunez et al. 2017). Slow pyrolysis results in a high yield of biochar (30–50 wt%) owing to the low heating rate and high residence time of the process, which favors a secondary cracking of vapors (Cheng and Li 2018). During the pyrolysis process, primary cracking, secondary decomposition, and the creation of oxygen functional groups begin approximately at 400–500°C. The ideal temperature for the onset of pyrolysis and the synthesis of char is reported to be 500°C (Yaashikaa et al. 2020).

Biomass mainly comprises cellulose, hemicellulose, and lignin, which follow different mechanisms for conversion into biochar. Cellulose decomposition takes place through the reduction in the degree of polymerization. The high heating rate of fast pyrolysis causes rapid volatilization and the production of levoglucosan. Along with biochar, bio-oil and syngas are also obtained here. Slow pyrolysis causes cellulose degradation due to the long residence time and low heating rate. Slow pyrolysis causes cellulose degradation owing to the long residence time and low heating rate. Hemicellulose also undergoes depolymerization, which follows a series of reactions that include decarboxylation, intramolecular rearrangement, depolymerization, and aromatization. During lignin decomposition, the β-O-4 lignin link breaks, releasing free radicals that absorb protons from other species, resulting in the formation of degraded products (Yaashikaa et al. 2020).

1.4.2   Hydrothermal Carbonization

A hydrothermal process is carried out in a closed reactor in the presence of water and at a temperature above 180°C. It is significantly influenced by temperature conditions; biochar, bio-oil, and gaseous products are obtained below 250°C, between 250–400°C, and above 400°C, respectively (Cheng and Li 2018). The hydrothermal process used to obtain biochar is termed hydrothermal carbonization (HTC). HTC is preferred to pyrolysis for treating biomass with high moisture content. It is also a cost-effective method, as it requires a lower temperature range. HTC product is known as hydrochar, to differentiate it from other biochar production processes, with an excellent yield of about 40–70% (Zhang et al. 2019; Yaashikaa et al. 2020). The reaction temperature, pressure, residence duration, and water-to-biomass ratio are the primary factors that influence the properties of hydrochar.

The process begins with the degradation of cellulose into oligomers, which further gives glucose rather than fructose. The hydrolyzed products give furfural and its derivatives by going through isomerization, dehydration, and fragmentation. These intermediates result in hydrochar by undergoing additional polymerization and condensation reactions and reverse aldol condensation and intermolecular dehydration (Zhang et al. 2019). The biochar produced from hydrothermal carbonization has more surface functional groups, and spherical microsized particles with a limited porosity are obtained. Other advantages of HTC include the facile decomposition of biomass and saving the energy required for drying biomass feed. Thus the process is preferred for treating biomass with high water content, such as aquatic plants (Cheng and Li 2018).

1.4.3 Hydrothermal Liquefaction

Hydrothermal liquefaction (HTL) is a thermochemical process that decomposes biomass macromolecules into energy-dense products by creating a highly reactive aqueous environment at high temperatures (573–623 K) and pressures (5–20 MPa) (Ibrahim et al. 2020). Similar to HTC, HTL is an energy-efficient process. Another advantage of HTL in biomass processing is that it recovers 70% of carbon present in feed as either biochar or bio-oil (Ponnusamy et al. 2020). During HTL, high pressure is maintained to keep water in the liquid state; this condition is considered subcritical.

The mechanism of HTL is described by three primary processes: depolymerization, decomposition, and recombination (Ibrahim et al. 2020). The final products obtained are biochar, bio-oil, or gaseous products, and their yield depends on the type of feedstock and operation parameters. The yield depends on the carbohydrate content of feed; the higher the carbohydrate content, the better the production rate. The operating parameters that affect biochar production are temperature conditions, liquid-to-solid, and residence time. At elevated temperature conditions, lignin degrades, which curbs the solid–solid conversion required for char formation. Thus biochar production decreases as temperature increases. Water, though it decreases the yield of biochar, increases the yield of bio-oil. The yield of biochar is reduced when residence time is increased. Though a shorter residence period causes incomplete degradation, it improves the biochar-associated polymerization reaction. The advantages of hydrothermal liquefaction for biochar production include high absorption capacity even with the low surface area due to the presence of volatile organic matter and functional groups and a high energy densification ratio (Ponnusamy et al. 2020).

1.4.4 Gasification

Gasification is a thermochemical process in which a carbon source is decomposed into a gaseous mixture under high-temperature conditions (>700°C) in the presence of oxidizing agents. The gasification agents can be air, oxygen, steam, or a mixture of these, and the product obtained is syngas containing CO, CO_2, CH_4, H_2, and traces of hydrocarbons (Zhang et al. 2019). The production of syngas depends significantly on the reaction temperature. The production of methane, carbon monoxide, and hydrogen increases with temperature rise, but carbon dioxide and hydrocarbons decrease

(Yaashikaa et al. 2020). Biochar is produced as a by-product of this process, and it has a lower yield of about 10 wt% of the biomass. The low production of biochar owes to the presence of oxidizing agents in the gasification process, which converts carbon into carbon monoxide. Also, the biochar obtained significantly comprises alkali and alkaline earth metals and polyaromatic hydrocarbons that impart high toxicity (Zhang et al. 2019).

The process of gasification starts with drying biomass to remove the moisture content without energy recovery. Drying is followed by oxidation and combustion to obtain the end products (Yaashikaa et al. 2020). Though each component of the biomass decomposes at a distinct temperature, the overall decomposition occurs between 400 and 500°C, where biochar is a major product. At a higher temperature, biochar undergoes secondary decomposition, resulting in syngas containing hydrogen and methane (Gabhane et al. 2020). Apart from temperature, the type of oxidizing agent, particle size, residence time, and gasification agent/biomass ratio are other important factors that influence the yield (Zhang et al. 2019).

1.4.5 TORREFACTION

Torrefaction, also referred to as mild pyrolysis, is a thermal decomposition process of biomass at low temperature and low heating rates. The decomposed biomass produces condensable and non-condensable gases with the end product of carbon-rich biochar (Pulka et al. 2019). With different decomposition processes, the carbon dioxide, oxygen, and moisture present in biomass are removed by utilizing inert atmospheric air in the absence of oxygen at a temperature of 300°C (Yaashikaa et al. 2020). Torrefaction helps in improving the quality of biomass by increasing its energy density by 30%, rendering it hydrophobic, modifying the particle size and surface area, increasing its carbon content, increasing its calorific value, and making it more suitable for gasification (Yaashikaa et al. 2020; Zhang et al. 2019; Pulka et al. 2019).

Torrefaction can be described with steam torrefaction, wet torrefaction, and oxidative torrefaction (Yaashikaa et al. 2020). The biomass is processed with steam at 260°C for a 10-min residence time in steam torrefaction. Wet torrefaction, also known as hydrothermal carbonization, occurs when biomass is exposed to water for 5–240 min at 180°C. In oxidative torrefaction, the biomass is treated with oxidizing agents. Torrefaction efficiency mainly depends on two factors: temperature and retention time. These two parameters must be adjusted for different types of biomasses in such a manner that the treatment process is cost-effective. Torrefaction is generally employed for the treatment of lignocellulose biomass as they tend to have good energy potential (Pulka et al. 2019).

1.4.6 FACTORS AFFECTING BIOCHAR SYNTHESIS

The yield and quality of biochar depend on the conditions during its production process. Temperature, quality and source of biomass, residence time, rate of heating, and conversion technique employed are some of the parameters that affect biochar synthesis. For instance, more biochar can be obtained from animal waste than from agricultural residues.

1.4.6.1 Source and Quality of Feedstock

During the production of biochar, even if the pyrolysis conditions are the same for different quality biomass, the biochar produced will not be the same due to structural and compositional differences in raw materials. Pyrolysis of high lignin biomass, for example, yields a high yield of biochar with a high carbon content (Cheng and Li 2018). Similarly, the moisture content of feedstock also alters the quality of biochar. While high moisture content reduces char formation and has more energy requirement, low moisture feedstock is more energy efficient for pyrolysis

(Yaashikaa et al. 2020). Another example is that hydrochar obtained from HTC of hexoses has a uniform micrometer size, whereas hydrochar from pentoses has a size between 100 and 500 nm (Cheng and Li 2018).

1.4.6.2 Temperature
The most crucial factor affecting biochar production is temperature; as the temperature increases, the yield of biochar is reduced. When the temperature rises, the thermal cracking of hydrocarbons present in biomass also increases. It leads to the formation of liquid and gaseous products along with a reduction in the production of biochar. Subsequently, at lower temperatures, the biomass is not completely decomposed, and thus the yield of biochar is high (Yadav and Jagadevan 2020). However, the physicochemical properties of biochar produced are positively correlated with temperature. When the temperature of pyrolysis is increased, the biochar obtained has higher pH, larger specific surface areas, higher pore volumes, and better stability (S. Li et al. 2020).

1.4.6.3 Residence Time
Longer residence time allows more heat transfer through feed, which affects the production of biochar (S. Li et al. 2020). When the residence time is increased, the prospects of biomass contents to be repolymerized and increased, which enhances the yield of biochar. The porosity of biochar also increases when residence time is increased (Yadav and Jagadevan 2020). High surface area and porosity can be obtained by slow pyrolysis (S. Li et al. 2020).

1.4.6.4 Conversion Techniques
Though the biochemical method of synthesis is more cost-effective and environmentally friendly, the yield of biochar is meager. Hence, the thermochemical technique is preferred (Yadav and Jagadevan 2020). The thermochemical mode conversion plays a vital role in the properties of products obtained. For example, biochar has more carbon content than hydrochar. Also, hydrochar has less surface area because decomposition products are retained on its surface and block the pores (Zhang et al. 2019).

1.4.7 CHARACTERIZATION TECHNIQUES

The biochar obtained from the processes previously discussed must be characterized for investigating and understanding their chemical structure, composition, morphology, and surface characteristics. These characterization techniques include transmission electron microscopy (TEM), scanning electron microscopy (SEM), nitrogen adsorption isotherm, Fourier transform infrared spectroscopy (FTIR), energy dispersive X-ray analysis (EDX), solid-state nuclear magnetic resonance (NMR), X-ray photoelectron spectroscopy (XPS), X-ray diffraction (XRD), and X-ray absorption spectroscopy (EXAFS). The significance of these characterization techniques for biochar is summarized in Figure 1.2.

1.5 METHODOLOGIES FOR FUNCTIONALIZATION OF BIOCHAR
The biochar obtained from agricultural waste and animal waste exhibits unique physical and chemical surface properties. It is extensively utilized for various applications such as the removal of dyes, heavy metals, organic pollutants, and pesticides. However, raw biochar has shown more unsatisfactory performance for the removal of contaminants. As a result, biochar can be functionalized or modified to increase its removal efficiency, especially for pollutants with higher concentrations. Extensive research has been conducted to improve the surface properties of

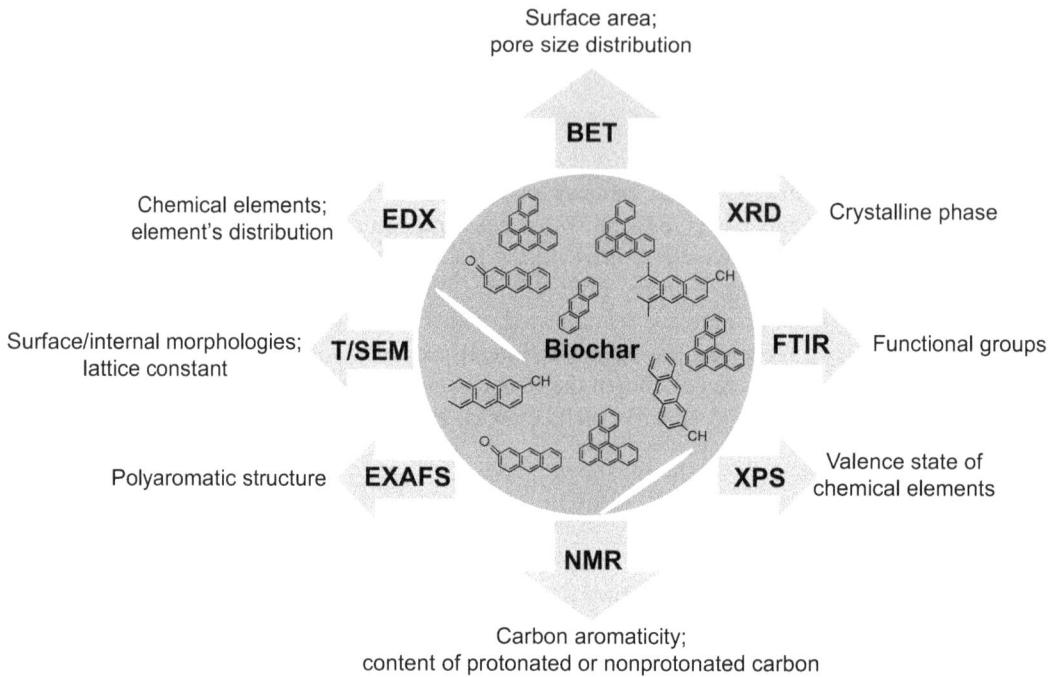

FIGURE 1.2   Characterization techniques for biochar.

Source: Reproduced with permission from J. Wang and Wang (2019).

biochar via physical, chemical, and biological methodologies in order to increase its adsorption capability (J. Wang and Wang 2019). The chemical techniques used for the functionalization of biochar include acid and alkaline treatment, oxidative functionalization and impregnation of metal oxide. The significance of the chemical modification of biochar is represented in Figure 1.3. Following that, the biochar modification or functionalization via chemical techniques is addressed subsequently.

1.5.1 Oxidative Functionalization of Biochar

Chemical oxidation of biochar leads to an increase in the number of oxygen-containing functional groups on the surface of biochar. The hydrophilicity of the product increases due to the presence of groups such as -COOH, -OH, which further causes a rise in the O/C molar ratio. The creation of OFGs (oxygen-containing functional groups) enhances the adsorption capacity of biochar, and thus it can be employed for the adsorption of heavy metal ions. The oxidative functionalization of biochar can be carried out with agents such as ozone, steam, CO_2, HCl, HNO_3, H_2O_2, H_3PO_4, etc. (Yang et al. 2019; Godwin et al. 2019).

The process of functionalization of biochar with agents such as ozone, steam, CO_2, etc., is referred to as gaseous activation. It involves treatment with these agents at a temperature above 700°C when they penetrate the char's interior structure and gasify the carbon atoms, causing the pores to open and enlarge. Along with the OFGs serving as adsorption sites, the product also has a high internal surface area. Steam modification leads to the formation of both meso- and micropores, while using CO_2 yields micropores on the surface, which are helpful for the adsorption of large and

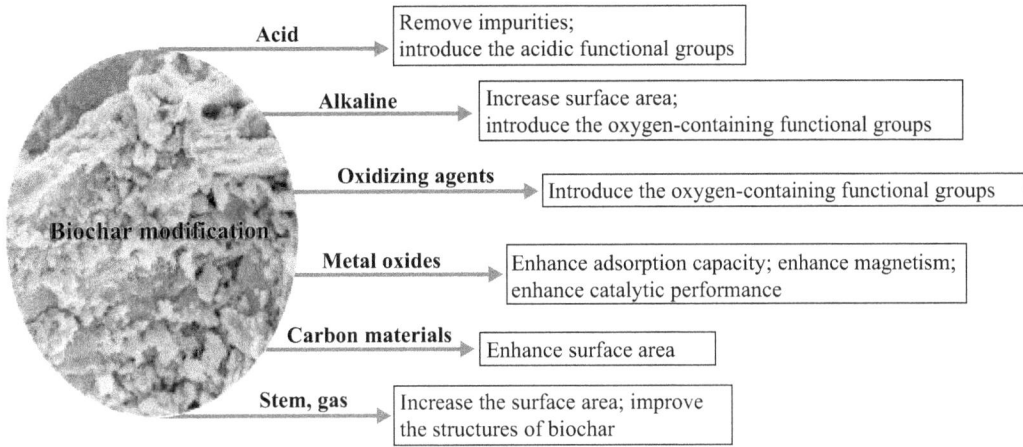

FIGURE 1.3 Chemical modification methods of biochar.

Source: Reproduced with permission from J. Wang and Wang (2019).

smaller molecules, respectively (Sajjadi, Chen, and Egiebor 2019). Acid-based modifications are discussed in the next section.

1.5.2 ACID TREATMENT

Acidic treatment involves the oxidation of biochar with agents like HCl, HNO_3, and H_3PO_4. The treatment of biochar with different acids results in varied characteristics. For example, biochar treated with HNO_3 includes more acidic oxygen-containing functional groups than biochar modified with HCl (Yang et al. 2019). HCl is a strong acid and therefore increases the amount of weak acidic OFGs on biochar. HCl dissolves salts present in pores and creates positive charges on the surface, which allows adsorption of organic and inorganic contaminants. Treatment with HNO_3 is a temperature-dependent process. HNO_3 enhances the surface area and pore size distribution and properties such as electrical conductivity and surface functionality. The treated biochar gets enriched in N-O bond-containing structures. Another frequently used agent is H_3PO_4, which is a low corrosivity agent and is generally employed to carbonize obtained biochar. The treatment of biochar with H_3PO_4 results in an environmentally friendly product. Due to the presence of oxygen-phosphorous functional groups, the activated biochar can adsorb positively charged ions. It can also adsorb inorganic contaminants and peptides from wastewater through chemical interactions, including surface adsorption and hydrogen bonding (Sajjadi, Chen, and Egiebor 2019).

1.5.3 ALKALI TREATMENT

The alkali treatment involves chemical treatment of biochar with agents such as KOH, NaOH, and NH_4OH, which reduce the functional groups present on the surface and enhances nonpolarity (Yang et al. 2019). It also helps in increasing the surface area and achieving a higher value of H/C and N/C ratios. When compared to acid-treated biochar, the alkali-modified biochar has a lower value of O/C, which indicates an increase in its hydrophobicity. Meanwhile, the increased N/C ratio indicates that more nitrogen-containing groups are present on the surface, which are also responsible for the biochar's fundamental nature (Godwin et al. 2019).

Alkali modification is generally carried out by soaking the biochar in the required concentration of alkaline solution at temperatures 25–100°C. Depending on the raw materials, the soaking time is between 6 and 24 h. The mechanism of functionalization and the characteristics of the final product depend on the type of reducing agent used. The production of KOH-modified biochar starts with KOH-impregnated biomass undergoing a series of reactions such as dehydration, cracking, polymerization, and, later, the conversion of lignocellulosic material to char. Reportedly, the char that results has high pore volume and surface area. It can be employed for adsorption for toxic metals due to the presence of phenolic and carboxylic groups. Comparatively, NaOH modification not only requires a lower amount of alkali but is also cheaper and less corrosive than KOH. The reaction leads to the formation and enhancement of pores on the surface. It has been reported that the micropores are converted to mesopores with a high amount of NaOH. It allows the adsorption of large molecules such as dyes on the surface of the functionalized biochar (Sajjadi, Chen, and Egiebor 2019).

1.5.4 METAL OXIDES

The functionalization or modification of biochar via metal salts or metal oxide can improve the surface characteristics of biochar for the adsorption of environmental pollutants. Common metals such as iron, manganese, aluminum, and magnesium were extensively used for the functionalization of biochar. The main objectives for the impregnation of metal salts into the biochar include:

1. *To improve the adsorption capacity for targeted pollutants*: For example, as the biochar surface is negatively charged, it lowers adsorption capacity toward anionic dyes. However, metal impregnation leads to enhancement in the surface characteristics of biochar and further elevates the adsorption capacity for anionic dyes.
2. *To recycle the biochar*: As the biochar particle size is too small, it's tedious to recover and recycle from the aqueous solution used for wastewater treatment. However, the magnetically modified iron salts may enhance the magnetic properties of biochar and subsequently make it easier to recollect and recycle.

The modification of biochar via metal salts can be executed in two ways:

1. By means of in situ pyrolysis, the metal salts are mixed with the feedstock and subjected to the pyrolysis process for the production of biochar.
2. The biochar is prepared via pyrolysis and then soaked in metal salt or metal oxide solution for physical mixing under certain conditions (J. Wang and Wang 2019).

Recently, Jung et al. (2016) introduced a novel technique for the preparation of Fe_3O_4-modified biochar using marine-derived microalgae as feedstock via electromagnetic technique. The obtained magnetic biochar/Fe_3O_4 nanocomposite has shown superior magnetic and surface characteristics, which resulted in the enhancement of adsorption capacity. Further, it was used for the adsorption of acid orange 7 azo dye. The maximum adsorption capacity was found to be 382 mg/g at 30°C. The nanocomposites demonstrated better adsorption and recyclability, as the induced magnetic characteristics aided in the facile recovery of nanocomposites (Jung et al. 2016). Liang et al. (2017) presented a creative synthesis of MnO_2-functionalized biochar (MBR) using aerobically digested swine manure as a feedstock. The obtained MBR showed a rough and larger surface area than that of pure biochar (BR). The adsorption performance of MBR was evaluated for removal of heavy metals such as Pb and Cd, and the results have shown superior adsorption capacity of MBR for Pb and Cd ions as 268 mg/g and 45.8 mg/g, respectively, than BR (127.75 and 14.41 mg/g) (Liang et al. 2017).

Similarly, considering the importance of the functionalization or modification of biochar via chemical techniques, as reported earlier, extensive research has been done for improving the overall adsorption performance of biochar via an enhancement in their surface properties for environmental remediation application. Some of the relevant applications of biochar for the treatment of textile industry effluent are discussed next.

1.6 APPLICATION OF BIOCHAR FOR TREATMENT OF TEXTILE INDUSTRIAL EFFLUENT

1.6.1 REMOVAL OF DYES

Dyes are colored organic substances. Many sectors, including leather, plastic, paper, textiles, and food, require dyes. The effluents of these industries contain a variety of polluted materials and dyes, which significantly influence water bodies. The release of these toxins into the environment has negative consequences for humans, microorganisms, and the aquatic ecosystem. The majority of dyes are poisonous, carcinogenic, or mutagenic, posing a health risk. Textile wastes containing synthetic dyes are dumped into the environment and cause severe environmental pollution to water bodies. Synthetic dyes are chemicals with azo chromophores or phthalocyanine chromophores (presence of nickel, copper, or other metals) and aromatic rings that are potentially poisonous, carcinogenic, and nonbiodegradable. Therefore, it is critical to remove dye from the water, and the search for efficient and cost-effective processes for dye removal is never ending (Zazycki et al. 2018; Srivatsav et al. 2020)

Several methods (chemical, physical, and biological) for removing dye molecules from wastewater are examined. Some of these methods are ion exchange, magnetic separation, advanced oxidation processes, coagulation-flocculation, biological treatment, filtration and adsorption (Srivatsav et al. 2020). Among them, adsorption, in which pollutants are transported from the solvent to the solid phase, is seen to be a promising approach. It is a desirable separation technique due to its high removal efficiency, ease of implementation and operation, excellent adsorbent regeneration capacity, and economic feasibility. Furthermore, no hazardous material is created during the adsorptive process in most cases. Activated carbon (AC) is one of the promising adsorbents for dye removal. However, the high cost of preparation and the difficulty of AC regeneration limit its potential applications.

At present, the focus of research has switched to the production of low-cost adsorbents. A low-cost adsorbent is one that is abundant in nature or generated from any industrial by-product or waste material. Biochar made from agricultural waste is considered a low-cost alternative to activated carbon. A number of studies have been reported to synthesize biochar from an alternative source such as sewage sludge, peanut shells, and oil distillation residue. These biochar's were utilized to clean the aqueous media of pollutants. Biochar made from raw pecan nutshells has also been shown to remove metals and dyes from an aqueous medium (Srivatsav et al. 2020; Roosta et al. 2015).

Activated biochar was found to remove the cationic dyes methylene blue (MB) and crystal violet (CV) by Abd-Elhamid et al. (2020). The biochar was prepared from rice straw, an agricultural residue. It is found that after pyrolysis, the obtained biochar becomes less active, which is termed p-biochar. This p-biochar was then activated using the wet attrition process to produce m-biochar (high active). The surface area and pore size distribution of the m-biochar were found to be about 104 m^2/g and 0.5–5.0 μm, respectively, which were much lower than those of p-biochar (223.4 m^2/g and 5–150 μm). The synthesis of oxygenated active groups in m-biochar during the wet attrition activation process is responsible for the substantial reduction in surface area and pore size. In the case of m-biochar, the equilibrium was swiftly attained in 15 min for MB dye and in 20 min for CV dye, with removal percentages of 94.45 and 92.70 for MB and CV dyes, respectively. Due to the weak adsorption ability of p-biochar, m-biochar has evolved as a superior sorbent for cationic dyes. In addition, the pH of the solution played a crucial influence in dye sorption, as the adsorption increases as the pH rises. At a higher pH, there are fewer H$^+$ ions in the solution, which improves cationic dye sorption on the biochar surface (Abd-Elhamid et al. 2020).

Caprariis et al. (2018) studied the influence of deep eutectic solvents (DES) on biochar activation. The biochar was made by pyrolyzing pine wood biomass in a quartz tube (2.5 cm ID and 50 cm L) for 1 hour at a temperature of 10°C/min up to 750°C with a nitrogen flow of 0.2 L/min. Later, both sodium hydroxide (NaOH) and DES were used to activate the biochar. The surface area of DES-treated biochar was found to be comparable to commercial activated carbon (800–1000 m^2/g) and higher than biochar modified by NaOH. Thus, the sorption capacity for DES-treated biochar is improved by 20% for methylene blue and rhodamine B dyes. Due to steric considerations and biochar surface functional groups, rhodamine B adsorption capability for all activated biochar is lower than that of methylene blue (De Caprariis et al. 2018). Zazycki et al. (2018) have developed cost-effective biochar prepared from pecan nutshells for the removal of anionic dye reactive red 141 (RR 141). In comparison to pure raw material (2.1 m²/g), the produced biochar has both micro- and mesoporous structures with a specific surface area of 93 m²/g. The sorption of RR 141 was favored in lower pH conditions (pH 2–3). This is because, in an acidic pH, the dye molecules' negatively charged groups (sulfonated groups) are attracted to the positive sites of the adsorbent. This is a common occurrence in the sorption of anionic dyes. In comparison to raw pecan nutshells (23%), produced biochar had an RR 141 removal rate of 85%. The highest sorption capacity of RR 141 was reported to be around 130 mg/g (Zazycki et al. 2018). Biochar and activated biochar from wasted *Auricularia auricula* substrate were reported by Su et al. (2021) for the removal of cationic azo dyes. Three different dyes, methylene blue (MB), crystal violet (CV), and rhodamine B (RB), were chosen for single and binary adsorptive systems. The substrate was carbonized at a temperature of 350°C and 650°C, respectively, for 3 h under N_2 flow. The development of physicochemical characteristics on the biochar surface was influenced by the synthesis temperature. The surface area of the biochar (made at 650°C) was 89.393 m²/g after activation with NaOH at a weight ratio of 1:1. In comparison to other produced adsorbents, activated biochar had the highest adsorption capacity for CV (735.73 mg/g) and was followed by MB (53.62 mg/g) and RB (32.33 mg/g). This was due to its higher surface area and increased oxygen comprising functional groups on the surface. Even after three cycles of adsorption-desorption, activated biochar was reusable and maintained 70% of initial adsorption. In the instance of a binary adsorption system, each dye had a competing effect on the adsorption of the others (Su et al. 2021). Chemically modified lignin-derived biochar for the elimination of methylene blue was reported by Liu, Li, and Singh (2021). For this, a variety of chemical modifiers such as MnO_2, $MnSO_4$, and $KMnO_4$ were utilized. MnO_2-biochar had a surface area of 349.65 m²/g and an average pore size of 3.892 nm, respectively. The maximum dye loading capacity was 248.96 mg/g, and the maximum removal rate was achieved at 99.73% for the said adsorbent. In comparison to other sorbents, the MnO_2-biochar surface had a rough texture and homogeneous deposition of MnO_2 particles, which may have increased the dye removal rate (X. J. Liu, Li, and Singh 2021).

1.6.2 Heavy Metal Removal

In today's world, water pollution is the supreme problem, and rapid industrialization is responsible for the generation of toxic metal ions that leads to severe environmental pollution. The major sources of heavy metals are metal finishing, mining, erosion from metal-rich soil, and electroplating industries. Therefore, for the safety of human health and the stability of the environment, the efficient separation of these heavy metals becomes a necessity. Considering the significance of environmental protection, several separation techniques such as membrane separation and adsorption are applied for heavy metal removal (X. Liu et al. 2019; Purkait, Kumar, and Maity 2009). Heavy metals particularly, lead (Pb), chromium (Cr), cadmium (Cd), copper (Cu) and nickel (Ni), mercury (Hg) are widely used for the production of color pigments of textile dyes. These heavy metals, when transferred to the environment, are highly toxic and can bioaccumulate in the human body, aquatic life, and natural water bodies and possibly be trapped in the soil (Enaime et al. 2020; Mondal, Nandi, and Purkait 2013).

Various research works concluded that for removing inorganic pollutants from wastewater like heavy metals, biochar had shown excellent effectivity in such applications. Specifically, such application depends on the biochar properties and target metal characteristics, which hugely controls the adsorption capacity of the biochar utilize. Due to the presence of numerous phenolic, hydroxyl, and carboxyl groups along with huge surface area, biochar thus has the benefit of being utilized as an effective sorption material for the removal of heavy metals (Zhao et al. 2020). Different biochar was prepared by Pan, Jiang, and Xu (2013) utilizing straws of rice, canola, peanut, and soybean to remove trivalent chromium through sorption. Depending on the acidic functional groups of the biochar, the adsorption capacity was found in the order of peanut > soybean > canola > rice. Zhao et al. (2020) studied the biochar properties obtained from raw materials such as chicken manure, rice straw, and sewage sludge. They studied their efficiency in the removal of metals such as Pb^{2+} and Zn^{2+}. The highest adsorption capacity toward Pb^{2+} was obtained for biochar derived from rice straw. The study confirmed that rice-straw-derived biochar has numerous functional groups (e.g., O-H and C=C/C=O), the highest surface negative charge, along with higher physical stability and alkaline pH. The study confirmed the possible mechanism of such heavy metal removal, supported by precipitation and ion exchange reaction mechanisms.

Zhou et al. (2013) investigated the production and utilization of low-cost biochars modified with chitosan toward the removal of heavy metals from an aqueous solution. The chitosan coating over the biochar was confirmed through various characterization techniques. The modified biochar was found to be helpful in improving soil properties along with its efficiency. Batch studies carried out showed the better efficiency of chitosan-modified biochar toward the removal of Cu^{2+}, Pb^{2+}, and Cd^{2+} from solution compared to the unmodified biochar. Through investigation, it was obtained that the sorption kinetics were slow on utilizing chitosan-modified biochar obtained from bamboo. Moreover, the lead sorption followed the Langmuir isotherm and had a relatively high sorption capacity of 14.3 mg/g biochar. The metal toxicity is reduced greatly due to the chitosan-modified biochar-based lead sorption. When compared to pristine chitosan-modified biochar, it was obtained that seed germination rate and seedling growth were the same for lead-sorbed chitosan-modified biochar. Moreover, a 60% reduction in lead uptake by plants was obtained when lead-sorbed chitosan-modified biochar was used. The work thus concludes the utilization of chitosan-modified biochar as an effective, low-cost adsorbent for controlling the contamination of heavy metals in the environment (Zhou et al. 2013).

Moreover, Inyang et al. (2012) prepared two different biochars through anaerobic digested biomass and then utilized it for adsorption of heavy metals from an aqueous solution. The biochars prepared were digested dairy waste biochar (DDWB) and digested whole sugar beet biochar (DWSB). From experiments, it was obtained that both of the prepared biochars are quite effective in removing heavy metals such as Cd^{2+}, Cu^{2+}, Pb^{2+}, and Ni^{2+} from aqueous solutions. Results concluded that DWSB was more efficient toward heavy metal adsorption of Ni^{2+} and Cd^{2+} in comparison to the DDWB. The study also revealed that the adsorption of lead onto both the prepared biochars mainly occurred through a surface precipitation mechanism. Such results were supported by batch experiments, mathematical model fittings, characterization of material studies such as XRD, SEM-EDX, and FTIR. The study further concludes that the adsorption capacity value of the two biochars was comparable to the commercially available activated carbons, which was > 200 mmol/g (Inyang et al. 2012).

Similarly, Shi et al. (2020) prepared a series of sustainable, cost-effective, and high-efficiency biochar utilizing waste glue residue for removal of Cr(VI). The work involved chemical modification of the synthesized biochar using HCl, $ZnCl_2$, and KOH for improved adsorption properties of the biochar. Furthermore, the ratio of modifier to glue waste and pyrolysis temperature being the controlling parameters were also investigated for the preparation of modified biochar. During the adsorption study of Cr(VI), various operating parameters such as the pH of the solution medium, adsorbent dosage, time of reaction, initial concentration of Cr(VI) were also studied. The biochar modified with $ZnCl_2$ was found to be more efficient, having the highest adsorption capacity of

325.54 mg/g, comparably higher than previously reported adsorbents. Moreover, the biochar was found to be effective about 90% after six run cycles during recycling experiments were conducted (Shi et al. 2020).

Li et al. (2020) investigated with hydrochar, which is a carbonaceous material, and studied its adsorption capacity for the toxic pollutant Cr (VI). In order to enhance the adsorption capacity of the hydrochar, it was modified utilizing the cohydrothermal carbonization technique carried out at 200°C for 7 h. The raw material used was bamboo sawdust, whereas the modifying agents were aluminum chloride ($AlCl_3$) and zinc chloride ($ZnCl_2$). In comparison to the pristine hydrochar, $ZnCl_2$-modified hydrochar was found to be more highly carbonized with increased surface area by 26 times; similarly, the $AlCl_3$-modified hydrochar was also found to have a higher surface area, about 4.3 times higher than the pristine hydrochar. Again, for $ZnCl_2$ and $AlCl_3$ modified hydrochar, the total pore volume was found to increase by 43 and 5.5 times, respectively. Through experimental studies, it was obtained that due to the modifications of $ZnCl_2$- and $AlCl_3$-modified hydrochar, the maximum adsorption capacity for Cr(VI) was enhanced by 3.4 and 2.8 times, respectively, when compared to pristine hydrochar. Moreover, the Langmuir isotherm fit the experimental adsorption data quite nicely, along with the adsorption kinetics following the pseudo-second-order model. The adsorption capacity was found to enhance at low pH value, indicating that electrostatic interaction along with an ion exchange mechanism plays a vital role during Cr(VI) removal (F. Li et al. 2020).

1.6.3 Adsorption Mechanism

The surface of biochar has been recognized for its heterogeneity, which has led to distinct adsorption mechanisms being established. Basically, the adsorption mechanism depends on the chemical properties of the adsorbent, the nature of the pollutant, and the interaction between the adsorbent surface and contaminant. The main routes for adsorption can be divided into a:

- Physical route, where the adsorbent surface is covered with the adsorbate.
- Pore filling route, where the condensation of adsorbate occurs within the pores of the adsorbent.
- Precipitation route, where the adsorbate forms a layer over the surface of the adsorbent.

In the case of organic pollutants such as dyes, the process of adsorption is executed via pore filling, electrostatic attraction, π–π staking, hydrophobic interactions, hydrogen bonding, and complexes adsorption.

In the case of removal of inorganic pollutants like heavy metals, the multitude mechanism may interfere alongside physical adsorption, surface precipitation, cationic and anionic attraction, and ion exchange mechanisms (Ahmad et al. 2014; Enaime et al. 2020). The proposed mechanism for the removal of organic and inorganic water contaminants by biochar is summarized in Figure 1.4.

The pore filling route of organic pollutants depends on the pore volumes, such as the micropore and mesopore volume of adsorbent. If the ionic radius is lower, then the penetration of pollutants on biochar is greater and results in a higher adsorption capacity of the biochar. However, the soluble pollutants can be adsorbed over hydrophobic adsorbents with hydrophobic functional sites. Generally, the surface of biochar is negatively charged due to the division of oxygen-rich functional groups. These functional sites possess electrostatic attraction between positively charged pollutants and biochar. Although the biochar synthesized at a higher temperature contains fewer oxygen-rich functional sites and makes biochar more aromatic and less polar in nature, this is no more effective for the efficient removal of organic pollutants. However, the adsorption can be commenced via hydrogen bonding (Ahmad et al. 2014; Qambrani et al. 2017; Enaime et al. 2020)

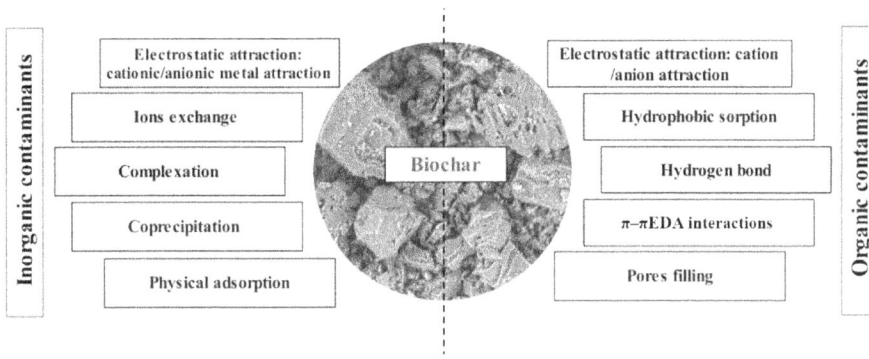

FIGURE 1.4 Proposed mechanism for removal of organic and inorganic water contaminants.
Source: Enaime et al. (2020).

1.7 CONCLUSION AND FUTURE PROSPECTIVE

Biochar as a renewable source has shown a significant potential for resolving various environmental issues, especially in wastewater treatment. It has been extensively utilized as an adsorbent with unique surface characteristics and high adsorption capacity for a wide range of pollutants. The present chapter explored the details related to biochar, its advantages, properties, feedstock, and applications for environmental remediation. The chapter also listed the available methods for the preparation and functionalization of biochar via chemical modification methods. Moreover, the factors affecting the properties and performance of biochar for wastewater treatments were also discussed. Based on the properties, the utilization of biochar and its modified forms for the efficient removal of heavy metals, dyes are critically reviewed. The major conclusion related to environmental remediations are as follows:

- A variety of feedstocks are available for biochar production; however, to accomplish the aims of efficient biochar synthesis, supply variables such as feedstock type, storage, collection, handling, and analysis of feedstocks must be carefully examined.
- It was found that the raw biochar has shown poor performance for the removal of toxic dyes and heavy metals. However, the activation or modification of biochar has resulted in improved surface characteristics of biochar with higher adsorption capacity and efficient removal of pollutants.
- The process parameters for biochar production, such as temperature, residence time, and other factors like source and type of feedstock and production technique, have dramatically affected the yield and properties of biochar.
- The functionalization of biochar can be significantly accomplished via chemical modification such as via metal salts or metal oxides, which have shown a greater adsorption capacity as well as recyclability.
- The goals of biochar production include fine-tuning biochar characteristics, decreasing environmental emissions during pyrolysis, improving process economics, and improving coproducts.

The major challenges related to biochar feedstocks are their availability, cost-effective transport, the uncertainty of cost and quality, a requirement of pretreatment processes. These challenges can be overcome via blending, comminution, pelleting, the combination of chemical and physical pretreatments, and a supply of more reliable, higher-volume, and cheaper biomass. The future perspective over the utilization of biochar includes further advancements in the biochar functionalization and identification of adsorption mechanisms for different types of pollutants. It was observed that very few studies have been conducted on the application of biochar to remove pesticides, drugs, and radioactive materials, which must be explored in the future. It is also important to establish and understand the relation between thermal conditions and the properties and behavior of biochar for the treatment of textile industry effluent. Moreover, the negative impacts from the utilization of biochar for wastewater treatment must be undertaken. In addition, efforts must be made to improve the stability of biochar as reflected by aromaticity. It was observed that when biochar was used for the wastewater treatment, a potential release of carbon and heavy metals occurred, especially biochar derived from sludge.

ACKNOWLEDGMENT

This work is partially supported by a grant (DST/TM/WTI/WIC/2K17/84(G)) from the DST (Department of Science and Technology), New Delhi. Any opinions, findings, and conclusions expressed in this paper are those of the authors and do not necessarily reflect the views of DST, New Delhi.

REFERENCES

Abd-Elhamid, A. I., Mohamed Emran, M. H. El-Sadek, Ahmed A. El-Shanshory, Hesham M. A. Soliman, M. A. Akl, and Mohamed Rashad. 2020. "Enhanced Removal of Cationic Dye by Eco-Friendly Activated Biochar Derived from Rice Straw." *Applied Water Science* 10 (1). Springer International Publishing: 1–11. doi:10.1007/s13201-019-1128-0.

Ahmad, Mahtab, Anushka Upamali Rajapaksha, Jung Eun Lim, Ming Zhang, Nanthi Bolan, Dinesh Mohan, Meththika Vithanage, Sang Soo Lee, and Yong Sik Ok. 2014. "Biochar as a Sorbent for Contaminant Management in Soil and Water: A Review." *Chemosphere* 99. Elsevier Ltd: 19–33. doi:10.1016/j.chemosphere.2013.10.071.

Baláž, Matej. 2014. "Eggshell Membrane Biomaterial as a Platform for Applications in Materials Science." *Acta Biomaterialia* 10 (9): 3827–3843. doi:10.1016/j.actbio.2014.03.020.

Caprariis, Benedetta De, Paolo De Filippis, Elisabetta Petrucci, and Marco Scarsella. 2018. "Activated Biochars Used as Adsorbents for Dyes Removal." *Chemical Engineering Transactions* 65: 103–108. doi:10.3303/CET1865018.

Chatterjee, Riya, Baharak Sajjadi, Wei Yin Chen, Daniell L. Mattern, Nathan Hammer, Vijayasankar Raman, and Austin Dorris. 2020. "Effect of Pyrolysis Temperature on PhysicoChemical Properties and Acoustic-Based Amination of Biochar for Efficient CO_2 Adsorption." *Frontiers in Energy Research* 8. doi:10.3389/fenrg.2020.00085.

Cheng, Feng, and Xiuwei Li. 2018. "Preparation and Application of Biochar-Based Catalysts for Biofuel Production." *Catalysts* 8 (9): 346. doi:10.3390/catal8090346.

Das, Pranjal P., Piyal Mondal, Anweshan, A. Sinha, P. Biswas, S. Sarkar, and Mihir K. Purkait. 2021. "Treatment of Steel Plant Generated Biological Oxidation Treated (BOT) Wastewater by Hybrid Process." *Separation and Purification Technology* 258: 118013. doi:10.1016/j.seppur.2020.118013.

Deepti, A. Sinha, P. Biswas, S. Sarkar, U. Bora, and M. K. Purkait. 2020. "Separation of Chloride and Sulphate Ions from Nanofiltration Rejected Wastewater of Steel Industry." *Journal of Water Process Engineering* 33 (December 2019). Elsevier: 101108. doi:10.1016/j.jwpe.2019.101108.

Dey, A., R. Singh, and M. K. Purkait. 2014. "Cobalt Ferrite Nanoparticles Aggregated Schwertmannite: A Novel Adsorbent for the Efficient Removal of Arsenic." *Journal of Water Process Engineering* 3 (C). Elsevier Ltd: 1–9. doi:10.1016/j.jwpe.2014.07.002.

Duarah, Prangan, Dibyajyoti Haldar, and Mihir Kumar Purkait. 2020. "Technological Advancement in the Synthesis and Applications of Lignin-Based Nanoparticles Derived from Agro-Industrial Waste Residues: A Review." *International Journal of Biological Macromolecules* 163. Elsevier B.V.: 1828–1843. doi:10.1016/j.ijbiomac.2020.09.076.

Enaime, Ghizlane, Abdelaziz Baçaoui, Abdelrani Yaacoubi, and Manfred Lübken. 2020. "Biochar for Wastewater Treatment-Conversion Technologies and Applications." *Applied Sciences (Switzerland)* 10 (10). doi:10.3390/app10103492.

Gabhane, Jagdish W., Vivek P. Bhange, Pravin D Patil, Sneha T. Bankar, and Sachin Kumar. 2020. "Recent Trends in Biochar Production Methods and Its Application as a Soil Health Conditioner: A Review." *SN Applied Sciences* 2. doi:10.1007/s42452-020-3121-5.

Garcia-Nunez, J. A., M. R. Pelaez-Samaniego, M. E. Garcia-Perez, I. Fonts, J. Abrego, R. J. M. Westerhof, and M. Garcia-Perez. 2017. "Historical Developments of Pyrolysis Reactors: A Review." *Energy and Fuels* 31: 5751–5775. doi:10.1021/acs.energyfuels.7b00641.

Ghodake, Gajanan Sampatrao, Surendra Krushna Shinde, Avinash Ashok Kadam, Rijuta Ganesh Saratale, Ganesh Dattatraya Saratale, Manu Kumar, Ramasubba Reddy Palem, et al. 2021. "Review on Biomass Feedstocks, Pyrolysis Mechanism and Physicochemical Properties of Biochar: State-of-the-Art Framework to Speed up Vision of Circular Bioeconomy." *Journal of Cleaner Production* 297. Elsevier Ltd: 126645. doi:10.1016/j.jclepro.2021.126645.

Godwin, Patrick M., Yuanfeng Pan, Huining Xiao, and Muhammad T. Afzal. 2019. "Progress in Preparation and Application of Modified Biochar for Improving Heavy Metal Ion Removal From Wastewater." *Journal of Bioresources and Bioproducts* 4 (1). Elsevier Masson SAS: 31–42. doi:10.21967/jbb.v4i1.180.

Ibrahim, Amr F. M., Kodanda Phani Raj Dandamudi, Shuguang Deng, and Y. S. Lin. 2020. "Pyrolysis of Hydrothermal Liquefaction Algal Biochar for Hydrogen Production in a Membrane Reactor." *Fuel* 265 (April). Elsevier: 116935. doi:10.1016/J.FUEL.2019.116935.

Inyang, Mandu, Bin Gao, Ying Yao, Yingwen Xue, Andrew R. Zimmerman, Pratap Pullammanappallil, and Xinde Cao. 2012. "Removal of Heavy Metals from Aqueous Solution by Biochars Derived from Anaerobically Digested Biomass." *Bioresource Technology* 110. Elsevier Ltd: 50–56. doi:10.1016/j.biortech.2012.01.072.

Jung, Kyung Won, Brian Hyun Choi, Tae Un Jeong, and Kyu Hong Ahn. 2016. "Facile Synthesis of Magnetic Biochar/Fe3O4 Nanocomposites Using Electro-Magnetization Technique and Its Application on the Removal of Acid Orange 7 from Aqueous Media." *Bioresource Technology* 220. Elsevier Ltd: 672–676. doi:10.1016/j.biortech.2016.09.035.

Kim, Jae Young, Shinyoung Oh, and Young Kwon Park. 2020. "Overview of Biochar Production from Preservative-Treated Wood with Detailed Analysis of Biochar Characteristics, Heavy Metals Behaviors, and Their Ecotoxicity." *Journal of Hazardous Materials* 384 (February). Elsevier: 121356. doi:10.1016/J.JHAZMAT.2019.121356.

Li, Feiyue, Andrew R. Zimmerman, Xin Hu, and Bin Gao. 2020. "Removal of Aqueous Cr(VI) by Zn- and Al-Modified Hydrochar." *Chemosphere* 260. doi:10.1016/j.chemosphere.2020.127610.

Li, Simeng, Celeste Y. Chan, Mohamadali Sharbatmaleki, Helen Trejo, and Saied Delagah. 2020. "Engineered Biochar Production and Its Potential Benefits in a Closed-Loop Water-Reuse Agriculture System." *Water (Switzerland)* 12 (10): 2847. doi:10.3390/w12102847.

Liang, Jie, Xuemei Li, Zhigang Yu, Guangming Zeng, Yuan Luo, Longbo Jiang, Zhaoxue Yang, Yingying Qian, and Haipeng Wu. 2017. "Amorphous MnO2 Modified Biochar Derived from Aerobically Composted Swine Manure for Adsorption of Pb(II) and Cd(II)." *ACS Sustainable Chemistry and Engineering* 5 (6): 5049–5058. doi:10.1021/acssuschemeng.7b00434.

Liu, Xiaolu, Ran Ma, Xiangxue Wang, Yan Ma, Yongping Yang, Li Zhuang, Sai Zhang, Riffat Jehan, Jianrong Chen, and Xiangke Wang. 2019. "Graphene Oxide-Based Materials for Efficient Removal of Heavy Metal Ions from Aqueous Solution: A Review." *Environmental Pollution* 252. Elsevier Ltd: 62–73. doi:10.1016/j.envpol.2019.05.050.

Liu, Xu Jing, Ming Fei Li, and Sandip K. Singh. 2021. "Manganese-Modified Lignin Biochar as Adsorbent for Removal of Methylene Blue." *Journal of Materials Research and Technology* 12. Elsevier Ltd: 1434–1445. doi:10.1016/j.jmrt.2021.03.076.

Mašek, Ondřej, Wolfram Buss, and Saran Sohi. 2018. "Standard Biochar Materials." *Environmental Science and Technology* 52 (17): 9543–9544. doi:10.1021/acs.est.8b04053.

Mohan, Dinesh, Ankur Sarswat, Yong Sik Ok, and Charles U. Pittman. 2014. "Organic and Inorganic Contaminants Removal from Water with Biochar, a Renewable, Low Cost and Sustainable Adsorbent—A Critical Review." *Bioresource Technology* 160. Elsevier Ltd: 191–202. doi:10.1016/j.biortech.2014.01.120.

Mondal, Dilip Kumar, Barun Kumar Nandi, and M. K. Purkait. 2013. "Removal of Mercury (II) from Aqueous Solution Using Bamboo Leaf Powder: Equilibrium, Thermodynamic and Kinetic Studies." *Journal of Environmental Chemical Engineering* 1 (4). Elsevier B.V.: 891–898. doi:10.1016/j.jece.2013.07.034.

Oni, Babalola Aisosa, Olubukola Oziegbe, and Obembe O. Olawole. 2019. "Significance of Biochar Application to the Environment and Economy." *Annals of Agricultural Sciences* 64 (2): 222–236. doi:10.1016/j.aoas.2019.12.006.

Pan, Jingjian, Jun Jiang, and Renkou Xu. 2013. "Adsorption of Cr(III) from Acidic Solutions by Crop Straw Derived Biochars." *Journal of Environmental Sciences (China)* 25 (10). The Research Centre for Eco-Environmental Sciences, Chinese Academy of Sciences: 1957–1965. doi:10.1016/S1001-0742(12)60305-2.

Pendolino, Flavio, and Nerina Armata. 2017. *Graphene Oxide in Environmental Remediation Process.* Springer. doi:10.1007/978-3-319-60429-9.

Ponnusamy, Vinoth Kumar, Senthil Nagappan, Rahul R. Bhosale, Chyi How Lay, Dinh Duc Nguyen, Arivalagan Pugazhendhi, Soon Woong Chang, and Gopalakrishnan Kumar. 2020. "Review on Sustainable Production of Biochar through Hydrothermal Liquefaction: Physico-Chemical Properties and Applications." *Bioresource Technology* 310 (August). Elsevier: 123414. doi:10.1016/J.BIORTECH.2020.123414.

Pulka, Jakub, Piotr Manczarski, Jacek A. Koziel, and Andrzej Białowiec. 2019. "Torrefaction of Sewage Sludge: Kinetics and Fuel Properties of Biochars." *Energies* 12 (3): 565. doi:10.3390/en12030565.

Purkait, M. K., V. Dinesh Kumar, and Damodar Maity. 2009. "Treatment of Leather Plant Effluent Using NF Followed by RO and Permeate Flux Prediction Using Artificial Neural Network." *Chemical Engineering Journal* 151 (1–3): 275–285. doi:10.1016/j.cej.2009.03.023.

Qambrani, Naveed Ahmed, Md Mukhlesur Rahman, Seunggun Won, Soomin Shim, and Changsix Ra. 2017. "Biochar Properties and Eco-Friendly Applications for Climate Change Mitigation, Waste Management, and Wastewater Treatment: A Review." *Renewable and Sustainable Energy Reviews* 79 (May). Elsevier Ltd: 255–273. doi:10.1016/j.rser.2017.05.057.

Rahman, M. T., Z. C. Guo, Z. B. Zhang, H. Zhou, and X. H. Peng. 2018. "Wetting and Drying Cycles Improving Aggregation and Associated C Stabilization Differently after Straw or Biochar Incorporated into a Vertisol." *Soil and Tillage Research* 175 (April 2017). Elsevier: 28–36. doi:10.1016/j.still.2017.08.007.

Roosta, M., M. Ghaedi, R. Sahraei, and M. K. Purkait. 2015. "Ultrasonic Assisted Removal of Sunset Yellow from Aqueous Solution by Zinc Hydroxide Nanoparticle Loaded Activated Carbon: Optimized Experimental Design." *Materials Science and Engineering C* 52. Elsevier B.V.: 82–89. doi:10.1016/j.msec.2015.03.036.

Sajjadi, Baharak, Wei Yin Chen, and Nosa O. Egiebor. 2019. "A Comprehensive Review on Physical Activation of Biochar for Energy and Environmental Applications." *Reviews in Chemical Engineering* 35 (6). doi:10.1515/revce-2017-0113.

Sajjadi, Baharak, Tetiana Zubatiuk, Danuta Leszczynska, Jerzy Leszczynski, and Wei Yin Chen. 2019. "Chemical Activation of Biochar for Energy and Environmental Applications: A Comprehensive Review." *Reviews in Chemical Engineering* 35. doi:10.1515/revce-2018-0003.

Shi, Yueyue, Rui Shan, Lili Lu, Haoran Yuan, Hong Jiang, Yuyuan Zhang, and Yong Chen. 2020. "High-Efficiency Removal of Cr(VI) by Modified Biochar Derived from Glue Residue." *Journal of Cleaner Production* 254. doi:10.1016/j.jclepro.2019.119935.

Smith, Sean C., and Debora F. Rodrigues. 2015. "Carbon-Based Nanomaterials for Removal of Chemical and Biological Contaminants from Water: A Review of Mechanisms and Applications." *Carbon* 91. Elsevier Ltd: 122–143. doi:10.1016/j.carbon.2015.04.043.

Sontakke, Ankush D., Pranjal P. Das, Piyal Mondal, and Mihir K. Purkait. 2021. "Thin-Film Composite Nanofiltration Hollow Fiber Membranes toward Textile Industry Effluent Treatment and Environmental Remediation Applications: Review." *Emergent Materials*, July. doi:10.1007/s42247-021-00261-y.

Sontakke, Ankush D., and M. K. Purkait. 2020. "Fabrication of Ultrasound-Mediated Tunable Graphene Oxide Nanoscrolls." *Ultrasonics Sonochemistry* 63. doi:10.1016/j.ultsonch.2020.104976.

Sontakke, Ankush D., and Mihir K. Purkait. 2021. "A Brief Review on Graphene Oxide Nanoscrolls: Structure, Synthesis, Characterization and Scope of Applications." *Chemical Engineering Journal* 420 (P1). Elsevier B.V.: 129914. doi:10.1016/j.cej.2021.129914.

Srivatsav, Prithvi, Bhaskar Sriharsha Bhargav, Vignesh Shanmugasundaram, Jayaseelan Arun, Kannappan Panchamoorthy Gopinath, and Amit Bhatnagar. 2020. "Biochar as an Eco-Friendly and Economical Adsorbent for the Removal of Colorants (Dyes) from Aqueous Environment: A Review." *Water (Switzerland)* 12 (12): 1–27. doi:10.3390/w12123561.

Su, Long, Haibo Zhang, Kokyo Oh, Na Liu, Yuan Luo, Hongyan Cheng, Guosheng Zhang, and Xiaofang He. 2021. "Activated Biochar Derived from Spent Auricularia Auricula Substrate for the Efficient Adsorption of Cationic Azo Dyes from Single and Binary Adsorptive Systems." *Water Science and Technology* 84 (1): 101–121. doi:10.2166/wst.2021.222.

Upadhyay, Ravi Kant, Navneet Soin, and Susanta Sinha Roy. 2014. "Role of Graphene/Metal Oxide Composites as Photocatalysts, Adsorbents and Disinfectants in Water Treatment: A Review." *RSC Advances* 4: 3823–3851. doi:10.1039/c3ra45013a.

Wang, Duo, Peikun Jiang, Haibo Zhang, and Wenqiao Yuan. 2020. "Biochar Production and Applications in Agro and Forestry Systems: A Review." *Science of the Total Environment* 723 (June). Elsevier: 137775. doi:10.1016/J.SCITOTENV.2020.137775.

Wang, Jianlong, and Shizong Wang. 2019. "Preparation, Modification and Environmental Application of Biochar: A Review." *Journal of Cleaner Production* 227. Elsevier Ltd: 1002–1022. doi:10.1016/j.jclepro.2019.04.282.

Wang, Xiaoqing, Zizhang Guo, Zhen Hu, and Jian Zhang. 2020. "Recent Advances in Biochar Application for Water and Wastewater Treatment: A Review." *PeerJ* 8: e9164. doi:10.7717/peerj.9164.

Yaashikaa, P. R., P. Senthil Kumar, Sunita Varjani, and A. Saravanan. 2020. "A Critical Review on the Biochar Production Techniques, Characterization, Stability and Applications for Circular Bioeconomy." *Biotechnology Reports* 28 (December). Elsevier: e00570. doi:10.1016/J.BTRE.2020.E00570.

Yadav, Krishna, and Sheeja Jagadevan. 2020. "Influence of Process Parameters on Synthesis of Biochar by Pyrolysis of Biomass: An Alternative Source of Energy." In *Recent Advances in Pyrolysis*. 1–14. IntechOpen Publisher. doi:10.5772/intechopen.88204.

Yang, Xue, Shiqiu Zhang, Meiting Ju, and Le Liu. 2019. "Preparation and Modification of Biochar Materials and Their Application in Soil Remediation." *Applied Sciences (Switzerland)* 9 (7). doi:10.3390/app9071365.

Yaranal, Naveenkumar Ashok, Senthilmurugan Subbiah, and Kaustubha Mohanty. 2020. "Environmental Technology & Innovation Identification, Extraction of Microplastics from Edible Salts and Its Removal from Contaminated Seawater." *Environmental Technology & Innovation*, xxxx. Elsevier B.V.: 101253. doi:10.1016/j.eti.2020.101253.

Zazycki, Maria A., Marcelo Godinho, Daniele Perondi, Edson L. Foletto, Gabriela C. Collazzo, and Guilherme L. Dotto. 2018. "New Biochar from Pecan Nutshells as an Alternative Adsorbent for Removing Reactive Red 141 from Aqueous Solutions." *Journal of Cleaner Production* 171. Elsevier Ltd: 57–65. doi:10.1016/j.jclepro.2017.10.007.

Zhang, Zhikun, Zongyuan Zhu, Boxiong Shen, and Lina Liu. 2019. "Insights into Biochar and Hydrochar Production and Applications: A Review." *Energy* 171 (March). Pergamon: 581–598. doi:10.1016/J.ENERGY.2019.01.035.

Zhao, Man, Yuan Dai, Miaoyue Zhang, Can Feng, Baojia Qin, Weihua Zhang, Nan Zhao, et al. 2020. "Mechanisms of Pb and/or Zn Adsorption by Different Biochars: Biochar Characteristics, Stability, and Binding Energies." *Science of the Total Environment* 717. Elsevier B.V.: 136894. doi:10.1016/j.scitotenv.2020.136894.

Zhou, Yanmei, Bin Gao, Andrew R. Zimmerman, June Fang, Yining Sun, and Xinde Cao. 2013. "Sorption of Heavy Metals on Chitosan-Modified Biochars and Its Biological Effects." *Chemical Engineering Journal* 231. Elsevier B.V.: 512–518. doi:10.1016/j.cej.2013.07.036.

2 Utilization of Modified Biochar for Wastewater Recycling and Management
An Overview

Kriti Shrivastava and Ankur Jain

CONTENTS

2.1 INTRODUCTION

United Nations Member States adopted the 2030 Agenda for Sustainable Development in 2015, which presents a common plan of action for the peace and prosperity of the people on this planet as well as for safeguarding the environment for upcoming generations. It includes the 17 Sustainable Development Goals (SDGs) and asks for urgent global action and coordination of all the developing and developed member states. The sustainable development goals, such as clean water and sanitation, sustainable cities and communities, climate action, life on land and life below water, all require

DOI: 10.1201/9781003203438-2

immediate attention to the problem of environment pollution. Continuous generation and discharge of complex organic and inorganic pollutants from various anthropogenic sources has constantly deteriorated the quality of natural waters both at the surface and beneath the ground, and during the past few decades, substantial research and investigations have been carried out in order to find a low-cost, environment-friendly, and sustainable method for the recycling and management of contaminated water and industrial effluents (Enaime et al. 2020; Mondal, Balomajumder, and Mohanty 2007; Gupta et al. 2009).

Several treatment methods such as membrane filtration, ultrafiltration, ion exchange, electrodialysis, chemical precipitation, reverse osmosis, coagulation, and flocculation can be employed for the removal of different toxic and emerging pollutants from industrial wastewater. But some of the major limitations associated with their wide application are the high setup cost, high operating and maintenance cost, continuous energy requirement, low proficiency, nonadaptability, and the generation of pollutants after the treatment process. If all these methods are compared, then the adsorption process is the method found to be environment-friendly and advantageous in many aspects, such as its high efficiency, easy operation, and cost-effectiveness, availability of adsorbent material and generation of zero pollutants (Kumi et al. 2020; Bergmann 2015; Enaime et al. 2020; Sounthararajah et al. 2015).

The adsorption method can be considered a simple and economical technology due to the fact that most of the adsorbents can be regenerated and reused. Many research investigations have found it be very effective and feasible for dye waste treatment (Ghaedi et al. 2011; Makrigianni et al. 2015; Ahmad et al. 2015; Manimekalai, Sivakumar, and Periyasamy 2015), the removal of heavy metal ions (Beni and Esmaeili 2019; Pyrzynska 2019; Álvarez-Merino, López-Ramón, and Moreno-Castilla 2005; Inam et al. 2016; Mondal, Balomajumder, and Mohanty 2007; Sounthararajah et al. 2015), the removal of antibiotics, pesticides, and complex organic pollutants (X. Fan et al. 2021; Gautam et al. 2015; Inam et al. 2016; Zhongtian Li, Dvorak, and Li 2012), industrial effluent treatment (Fathy, El-Shafey, and Khalil 2013; Wani and Patil 2017; Patil et al. 2019), gray water treatment and turbidity removal (Singh 2017; Fathima et al. 2016; Christopher 2012), and radioactive wastewater treatment (Zizhen et al. 2021). Biochar is an effective adsorbent material due to its large surface area, high porosity, oxygen-containing surface functional groups, and low cost, and hence it has the potential to absorb the various existing and emerging contaminants present in industrial wastewater (L. Li et al. 2019).

Biochar is a stable carbon-rich material obtained as the solid by-product of pyrolysis (Mohan, Sarswat et al. 2014). Its extensive porous structure and surface functional groups make it a suitable adsorbent for such applications as soil remediation, wastewater recycling, anaerobic digestion, catalysis, and electrochemical devices (Figure 2.1).

Enormous research has been carried out to establish the ability of biochar for the absorption of organic matter and contaminants, metals and their complex compounds, filtration of suspended solids, promoting the growth of microorganisms, holding water and nutrients with the simultaneous increase in carbon content of the soil (Pokharel, Acharya, and Farooque 2020; Wang et al. 2020; X. Fan et al. 2021). Biochar can be prepared from a variety of raw material that includes the biomass not only from different sources but also from plastic waste and organic waste like sewage and animal manures (L. Li et al. 2019). It is lesser carbonized than activated carbon and contains more oxygen, and hydrogen remains in the structure and therefore adsorbs hydrocarbons along with other organic and inorganic ions. Biochar can replace activated carbon obtained from other common sources like wood, coconut shell, and coal as a low-cost adsorbent for both pathogens and contaminants (Sharma et al. 2018).

Biochars have a remarkable ability to adsorb different synthetic organic chemicals (SOC) such as pesticides, herbicides, detergents, polycyclic aromatic hydrocarbons, nitrosamines, phenolic compounds, trihalomethanes, and other pollutants and natural organic matter (NOM), which is basically composed of partially and completely decomposed plants and animal remains in the environment. Use of chlorine as a disinfectant can generate the formation of carcinogenic compounds like chlorophenols and halomethanes.

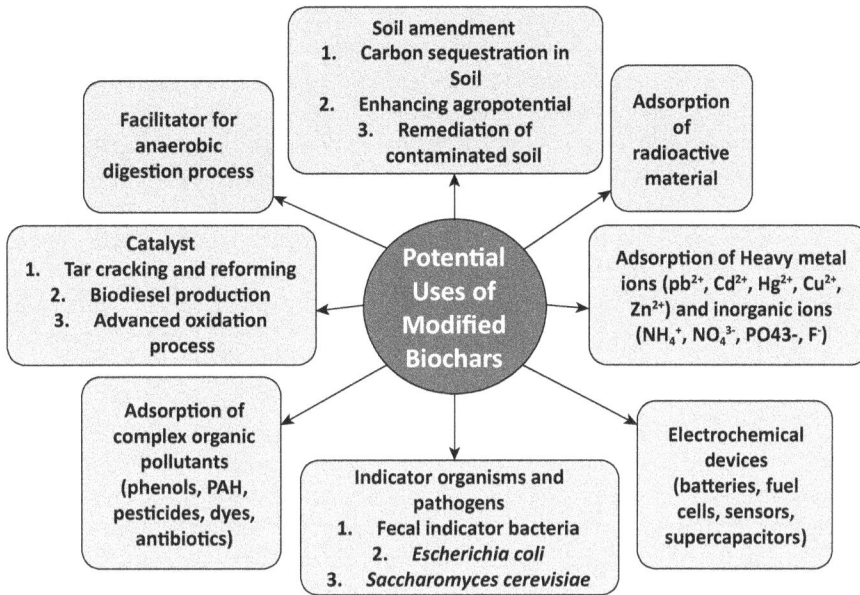

FIGURE 2.1 Potential uses of modified biochar.

Among the other several methods available for the control of organic contamination in water, the use of biochar has been proved to be a wide-spectrum technology that can be employed for the removal of different categories of pollutants at the same time. Therefore a constant increase has been observed in the utilization of activated carbon for water treatment purposes all over the world (A. Da̧browski 2001).

Granular activated carbon (GAC) adsorption is an effective treatment technology for the removal of organics during potable water treatment and thereby improves its taste, color, and odor. It is also found to be helpful in the removal of trace heavy metals such as Cd, Cr, Hg, Cu, Fe, V, Zn, and Ni from the water. The important properties associated with the use of GAC are their tremendous adsorptive capacity and selectivity, ability to be regenerated, stability during thermal treatment, and easy storage and transport. Therefore, nowadays, activated carbons are being used on a much larger scale than ever before. GAC application methods have been extensively developed for municipal and industrial wastewater treatment with a limited approach for the regeneration and disposal of spent adsorbent (Van Der Hoek, Hofman, and Graveland 1999).

This chapter is aimed at providing an overview of the most significant developments in the application of modified biochars in the application of wastewater recycling and management. A brief account of the wide variety of available feedstock for biochar production and different methods of preparation is taken into consideration. The physical and chemical properties of biochar and methods of modification are discussed in order to enhance the selectivity and efficiency of biochar treatment. Environmental protection and sustainability issues associated with the biochar applications also require the careful attention of researchers in order to develop this method as a holistic approach of water remediation.

2.2 FEEDSTOCK FOR BIOCHAR PREPARATION

The raw material required for biochar production is biomass, which can be obtained from various sources and is abundantly available at low cost. It mainly includes agricultural waste, municipal solid waste, forestry residue, etc. Solid waste can be converted into biomass, and this could be a viable solution to the vast problem of solid waste management (Santos et al. 2016). Pyrolysis is commonly used process for the conversion of biodegradable solid waste into biochar. The physical and chemical properties of biochar depends on pyrolysis conditions i.e., temperature, residence time, reactor type and rate, and chemical composition of feedstock (Huang, Lee, and Huang 2021).

The potential sources of biochar could be forestry residues and biomass, agricultural residues like corn stover, husk, sawdust or municipal waste, e.g., kitchen waste. Biomass with high volatile content gives a very low yield of biochar (Y. van Fan, Klemeš, and Lee 2021). The biochar parameters most affected by feedstock properties were, e.g., total organic carbon, fixed carbon, mineral elements of biochar, and potential total C sequestration (L. Zhao et al. 2013). Lignocellulosic plants as a feedstock give a high biochar yield as compared to other feedstock because lignin preferentially forms char during pyrolysis (Antal 2003). So it was observed that biochars produced from woodchip had a higher C/N ratio and specific surface area (SSA) than the dairy manure biochar produced at the same temperature (Lei and Zhang 2013).

Cellulose biomass is preferred over lignin biomass when pyrolyzed at the same temperature because the micropore area of cellulose biochar was found to be greater than that of the lignin biochar. The pinewood biochar, which is obtained by high temperature pyrolysis, usually contains a larger surface area and total pore volume due to greater degree of lignin. It reflects the role of high temperature in the development of well developed pore structure (Jianfa Li et al. 2014). Freshly harvested biomass like animal waste, sewage sludge, vegetable wastes, and algae have higher moisture content (>30%) and are known as wet biomasses.

But biomass of wood species and agricultural residues have lower moisture content (<30%) and is therefore called dry biomass (Sharma et al. 2018). Some of the examples of feedstock used for biochar preparation are pine sawdust biochars (Lou et al. 2016), *Miscanthus saccharifloru* (Shim et al. 2015), broiler litter (Uchimiya et al. 2010), potato leaves and stems (J. Zhao et al. 2019), rice husk (Kaetzl et al. 2019), siderite (M. Li et al. 2019), bamboo (Q. S. Liu et al. 2010), acenaphthylene (Turner and Thomas 1999), acorn (Ghaedi et al. 2011), plastic wastes (Manimekalai, Sivakumar, and Periyasamy 2015), oil-palm-stone (Jia Guo 2000), peanut shells (Fatimah et al. 2018), sewage sludge (X. Fan et al. 2021), etc. (Figure 2.2).

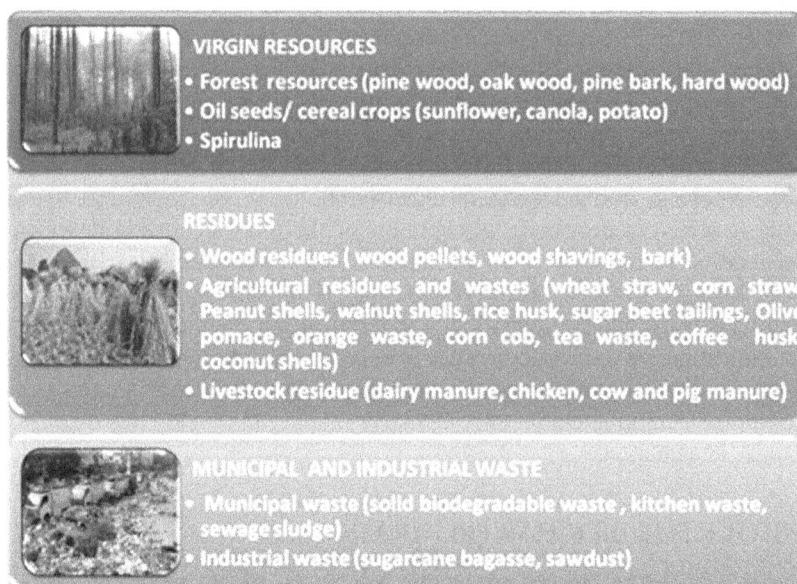

VIRGIN RESOURCES
- Forest resources (pine wood, oak wood, pine bark, hard wood)
- Oil seeds/ cereal crops (sunflower, canola, potato)
- Spirulina

RESIDUES
- Wood residues (wood pellets, wood shavings, bark)
- Agricultural residues and wastes (wheat straw, corn straw, Peanut shells, walnut shells, rice husk, sugar beet tailings, Olive pomace, orange waste, corn cob, tea waste, coffee husk, coconut shells)
- Livestock residue (dairy manure, chicken, cow and pig manure)

MUNICIPAL AND INDUSTRIAL WASTE
- Municipal waste (solid biodegradable waste , kitchen waste, sewage sludge)
- Industrial waste (sugarcane bagasse, sawdust)

FIGURE 2.2   Common feedstock for biochar preparation.

During pyrolysis, the complex organic compounds present in biomass, i.e., cellulose, hemicellulose, lignin, are converted into carbon-rich mass, and all the other impurities are removed. After the removal of different gaseous and volatile by-products, the O/C and H/C ratios decrease, and the aromatic and carbonaceous content increases. At higher temperatures, these effects become more pronounced, and the hydrophobic character, stability, and porous structure of the biomass are enhanced (Enaime et al. 2020). Biomass can be carbonized by slow or fast pyrolysis in which the thermal destruction of biomass takes place in the absence of oxygen and the rate of heating, temperature, residence time decide the characteristics of biochar (Godwin et al. 2019).

2.3 METHODS FOR BIOCHAR PRODUCTION AND MODIFICATION

2.3.1 METHODS FOR BIOCHAR PRODUCTION

Biochar can be produced as a coproduct from several different processes. Commonly, biochar is produced by pyrolysis or carbonization of biomass due to the high charcoal yields, but other technologies cannot be underestimated (Manyà 2012). Some of the conventional techniques used for biochar production are mentioned in Figure 2.3.

Biochar production can be an intermediate step in the preparation of activated carbon. After the carbonization of the raw material (such as agriculture residue) under inert atmosphere or limited oxygen atmosphere, biochar is produced, and then it is activated by physical or chemical activation processes to form activated carbon (Qian et al. 2015). Activated carbon is a versatile material for the various water and wastewater treatment applications due to well developed internal pore structure. Acid-treated activated carbon is found effective for the higher uptake of metal ions, while base-treated activated carbon has shown higher uptake of anionic species from aqueous solutions (Bhatnagar et al. 2013).

2.3.1.1 Conventional Pyrolysis

This is a traditional heating method for biochar production in which heat is transferred from an external source to the biomass via conduction, radiation, and convection. Due to such heat transfer, temperature at the surface of the feedstock is very high and decreases toward the center of the feedstock. The peak temperature is the maximum temperature attained during the process,

CONVENTIONAL METHODS FOR BIOCHAR PRODUCTION			
PYROIYSIS	**HYDROTHERMAL CARBONIZATION (HTC)**	**GASIFICATION**	**TORREFACTION**
1. Fast pyrolysis 2. Slow pyrolysis 3. Thermal destruction of biomass in the absence of oxygen 4. Enrich carbon content and eliminate noncarbon specie	1. Used for high energy density material 2. Can be used for wet feedstock 3. Energy-intensive drying process is not required 4. High conversion efficiency 5. High yield at low operational temperature	1. Thermochemical conversion of feedstock into syngas 2. Tar and char are two undesirable products 3. Properties of char depend upon gasification temperature, reactor design, nature of feedstock, and gasifying reagent	1. Mainly used for Biofuel production 2. Enables energy densification and homogenization of biomass 3. Char contains high amount of oxygen containing functional groups 4. Involves slow heating at lower temperatures

FIGURE 2.3 Methods of biochar production,

and an increase in the peak temperature causes an increase in the fixed carbon content of biochar. A tube-type furnace and muffle furnace are used in slow pyrolysis, and the advantages are simple operation and high yield of biochar. During slow pyrolysis, the temperature remains between 250 and 700°C, and the retention time could be several hours. The other technique is fast pyrolysis in which temperature is about 500°C, the heating rate is about 20–50°C/min, and the retention time is a few seconds. Fixed-bed and fluidized-bed reactors are used, and the major product is tar with a very low yield of biochar. The major factors affecting product quality are the heating rate and residence time in contact with vapors. Therefore, biomass with high moisture content can be suitably used in this process. The limiting factors of this process are thermal conductivity of feedstock and convection current in the system (Zizhen et al. 2021; Manyà 2012; Jianfa Li et al. 2014).

Generally, biochars produced at lower temperature have lower surface area, low pore volume, and functional groups containing more oxygen at the surface. These can be effectively applied for the removal of inorganic pollutants while the biochars produced at high temperature show more hydrophobic and aromatic character and can be effectively applied to the removal of organic contaminants (Justine Cox et al. 2012).

2.3.1.2 Dry Torrefaction

In this process, biomass is heated in an inert atmosphere keeping the temperature range at 200–300°C with residence time ranging from 30 min to 2–3 h. Sometimes it is called mild pyrolysis. Generally, a loss of 30% is observed in mass, while 10% of the energy is lost in gaseous form. As a result, energy density of the torrefied product increases. This method is employed to improve the physiochemical and thermochemical properties of biomass for its conversion into fuel for the combustion process (Kaliaguine and Bardestani 2019; Sharma et al. 2018; Enaime et al. 2020).

2.3.1.3 Hydrothermal Carbonization (HTC)

HTC is a thermochemical process for converting biomass into a solid hydrochar having high carbon content. The HTC process is also called wet torrefaction or hydrothermal treatment. Inside a hydrothermal reactor, temperature is kept at 180–260°C, and heating is done at pressure 2–6 MPa in an oxygen-limited environment. Biomass is submerged into water, and, due to the low pyrolysis temperature, a mild reaction takes place (Zizhen et al. 2021; Pokharel, Acharya, and Farooque 2020). At temperatures higher than 260°C, the reaction becomes severe, and HTC can be classified as hydrothermal gasification (HTG) or hydrothermal vaporization (HTV) and hydrothermal liquefaction (HTL). In both processes, the desired product is gaseous or liquid fuel, and hence feedstock with high moisture content can also be used (Sharma et al. 2018).

2.3.1.4 Gasification

In this process, partial oxidation of feedstock takes places at high temperature (600–1200°C) with small residence time (10–12 s). The product contains a large amount of inorganic metals due to ash formation, and sometimes highly toxic polyaromatic hydrocarbons (PAHs) can also be present depending on the ash content and composition of the feedstock. This raises safety concerns for the application of this biochar for soil amendment. Because of the high operating temperature and partial oxidation of biomass, the yield of biochar is very small, and the main product is a mixture of gases (CO, H_2, CO_2) known as syngas (L. Li et al. 2019; X. Wang et al. 2020). The process is characterized by high residence time in fluidized bed and thermal cracker for short residence time (Hale et al. 2012). Commonly, air is used as the gasification agent, and, due to partial combustion of the fuel, a combustible gas with a low heating value of 3.5–10.0 MJ Nm^3 is generated, which can be used as a fuel for a boiler, gas turbine, or gas engine. The quality of the gas can be further improved by using other oxidizing agents such as steam, carbon dioxide, or a mixture of oxygen and steam. This enhances the amount of carbon monoxide and hydrogen, and now the gas can be used in various fields like chemical synthesis, fuel cell feed, hydrogen production, etc. due to the high operating temperature and the partial oxidizing atmosphere.

2.3.2 Modification of Biochar

Surface modifications of the biochar can be considered a promising and attractive method for the selective and effective removal of organic and inorganic contaminants during wastewater treatment. Modification basically involves oxidation and further grafting onto the activated carbon surface by chemical, electrochemical, plasma, and/or microwave method to introduce functional groups (e.g., carboxylic acid, amine, etc.) (Bhatnagar et al. 2013). Table.2.1 provides a detailed input of different agents and methods for biochar modification.

2.3.2.1 Acid and Alkaline Treatment

Acid treatment of biochar oxidizes the porous carbon surface and increases the acidic property, removes the mineral elements, and improves the hydrophilic character. Nitric acid and sulfuric acid

TABLE 2.1
Methods of Biochar Modification

S. No.	Modification treatment method	Modification agent	Change in biochar characteristics	Application in wastewater treatment	References
1.	Acidic treatment	Nitric acid, sulfuric acid	Improves acidic character and hydrophilic nature of surface, removes minerals	Removal of heavy metal ions	Álvarez-Merino, López-Ramón, and Moreno-Castilla 2005; Turner and Thomas 1999
2.	Alkaline treatment	Ammonia, urea	Produces positive porous surface structure, formation of nitrogen containing functional groups on surface, better adsorbate–adsorbent interactions	Adsorption of organic species (phenols, p-chlorophenol)	Park and Jang 2002; Stavropoulos, Samaras, and Sakellaropoulos 2008
3.	Impregnation	Ag, Cu, Al, Fe	Enhanced adsorption potential toward fluoride, cyanide, and heavy metals like arsenic	Removal of fluoride, arsenic and dissolved organic matter in water	Mondal, Balomajumder, and Mohanty 2007; Leyva Ramos, Ovalle-Turrubiartes, and Sanchez-Castillo 1999
4.	Steam activation	Steam	To increase structural porosity and remove impurities of incomplete combustion	Adsorption of heavy metals and antibiotics	Shim et al. 2015; Lou et al. 2016
5.	Microwave treatment	Microwave heating	Formation of stable aromatic and elemental carbon, polar functional groups on surface, highly porous structure	Removal of heavy metals and dyes	Wahi et al. 2017; Q. S. Liu et al. 2010
6.	Ozone treatment	Ozone	Formation of acidic surface oxygen groups	Mercury sorption	Manchester et al. 2008; Jaramillo, Álvarez, and Gómez-Serrano 2010
7.	Plasma treatment	Plasma in vacuum/ limited air	Formation of weakly acidic functional group, oxygen-containing functional group	Metal ions and organic compounds	Lee et al. 2005; W. Zhang et al. 2012

(Continued)

TABLE 2.1
(Continued)

S. No.	Modification treatment method	Modification agent	Change in biochar characteristics	Application in wastewater treatment	References
8.	Biological modification	Bacteria	Biologically activated carbon used in activated carbon chamber	Removal of wide variety of toxic aquatic pollutants, natural organic matter and organic micropollutants	Zhongtian Li, Dvorak, and Li 2012; Van Der Hoek, Hofman, and Graveland 1999
9.	Magnetization	Fe^{3+}/Fe^{2+} solution	Magnetic components impregnated in biochar, which makes the separation of biochar from aqueous phase easy and also enhances the sorption ability	Pb^{2+} and Cd^{2+} removal, remediation of Cr(VI)	Z. Liu, Zhang, and Sasai 2010; Mohan, Kumar et al. 2014; Yu et al. 2013

Other methods of chemical surface modification using surfactants

S. No.	Modification treatment method	Modification agent	Change in biochar characteristics	Application in wastewater treatment	References
10.	Sodium dodecyl-sulfate (SDS) and sodium diethyl dithiocarbamate (SDDC)			Removal of heavy metals from industrial phosphoric acid	Monser, Ben, and Ksibi 1999
11.	Tetrabutyl ammonium iodide (TBAI) and SDDC			Removal of Cu, Zn, Cr, and CN^- from metal finishing wastewater	Monser 2002
12.	Tris-(hydroxymethyl) amino methane (AC–TRIS)			Selective separation of Au(III)	Albishri and Marwani 2011
13.	Zincon-modified activated carbon (AC–ZCN)			Preconcentration of trace Cr(III) and Pb(II)	Zhenhua Li et al. 2009
14.	Cationic surfactant (cetyltrimethylammonium chloride (CTAC))			Removal of bromate ion (BrO_3^-)	W. Chen et al. 2012
15.	Quaternary ammonium-/epoxide-forming compounds (QAE)			Perchlorate removal	Hou et al. 2012
16.	Ammonium pyrrolidine dithiocarbamate			Removal of different metal ions	Soylak and Doan 1996
17.	5,5-diphenylimidazoli- dine-2,4-dione (phenytoin)			Removal of different metal ions	Ghaedi et al. 2008
18.	Chelating agent pyrocatechol violet			Removal of Cu, Mn, Co, Cd, Pb, Ni, and Cr	Narin, Soylak, and Elc 2000
19.	Pyridyl azo resorcinol			Removal of Cu, Co, Cd, Cr, Ni, Pb, and, V	Chakrapani et al. 1998
20.	Heating from ambient temperature to 900°C in SO_2			To increase the adsorption efficiency	Ansari 2009

are the most commonly used for this purpose. The presence of acidic functional groups (i.e., oxygen functional groups containing proton donors) on surfaces generates favorable conditions for the selective adsorption of heavy metals due to the tendency of metal ions to form complexes with the negatively charged groups. On the other side, alkaline treatment produces a positive surface charge, which in turn helps in the efficient adsorption of negatively charged species (Stavropoulos, Samaras, and Sakellaropoulos 2008; Park and Jang 2002).

Alkaline treatment is done to introduce porous structure with basic. Generally, it is performed by heating the biochar at high temperature in inert hydrogen or ammonia atmosphere. Nitrogen-containing groups are introduced at the surface, and it enhances the interaction between porous carbon and acid molecules by dipole–dipole, hydrogen bonding, and covalent interactions. Additionally, under alkaline conditions, hydroxide ions react with the surface functional groups, and the adsorption of organic contaminants is enhanced (Godwin et al. 2019; Makrigianni et al. 2015).

Biochars treated by urea developed basic characteristics at the surface, and enhancement of phenol removal capacity was observed, which can be due to the increased nitrogen content of these samples. Activated carbons treated by oxygen and nitric acid presented significant acidic character and a lower phenol adsorption capacity. The oxidation treatment affected the pore structure development in different modes: oxygen gasification slightly affected the surface area and pore volumes of the produced samples, while the highest reduction in pore structure was observed for the samples treated by nitric acid solutions. As a result, urea and possibly other nitrogenous compounds may be used as effective media for the enhancement of phenol adsorption potential in activated carbons (Stavropoulos, Samaras, and Sakellaropoulos 2008).

2.3.2.2 Steam Activation

Steam activation is commonly used to introduce porous structure and oxygen-containing functional groups (e.g., carboxylic, carbonyl, ether and phenolic hydroxyl groups) onto BC surfaces; hence the increasing hydrophilicity of biochar despite steam being a weaker oxidant. Steam changes the properties of BCs by removing the trapped products of incomplete combustion during pyrolysis and oxidizes the carbon surface by generating mainly H_2, CO, and CO_2. Steam removes carbon atoms from the surface network and helps in opening clogged pores and creating new pores. In addition, the polarity index $((O + N)/C)$ decreased from 0.154 to 0.138 because of both an increase in the C content and a decrease in the O content after the activation process (Shim et al. 2015). More basic surface functional groups are introduced to biochar, which is generally used for adsorption of hydrocarbons. There is increase in surface hydrophobicity due to removal of hydrophilic groups such as carbonyl and ether groups (Ahmed et al. 2016) and consequently the capacity for the adsorption of complex organic contaminants is also increased (Uchimiya et al. 2010; Lou et al. 2016; Shim et al. 2015).

2.3.2.3 Plasma Treatment

Plasma treatment of biochar performs the oxidation of biochar surface when it is exposed to plasma under vacuum or atmospheric pressure in the presence of controlled air or oxygen. This process significantly changes the surface characteristics of biochar. Plasma oxidation increases the surface acidity by the chemical addition of oxygen to the carbon surfaces due to the aggressive reaction of oxygen-free radicals with carbon atoms located at the peripheral layers. Plasma treatment can not only alter the amount and property of surface functional groups but also change the physical structure of the activated carbon. These changes are related to the change in adsorption ability of biochar for metal ions (Bhatnagar et al. 2013).

Low-temperature oxygen plasma was used to modify activated carbons (ACs), and the adsorption capacity toward dibenzothiophene (DBT) in a model diesel fuel was enhanced. The oxygen plasma made a carbon surface rich in oxygen-containing groups and significantly improved their adsorption capacities. Additional advantage of this method is very little mass loss during modification as compared to conventional thermal oxidation treatment, which resulted in huge mass loss (W. Zhang et al. 2012). Similar studies were done using helium–oxygen (He-O_2) plasma, which was generated in a DBD (dielectric barrier discharge) reactor of planar type that can easily produce spatially uniform plasma with a little amount of oxygen. This plasma can treat the activated carbon surface more uniformly than the pure oxygen plasma. Another advantage of this method is the additive oxygen in the mixture gas that enables the discharge to create oxygen radicals that react on the carbon surface and create weakly acidic functional groups helpful in the adsorption of metal ions (Lee et al. 2005).

2.3.2.4 Surface Oxidation by Ozone

Surface oxidation is a very common low-cost method for the modification of biochar surface and enhancement of adsorption capacities. Reagents that can be used for this purpose are molecular oxygen, ozone, hydrogen peroxide, nitric acid, and permanganate. It provides hydrophilic character to the surface. Ozone treatment causes different structural changes as compared to air or hydrogen peroxide oxidation, and a transient increase in the affinity of surface toward mercury adsorption has been noticed due to the introduction of high-activity oxygen-containing groups (Manchester et al. 2008). But this decreases exposure to atmospheric moisture or heating. Spectroscopic and wet chemical assays suggest the presence of epoxides or secondary ozonides as the active but labile oxidizing groups. Compared to the alkaline treatment of biochars, treatment of the activated carbons with ozone yields a more basic surface (Manchester et al. 2008; Jaramillo, Álvarez, and Gómez-Serrano 2010). Experimental results revealed that the adsorption capacity sharply decreased as the number of sulfonic groups in the aromatic ring increased. As the concentration of oxygenated electron-withdrawing groups on the carbon surface increased, a significant reduction in adsorption capacity of aromatic sulfonic compounds was observed (Bhatnagar et al. 2013).

The point of zero charge (pH_{PZC}) is the pH value required for the surface of adsorbent to have a net neutral charge. The pH of the solution affects the adsorption capacity and overall adsorption process (Bergmann 2015). Ozone-treated activated carbons can be subjected to thermal treatment for the control of the acidic–basic character and control over the surface chemistry of activated carbon is possible. The ozone treatment of the activated carbons changes the surface characteristic from basic to acidic. pH_{PZC} is as low as 3.6. But when the heat treatment is applied to the ozone-treated product, the basic strength of the carbon surface increases with the increasing heat treatment temperature. So now pH_{PZC} is as high as 10.3 (Jaramillo, Álvarez, and Gómez-Serrano 2010).

2.3.2.5 Microwave Treatment

Microwave treatment comes with several advantages over conventional methods of biochar production. It provides rapid, uniform, and selective heating of biochar via microwave radiation. Direct contact between microwave source and biomass is not required, and the product has better surface physical properties like porosity and surface area (Wahi et al. 2017). Moisture content of the feedstock helps to enhance the heating rate during microwave pyrolysis contrary to the conventional pyrolysis method in which process slows down due the presence of moisture in the feedstock. Microwave treatment helps in evaporation of the moisture content of the feedstock before removal of volatile organic content and thereby increases the porosity of the biochar surface (Jing Li et al. 2016).

Being a good absorber of microwave energy, heat is conducted through dipole rotation and ionic conduction. Major advantages are rapid increase in temperature, uniform heat distribution, and energy savings. Significant structural changes can be obtained within a short time due to a distinct mechanism of microwave heating. During conventional pyrolysis, the heat is transferred from outside to the interior part by conduction and convection, causing a thermal gradient from the surface toward the interior side. On the other hand, microwave heating produces a thermal gradient in the opposite direction generating internal and volumetric heating (Jia Guo 2000).

Bamboo-based activated carbon was modified by microwave heating in the presence of N_2 atmosphere, and a gradual decrease in the surface acidic groups and increase in the basicity were observed, causing an increase in the pH_{PZC} value (Q. S. Liu et al. 2010).

2.3.2.6 Biological Modification

Biologically modified activated carbon can be prepared by trapping the bacteria within the surface carbon network, and these bacteria multiply under an ideal environment of temperature and organic nutrients for growth. The life of biologically activated carbon-bed can be increased by preozonation. The attached microorganisms then convert the biodegradable portion to biomass, carbon dioxide, and waste products (Bhatnagar et al. 2013). Granular activated carbon (GAC) can be converted into biologically activated carbon (BAC) by promoting biofilm growth in the reactor, and this has been effectively used for the removal of 17b-estradiol (E2) and estrone (E1), which is an E2 biodegradation intermediate (Zhongtian Li, Dvorak, and Li 2012). Biological activated carbon filtration (BACF) is found to be very useful in the domestic water treatment process for the removal of natural organic matter (NOM) and particularly for the organic micropollutants (Van Der Hoek, Hofman, and Graveland 1999).

2.3.2.7   Chemical Impregnation

Impregnation refers to the fine distribution of chemicals and/or metal particles in the pores of activated carbon. For this purpose, metals such as silver, copper, aluminum, and iron can be used, and it has shown significant increase in surface adsorption capacity. The adsorption potential of biochar for fluoride, cyanide, and heavy metals like arsenic in water was found to be enhanced by this method (Bhatnagar et al. 2013). Generally, simple metal ions do not adsorb easily onto biochar surface due to their good solvation in aqueous solutions and the hydrophobic nature of the carbon surface. Adsorption of inorganic contaminants can also be increased by impregnating the activated carbon with suitable chemicals, which causes chemisorption of ions by chemical reactions, e.g., chelation, neutralization, redox, hydrolysis, precipitation, and catalytic reactions. pH is the most important factor affecting the adsorption in aqueous phase in the case of inorganics, weak organic acid, and bases because their dissociation is highly pH dependent. AC impregnated with organic compounds containing active groups like -SH or -NH can be more effective for the elimination of heavy metals from the effluents (Ansari 2009). Aluminum-impregnated biochar was used for the removal of fluoride from water (Leyva Ramos, Ovalle-Turrubiartes, and Sanchez-Castillo 1999), and urea impregnation increased phenol adsorption (Stavropoulos, Samaras, and Sakellaropoulos 2008). Impregnation with K_2CO_3 causes increased capability for the adsorption of cationic dye like methylene green 5 (Tran, You, and Chao 2017), and Fe^{3+}-impregnated AC was used for the removal of arsenic, iron, and manganese from groundwater (Mondal, Balomajumder, and Mohanty 2007). Sewage sludge biochar was modified by the metal-loaded method, and biochar samples were prepared by impregnating the biochar with Fe, Mn, and Al. All impregnated biochar samples showed remarkable adsorption of antibiotics tetracycline (TC), sulfamethoxazole (SMZ), and amoxicillin (AMC). Surface studies revealed the possible adsorption mechanism to be pores filling, van der Waals forces, and H-bonding (X. Fan et al. 2021).

2.3.2.8   Magnetization of Biochar

To deal with the problems of biochar separation from aqueous solution, magnetization of biochar can be done, which helps in easy removal of biochar from the aqueous phase and its further regeneration after the treatment is complete (Z. Zhang et al. 2019). Mohan, Kumar et al. (2014) produced magnetic biochar using Fe^{3+}/Fe^{2+} solution. By increasing the iron content up to 80.6%, the biochar was effectively magnetized. Biochar showed significantly high sorption capacity for Pb^{2+} and Cd^{2+} removal from solutions. Microwave heating can be effectively used for the synthesis of magnetic biochar. Yu et al. (2013) prepared magnetic biochar by mixing it with ammoniacal solution of Fe^{3+}/Fe^{2+} and irradiating the sample with ultrasound at 60°C. It resulted in an increase in the number of surface carboxyl functional groups, causing more negative charge, and consequently the sorption rate and capacity for heavy metal ions were found to be improved.

Sorption ability is increased in the presence of magnetic components like Fe_2O_3, Fe_3O_4, FeO, and Fe^0 in biochar. Fe^0 contributes to the enhanced Pb(II) adsorption by direct reduction, while Fe_3O_4 helps in the remediation of Cr(VI) contaminated water. This can be attributed to the presence of Fe(II) and Fe(III) in Fe_3O_4 in the octahedral state acting as an active site of chemical sorption or reduction. Surface morphology of magnetic biochar is affected by peak pyrolysis temperature, and at high temperature, the Fe_3O_4 in magnetic biochar is transformed into FeO (Y. di Chen et al. 2018). Z. Liu, Zhang, and Sasai (2010) investigated arsenate adsorption from water using Fe_3O_4-loaded biochar prepared from waste biomass (pinewood sawdust) in the presence of $FeCl_3$ as the metallic solution to provide the magnetic effect. S. Wang et al. (2015) studied arsenic removal by a magnetically modified biochar produced using natural hematite and pinewood. Meng et al. (2015) used chitosan as the raw material to produce $MnFe_2O_4$-loaded magnetic biochar by a microwave-assisted hydrothermal method in which microwave heating was done at the temperature of 120°C for 10 min. The magnetic biochars were used for adsorption of Cu^{2+} ions from synthetic wastewater.

2.4 SURFACE CHARACTERISTICS OF BIOCHAR FOR WASTEWATER TREATMENT

Feedstock properties and production conditions decide the surface characteristics of biochar. Feedstock properties influences the biochar parameters like total organic carbon, fixed carbon, and mineral elements of biochar, while the highest treatment temperature (HTT) affects the surface area and pH. HTT also determines the biochar recalcitrance while the potential total C sequestration (product of recalcitrance and pyrolysis carbon yield) depends more on the nature of feedstock (L. Zhao et al. 2013). Adsorption capacity is determined by the porous structure of the surface while the presence of functional groups on the surface affects adsorbate–adsorbent interactions. The surface-active sites allows interaction of functional groups with other hetero-atoms or molecules in the adsorbate (Park and Jang 2002).

Biochar surface can have various functional groups, and they can be classified into several groups: O-containing groups (e.g., carboxylic groups, phenolic groups, lactonic groups), N-containing groups (e.g., amine-N groups, pyrrolic-N groups, graphitic-N groups), S-containing groups (e.g., sulfonic groups), and other functional groups (Huang, Lee, and Huang 2021). The textural properties of the adsorbents are decided by its pore size, total pore volume, and superficial area. The pores allow the adsorbate to be retained on the surface and to diffuse in the internal pores of the adsorbent (intraparticle diffusion). Therefore, pore size greatly affects the adsorption efficiency of adsorbent by controlling the rate of diffusion of adsorbate in the internal pores of the adsorbent. Small molecules are adsorbed at the micropores, while larger molecules are retained by the mesopores of adsorbent. The greater number of micropores on material provides higher surface area (Bergmann 2015).

Physicochemical properties of biochar depends on its surface area, porosity, elementary composition, and molar ratios such as H/C, O/C, and N/C. Steam activation can be used to enhance the surface area of biochar, decreasing the H/C ratio (aromaticity) and reducing polarity at the surface (Ahmed et al. 2016). The physical activation of biochar is less effective compared to acidic, alkaline, or salt modified biochars in terms of average surface area. Acid-modified biochars possess high H/C and O/C ratios compared to the primary BCs, while the N/C ratio can vary. Alkaline treatment produces a larger surface area with a higher ratio of surface aromaticity (H/C) and higher N/C ratio with a lower value of O/C compared to acid-modified BCs. More nitrogen-containing groups, which are responsible for the basic properties, are present on the modified BC surface. Additionally, the reaction of alkaline materials (containing OH- ion or $-NH_2$ group) with surface functional groups increases sorption of negatively charged species and organic contaminants from water or wastewater (Ahmed et al. 2016). The impregnation method improves physical and chemical properties of biochars forming new sites like composites with the distribution of nanosized particles on their surface. These modifications effectively enhances the sorption capacity of biochar for a variety of contaminants from water and wastewater (Godwin et al. 2019).

When a gas, vapor, or liquid comes in contact with a solid surface, some of the molecules are taken up by solid. They can enter the inside of the solid, which is known as absorption, or remain on the outside attached to the surface, which is called adsorption. When the phenomena occur simultaneously, the process is termed sorption. The solid surface involved in sorption is called adsorbent, and the gas, vapor, or solute taken up on the surface is called adsorbate. It is a reversible reaction in case of physical adsorption due to the weak van der Waals attraction between adsorbate and adsorbent, but it becomes irreversible in case of chemical adsorption, which involves stronger chemical forces between the adsorbate and adsorbent surface (Ansari 2009). Figure 2.4 represents sorption mechanism for metal ions and organic contaminants on biochar surface.

Chemical composition and morphology of biochar surface can be understood by biochar characterization, and it includes the techniques of SEM (scanning electron microscopy), TEM (transmission, electron microscopy), XRD (X-ray diffraction), XPS (X-ray photoelectron spectroscopy), FTIR (Fourier transform infrared spectrometry), X-ray spectroscopy, STEM (scanning transmission electron microscopy), and EXAFS (X-ray adsorption spectroscopy). The surface area, pore volume, and diameter at the biochar surface can be measured with Brunauer–Emmett–Teller (BET) adsorption model in which the nitrogen adsorption isotherm through multilayer adsorption capacity is determined under different nitrogen partial pressures (Mohan, Kumar et al. 2014; Ghaedi et al. 2011). The technique of XRD is used for determination of angle and intensity of diffracted biochar beams in which sharp narrow peaks identify graphitized carbon, and broad peaks represent non-graphitized carbon (L. Li et al. 2019; X. Fan et al. 2021). In FTIR spectrum, different functional groups are identified on the biochar surface. The different vibration position represents different

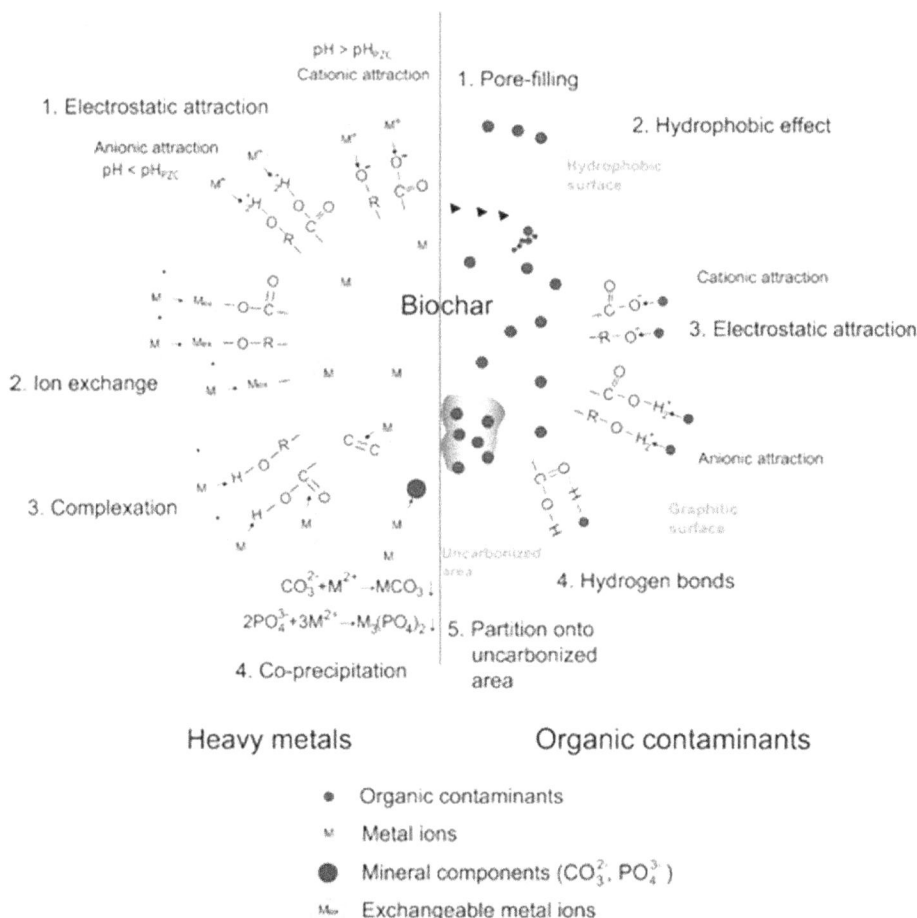

FIGURE.2.4 Biochar sorption mechanism.

Source: X. Wang et al. (2020).

functional groups in the infrared spectrometry. XPS studies the distribution of photoelectrons with energy and identifies the valence state of chemical elements. SEM shows the surface morphologies of biochar through electron beam, and TEM is used for similar applications with high resolution and to find the lattice constant of biochar (Manimekalai, Sivakumar, and Periyasamy 2015; Jia Guo 2000). STEM characterizes biochar at the atomic levels, while EDS is used to determine the molar fraction of elements on the surface of biochar. EXAFS characterizes the polyaromatic structure of biochar, and surface pH can be measured by a pH probe. An element analyzer can be used for measuring carbon and nitrogen content (Kumi et al. 2020).

2.5 WASTEWATER RECYCLING AND MANAGEMENT USING MODIFIED BIOCHAR

Biochar can perform various functions during wastewater treatment process (Figure 2.5), and not only can this enhance the treatment efficiency, but also value-added byproducts can be recovered afterward. The principle mechanisms involved in the removal of contaminants can be adsorption, buffering, and immobilization of microbial cells. Suitably modified biochar can be used for the selective adsorption of various emerging and existing contaminants from WWTP (wastewater treatment plant) effluent before its discharge into the environment.

2.5.1 REMOVAL OF INORGANIC POLLUTANTS

Removal of inorganic ions by biochar upon on the valence state of the ion as well as the solution pH. Removal of heavy metal ions by surface adsorption may involve (1) electrostatic forces of attraction between heavy metals ions in solution and the biochar surface; (2) ion exchange between heavy

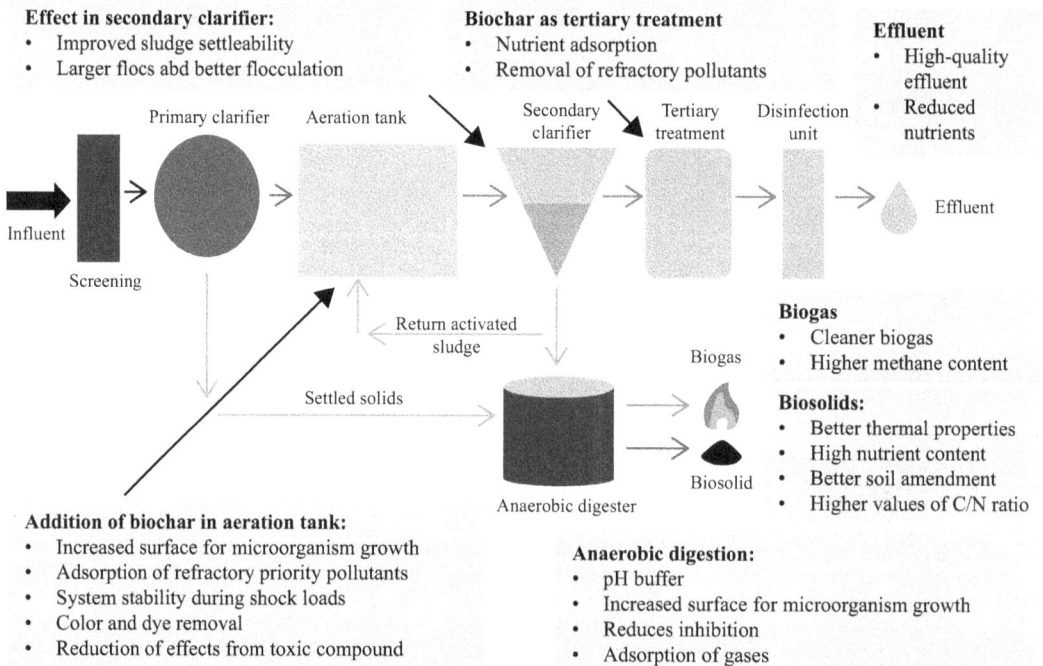

FIGURE.2.5 Functions of biochar during wastewater treatment.

Source: Pokharel, Acharya, and Farooque (2020).

metals ions and protons or other lighter group 1 and 2 metal ions on biochar surface; (3) complexation of metal ions with unsaturated surface functional groups; (4) coprecipitation on the surface to form insoluble compounds. Solution pH can strongly influence the surface charge of biochar. pH_{PZC} is the pH value of solution in contact with biochar surface at which the net surface charge becomes zero. Therefore, the surface is positively charged at solution $pH < pH_{PZC}$, and it can bind metal anions such as $HAsO_4^{2-}$ and $HCrO_4^-$ (H. Li et al. 2017). The surface is negatively charged at solution $pH > pH_{PZC}$ and combines with metal cations such as Hg^{2+}, Pb^{2+}, and Cd^{2+}. Biochar generated by carbonization of biomass contains numerous cations on the surface (Na^+, K^+, Mg^+, Ca^{++}) that can be exchanged heavy metal ions. Mineral components in the biochar surface acts as other sorption sites and helps in the adsorption of heavy metals by precipitation (X. Wang et al. 2020). Several investigations have been made to study removal of inorganic ions from wastewater.

Zinc is a common pollutant in different industrial effluents. The effect of surface oxidation, solution pH, and ionic strength of biochar on the static and dynamic adsorption of Zn(II) ions was studied by Álvarez-Merino et al. 2005 using commercial granular activated carbon, activated carbon cloth, and column beds. The oxidation process causes an increase in the number of total acid sites, mainly carboxyl acid groups, and a large decrease in the pH_{PZC} (pH at the point of zero charge) of the samples. Oxidation largely increased Zn(II) uptake under static conditions because of the electrostatic adsorbate–adsorbent interactions between the negatively charged surface and positively charge metal ions. Sodium titanate nanofibrous material was used for the removal of Cu, Cd, Ni, Pb, and Zn individually (single metal system) and together (mixed metals system) from water. The study revealed high potential of titanate material for removing heavy metals from polluted water when used with granular activated carbon at a very low proportion in fixed-bed columns (Sounthararajah et al. 2015).

The chemically modified activated carbon was used to study the mechanism and rate of Cr(VI) adsorption in aqueous solution. The possible mechanisms of total chromium removal were found to be highly dependent on the surface oxygen complexes at low pH. Hydrochloric acid treatment produced increased numbers of various surface oxygen complexes, thus increasing the adsorption active sites and improving the reduction rates of Cr(VI) (Park and Jang 2002).

Modification of activated carbon was done by DBD treatment using a helium plasma mixed with oxygen in iron cation solutions. The increase in adsorptivity of plasma-treated activated carbon is about 3.8 times higher than that of the untreated one. It is found that after the helium-oxygen plasma treatment, the surface area decreases due to wall destruction and pore blockage, but acidic functional groups have been generated enhancing the absorptivity by giving more hydrogen ions (Lee et al. 2005).

Use of surface-modified adsorbents is also becoming common for the development of cheaper arsenic removal technique. Adsorbents are developed from different feedstock, and the modification method like Cu-impregnated coconut husk carbon, iron-oxide-coated polymeric materials, iron-oxide-coated sand, iron-oxide-coated cement, bead cellulose loaded with iron oxy hydroxide (P. Mondal, Majumder, and Mohanty 2006). Removal of arsenic from groundwater was investigated by the adsorption onto Fe^{3+}-impregnated granular activated carbon (GAC-Fe) in the presence of Fe^{2+}, Fe^{3+}, and Mn^{2+}, and the arsenic concentration in the treated water was reduced below 10 and 50 ppb. GAC-Fe may be used as a cheap adsorbent for arsenic removal from drinking water (Mondal, Balomajumder, and Mohanty 2007). Hematite-modified magnetic biochar was also used for arsenic removal, and apart from its greater ability to remove As from aqueous solution, it can be easily isolated and removed with external magnets and can be conveniently used in various arsenic contaminant removal applications (S. Wang et al. 2015). Ozone-treated biochar was shown to have mercury capture capacity higher by a factor of 134, but the activity was easily destroyed when exposed to the atmosphere, water vapor, or even mild heating. The very high mercury adsorption capacity is due to the presence of labile epoxides or secondary ozonide (Manchester et al. 2008).

Adsorption of copper ions was evaluated by using steam-activated biochar (BC) produced from *Miscanthus sacchariflorus*. Although the surface area of activated biochar was quiet high, the Cu sorption capacities remains nearly same as compared to as-received biochar (Shim et al. 2015).

The adsorption efficiency of activated carbon prepared from plantain peels (a biowaste from confectioneries) and adsorption of lead (II) and 3,7-bis (dimethylamino)-phenothiazin-5-ium chloride was studied as common water pollutants. Results have indicated potential of this modified biochar to be used as adsorbent for removal of heavy metal pollutants from textile wastewater (Inam et al. 2016). Biochars were produced from *Miscanthus sacchariflorus*, and high pyrolysis temperature greatly increased the Cd sorption capacity of biochar. A possible mechanism of removal could be surface sorption or a precipitation reaction depending on pH (Kim et al. 2013).

Biochar was treated with saturated $FeCl_3$ solution followed by neutralization with NaOH solution, and then this modified biochar was used to observe phosphorus adsorption from wastewater. It was found to contain one-third of the phosphorus amount present in the commonly used agricultural fertilizer and to promote easier absorption of phosphorus by plants. This method can be utilized for recycling of phosphorus from wastewater (Kopecký et al. 2020). Similarly, tailor-made biochars with large phosphate adsorption capacity can be used to tackle the problem of eutrophication in water bodies. Pine sawdust was used to prepare biochars, followed by steam activation resulting in a negatively charged surface and thereby representing the limited capacity of phosphate removal. This observation suggests that a high surface area and oxygen functionality could not always enhance adsorption capacity. This biochar shows effective adsorption of cationic adsorbates. Therefore a selective and critical approach is required to apply biochar for phosphate removal (Lou et al. 2016). The adsorption isotherms of fluoride from an aqueous solution on plain and Al-impregnated activated carbon were studied by Ramos et al. (1999), and they found the impregnated carbon with fluoride adsorption capacity 3–5 times greater than that of the plain carbon.

2.5.2 REMOVAL OF ORGANIC POLLUTANTS

Adsorption of organic contaminants on biochar surface takes place by the mechanisms of pore filling, hydrophobic and electrostatic interaction, and hydrogen bonds depending on the properties of both the contaminant and biochar. At higher pyrolysis temperatures, greater carbonization of biochar is achieved, and it causes well developed micropores and larger surface area, leading to an enhanced sorption rate. On the other hand, biochars produced at intermediate temperatures (250–350°C) show slower sorption rates due to the presence of highly condensed organic phases causing difficult pore filling. Electrostatic interaction is considered the mechanism of adsorption for polar organic contaminants (Inyang et al. 2014). For example, the methyl violet represented adsorption due to electrostatic interaction between dye molecules containing carboxylate anion and phenolic hydroxyl groups on biochar surface (X. Wang et al. 2020).

Organic contaminants in water can be natural organic matter (NOM), pesticides, antibiotics, dyes, phenols, and other complex organic compounds. Removal of a broad spectrum of organic pollutants has been studied by researchers worldwide. Biological activated carbon filtration (BACF) is a very reliable technique for the removal of organic matter and pesticides. Specific advantages are the very long lifetime of the carbon with an excellent pesticide removal capacity. A very good residual capacity for pesticide removal has been shown by filters even after an operational period of three years. This method can be accompanied by other advance methodologies like reverse osmosis, ozonization, and slow sand filtration to increase the effectiveness of the water treatment process (Van Der Hoek, Hofman, and Graveland 1999).

Zheng et al. 2010 investigated the sorption of two triazine pesticides, atrazine and simazine, on biochar. Under different sorption conditions, the sorption ability of atrazine was greater than simazine. Uchimiya et al. 2010 produced biochar by pyrolyzing broiler litter at 350 and 700°C and then using it to remove deisopropylatrazine, which is a stable metabolite of atrazine from water. The biochar prepared at 700°C was found to have greater surface area due to more micropores, and hence it displays greater aromatic character, which is required for effective removal of target contaminants. The biochar prepared below 500°C was found to have comparatively lower adsorption ability for similar reasons. Tetracyclines (TCs) and sulfonamides (SAs) are two commonly used antibiotics used as field additives in intensive agricultural practices. Their extensive use is hazardous to both the environment and human health. Under weakly acidic conditions, high-temperature biochar displayed high sorption capacity, which can be due to strong π–π interactions between the antibiotic molecules and aromatic groups on the biochar surface and micropore filling due to the smaller size of SAs (Peiris et al. 2017). J. Zhao et al. (2019) prepared magnetic biochar coated with humic acid from potato stems, and adsorption of three antibiotic fluoroquinolones (FQs) (enrofloxacin (ENR), norfloxacin (NOR), and ciprofloxacin (CIP)) was studied. High FQs removal efficiency was observed due to the mechanism of hydrophobic and electrostatic interactions and formation of hydrogen bonds.

Similar studies were carried out to remove estrogenic compounds in drinking water sources that pose potential threats to human health. Granular activated carbon was tested for its ability to remove 17b-estradiol (E2), and a high adsorption capacity was observed. To further enhance E2 removal, the GAC reactor was converted to a biologically active carbon (BAC) reactor by promoting biofilm growth in the reactor (Zhongtian Li, Dvorak, and Li 2012).

Urea-treated samples with a basic character and high nitrogen content presented the highest phenol uptake capacity, while samples treated with nitric acid or oxygen gasified have shown acidic surface properties with low phenol uptake capacity. The presence of nitrogen on biochar surface causes favorable adsorbate–adsorbent interactions and increases phenol uptake (Stavropoulos, Samaras, and Sakellaropoulos 2008). To obtain low-sulfur liquid transportation fuel, deep desulfurization of transportation fuels is done by using plasma modified AC. Low-temperature oxygen plasma modifies carbon surfaces for affective adsorption of dibenzothiophene (DBT) by increasing oxygen-containing acidic functional groups at the surface and also avoids the severe mass loss during modification (W. Zhang et al. 2012).

Direct discharge of textile industry effluent causes the contamination of natural waters with a complex mixture of dyes that are difficult to remove from water. These dyes possess their toxic effects on living beings, and also the by-products enter the local food chain. They alter the chemical properties of natural water by changing the solution pH, color, and chemical oxygen demand, and they obstruct the microbial growth. They hinder the photosynthetic activity due to less penetration of solar radiation in water bodies. Their immediate and effective removal is required from the contaminated sources once they enter into the natural environment. It was seen that extended micropore volume enhances adsorption of methylene blue on the activated carbons. A faster adsorption rate is also observed by the kinetic studies (Q. S. Liu et al. 2010).

Biochar was prepared from green tea dredge by pyrolysis, and it was found effective to remove methylene blue dye from aqueous solutions. Since the pH does not affect MB removal, treatment can be done without pH adjustment. Due to significant reduction in the dye concentration, green tea dredge can be considered a low-cost adsorbent for the removal of basic dyes (Ahmad et al. 2015). Acid-treated pyrolytic tire char (PTC) was studied for adsorption of both the phenol and methylene blue dye from aqueous solutions. Phenol uptake can be attributed to the π–π interactions that cause hydrogen bonding of phenol with surface groups, while electrostatic attraction between the carboxylic groups of biochar surface and cationic dye is responsible for the adsorption of MB (Makrigianni et al. 2015). Activated carbon prepared from acorn was studied for the removal of brilliant green dye by the adsorption technique, and more than 90% removal efficiency was obtained within 30 min at an adsorbent dose of 2 g/100 mL for initial dye concentration of 25 mg/L (Ghaedi et al. 2011).

Plastic waste (mostly PP, PVC, and PET) can be utilized for the production of activated carbon (PWAC), and it was studied for the removal of acid red-114 from textile industry effluent. The rate of adsorption decreases with an increase in temperature, which reflects the physical adsorption mechanism (Manimekalai, Sivakumar, and Periyasamy 2015). Preparation of the biosorbents from *Lantana camara* weed by sulfuric acid activation was carried out, and it was used as an adsorbent for the removal of an acidic dye tartrazine from aqueous solutions. The adsorption was found to be spontaneous, endothermic and favorable (Gautam et al. 2015). Activated carbon (AC) was synthesized from golden shower (GS) through a three-stage process: the first step is the hydrothermal carbonization of GS to produce hydrochar, followed by pyrolysis of hydrochar to produce biochar, and subsequent chemical activation of biochar with K_2CO_3 at the end. Surface adsorption was found to be irreversible, and oxygenation of the ACs' surface through a hydrothermal process with acrylic acid causes a decrease in MG5 adsorption (Tran, You, and Chao 2017).

Biochar application can be utilized for the removal of indicator organisms and pathogens. Biochar filters for removal of microorganism from water have been studied extensively (X. Wang et al. 2020). Kaetzl et al. (2019) prepared low-cost filter materials from rice-husk biochar and evaluated their potential and limitation for wastewater treatment. These filters have shown superior performance as compared to standard sand filters. When the treated wastewater was used for lettuce irrigation in a pot test, very little contamination was found with fecal indicator bacteria. Perez-Mercado et al. (2019) successfully removed *Saccharomyces cerevisiae* by using biochar as a filter medium from diluted wastewater under the condition of on-farm irrigation. A greater number of micropores with smaller pore size causes effective removal of bacteria. Mohanty et al. (2014) improved sand biofilters with biochar, and hence the bacteria removal capacity was enhanced. It can filter three times more *Escherichia coli* and also can prevent their mobility during continuous or intermittent flows.

Research on the application of biochar in the treatment of radioactive wastewater has been increasing year by year, especially the improvement of the adsorption effect of biochar modification methods (Zizhen et al. 2021). Uranium (U) is a toxic and radioactive element. Excessive amounts of aqueous U(VI) generated from uranium mining, processing, and the nuclear industry cause severe and irreversible damage to the environment.

The aerial root of *Ficus microcarpa* (FMAR), which is a biowaste material, was used to adsorb U(VI) from aqueous solutions. Potassium permanganate ($KMnO_4$)-modified FMAR biochar was compared with raw unmodified biochar with respect to U(VI) adsorption. The results indicated higher adsorption capability of the modified FMAR biochar. It was seen that $KMnO_4$-modified FMAR biochar has a good potential to serve as an environment-friendly adsorbent for the removal of U(VI) from solution (N. Li et al. 2019).

The magnetic biochar composites were applied for the adsorption of U(VI). They are fabricated by direct heating of siderite and rice husk under N_2 condition by a low-cost, effective, and eco-friendly method. Due to a porous structure with larger specific surface area, high adsorption properties of magnetic biochar composites were displayed for U(VI), and it is found to be affected by Na_2CO_3. U(VI) adsorbed onto magnetic biochar was reduced to U(IV) by Fe_3O_4. Studies revealed inner-sphere surface complexation of U(VI) on magnetic biochar. Therefore the magnetic biochar could be regarded as an alternative adsorbent in many environmental applications to reduce the risks of U(VI) contamination (M. Li et al. 2019).

2.5.3 REGENERATION OF SPENT BIOCHAR

The adsorption capacity of modified BCs is limited due to the limited number of active sites available on the surface; therefore a prolonged usage of biochars during wastewater treatment results in the establishment of thermodynamic equilibrium between the BCs and contaminants, and adsorption capacity becomes limited. This generates the need for biochar regeneration.

Desorption is done for biochars having saturated surfaces, and it can be performed with solutions of KNO_3, HNO_3, or $NaNO_3$ at different concentrations. These types of desorbents can supply a significant quantity of cations, which displace heavy metal ions adsorbed on the biochar surface. The effectiveness of desorption can be increased at low pH (Southtararajah et al. 2015). Several other techniques are also available for regeneration of biochars such as partial pressure reduction, heat treatment, use of inert gas or fluid to purge, and simply changing the pH. Commonly used solvents for regeneration purpose are acetic acid, NaOH, HCl, NaCl, EDTA, etc. (Godwin et al. 2019; Southtararajah et al. 2015).

2.6 SUSTAINABILITY ISSUES OF BIOCHAR APPLICATION

Although the effectiveness of biochar applications is well established, the sustainability issues associated with commercial large-scale applications are not properly addressed (Y. van Fan, Klemeš,

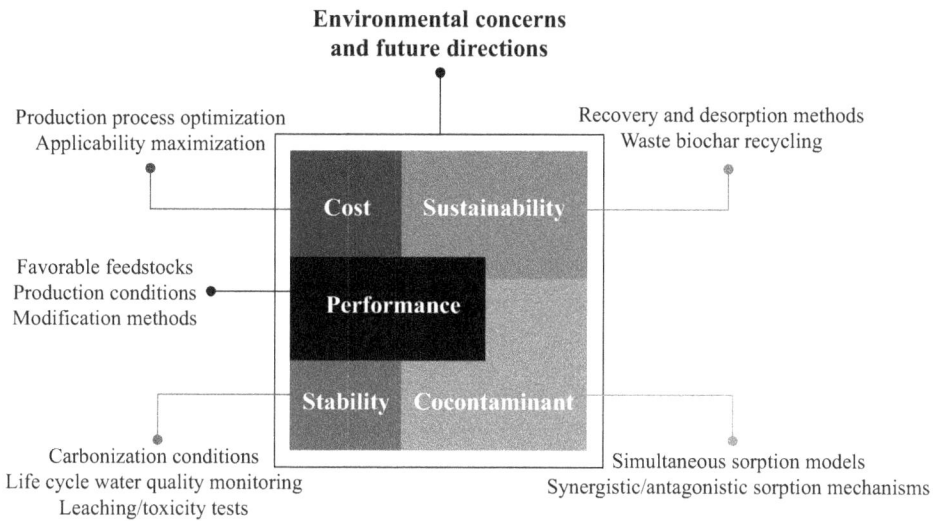

Environmental concerns and future directions

Production process optimization
Applicability maximization

Recovery and desorption methods
Waste biochar recycling

Cost   Sustainability

Favorable feedstocks
Production conditions
Modification methods

Performance

Stability   Cocontaminant

Carbonization conditions
Life cycle water quality monitoring
Leaching/toxicity tests

Simultaneous sorption models
Synergistic/antagonistic sorption mechanisms

FIGURE 2.6   Environmental concerns and future directions.

Source: X. Wang et al. (2020).

and Lee 2021). The technique of biochar application is still in the testing stage and not yet applied on commercial levels (Figure 2.6). Many countries in the world still lack a proper technology for biochar production, as well as industrial facilities to produce and consume biochars, and several environmental concerns are still related with the practical applications of biochar in the field of wastewater treatment. More focused and substantial research is required in this field in order to address the potential environmental problems associated with the large-scale production and application of biochar adsorbent (X. Wang et al. 2020).

The modification methods of biochar include many disadvantages like the high cost of modification process and leaching of the modifying chemicals used in the water being treated. Very little information is available about the usage of surface-modified ACs on the level of column, pilot, or full scale because most of the studies are limited to only batch tests. Therefore, for the wider range of biochar applications, performance assessment of modified biochar is required at column, pilot, or full scale.

Another important issue related to the biochar application is their real-time behavior when used for the treatment of real wastewater, groundwater, or surface water. The potential of modified biochar can be altered due to the presence of different coexisting ions. Also, very little information about cost and regeneration of such modified ACs is available. It is important to find out that, during the regeneration of biochar, the adsorbed pollutant is also getting released or retained in the surface. Furthermore, no clear information is available regarding the disposal of waste biochar generated after several wastewater treatment cycles. Researchers should focus their efforts toward the development of simple and efficient biochars that are also easily regenerated.

Sometimes biochar may incorporate toxic chemicals generated during the pyrolysis of feedstock. Hale et al. (2012) found that biochar may contain polycyclic aromatic hydrocarbons or dioxins depending on the production conditions. During pyrolysis, organic compounds contained in the biomass are partly cracked into smaller and unstable fragments that contain highly reactive free radicals combinations from which stable polyaromatic hydrocarbons known as toxic contaminants can form. Dioxins are mostly formed on solid surfaces when the pyrolysis temperature is 200–400°C and the pyrolysis times is in seconds. Biochar source, production temperature, pyrolysis time, and aging all affect the level of toxins in the resulting biochars (Hale et al. 2012).

During sorption studies, the main focus remains on the study of single contaminants in aqueous solutions. But during real water applications on a large scale, several contaminants can coexist, and they can produce ionic interference by competing for active binding sites, and thereby removal efficiency is reduced. Due to the limited availability of research work in this field, the establishment of simultaneous sorption models is required (X. Wang et al. 2020).

Biochar production requires processing of feedstock like grinding and cleaning followed by drying and pyrolysis of biomass. Modification steps are also required for an ideal sorption effect. Therefore, all these preparatory methods inevitably increase the production cost. So further studies are required to optimize the production process while maximizing the applicability of biochar and minimizing the production cost.

Biochar can be considered a sustainable material to be applied as activated carbon, absorbent material, catalyst, and additive. But it usually depends on the selected conditions and applications scenarios. While enhancing the sustainability of the biochar application, its efficiency needs to be optimized through use of the surface area and attached functional groups. Suitably tailored and engineered biochar can be developed for area-specific and industry-specific needs (Y. van Fan, Klemeš, and Lee 2021).

2.7 CONCLUSION

In recent years, a rapid growth has been witnessed in the human population as well as in industrial and commercial activities. There has been a high load on environmental resources to meet the exponential demands in every sector of life. It is a great challenge for governments worldwide to provide clean and safe drinking water to their populations. Public and private water supply cannot rely only upon the fresh natural water sources, which are also contaminated to serious levels. Under such circumstances, the treatment and management of wastewater is the only viable solution to the problem of the growing water crisis.

Adsorption of common pollutants and selective removal of specific contaminants is a promising method for the recycling of wastewater, and it can be easily adopted in municipal water treatment plants, sewage treatment plants, and industrial effluent treatment plants. Another advantage is the regeneration of spent adsorbent, which makes it more sustainable and cost-effective. In addition, it requires less manpower and low technical skills.

Biochar is environment-friendly adsorbent material, and it can be produced at a very low cost using abundantly available different varieties of feedstock. Several researches have indicated that by a suitable modification method or merely by changing the method of biochar production, desired contaminants can be adsorbed from natural and wastewater. Developing countries are constantly facing the need for a low-cost and low-energy-intensive technique that can be applied for the large-scale treatment of water in remote villages and hilly areas. Pilot-scale studies should be promoted to find out the technical and commercial feasibility of the biochar adsorption technique at commercial levels.

This technique can be molded to fit area-specific requirements of water and wastewater treatment and availability of feedstock. Biochar can be produced from the agricultural and farm residues available in rural areas and selectively modified for the removal of a broad category of low-level contamination from water. However, in the urban areas, municipal waste can be utilized for the same purpose, and biochar can be modified for the removal of specific contaminants present in that region that may be due to geographical conditions, e.g., the presence of fluoride, arsenic, and calcium in the water, or due to the presence of particular industries around that area, e.g., presence of chromium ions, phenols, antibiotics, or fecal microorganisms, among others. Industry-specific ready-to-use biochars can be developed for the purpose of water treatment at effluent treatment plants (ETPs) for the selective removal of the major industry-specific contaminants.

The problem of solid waste management can also be targeted by this technology, and this is an additional advantage. A great scope of research and investigation lies in the development of biochar adsorbents for the small-scale removal of a broad category of common pollutants as well as the large-scale removal of specific severe pollutants to find ambient common solutions for the problems of water pollution and the water crisis.

REFERENCES

Ahmad, Muhammad, Robert Thomas Bachmann, Misbahul Ain Khan, Robert G.J. Edyvean, Umar Farooq, and Muhammad Makshoof Athar. 2015. "Dye Removal Using Carbonized Biomass, Isotherm and Kinetic Studies." *Desalination and Water Treatment* 53 (8): 2289–2298. https://doi.org/10.1080/19443994.2013.867818.

Ahmed, Mohammad Boshir, John L. Zhou, Huu H. Ngo, Wenshan Guo, and Mengfang Chen. 2016. "Progress in the Preparation and Application of Modified Biochar for Improved Contaminant Removal from Water and Wastewater." *Bioresource Technology* 214: 836–851. https://doi.org/10.1016/j.biortech.2016.05.057.

Albishri, Hassan M., and Hadi M. Marwani. 2011. "Chemically Modified Activated Carbon with Tris (Hydroxymethyl) Aminomethane for Selective Adsorption and Determination of Gold in Water Samples." *Arabian Journal of Chemistry* 9: S252–S258. https://doi.org/10.1016/j.arabjc.2011.03.017.

Álvarez-Merino, Miguel A., Victoria López-Ramón, and Carlos Moreno-Castilla. 2005. "A Study of the Static and Dynamic Adsorption of Zn(II) Ions on Carbon Materials from Aqueous Solutions." *Journal of Colloid and Interface Science* 288 (2): 335–341. https://doi.org/10.1016/j.jcis.2005.03.025.

Ansari, R. 2009. "Activated Charcoal: Preparation, Characterization and Applications: A Review Article." *International Journal of ChemTech Research* 1 (4): 859–864.

Antal, Michael Jerry. 2003. "The Art, Science, and Technology of Charcoal Production †." *Industrial & Engineering Chemistry Research* 42: 1619–1640. https://doi.org/10.1021/ie0207919.

Beni, Ali Aghababai, and Akbar Esmaeili. 2019. "Biosorption, an Efficient Method for Removing Heavy Metals from Industrial Effluents: A Review Ali." *Environmental Technology & Innovation* 17 (February): 100503. https://doi.org/10.1016/j.eti.2019.100503.

Bergmann, Carlos P. 2015. *Carbon Nanomaterials as Adsorbents for Environmental and Biological Applications*. Edited by Carlos P. Bergmann and Fernando Machado Machado. Switzerland: Springer International Publishing AG. www.springer.com/series/8633.

Bhatnagar, Amit, William Hogland, Marcia Marques, and Mika Sillanpää. 2013. "An Overview of the Modification Methods of Activated Carbon for Its Water Treatment Applications." *Chemical Engineering Journal* 219: 499–511. https://doi.org/10.1016/j.cej.2012.12.038.

Chakrapani, G. Ł., D. S. R. Murty, P. L. Mohanta, and R. Rangaswamy. 1998. "Sorption of PAR—Metal Complexes on Activated Carbon as a Rapid Preconcentration Method for the Determination of Cu, Co, Cd, Cr, Ni, Pb and V in Ground Water." *Journal of Geochemical Exploration* 63: 145–152.

Chen, Wei-fang, Ze-ya Zhang, Qian Li, and Hong-yan Wang. 2012. "Adsorption of Bromate and Competition from Oxyanions on Cationic Surfactant-Modified Granular Activated Carbon (GAC)." *Chemical Engineering Journal* 203: 319–325. https://doi.org/10.1016/j.cej.2012.07.047.

Chen, Yi di, Shih Hsin Ho, Dawei Wang, Zong su Wei, Jo Shu Chang, and Nan Qi Ren. 2018. "Lead Removal by a Magnetic Biochar Derived from Persulfate-ZVI Treated Sludge Together with One-Pot Pyrolysis." *Bioresource Technology* 247 (September 2017): 463–470. https://doi.org/10.1016/j.biortech.2017.09.125.

Christopher, David. 2012. "The Effect of Granular Activated Carbon Pretreatment and Sand Pretreatment on Microfiltration of Greywater." *All Theses* 1463: 1–150.

Cox, Justine, Adriana Downie, Mark Hickey, Abigail Jenkins, Rebecca Lines-Kelly, Anthea McClintock, Janine Powell, Bhupinder Pal Singh, and Lukas van Zwieten. 2012. *Biochar in Horticulture: Prospects for the Use of Biochar in Australian Horticulture*. Sydney, Australia: NSW Trade and Investment.

Dabrowski, A. 2001. "Adsorption – From Theory to Practice." *Advances in Colloid and Interface Science* 93 (2001): 135–224.

Enaime, Ghizlane, Abdelaziz Baçaoui, Abdelrani Yaacoubi, and Manfred Lübken. 2020. "Biochar for Wastewater Treatment-Conversion Technologies and Applications." *Applied Sciences (Switzerland)* 10 (10). https://doi.org/10.3390/app10103492.

Fan, Xiulei, Zheng Qian, Jiaqiang Liu, Nan Geng, Jun Hou, and Dandan Li. 2021. "Investigation on the Adsorption of Antibiotics from Water by Metal Loaded Sewage Sludge Biochar." *Water Science and Technology* 83 (3): 739–750. https://doi.org/10.2166/wst.2020.578.

Fan, Yee van, Jiří Jaromír Klemeš, and Chew Tin Lee. 2021. "Environmental Performance and Techno-Economic Feasibility of Different Biochar Applications: An Overview." *Chemical Engineering Transactions* 83 (February): 469–474. https://doi.org/10.3303/CET2183079.

Fathima, Nazia, Uzma Baig, Smita Asthana, and D. Sirisha. 2016. "Removal of Turbidity of Waste Water by Adsorption Technology." *International Journal of Innovative Research in Science, Engineering and Technology* 5 (11): 20010–20016. https://doi.org/10.15680/IJIRSET.2016.0511066.

Fathy, Nady A., Ola I. El-Shafey, and Laila B. Khalil. 2013. "Effectiveness of Alkali-Acid Treatment in Enhancement the Adsorption Capacity for Rice Straw: The Removal of Methylene Blue Dye." *ISRN Physical Chemistry* 2013: 1–15. https://doi.org/10.1155/2013/208087.

Fatimah, Is, Bonusa Nabila Huda, Ilmi Lucyawati Yusuf, and Budi Hartono. 2018. "Enhanced Adsorption Capacity of Peanut Shell toward Rhodamine b via Sodium Dodecyl Sulfate Modification." *Rasayan Journal of Chemistry* 11 (3): 1166–1176. https://doi.org/10.31788/RJC.2018.1134021.

Gautam, Ravindra Kumar, Pavan Kumar Gautam, Sushmita Banerjee, Vandani Rawat, Shivani Soni, Sanjay K. Sharma, and Mahesh Chandra Chattopadhyaya. 2015. "Removal of Tartrazine by Activated Carbon Biosorbents of Lantana Camara: Kinetics, Equilibrium Modeling and Spectroscopic Analysis." *Journal of Environmental Chemical Engineering* 3 (1): 79–88. https://doi.org/10.1016/j.jece.2014.11.026.

Ghaedi, M., F. Ahmadi, Z. Tavakoli, and M. Montazerozohori. 2008. "Three Modified Activated Carbons by Different Ligands for the Solid Phase Extraction of Copper and Lead." *Journal of Hazardous Materials* 152: 1248–1255. https://doi.org/10.1016/j.jhazmat.2007.07.108.

Ghaedi, M., H. Hossainian, M. Montazerozohori, A. Shokrollahi, F. Shojaipour, M. Soylak, and M. K. Purkait. 2011. "A Novel Acorn Based Adsorbent for the Removal of Brilliant Green." *Desalination* 281 (1): 226–233. https://doi.org/10.1016/j.desal.2011.07.068.

Godwin, Patrick M., Yuanfeng Pan, Huining Xiao, and Muhammad T. Afzal. 2019. "Progress in Preparation and Application of Modified Biochar for Improving Heavy Metal Ion Removal From Wastewater." *Progress in the Preparation and Application of Modified Biochar for Improving Heavy Metal Ion Removal from Wastewater* 4 (1): 31–42. https://doi.org/10.21967/jbb.v4i1.180.

Guo, Jia, and Aik Chong Lua. 2000. "Preparation of Activated Carbons from Oil-Palm-Stone Chars by Microwave-Induced Carbon Dioxide Activation." *Carbon* 38: 1985–1993. https://doi.org/10.1260/026361703772776457.

Gupta, V. K., P. J. M. Carrott, M. M. L. Ribeiro Carrott, and Suhas. 2009. "Low-Cost Adsorbents: Growing Approach to Wastewater Treatmenta Review." *Critical Reviews in Environmental Science and Technology* 39 (10): 783–842. https://doi.org/10.1080/10643380801977610.

Hale, Sarah E., Johannes Lehmann, David Rutherford, Andrew R. Zimmerman, Robert T. Bachmann, Victor Shitumbanuma, Adam O'Toole, Kristina L. Sundqvist, Hans Peter H. Arp, and Gerard Cornelissen. 2012. "Quantifying the Total and Bioavailable Polycyclic Aromatic Hydrocarbons and Dioxins in Biochars." *Environmental Science and Technology* 46 (5): 2830–2838. https://doi.org/10.1021/es203984k.

Hou, Pin, Fred S. Cannon, Nicole R. Brown, Timothy Byrne, Xin Gu, and Cesar Nieto. 2012. "Granular Activated Carbon Anchored with Quaternary Ammonium/Epoxide-Forming Compounds to Enhance Perchlorate Removal from Groundwater." *Carbon* 53: 197–207. https://doi.org/10.1016/j.carbon.2012.10.048.

Huang, Wei Hao, Duu Jong Lee, and Chihpin Huang. 2021. "Modification on Biochars for Applications: A Research Update." *Bioresource Technology* 319 (August 2020). https://doi.org/10.1016/j.biortech.2020.124100.

Inam, Edu I., Ubong J. Etim, Ememobong G. Akpabio, and Saviour A. Umoren. 2016. "Simultaneous Adsorption of Lead (II) and 3,7-Bis(Dimethylamino)-Phenothiazin-5-Ium Chloride from Aqueous Solution by Activated Carbon Prepared from Plantain Peels." *Desalination and Water Treatment* 57 (14): 6540–6553. https://doi.org/10.1080/19443994.2015.1010236.

Inyang, Mandu, Bin Gao, Andrew Zimmerman, Ming Zhang, and Hao Chen. 2014. "Synthesis, Characterization, and Dye Sorption Ability of Carbon Nanotube–Biochar Nanocomposites." *Chemical Engineering Journal* 236: 39–46. https://doi.org/10.1016/j.cej.2013.09.074.

Jaramillo, J., P. M. Álvarez, and V. Gómez-Serrano. 2010. "Preparation and Ozone-Surface Modification of Activated Carbon. Thermal Stability of Oxygen Surface Groups." *Applied Surface Science* 256 (17): 5232–5236. https://doi.org/10.1016/j.apsusc.2009.12.109.

Kaetzl, Korbinian, Manfred Lübken, Gülkader Uzun, Tito Gehring, Edith Nettmann, Kathrin Stenchly, and Marc Wichern. 2019. "On-Farm Wastewater Treatment Using Biochar from Local Agroresidues Reduces Pathogens from Irrigation Water for Safer Food Production in Developing Countries." *Science of the Total Environment* 682: 601–610. https://doi.org/10.1016/j.scitotenv.2019.05.142.

Kaliaguine, S., and R. Bardestani. 2019. "The Application of Pyrolysis Biochar for Wastewater Treatment." https://dc.engconfintl.org/pyroliq_2019/14/.

Kim, Woong Ki, Taeyong Shim, Yong Seong Kim, Seunghun Hyun, Changkook Ryu, Young Kwon Park, and Jinho Jung. 2013. "Characterization of Cadmium Removal from Aqueous Solution by Biochar Produced from a Giant Miscanthus at Different Pyrolytic Temperatures." *Bioresource Technology* 138: 266–270. https://doi.org/10.1016/j.biortech.2013.03.186.

Kopecký, Marek, Ladislav Kolář, Petr Konvalina, Otakar Strunecký, Florina Teodorescu, Petr Mráz, Jiří Peterka, et al. 2020. "Modified Biochar-A Tool for Wastewater Treatment." *Energies* 13 (20): 1–13. https://doi.org/10.3390/en13205270.

Kumi, Andy G., Mona G. Ibrahim, Mahmoud Nasr, and Manabu Fujii. 2020. "Biochar Synthesis for Industrial Wastewater Treatment: A Critical Review." *Materials Science Forum* 1008 MSF: 202–212. https://doi.org/10.4028/www.scientific.net/MSF.1008.202.

Lee, Dongsoo, Sang Hee Hong, Kwang Hyun Paek, and Won Tae Ju. 2005. "Adsorbability Enhancement of Activated Carbon by Dielectric Barrier Discharge Plasma Treatment." *Surface and Coatings Technology* 200 (7): 2277–2282. https://doi.org/10.1016/j.surfcoat.2004.11.027.

Lei, Ouyang, and Renduo Zhang. 2013. "Effects of Biochars Derived from Different Feedstocks and Pyrolysis Temperatures on Soil Physical and Hydraulic Properties." *Journal of Soils and Sediments* 13: 1561–1572. https://doi.org/10.1007/s11368-013-0738-7.

Li, Hongbo, Xiaoling Dong, Evandro B. da Silva, Letuzia M. de Oliveira, Yanshan Chen, and Lena Q. Ma. 2017. "Mechanisms of Metal Sorption by Biochars: Biochar Characteristics and Modifications." *Chemosphere* 178: 466–478. https://doi.org/10.1016/j.chemosphere.2017.03.072.

Li, Jianfa, Yimin Li, Yunlu Wu, and Mengying Zheng. 2014. "A Comparison of Biochars from Lignin, Cellulose and Wood as the Sorbent to an Aromatic Pollutant." *Journal of Hazardous Materials* 280: 450–457. https://doi.org/10.1016/j.jhazmat.2014.08.033.

Li, Jing, Jianjun Dai, Guangqing Liu, Hedong Zhang, Zuopeng Gao, Jie Fu, Yanfeng He, and Yan Huang. 2016. "Biochar from Microwave Pyrolysis of Biomass: A Review." *Biomass and Bioenergy* 94: 228–244. https://doi.org/10.1016/j.biombioe.2016.09.010.

Li, Longcheng, Dongsheng Zou, Zhihua Xiao, Xinyi Zeng, Liqing Zhang, Longbo Jiang, Andong Wang, Dabing Ge, Guolin Zhang, and Fen Liu. 2019. "Biochar as a Sorbent for Emerging Contaminants Enables Improvements in Waste Management and Sustainable Resource Use." *Journal of Cleaner Production* 210: 1324–1342. https://doi.org/10.1016/j.jclepro.2018.11.087.

Li, Mengxue, Haibo Liu, Tianhu Chen, Chen Dong, and Yubing Sun. 2019. "Synthesis of Magnetic Biochar Composites for Enhanced Uranium(VI) Adsorption." *Science of the Total Environment* 651: 1020–1028. https://doi.org/10.1016/j.scitotenv.2018.09.259.

Li, Nuo, Meiling Yin, Daniel C. W. Tsang, Shitong Yang, Juan Liu, Xue Li, Gang Song, and Jin Wang. 2019. "Mechanisms of U(VI) Removal by Biochar Derived from Ficus Microcarpa Aerial Root: A Comparison between Raw and Modified Biochar." *Science of the Total Environment* 697: 134115. https://doi.org/10.1016/j.scitotenv.2019.134115.

Li, Zhenhua, Xijun Chang, Zheng Hu, Xinping Huang, Xiaojun Zou, Qiong Wu, and Rong Nie. 2009. "Zincon-Modified Activated Carbon for Solid-Phase Extraction and Preconcentration of Trace Lead and Chromium from Environmental Samples." *Journal of Hazardous Materials* 166: 133–137. https://doi.org/10.1016/j.jhazmat.2008.11.006.

Li, Zhongtian, Bruce Dvorak, and Xu Li. 2012. "Removing 17β-Estradiol from Drinking Water in a Biologically Active Carbon (BAC) Reactor Modified from a Granular Activated Carbon (GAC) Reactor." *Water Research* 46 (9): 2828–2836. https://doi.org/10.1016/j.watres.2012.03.033.

Liu, Qing Song, Tong Zheng, Nan Li, Peng Wang, and Gulizhaer Abulikemu. 2010. "Modification of Bamboo-Based Activated Carbon Using Microwave Radiation and Its Effects on the Adsorption of Methylene Blue." *Applied Surface Science* 256 (10): 3309–3315. https://doi.org/10.1016/j.apsusc.2009.12.025.

Liu, Zhengang, Fu Shen Zhang, and Ryo Sasai. 2010. "Arsenate Removal from Water Using Fe3O4-Loaded Activated Carbon Prepared from Waste Biomass." *Chemical Engineering Journal* 160 (1): 57–62. https://doi.org/10.1016/j.cej.2010.03.003.

Lou, Kangyi, Anushka Upamali Rajapaksha, Yong Sik Ok, and Scott X. Chang. 2016. "Pyrolysis Temperature and Steam Activation Effects on Sorption of Phosphate on Pine Sawdust Biochars in Aqueous Solutions." *Chemical Speciation and Bioavailability* 28 (1–4): 42–50. https://doi.org/10.1080/09542299.2016.1165080.

Makrigianni, Vasiliki, Aris Giannakas, Yiannis Deligiannakis, and Ioannis Konstantinou. 2015. "Adsorption of Phenol and Methylene Blue from Aqueous Solutions by Pyrolytic Tire Char: Equilibrium and Kinetic Studies." *Journal of Environmental Chemical Engineering* 3 (1): 574–582. https://doi.org/10.1016/j.jece.2015.01.006.

Manchester, Shawn, Xuelei Wang, Indrek Kulaots, Yuming Gao, and Robert H. Hurt. 2008. "High Capacity Mercury Adsorption on Freshly Ozone-Treated Carbon Surfaces." *Carbon* 46 (3): 518–524. https://doi.org/10.1016/j.carbon.2007.12.019.

Manimekalai, T. K., N. Sivakumar, and S. Periyasamy. 2015. "Catalytic Pyrolysis of Plastic Waste into Activated Carbon and Its Adsorption Studies on Acid Red 114 as Model Organic Pollutant." *International Journal of ChemTech Research* 8 (8): 333–348.

Manyà, Joan J. 2012. "Pyrolysis for Biochar Purposes: A Review to Establish Current Knowledge Gaps and Research Needs." *Environmental Science and Technology* 46 (15): 7939–7954. https://doi.org/10.1021/es301029g.

Meng, Yuying, Deyang Chen, Yitao Sun, Dongling Jiao, Dechang Zeng, and Zhongwu Liu. 2015. "Adsorption of Cu 2+ Ions Using Chitosan-Modified Magnetic Mn Ferrite Nanoparticles Synthesized by Microwave-Assisted Hydrothermal Method." *Applied Surface Science* 324: 745–750. https://doi.org/10.1016/j.apsusc.2014.11.028.

Mohan, Dinesh, Hemant Kumar, Ankur Sarswat, M. Alexandre-Franco, and Charles U. Pittman. 2014. "Cadmium and Lead Remediation Using Magnetic Oak Wood and Oak Bark Fast Pyrolysis Bio-Chars." *Chemical Engineering Journal* 236: 513–528. https://doi.org/10.1016/j.cej.2013.09.057.

Mohan, Dinesh, Ankur Sarswat, Yong Sik Ok, and Charles U. Pittman. 2014. "Organic and Inorganic Contaminants Removal from Water with Biochar, a Renewable, Low Cost and Sustainable Adsorbent—A Critical Review." *Bioresource Technology* 160: 191–202. https://doi.org/10.1016/j.biortech.2014.01.120.

Mohanty, Sanjay K., Keri B. Cantrell, Kara L. Nelson, and Alexandria B. Boehm. 2014. "Efficacy of Biochar to Remove Escherichia Coli from Stormwater under Steady and Intermittent Flow." *Water Research* 61: 288–296. https://doi.org/10.1016/j.watres.2014.05.026.

Mondal, Prasenjit, Chandrajit Balomajumder, and Bikash Mohanty. 2007. "A Laboratory Study for the Treatment of Arsenic, Iron, and Manganese Bearing Ground Water Using Fe3+ Impregnated Activated Carbon: Effects of Shaking Time, PH and Temperature." *Journal of Hazardous Materials* 144 (1–2): 420–426. https://doi.org/10.1016/j.jhazmat.2006.10.078.

Mondal, P., C. B. Majumder, and B. Mohanty. 2006. "Laboratory Based Approaches for Arsenic Remediation from Contaminated Water: Recent Developments." *Journal of Hazardous Materials* 137 (1): 464–479. https://doi.org/10.1016/j.jhazmat.2006.02.023.

Monser, Lotfi. 2002. "Modified Activated Carbon for the Removal of Copper, Zinc, Chromium and Cyanide from Wastewater." *Separation and Purification Technology* 26: 137–146.

Monser, Lotfi, Mohamed Ben, and Mohamed Ksibi. 1999. "Purification of Wet Phosphoric Acid Using Modified Activated Carbon." *Chemical Engineering and Processing: Process Intensification* 38: 267–271.

Narin, I., Mustafa Soylak, and Latif Elc. 2000. "Determination of Trace Metal Ions by AAS in Natural Water Samples after Preconcentration of Pyrocatechol Violet Complexes on an Activated Carbon Column." *Talanta* 52: 1041–1046.

Park, Soo Jin, and Yu Sin Jang. 2002. "Pore Structure and Surface Properties of Chemically Modified Activated Carbons for Adsorption Mechanism and Rate of Cr(VI)." *Journal of Colloid and Interface Science* 249 (2): 458–463. https://doi.org/10.1006/jcis.2002.8269.

Patil, Chandrashekhar S., Datta B. Gunjal, Vaibhav M. Naik, Namdev S. Harale, Suryabala D. Jagadale, Abhijit N. Kadam, Pramod S. Patil, Govind B. Kolekar, and Anil H. Gore. 2019. "Waste Tea Residue as a Low Cost Adsorbent for Removal of Hydralazine Hydrochloride Pharmaceutical Pollutant from Aqueous Media: An Environmental Remediation." *Journal of Cleaner Production* 206 (January): 407–418. https://doi.org/10.1016/j.jclepro.2018.09.140.

Peiris, Chathuri, Sameera R. Gunatilake, Todd E. Mlsna, Dinesh Mohan, and Meththika Vithanage. 2017. "Biochar Based Removal of Antibiotic Sulfonamides and Tetracyclines in Aquatic Environments: A Critical Review." *Bioresource Technology* 246: 150–159. https://doi.org/10.1016/j.biortech.2017.07.150.

Perez-Mercado, Luis Fernando, Cecilia Lalander, Abraham Joel, Jakob Ottoson, Sahar Dalahmeh, and Björn Vinnerås. 2019. "Biochar Filters as an On-Farm Treatment to Reduce Pathogens When Irrigating with Wastewater-Polluted Sources." *Journal of Environmental Management* 248 (February): 109295. https://doi.org/10.1016/j.jenvman.2019.109295.

Pokharel, Abhishek, Bishnu Acharya, and Aitazaz Farooque. 2020. "Biochar-Assisted Wastewater Treatment and Waste Valorization." *Applications of Biochar for Environmental Safety*: 1–19. https://doi.org/10.5772/intechopen.92288.

Pyrzynska, Krystyna. 2019. "Removal of Cadmium from Wastewaters with Low-Cost Adsorbents." *Journal of Environmental Chemical Engineering* 7 (1): 102795. https://doi.org/10.1016/j.jece.2018.11.040.

Qian, Kezhen, Ajay Kumar, Hailin Zhang, Danielle Bellmer, and Raymond Huhnke. 2015. "Recent Advances in Utilization of Biochar." *Renewable and Sustainable Energy Reviews* 42: 1055–1064. https://doi.org/10.1016/j.rser.2014.10.074.

Ramos, R. L., J. Ovalle-Turrubiartes, and M. A. Sanchez-Castillo. 1999. "Adsorption of Fluoride from Aqueous Solution on Aluminum-Impregnated Carbon." *Carbon* 37: 609–617. https://doi.org/10.1016/S0008-6223(98)00231-0.

Santos, M. T., J. F. Puna, A. M. Barreiros, and M. Matos. 2016. "Agricultural Wastes for Wastewater Treatment." In *CYPRUS2016 4th International Conference on Sustainable Solid Waste Management*, Limassol, 1–11. http://uest.ntua.gr/cyprus2016/proceedings/pdf/Santos_Puna_Agriculture_wastes_for_wastewater_treatment.pdf

Sharma, Monika, Joginder Singh, Chinnappan Baskar, and Ajay Kumar. 2018. "A Comprehensive Review on Biochar Formation and Its Utilization for Wastewater Treatment." *Pollution Research* 37 (June): S1–18.

Shim, Taeyong, Jisu Yoo, Changkook Ryu, Yong Kwon Park, and Jinho Jung. 2015. "Effect of Steam Activation of Biochar Produced from a Giant Miscanthus on Copper Sorption and Toxicity." *Bioresource Technology* 197: 85–90. https://doi.org/10.1016/j.biortech.2015.08.055.

Singh, Sukhdeep. 2017. "Turbidity Removal from Water by Use of Different Additive Materials." *International Journal of Scientific Engineering and Science* 1 (10): 55–57.

Sounthararajah, D. P., P. Loganathan, J. Kandasamy, and S. Vigneswaran. 2015. "Adsorptive Removal of Heavy Metals from Water Using Sodium Titanate Nanofibres Loaded onto GAC in Fixed-Bed Columns." *Journal of Hazardous Materials* 287: 306–316. https://doi.org/10.1016/j.jhazmat.2015.01.067.

Soylak, M., and M. Doan. 1996. "Column Preconcentration of Trace Amounts of Copper on Activated Carbon from Natural Water Samples." *Analytical Letters* 29 (4): 635–643. https://doi.org/10.1080/00032719608000426.

Stavropoulos, G. G., P. Samaras, and G. P. Sakellaropoulos. 2008. "Effect of Activated Carbons Modification on Porosity, Surface Structure and Phenol Adsorption." *Journal of Hazardous Materials* 151 (2–3): 414–421. https://doi.org/10.1016/j.jhazmat.2007.06.005.

Tran, Hai Nguyen, Sheng Jie You, and Huan Ping Chao. 2017. "Fast and Efficient Adsorption of Methylene Green 5 on Activated Carbon Prepared from New Chemical Activation Method." *Journal of Environmental Management* 188 (2017): 322–336. https://doi.org/10.1016/j.jenvman.2016.12.003.

Turner, J. A., and K. M. Thomas. 1999. "Temperature-Programmed Desorption of Oxygen Surface Complexes on Acenaphthylene-Derived Chars: Comparison with Oxygen K-Edge XANES Spectroscopy." *Langmuir* 15 (19): 6416–6422. https://doi.org/10.1021/la981165c.

Uchimiya, Minori, Lynda H. Wartelle, Isabel M. Lima, and K. Thomas Klasson. 2010. "Sorption of Deisopropylatrazine on Broiler Litter Biochars." *Journal of Agricultural and Food Chemistry* 58 (23): 12350–12356. https://doi.org/10.1021/jf102152q.

Van Der Hoek, J. P., J. A. M. H. Hofman, and A. Graveland. 1999. "The Use of Biological Activated Carbon Filtration for the Removal of Natural Organic Matter and Organic Micropollutants from Water." *Water Science and Technology* 40 (9): 257–264. https://doi.org/10.1016/S0273-1223(99)00664-2.

Wahi, Rafeah, Nur Fakhirah Qurratu ain Zuhaidi, Yusralina Yusof, Jamliah Jamel, Devagi Kanakaraju, and Zainab Ngaini. 2017. "Chemically Treated Microwave-Derived Biochar: An Overview." *Biomass and Bioenergy* 107: 411–421. https://doi.org/10.1016/j.biombioe.2017.08.007.

Wang, Shengsen, Bin Gao, Andrew R. Zimmerman, Yuncong Li, Lena Ma, Willie G. Harris, and Kati W. Migliaccio. 2015. "Removal of Arsenic by Magnetic Biochar Prepared from Pinewood and Natural Hematite." *Bioresource Technology* 175: 391–395. https://doi.org/10.1016/j.biortech.2014.10.104.

Wang, Xiaoqing, Zizhang Guo, Zhen Hu, and Jian Zhang. 2020. "Recent Advances in Biochar Application for Water and Wastewater Treatment: A Review." *PeerJ* 8: e9164. https://doi.org/10.7717/peerj.9164.

Wani, Pawan R., and Sonali B. Patil. 2017. "Treatment of Dairy Waste Water by Using Groundnut Shell as Low Cost Adsorbant." *International Journal of Innovative Research in Science, Engineering and Technology* 6 (7): 14941–14948. https://doi.org/10.15680/IJIRSET.2017.0607353.

Yu, Jun Xia, Li Yan Wang, Ru An Chi, Yue Fei Zhang, Zhi Gao Xu, and Jia Guo. 2013. "Competitive Adsorption of Pb 2+ and Cd 2+ on Magnetic Modified Sugarcane Bagasse Prepared by Two Simple Steps." *Applied Surface Science* 268: 163–170. https://doi.org/10.1016/j.apsusc.2012.12.047.

Zhang, Wei, Haiyong Liu, Qibin Xia, and Zhong Li. 2012. "Enhancement of Dibenzothiophene Adsorption on Activated Carbons by Surface Modification Using Low Temperature Oxygen Plasma." *Chemical Engineering Journal* 209: 597–600. https://doi.org/10.1016/j.cej.2012.08.050.

Zhang, Zhifeng, Huijuan Chen, Wenmei Wu, Wenting Pang, and Guiqin Yan. 2019. "Efficient Removal of Alizarin Red S from Aqueous Solution by Polyethyleneimine Functionalized Magnetic Carbon Nanotubes." *Bioresource Technology* 293 (August): 122100. https://doi.org/10.1016/j.biortech.2019.122100.

Zhao, Jing, Guiwei Liang, Xiaoli Zhang, Xuewei Cai, Ruining Li, Xiaoyun Xie, and Zhaowei Wang. 2019. "Coating Magnetic Biochar with Humic Acid for High Efficient Removal of Fluoroquinolone Antibiotics in Water." *Science of the Total Environment* 688: 1205–1215. https://doi.org/10.1016/j.scitotenv.2019.06.287.

Zhao, Ling, Xinde Cao, Ondřej Mašek, and Andrew Zimmerman. 2013. "Heterogeneity of Biochar Properties as a Function of Feedstock Sources and Production Temperatures." *Journal of Hazardous Materials* 256–257: 1–9. https://doi.org/10.1016/j.jhazmat.2013.04.015.

Zheng, Wei, Mingxin Guo, Teresa Chow, Douglas N. Bennett, and Nandakishore Rajagopalan. 2010. "Sorption Properties of Greenwaste Biochar for Two Triazine Pesticides." *Journal of Hazardous Materials* 181 (1–3): 121–126. https://doi.org/10.1016/j.jhazmat.2010.04.103.

Zizhen, Bao, Liang Dong, Li Honghui, Liu Yuchen, and Cui Anxi. 2021. "Advances on the Application of Biochar in Radioactive Wastewater Treatment." In *ICESCE 2021, E3S Web of Conferences*, 1–6. https://doi.org/10.1051/e3sconf/202126702016.

3 Removal of Emerging Contaminants by Biochar
An Eco-friendly Approach for a Sustainable Environment

Riti Thapar Kapoor, V. P. Sharma, and Maulin P. Shah

CONTENTS

3.1 INTRODUCTION

Clean and safe water is a fundamental requirement of human beings for the multifaceted development of society and a flourishing economy. Global economic and societal development have led to increased water demand along with a water deficit. Water resources are under pressure due to the population growth, urbanization, industrialization, and pollution particularly in developing countries. Rampant industrialization and extensive agriculture practices are mainly responsible for deterioration of water quality, and it is a matter of global concern. Millions of people die every year due to diseases communicated through consumption of water contaminated by deleterious contaminants. The strain on the water system will grow by the year 2050 when the world population will increase to ten billion. Water security means a sustainable and adequate quantity and quality of water, which is essential to human life, food and energy security, health and well-being, and economic prosperity. It has been predicted that by the year 2030 with current water utilization practices, there may be a 40% water deficit at the global level (Al-Jabri et al. 2021).

DOI: 10.1201/9781003203438-3

The priority for safe water has been highlighted by various developed and developing nations through sustainable development goals at various scientific platforms. The United Nations considers clean water and sanitation to be a priority objective, and one of its goals is to ensure universal access to safe and affordable drinking water. The direct disposal of untreated industrial effluents into water bodies has an adverse effect on the aquatic ecosystem. It creates mutagenic effects on aquatic organisms due to the presence of heavy metals and aromatics (Dutta et al. 2021). The disposal of contaminates in the environment adversely affect water quality, aquatic life, and human health. The wastewater contaminates freshwater and the coastal ecosystem, threatening food security and access to safe drinking water and becoming a major health and environmental challenge. Due to the hazardous effects of municipal and industrial effluent on the environment, treatment of wastewater and proper disposal of sludge are indispensable from an environmental safety point of view.

Several conventional technologies have been used for the removal of wastewater pollutants such as coagulation-flocculation, adsorption, membrane filtration, reverse osmosis, chemical precipitation, ion exchange, electrochemical treatment, solvent extraction, and flotation (Vunain et al. 2019). However, these technologies suffer from various disadvantages stretching from inefficiency to removing pollutants at low concentrations and to completely converting pollutants into less toxic by-products and the use of chemicals, in addition to the processes being complex and time-consuming and having high maintenance and operation costs (Enaime et al. 2020). Some of these mentioned methods are energetically and operationally intensive, and they require engineering expertise and infrastructure. Although nanotechnology exhibits promising outcomes in wastewater treatment, many barriers stand between the promises and their delivery. The most common barriers include nanomaterial toxicity, cost-effectiveness, and social acceptance. Hence, there is a need for alternative wastewater treatment methods that are low carbon-emitting and high on resource recycling and that consume less energy and promote biorefinery and circular economy concepts. One such alternative option could be the use of biochar to treat industrial, agricultural, and domestic wastewater. The present chapter highlights a systematic overview of the properties of biochar and its application in wastewater treatment by the removal of emerging organic and inorganic contaminants.

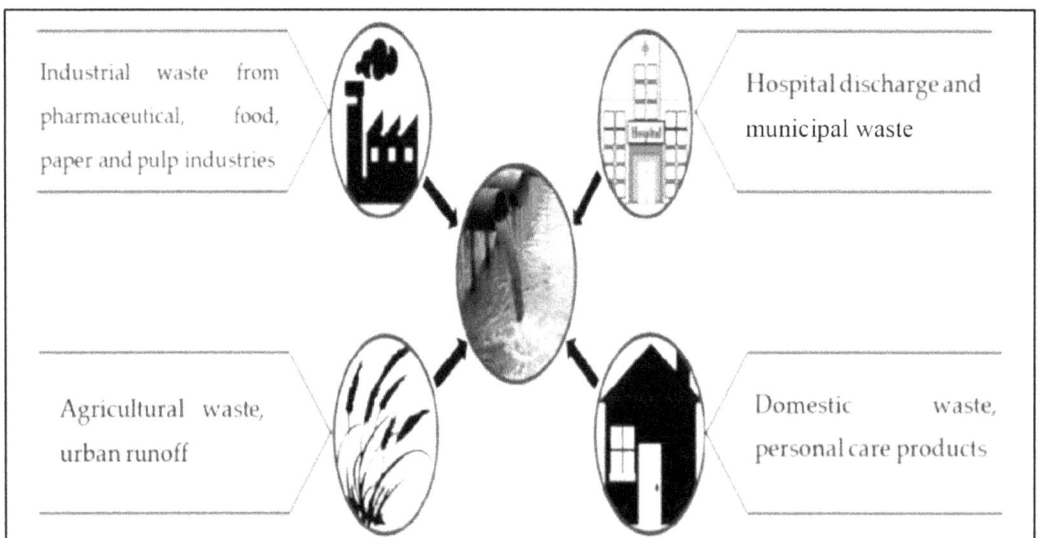

FIGURE 3.1   Various sources of wastewater.

3.2 BIOCHAR—A SUSTAINABLE TECHNOLOGY

Biochar is a carbon-rich material produced from waste biomass by pyrolysis under conditions of limited oxygen (Wang and Wang 2019). Biochar is a promising alternative to activated carbon due to its large surface area, porous structure, and high variability of surface functional groups (Rajec et al. 2016). The low cost of biochar production, its wide availability, and its role in soil and waste-water remediation are getting attention (Tang et al. 2019). Biochar enhances the physical properties of soil by improving its water holding capacity, moisture level, and oxygen content. Biochar can be used for improvement of soil properties and increasing crop yield (Yu et al. 2019), which ultimately contributes to soil carbon sequestration and reduction of greenhouse gases (Kumar et al. 2021). Biochar has the potential to reduce surface runoff and soil erosion (Ayaz et al. 2021). Biochar particles can bond with soil mineral surface through phenolic and carboxylic functional groups and thus improve the stability of soil aggregation and structure (Tanure et al. 2019). Bayabil et al. (2015) reported that biochar amendment in soil can improve soil properties and reduce runoff, erosion, and waterlogging.

Biochar is a carbon-rich pyrolysis product of feedstock (Yang et al. 2020) and its capacity to remove multiple contaminants from aqueous solution makes biochar an ideal adsorbent for waste-water treatment. The term "biochar" is the abbreviation for biocharcoal; it was called biochar during the first International Biochar Conference held in Australia in 2007. Due to the large surface area and pore volume, rich organic carbon and mineral contents, and diverse functional groups, biochar shows prominent sorption ability for both inorganic and organic contaminants in aqueous solution (Ahmad et al. 2014). Figure 3.2 shows a significant increase in the number of published scientific papers on the biochar topic, increasing from less than 50 at the beginning of the year 2000 to more than 3500 in 2020 (Conte et al. 2021). This information was collected from Web of Science by using the keyword "biochar."

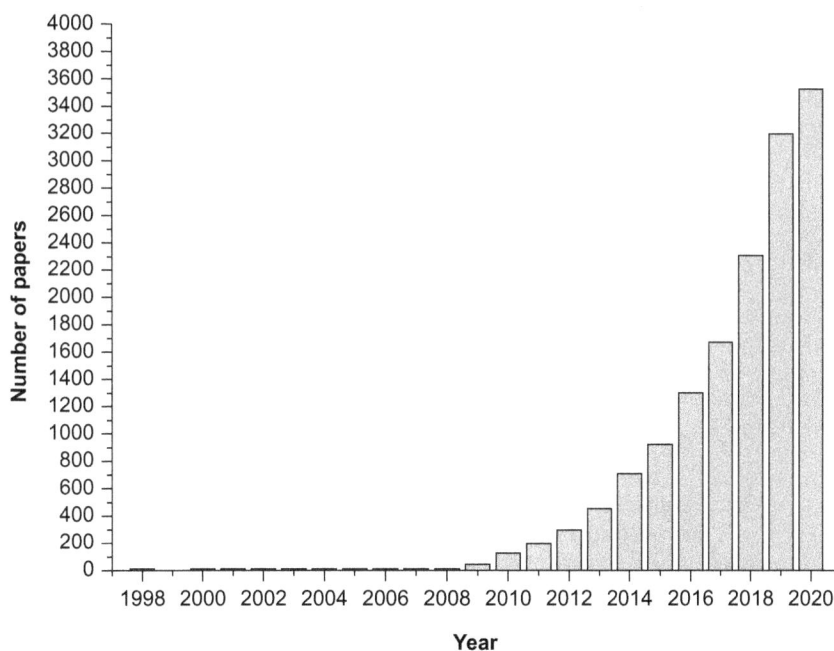

FIGURE 3.2 Publication of scientific research papers on the biochar topic in the last two decades

Source: Web of Science by using the keyword "biochar."

3.3 PRODUCTION OF BIOCHAR FROM WASTE BIOMASS

Agricultural and forest residues such as woodchips, straw, shell, and bagasse are the most widely used biomasses for the preparation of biochar because of their availability at almost zero cost as a natural resource (Nanda et al. 2016). The utilization of biomass rich in cellulose, hemicellulose, lignin, and other organic matters to produce biochar plays a significant role in reducing greenhouse gases emission, carbon sequestration, absorbing contaminants, and soil remediation (Yaashik et al. 2020). It is estimated that around 822 million tonnes of rice husk wastes are generated at the global level every year, which can be used for production of biochar (Donner et al. 2021). The common methods used for the disposal of these agricultural wastes are composting or open burning, which can produce greenhouse gases (CO_2, CH_4, and N_2O) and other toxic gases such as H_2S, SO_2, and NH_3 (Yang et al. 2021). When biochar is used in the agricultural field, it can hold nutrients in the soil and improve agricultural yield and sustainability, sequestering carbon and reducing carbon emission (Hu et al. 2020). Lima et al. (2020) reported that the use of 1 m^3 of *Dillenia excelsa* wood wastes is equivalent to reducing 1687 kg of CO_2 emission. The use of organic wastes in the form of biochar for agricultural production can return nutrients to the soil and play a significant role in development of the circular economy (Kwoczynski and Cmelík 2020).

Thermochemical conversion technologies such as pyrolysis, gasification, torrefaction, and hydrothermal carbonization are used for biochar production (Yaashikaa et al. 2019; Chen et al. 2020). The differences between these methods are in biomass type and process conditions like temperature, pressure, heating rate, residence time, etc. (Table 3.1).

The physical (pH, surface area, electrical conductivity, and pore size), chemical (functional groups, cation exchange capacity and nutrition) and biological properties of biochar are affected by the feedstock, preparation conditions, and processes. Pyrolysis under a low oxygen condition is one of the most common methods for biochar production. During the pyrolysis process, biochar is produced with the generation of liquid (bio-oil) and gas (syngas) which can be used as energy

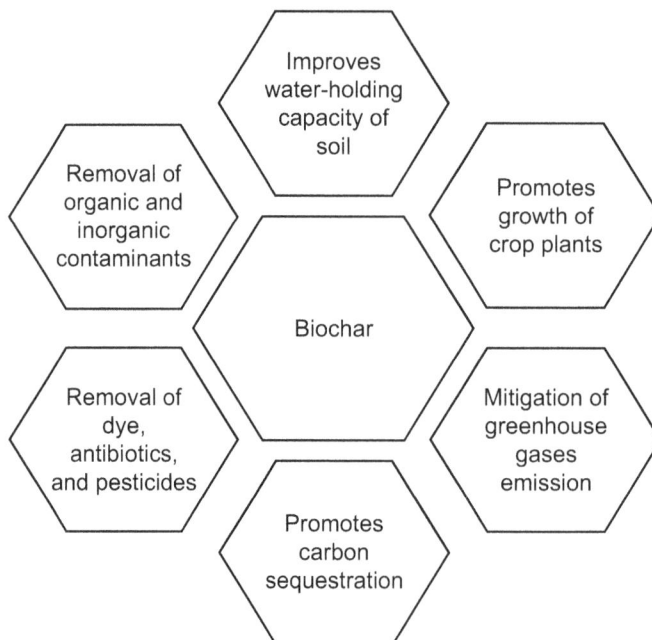

FIGURE 3.3 Application of biochar for wastewater treatment and improvement in soil health.

TABLE 3.1
Utilization of Different Thermal Carbonization Technologies

Thermal carbonization technologies	Temperature/power range	Residence time	End product	Benefits
Pyrolysis	300–850°C	1–3 h	Biochar	Simple, robust, cost-effective, applicable to small-scale biochar production
Microwave-assisted pyrolysis	400–500 W	1–10 min	Biochar and biofuel	Fast, selective, and efficient heating
Hydrothermal carbonization	120–260°C	1–16 h	Hydrochar	Suitable for feedstock with high moisture content
Gasification	> 800°C	10–20 s	Syngas	Biochar production is lower compared to pyrolysis, but the biochar contains high amount of alkali salts.

substances to meet the energy crisis (Chi et al. 2020). The proportion of these products depend on the temperature, pressure, residence time, etc. (Tomczyk et al. 2020). Pyrolysis is classified into three types: slow, intermediate, and fast pyrolysis (Chi et al. 2020). Biomass pyrolysis for a long time (a few hours to a few days) at a temperature lower than 400°C with a very low heating rate under 10°C/min can yield 30% biochar. The process is known as slow pyrolysis. In fast pyrolysis, temperature is between 500 and 700°C, the heating rate is faster, between 10 and 100°C/s, and the resident time is less than 2 s. The biochar obtained by this method has higher oxygen content and lower calorific value. Such conditions exhibit more production of bio-oil and less yield of biochar (Gabhane et al. 2020). The slow pyrolysis is used to produce biochar for its high solid yield, whereas fast pyrolysis at a faster heating rate and high temperature can produce more liquid and gas like bio-oil, CO, H_2, and syngas (Yaashik et al. 2020). Intermediate pyrolysis conditions are between slow and fast pyrolysis, but it is rarely used for the production of biochar. The physical or chemical modification of biochar improves its properties for its application in different fields (Zhuang et al. 2020). Biochar modification can be achieved through acid or alkaline treatment, oxidation, sulfuration, nitrogenation, and physical activation. Modified biochar shows excellent properties and can be used for soil amendment, as an electrochemical material, catalyst, and adsorbent (Huang et al. 2021).

3.4 FACTORS AFFECTING BIOCHAR ACTIVITY

Various physical and chemical factors affect the properties and activity of biochar such as feedstock, temperature, heating rate, and particle size.

3.4.1 FEEDSTOCK

Feedstock has a significant effect on biochar properties and elemental composition. The feedstock materials containing higher nutrients, such as animal manure, can produce biochar with high nutrient content compared to plant feedstock, which is composed of cellulose, lignin, and inorganic compounds. The degradation of lignocellulose, hemicellulose, and cellulose occurs at 120°C, 200–260°C, and 240–350°C, respectively, whereas lignin degradation requires 280–350°C temperatures

(Downie et al. 2009). The presence of functional groups in biochar is determined by the type of feedstock and temperature used during pyrolysis (Uchimiya et al. 2013). The feedstock composition and pyrolysis conditions may affect the physical properties of biochar such as pH, pore structure, surface area, organic carbon content, and adsorption properties.

3.4.2 TEMPERATURE

The pyrolysis temperature is one of the key factors that determines the yield and quality of biochar. Biochar obtained at high temperature is characterized by large surface area and micropores with a higher part of the stable fraction, whereas a low temperature provides biochar with less adsorption capacity. Day et al. (2005) reported that the surface area of biochar was significantly increased from 120 to 460 m^2/g with an increase in temperature from 400 to 900°C. The increase in temperature results in a decrease of volatile compounds in the feedstock material but enhances pH, surface area, and ash content of the biochar (Yorgun and Yildiz 2015). The change in pyrolysis temperature showed different adsorption capacities of biochar for organic compounds like naphthalene, benzene, nitrobenzene, dinitrobenzene, and catechol.

3.4.3 HEATING RATE

The heating rate shows significant effect on the structure of biochar and yield of bio-oil and syngas. The low pyrolysis temperature and heating rate led to the formation of a biochar, whereas a high pyrolysis temperature but low heating rate with longer residence time favor the conversion of biomass into a gaseous product. The intermediate pyrolysis temperature, high heating rate, and less vapor residence time produce a liquid product.

3.4.4 SIZE OF PARTICLES

The particle size of biomass material is an important parameter that affects heat and mass transfer rate and the extent of secondary reactions within the particle. The small particle size is preferred for the fast pyrolysis process as the biomass heats up uniformly and volatile matter is released, leading to higher production of bio-oil and gas product as compared to large particles.

3.5 PHYSICAL PROPERTIES OF BIOCHAR

The density and porosity of the biochar affect its mobility in nature, its interaction with the hydrologic cycle of the soil and its suitability as a natural niche for small soil microbes. Hydrothermal treatment or oxygenation is another method to enhance biochar porosity. Biochar has a large specific surface area with high cation exchange capacity and porosity, which increases water holding capacity and nutrient uptake ability in soil. The biochar capacity for water and gases adsorption depends on the size of the pores. The biochar pores may be micropores (pore size 0.0001–0.05 μm), medium pores (0.002–0.05 μm), and large pores (0.05–1000 μm). The biochar pore structure is made up of micropores that occupy more than 80% of the total pore volume, whereas the quantity of micropores in untreated agro-waste biomass is less than 10% (Weber and Quicker 2018). Particle size is inversely related to reaction temperature as high heating temperature produces smaller biomass particles, whereas long residence time and low heating rate generate large biomass particles. The mechanical strength of biochar depends on its density.

The biochar adsorption capacity relies on the number of positively and negatively charged groups on the surface and porosity, which is distributed by volume size. As compared to inactive biochar, phosphoric-acid-treated, pine-tree-sawdust biochar enhanced microporosity and P-O-P bonds incorporation into biochar matrix and showed a 20% increase in lead adsorption (Hu et al. 2020).

3.6   CHEMICAL PROPERTIES OF BIOCHAR

Biochar has low volatile matter content as compared to biowastes. Biochar is formed from both aliphatic and aromatic substances with high carbon and low oxygen content (Ortiz et al. 2020). Untreated wood has a carbon and oxygen content of more than 50 and 40%, respectively. During biochar production at high temperature, carbon content can reach more than 95%, and oxygen contents are less than 5%. The hydrogen content of wood varies from 5 to 7%, and it is reduced to less than 2% during pyrolysis above 700°C (Weber and Quicker 2018).

The primary biomass is slightly acidic or alkaline with a pH between 5 and 7.5 (Weber and Quicker 2018). Pyrolysis promotes the separation of acidic functional groups of the biochar such as carboxyl, hydroxyl, or carbonyl groups early in the carbonization process, which enhances the pH of the biochar. The increase in biochar electrical conductivity with an increase in temperature may be related with the loss of biomass volatiles during carbonization, the increase in cation exchange capacity, and nutrients (Chandra and Bhattachary 2019). The functional groups on the biochar surface are mainly negatively charged and thus can hardly adsorb anions. The presence of a large number of minerals such as K, Ca, Na, and Mg in biochar promotes the formation of surface functional groups, resulting a high cation exchange capacity for biochar (Li et al. 2019a).

3.7   APPLICATIONS OF BIOCHAR

3.7.1   USE OF BIOCHAR FOR WASTEWATER TREATMENT

The industrial wastewater comes from various sources such as mining, smelting, battery manufacturing, chemical industry, leather and dye manufacturing units. Heavy metals present in wastewater are not biodegradable and tend to accumulate in living organisms through the food chain. Long-term exposure to heavy metals can cause serious health threats such as respiratory ailments, endocrine disorders, cancer, and other diseases in human beings even at low concentrations (Ahmed et al. 2016). Biochar can effectively remove metal ions by adsorption, ion exchange, electrostatic interaction, complexation, precipitation, etc. (Dai et al. 2017). Thus biochar has become an important solution to remediate pollutants present in different industrial and agricultural sectors for improving environmental quality (Wang et al. 2017).

Liang et al. (2021) found biochar prepared by the pyrolysis of water hyacinth at different temperature (250–550°C) significantly removed cadmium ions from wastewater. Biochar produced at 450°C showed the complete removal of cadmium ions from the aqueous solution within 1 h. The adsorption capacity of arsenite was enhanced from 5.7 to 7.0 mg/g through the surface modification of biochar by $Zn(NO_3)_2$ impregnation (Van Vinh et al. 2015). Biochar produced from paper mill sludge was used to adsorb arsenate, and a 34.1 mg/g adsorption capacity was observed (Cho et al. 2017). The removal efficiency of lead by fresh and dehydrated banana peel biochar was 359 and 193 mg/g, respectively (Zhou et al. 2017). Biochar prepared from sewage sludge adsorbed 70% of chromium from the aqueous solution (Agrafioti et al. 2013).

Wang et al. (2015) used 3.65% manganese to impregnate biochar to enhance lead removal capacity from 6 to 99% at pH 5. The adsorption capacity of modified biochar at 25°C was five times higher as compared to pristine biochar. Li et al. (2019b) prepared $KMnO_4$ modified biochar and used it for uranium removal. Zhang et al. (2020) used chitosan-modified magnetic biochar to enhance chromium removal from aqueous solution, and maximum adsorption capacity was increased from 76 to 127 mg/g. Shi et al. (2020) developed $ZnCl_2$-modified glue residue biochar for Cr(VI) sorption and exhibited 325.5 mg/g adsorption capacity. Liu et al. (2014) used zinc borate to prepare lotus stalks biochar for removal of nickel. Adsorption of Ni(II) on lotus stalk biochar was enhanced 3–10 times compared with the biochar without adding zinc borate. Dos Reis et al. (2016) produced biochar from sewage sludge by pyrolysis at 500°C, followed by HCl treatment. The biochar displayed a very high adsorption capacity for hydroquinone, and it was 1218.3 mg/g.

3.7.2 Removal of Pesticide Contaminants by Biochar

The widespread application of pesticides has caused the pollution of soil and water. Pesticides are toxic, persistent, and carcinogenic compounds that pose a serious threat to the environment and human health. Excessive use of pesticides causes toxicity on nontarget organisms and destruction to ecological balance and human health (Zhong et al. 2018). Zhang et al. (2018) produced maize straw biochar at 300, 500, and 700°C and used it for thiacloprid sorption. Zheng et al. (2010) investigated the sorption of two triazine pesticides, atrazine and simazine, on biochar. The adsorption ability of atrazine was 451–1158 mg/g and 243–1066 mg/g for simazine.

Zhang et al. (2013) used pig manure as biomass for preparation of biochar at different temperatures (350 and 700°C) for the removal of carbaryl and atrazine.

The decomposition rate of carbaryl and atrazine pesticides by biochar prepared at 700°C after 12 h was 72 and 28%, respectively. Wu et al. (2019) added biochar to oxyfluorfen containing soil and investigated adsorption, degradation, and bioavailability of oxyfluorfen. The degradation of oxyfluorfen in biochar-treated soil was greater compared to untreated soil, and bioavailability of oxyfluorfen was significantly reduced (18–63%). Rice straw and phosphoric-acid-modified rice straw biochar showed significantly high adsorption for imidacloprid and atrazine from agricultural wastewater (Mandal and Singh 2017). Soybean and corn straw biochar showed high atrazine removal (Liu et al. 2015). Devi and Saroha (2014) used zero-valent iron magnetic paper mill sludge biochar for removal of pentachlorophenol from the effluent.

3.7.3 Removal of Nitrogen, Phosphorus, and Hydrogen Sulfide Gas

Ammonium, nitrate, and phosphate are the common forms of reactive nitrogen and phosphorus in wastewater and can lead to eutrophication (Xu et al. 2018). Biochar can absorb nitrogen and phosphorus present in the aqueous phase (Xue et al. 2016). Oxidized maple wood biochar showed higher ammonium adsorption capacity in comparison to maple wood biochar (Wang et al. 2016). Digested sugar beet tailing biochar showed 73% phosphate removal ability (Yao et al. 2011). Biochar derived from tomato plants enriched with magnesium during their growth showed increased adsorption of phosphate up to 100 mg/g from aqueous solution (Yao et al. 2013). The biochar produced from wood waste pretreated with magnesium oxide was used to recover ammonium and phosphate (Xu et al. 2018). Antunes et al. (2017) found that the phosphorus adsorption capacity of biochar was almost seven times higher than activated carbon. Yang et al. (2016) reported that biochar exhibited an excellent capacity of H_2S removal.

Biochar has a porous surface area that allows it to act as a biofilter for municipal wastewater treatment. An aluminum-impregnated biochar can effectively remove arsenate and other pollutants such as lead, zinc, copper, and phosphate in a polluted urban water runoff (Liu et al. 2019). Biochar can be directly used or combined with biofilter and other technologies for municipal wastewater treatment, which shows the recovery of labile nitrogen and phosphorus (Cole et al. 2017). Manyuchi et al. (2018) reported that chemical oxygen demand, total suspended solids, and total Kjeldahl nitrogen of wastewater were reduced to 90, 89, and 64%, respectively, after being passed through the biochar biofilter.

The engineered biochar loaded with aluminum oxyhydroxide was applied to recycle and reuse phosphorus from secondary treated wastewater (Zheng et al. 2019). Phosphorus adsorbed on engineered biochar can be used as a slow release fertilizer for crop production. Biochar can adsorb 20–43% ammonium and 19–65% phosphate from flushed dairy manure within 24 h (Ghezzehei et al. 2014).

Fan et al. (2019) reported NH_4^+ sorption by hydrous bamboo biochar. Walnut shell and sewage sludge were copyrolyzed to prepare biochar for phosphate adsorption from eutrophic water (Yin et al. 2019). Ajmal et al. (2020) compared phosphate removal efficiency from wastewater by biochar before and after magnetic modification. They found that the adsorption ability of magnetic biochar was twice that of unmodified biochar. Tang et al. (2019) found biochar derived at 450°C showed the highest ammonium removal capacity due to its greater surface area and high functional group density. Blum et al. (2018) reported that wastewater released from residential units not connected to any municipal sewage treatment plant can be treated with biochar.

3.7.4 Removal of Antibiotic Contaminants by Biochar

The presence of antibiotic contaminants in the environment is a matter of great concern for human health. Antibiotics cannot be degraded easily, and residual antibiotics may increase the resistance of aquatic microbes, which may lead to an adverse impact on the environment through bioaccumulation and biomagnification (Carvalho and Santos 2016). The removal of tetracycline by $ZnCl_2$/$FeCl_3$–solution-doped sawdust biochar showed more than 89% removal rate after three cycles (Zhou et al. 2017). Zhao et al. (2019) prepared humic-acid-coated magnetic biochar derived from potato stems and leaves to adsorb enrofloxacin, norfloxacin, and ciprofloxacin. The adsorption capacities were 8.4, 10, and 11.5 mg/g for enrofloxacin, norfloxacin, and ciprofloxacin.

Fan et al. (2018) pyrolyzed rice straw at different temperature for the removal of tetracycline antibiotic, and biochar obtained at high temperature showed the maximum adsorption capacity of 51 mg/g due to its large specific surface area, aromatic structure, and graphite carbon. Peiris et al. (2017) used biochar to remove tetracycline. Liu et al. (2019) modified biochar by chitosan and FeS_x for tetracycline removal, and they found that the maximum capacity of tetracycline removal by biochar and modified biochar was 52 and 193 mg/g, respectively. Iron- and zinc-doped sawdust biochar exhibited tetracycline removal from aqueous solution (Zhou et al. 2017).

Sun et al. (2018) reported that biochar can remove antibiotics such as sulfonamide and tetracycline. The mechanism of removal of sulfonamide and tetracycline was dependent on the groups available on the surface area rings and electron donor–acceptor interactions (Peiris et al. 2017). Sulfamethoxazole is one of the typical sulfonamide antibiotics widely used for both human beings and animals. Yao et al. (2018) reported that sulfamethoxazole adsorption on the digested bagasse biochar was affected by the pH value of solution.

3.7.5 Removal of Organic Contaminants by Biochar

The organic contaminants such as dye and phenolic compounds have attracted much attention because of their complex aromatic structure, toxicity, and nonbiodegradable nature. Organic pollutants are toxic in nature, can reduce dissolved oxygen content in water, and cause harm to the aquatic ecosystem and human health (Ahmed et al. 2016). The presence of dye in the environment has adverse effects on human health such as carcinogenic, mutagenic, allergic effects, as well as dermatitis and kidney diseases (Ardila-Leal et al. 2021).

Dyes are extensively used in the textile, paper, food, and printing industries. The presence of colored compounds in water can be easily recognized, and they reduce light penetration into the water and cause an adverse impact on aquatic organisms, which may die due to oxygen deprivation. *Gliricidia* biochar can be used for removal of crystal violet dye from an aqueous medium, and this process depends on the pH, surface area, and pore volume of biochar (Wathukarage et al. 2017). Bagasse biochar was used to adsorb lead from battery manufacturing industry effluent, and it showed a maximum adsorption capacity of 12.7 mg/g; the adsorptive process was dependent on pH, exposure time, and dosage (Poonam and Kumar 2018). Fan et al. (2017) used municipal sludge to prepare biochar for removal of methylene blue; the removal rate of methylene blue was 60% after three cycles. Khataee et al. (2017) reported the ultrasound-catalyzed degradation of RB69 by TiO_2-biochar. The removal efficiency of RB69 was 98% by ultrasound/TiO_2-biochar, which was higher than that of ultrasound/biochar, which showed only 64% removal. Lu et al. (2019) used TiO_2 biochar for methyl orange decolorization, and they observed 97% decolorization.

Phenols are organic contaminants that can affect the taste of fish, shrimp, and drinking water even present at low concentration. Phenols are harmful to the human body as they have an aromatic ring structure and strong adsorption affinity, and they cannot be easily degraded by the microbes. Mohammed et al. (2018) reported that biochar can remove phenolic compounds. Thang et al. (2019) prepared biochar by using chicken manure for removal of phenol and 2,4-dinitrophenol, and they could completely remove both of the complexes within 90 min. The maximum adsorption capacity of biochar for phenol and DNP were 106 and 148 mg/g, respectively, which was due to high surface area with more pore structure and a large number of functional groups on biochar surface. Lawal et al. (2021) used oil palm leaves biochar for removal of phenol, and they observed 63 mg/g adsorption capacity for phenol. Zhang et al. (2020b) used zinc oxide with biochar for removal of phenolic compounds from aqueous solution. The composite exhibited strong stability and 95% photodegradation efficiency up to five cycles.

3.8 POTENTIAL RISKS OF BIOCHAR AND BIOCHAR-BASED COMPOSITE MATERIALS

A large number of toxic substances such as polycyclic aromatic hydrocarbons, dioxin, and acrolein have been formed during biochar production (Zheng et al. 2019). Lyu et al. (2016) reported that biochar prepared at low temperature exhibited higher biological toxicity. The concentration of polycyclic aromatic hydrocarbons in biochar was positively correlated with pyrolysis temperature. The concentration of polycyclic aromatic hydrocarbons in ordinary biochar was 0.07–4 μg/g, but it was significantly enhanced to 17–27 μg/g when biochar was prepared from paper mill sludge and wood (Devi and Saroha 2014). Polychlorinated dibenzo-para-dioxins and polychlorinated dibenzofurans are highly toxic compounds; their concentration was 610 and 67 pg/g at 300 and 700°C, respectively (Lyu et al. 2016).

Zhang et al. (2019) found that 1 and 3% biochar-amended soil can effectively reduce the toxicity of mesotrione herbicide to earthworms. However, 10% biochar-amended soil significantly inhibited the growth of earthworms though damage in DNA.

Zheng et al. (2019) reported that biochar composed from sludge and animal manure was rich in heavy metals such as lead, cadmium, copper, nickel, and chromium as compared with green biomass, which limits its long-term application. Biochar can effectively remove water pollutants in the primary stage of the experiment. Long-term experiments and risk assessment are needed before its application for water or soil treatment as some reports revealed that biochar releases toxic substances and may show side effects on organisms. Liao et al. (2014) reported that biochar can generate free radicals, which may inhibit seed germination and growth of crop plants. Hence there is an urgent need to identify potential hazards due to the application of biochar to reduce its adverse effects on human health and environment.

3.9 CONCLUSION AND FUTURE PERSPECTIVES

Biochar is an efficient and low-cost adsorbent, which can be produced from waste biomass such as agricultural and forestry residues, sewage sludge, and solid organic municipal wastes. The biochar potential for removal of pollutants released from industrial wastewater and from the municipal and agricultural sectors has been well demonstrated in laboratory experiments, but they are devoid of field trials. Most of the studies are lacking reports on the stability of biochar or biochar-based composites and on their toxicity for aquatic and soil microbes. The exact mechanism for removal of contaminants by biochar and its relationship with biomass properties and preparation methods are still not clear. It is essential to recognize 3-D aspects of biochar adsorption and the optimum dose of biochar for wastewater treatment and crop growth. Engineers need to develop simple and portable reactors for pyrolysis, which could minimize biochar production cost. There are many constraints for wide-scale utilization of biochar but by addressing the knowledge gaps and constraints, biochar technology can offer a sustainable solution to environmental problems. Further, in-depth investigations are required to develop low-cost, efficient biochar modification technologies that are useful for field experiments under different environmental conditions.

ACKNOWLEDGMENTS

The authors extend their appreciation to Amity University, Noida, Indian Institute of Toxicology Research (CSIR), Lucknow and Enviro Technology Limited, Ankleshwar, Gujarat for providing all the necessary facilities and valuable support.

REFERENCES

Agrafioti, E., G. Bouras, D. Kalderis, and E. Diamadopoulos, 2013. Biochar production by sewage sludge pyrolysis. *J. Anal. Appl. Pyrol.* 101: 72–78.

Ahmad, M., A.U. Rajapaksha, J.E. Lim, M. Zhang, N. Bolan, D. Mohan, M. Vithanage, S.S. Lee, and Y.S. Ok, 2014. Biochar as a sorbent for contaminant management in soil and water: a review. *Chemosphere.* 99: 19–33.

Ahmed, M.B., J.L. Zhou, H.H. Ngo, W. Guo, and M. Chen, 2016. Progress in the preparation and application of modified biochar for improved contaminant removal from water and wastewater. *Bioresour. Technol.* 214: 836–851.

Ajmal, Z., A. Muhmood, R. Dong, and S. Wu, 2020. Probing the efficiency of magnetically modified biomass-derived biochar for effective phosphate removal. *J. Environ. Manage.* 253: 109730.

Al-Jabri, H., P. Das, S. Khan, M. Thaher, and M. AbdulQuadir, 2021. Treatment of wastewaters by microalgae and the potential applications of the produced biomass—a review. *Water.* 13: 27.

Antunes, E., J. Schumann, G. Brodie, M.V. Jacob, and P.A. Schneider, 2017. Biochar produced from biosolids using a single-mode microwave: characterisation and its potential for phosphorus removal. *J. Environ. Manag.* 196: 119–126.

Ardila Leal, L.D., R.A. Poutou Pinales, A.M. Pedroza Rodríguez, and B.E. Quevedo Hidalgo, 2021. A brief history of colour, the environmental impact of synthetic dyes and removal by using laccases. *Molecules.* 26: 3813.

Ayaz, M., D. Feizien, V. Tilvikien, K. Akhtar, U. Stulpinait, and R. Iqbal, 2021. Biochar role in the sustainability of agriculture and environment. *Sustainability.* 13: 1330. https://doi.org/10.3390/su13031330.

Bayabil, H.K., C.R. Stoof, J.C. Lehmann, B. Yitaferu, and T.S. Steenhuis, 2015. Assessing the potential of biochar and charcoal to improve soil hydraulic properties in the humid Ethiopian Highlands: the Anjeni watershed. *Geoderma.* 243–244: 115–123.

Blum, K.M., C. Gallampois, P.L. Andersson, G. Renman, A. Renman, and P. Haglund, 2018. Comprehensive assessment of organic contaminant removal from on-site sewage treatment facility effluent by char-fortified filter beds. *J. Hazard Mater.* 361: 111.

Carvalho, I.T., and L. Santos, 2016. Antibiotics in the aquatic environments: a review of the European scenario. *Environ. Int.* 94: 736–757. http://doi.org/10.1016/j.envint.2016.06.025.

Chandra, S., and J. Bhattachary, 2019. Influence of temperature and duration of pyrolysis on the property heterogeneity of rice straw biochar and optimization of pyrolysis conditions for its application in soils. *J. Clean. Prod.* 215: 1123–1139.

Chen, L.C., C. Wen, W.Y. Wang, T.Y. Liu, E.Z. Liu, H.W. Liu, and Z.X. Li, 2020. Combustion behavior of bio-chars thermally pretreated via torrefaction, slow pyrolysis, or hydrothermal carbonisation and co-fired with pulverised coal. *Renew. Energ.* 161: 867–877.

Chi, N.T.L., A. Susaimanickam, T.S. Ahamed, S.S. Kumar, and A. Pugazhendhi, 2020. A review on biochar production techniques and biochar based catalyst for biofuel production from algae. *Fuel.* 287: 119411.

Cho, D.W., G. Kwon, K. Yoon, Y.F. Tsang, Y.S. Ok, E.E. Kwon, and H. Song, 2017. Simultaneous production of syngas and magnetic biochar via pyrolysis of paper mill sludge using CO_2 as reaction medium. *Energy Convers. Manag.* 145: 1–9.

Cole, A.J., N.A. Paul, R.N. De, and D.A. Roberts, 2017. Good for sewage treatment and good for agriculture: algal based compost and biochar. *J. Environ. Manag.* 200: 105.

Conte, P., R. Bertani, P. Sgarbossa, P. Bambina, H.P. Schmidt, R. Raga, G. Lo Papa, D.F. Chillura Martino, and P. Lo Meo, 2021. Recent developments in understanding biochar's physical-chemistry. *Agronomy.* 11: 615. https://doi.org/10.3390/agronomy11040615.

Dai, L., L. Fan, Y. Liu, R. Ruan, Y. Wang, Y. Zhou, Y. Zhao, and Z. Yu, 2017. Production of bio-oil and biochar from soapstock via microwave-assisted co-catalytic fast pyrolysis. *Bioresour. Technol.* 225: 1–8.

Day, D., R.J. Evans, J.W. Lee, and D. Reicosky, 2005. Economical CO_2, SOx, and NOx capture from fossil-fuel utilization with combined renewable hydrogen production and large-scale carbon sequestration. *Energy.* 30(14): 2558–2579.

Devi, P., and A.K. Saroha, 2014. Synthesis of the magnetic biochar composites for use as an adsorbent for the removal of pentachlorophenol from the effluent. *Bioresour. Technol.* 169: 525–531.

Donner, M., A. Verniquet, J. Broeze, K. Kayser, and H. de Vries, 2021. Critical success and risk factors for circular business models valorizing agricultural waste and by-products. *Resour. Conserv. Recycl.* 165: 105236.

dos Reis, G.S., M.A. Adebayo, C.H. Sampaio, E.C. Lima, P.S. Thue, I.A.S. de Brum, S.L.P. Dias, and F.A. Pavan, 2016. Removal of phenolic compounds from aqueous solutions using sludge-based activated carbons prepared by conventional heating and microwave-assisted pyrolysis. *Wat. Air Soil Poll.* 228: 33.

Downie, A., A. Crosky, and P. Munroe, 2009. Physical properties of biochar. In: Lehmann J., Joseph S., editors. *Biochar for environmental management: science and technology.* London, UK: Earthscan, pp. 13–32.

Dutta, S., B. Gupta, S.K. Srivastava, and A.K. Gupta, 2021. Recent advances on the removal of dyes from wastewater using various adsorbents: a critical review. *Mater. Adv.* 2: 4497–4531.

Enaime, G., A. Baçaoui, A. Yaacoubi, and M. Lubken, 2020. Biochar for wastewater treatment—conversion technologies and applications. *Appl. Sci.* 10: 3492.

Fan, R., C. Chen, J. Lin, J. Tzeng, C. Huang, C. Dong, and C.P. Huang, 2019. Adsorption characteristics of ammonium ion onto hydrous biochars in dilute aqueous solutions. *Bioresour. Technol.* 272: 465–472.

Fan, S.S., Y. Wang, Y. Li, Z. Wang, Z.X. Xie, and J. Tang, 2018. Removal of tetracycline from aqueous solution by biochar derived from rice straw. *Environ. Sci. Pollut. Res.* 25(29): 29529–29540.

Fan, S.S., Y. Wang, Z. Wang, J. Tang, and X.D. Li, 2017. Removal of methylene blue from aqueous solution by sewage sludge-derived biochar: adsorption kinetics, equilibrium, thermodynamics and mechanism. *J. Environ. Chem. Eng.* 5(1): 601–611.

Gabhane, J.W., V.P. Bhange, P.D. Patil, S.T. Bankar, and S. Kumar, 2020. Recent trends in biochar production methods and its application as a soil health conditioner: a review. *SN Appl. Sci.* 2: 1307.

Ghezzehei, T.A., D.V. Sarkhot, and A.A. Berhe, 2014. Biochar can be used to capture essential nutrients from dairy wastewater and improve soil physico-chemical properties. *Solid Earth.* 5: 953–962.

Hu, B.W., Y.J. Ai, J. Jin, T. Hayat, A. Alsaedi, L. Zhuang, and X.K. Wang, 2020. Efficient elimination of organic and inorganic pollutants by biochar and biochar-based materials. *Biochar.* 2: 47–64.

Huang, W.H., D.J. Lee, and C.P. Huang, 2021. Modification on biochars for applications: a research update. *Bioresour. Technol.* 319: 124100.

Khataee, A., B. Kayan, P. Gholami, D. Kalderis, and S. Akay, 2017. Sonocatalytic degradation of an anthraquinone dye using TiO_2 biochar nanocomposite. *Ultrason. Sonochem.* 39: 120–128.

Kumar, G.R., G. Mandavi, K. Mishra Rakesh, C. Preeti, M.K. Awasthi, S.R. Sharan, G.B. Shekhar, and P. Ashok, 2021. Biochar for remediation of agrochemicals and synthetic organic dyes from environmental samples: a review. *Chemosphere.* 272.

Kwoczynski, Z., and J. Cmelík, 2020. Characterization of biomass wastes and its possibility of agriculture utilization due to biochar production by torrefaction process. *J. Clean. Prod.* 280: 124302.

Lawal, A.A., M.A. Hassan, M.A.A. Farid, T.A.T.Y. Anuar, M.H. Samsudin, M.Z.M. Yusoff, M.R. Zakaria, M.N. Mokhtar, and Y. Shirai, 2021. Adsorption mechanism and effectiveness of phenol and tannic acid removal by biochar produced from oil palm frond using steam pyrolysis. *Environ. Pollut.* 269: 116197.

Li, J., Z. Hu, F. Li, J. Fan, J. Zhang, F. Li, and H. Hu, 2019a. Effect of oxygen supply strategy on nitrogen removal of biochar-based vertical subsurface flow constructed wetland: intermittent aeration and tidal flow. *Chemosphere.* 223: 366–374.

Li, N., M.L. Yin, D.C.W. Tsang, S.T. Yang, J. Liu, X. Li, G. Song, and J. Wang, 2019b. Mechanisms of U(VI) removal by biochar derived from *Ficus microcarpa* aerial root: a comparison between raw and modified biochar. *Sci. Total. Environ.* 697: 134115.

Liang, L., F. Xi, W. Tan, Xu, Meng, B. Hu, and X. Wang, 2021. Review of organic and inorganic pollutants removal by biochar and biochar-based composites. *Biochar.* 3: 255–281.

Liao, S.H., B. Pan, H. Li, D. Zhang, and B.S. Xing, 2014. Detecting free radicals in biochars and determining their ability to inhibit the germination and growth of corn, wheat and rice seedlings. *Environ. Sci. Technol.* 48(15): 8581–8587.

Lima, M.D.R., E.P.S. Patrício, U. de Oliveira Barros Junior, M.R. de Assis, C.N. Xavier, L. Bufalino, P.F. Trugilho, P.R.G. Hein, and T. de Paula Protasio, 2020. Logging wastes from sustainable forest management as alternative fuels for thermochemical conversion systems in Brazilian Amazon. *Biomass. Bioenergy.* 140: 105660.

Liu, H., S. Liang, J. Gao, H.H. Ngo, W. Guo, Z. Guo, and Y. Li, 2014. Development of biochars from pyrolysis of lotus stalks for Ni(II) sorption: using zinc borate as flame retardant. *J. Anal. Appl. Pyrolysis.* 107: 336–341.

Liu, H., Y.F. Wei, J.M. Luo, T. Li, D. Wang, S.L. Luo, and J.C. Crittenden, 2019. 3D hierarchical porous-structured biochar aerogel for rapid and efficient phenolic antibiotics removal from water. *Chem. Eng. J.* 368: 639–648.

Liu, P., R. Jiang, W.C. Zhou, H. Zhu, W. Xiao, D.H. Wang, and X.H. Mao, 2015. g-C3N4 modified biochar as an adsorptive and photocatalytic material for decontamination of aqueous organic pollutants. *Appl. Surf. Sci.* 358: 231–239.

Lu, L.L., R. Shan, Y.Y. Shi, S.X. Wang, and H.R. Yuan, 2019. A novel TiO_2/biochar composite catalysts for photocatalytic degradation of methyl orange. *Chemosphere.* 222: 391–398.

Lyu, H.H., Y.H. He, J.C. Tang, M. Hecker, Q.L. Liu, P.D. Jones, G. Codling, and J.P. Giesy, 2016. Effect of pyrolysis temperature on potential toxicity of biochar if applied to the environment. *Environ. Pollut.* 218: 1–7.

Mandal, A., and N. Singh, 2017. Optimization of atrazine and imidacloprid removal from water using biochars: designing single or multi-staged batch adsorption systems. *Int. J. Hyg. Environ. Health.* 220: 637–645.

Manyuchi, M.M., C. Mbohwaa, and E. Muzenda, 2018. Potential to use municipal waste bio char in wastewater treatment for nutrients recovery. *Phys. Chem. Earth.* 107: 92–95.

Mohammed, N.A.S., R.A. Abu-Zurayk, I. Hamadneh, and A.H. Al-Dujaili, 2018. Phenol adsorption on biochar prepared from the pine fruit shells: equilibrium, kinetic and thermodynamics studies. *J. Environ. Manag.* 226: 377–385.

Nanda, S., A.K. Dalai, I. Gokalp, and J.A. Kozinski, 2016. Valorization of horse manure through catalytic supercritical water gasification. *Waste Manag.* 52: 147–158.

Ortiz, L.R., E. Torres, D. Zalazar, H.L. Zhang, R. Rodriguez, and M. German, 2020. Influence of pyrolysis temperature and bio-waste composition on biochar characteristics. *Renew. Energ.* 155: 837–847.

Peiris, C., S.R. Gunatilake, T.E. Mlsna, D. Mohan, and M. Vithanage, 2017. Biochar based removal of antibiotic sulfonamides and tetracyclines in aquatic environments: a critical review. *Bioresour. Technol.* 246: 150–159.

Poonam, S.K. Bharti, and N. Kumar, 2018. Kinetic study of lead (Pb2þ) removal from battery manufacturing wastewater using bagasse biochar as biosorbent. *Appl. Water Sci.* 8: 119.

Rajec, P., O. Rosskopfova, M. Galambos, V. Fristak, G. Soja, A. Dafnomili, F. Noli, A. Dukic, and L. Matovic, 2016. Sorption and desorption of pertechnetate on biochar under static batch and dynamic conditions. *J. Radioanal. Nucl. Chem.* 310: 253–261.

Shi, Y., R. Shan, L. Lu, H. Yuan, H. Jiang, Y. Zhang, and Y. Chen, 2020. High-efficiency removal of Cr(VI) by modified biochar derived from glue residue. *J. Clean. Prod.* 254: 119935

Sun, P., Y. Li, T. Meng, R. Zhang, M. Song, and J. Ren, 2018. Removal of sulphonamide antibiotics and human metabolite by biochar and biochar/H2O2 in synthetic urine. *Water Res.* 147: 91–100.

Tang, Y., M.S. Alam, K.O. Konhauser, D.S. Alessi, S. Xu, W. Tian, and Y. Liu, 2019. Influence of pyrolysis temperature on production of digested sludge biochar and its application for ammonium removal from municipal wastewater. *J. Clean. Prod.* 209: 927–936.

Tanure, M.M.C., L.M. Da Costa, H.A. Huiz, R.B.A. Fernandes, P.R. Cecon, J.D. Pereira Junior, and J.M.R. DaLuz, 2019. Soil water retention, physiological characteristics, and growth of maize plants in response to biochar application to soil. *Soil Tillage Res.* 192: 164–173.

Thang, P.Q., K. Jitae, B.L. Giang, N.M. Viet, and P.T. Huong, 2019. Potential application of chicken manure biochar towards toxic phenol and 2,4-dinitrophenol in wastewaters. *J. Environ. Manage.* 251: 109556.

Tomczyk, A., Z. Sokołowska, and P. Boguta, 2020. Biochar physicochemical properties: pyrolysis temperature and feedstock kind effects. *Rev. Environ. Sci. Biotechnol.* 19: 191–215.

Uchimiya, M., T. Ohno, and Z. He, 2013. Pyrolysis temperature-dependent release of dissolved organic carbon from plant, manure, and biorefinery wastes. *J. Anal. Appl. Pyrolysis.* 104: 84–94

Van Vinh, N., M. Zafar, S. Behera, and H.S. Park, 2015. Arsenic (III) removal from aqueous solution by raw and zinc-loaded pine cone biochar: equilibrium, kinetics, and thermodynamics studies. *Int. J. Environ. Sci. Technol.* 12: 1283–1294.

Vunain, E., E.F. Masoamphambe, P.M.G. Mpeketula, M. Monjerezi, and A. Etale, 2019. Evaluation of coagulating efficiency and water borne pathogens reduction capacity of *Moringa oleifera* seed powder for treatment of domestic wastewater from Zomba, Malawi. *J. Environ. Chem. Eng.* 7: 103–118.

Wang, B., B. Gao, and J. Fang, 2017. Recent advances in engineered biochar productions and applications. *Crit. Rev. Environ. Sci. Technol.* 47: 2158–2207.

Wang, B., J. Lehmann, K. Hanley, R. Hestrin, and A. Enders, 2016. Ammonium retention by oxidized biochars produced at different pyrolysis temperatures and residence times. *RSC Adv.* 6: 41907–41913.

Wang, H.Y., B. Gao, S.S. Wang, J. Fang, Y.W. Xue, and K. Yang, 2015. Removal of Pb(II), Cu(II), and Cd(II) from aqueous solutions by biochar derived from $KMnO_4$ treated hickory wood. *Bioresour. Technol.* 197: 356–362.

Wang, S.Z., and J.L. Wang, 2019. Activation of peroxymonosulfate by sludge derived biochar for the degradation of triclosan in water and wastewater. *Chem. Eng. J.* 356: 350–358.

Wathukarage, A., I. Herath, M.C.M. Iqbal, and M. Vithanage, 2017. Mechanistic understanding of crystal violet dye sorption by woody biochar: implications for wastewater treatment. *Environ. Geochem. Health.* 1–15.

Weber, K., and P. Quicker, 2018. Properties of biochar. *Fuel.* 217: 240–261.

Wu, C., X.G. Liu, X.H. Wu, F.S. Dong, J. Xu, and Y.Q. Zheng, 2019. Sorption, degradation and bioavailability of oxyfluorfen in biochar-amended soils. *Sci. Total. Environ.* 658: 87–94.

Xu, K., F. Lin, X. Dou, M. Zheng, W. Tan, and C. Wang, 2018. Recovery of ammonium and phosphate from urine as value-added fertilizer using wood waste biochar loaded with magnesium oxides. *J. Clean. Prod.* 187.

Xue, L.H., B. Gao, Y.S. Wan, J.N. Fang, S.S. Wang, Y.C. Li, R. Munoz-Carpena, and L.Z. Yang, 2016. High efficiency and selectivity of MgFe-LDH modified wheatstraw biochar in the removal of nitrate from aqueous solutions. *J. Taiwan Inst. Chem. Eng.* 63: 312–317.

Yaashikaa, P.R., P.S. Kumar, S.J. Varjani, and A. Saravanan, 2019. Advances in production and application of biochar from lignocellulosic feedstocks for remediation of environmental pollutants. *Bioresour. Technol.* 292: 122030.

Yaashikaa, P.R., P.S. Kumar, S. Varjani, and A. Saravanan, 2020. A critical review on the biochar production techniques, characterization, stability and applications for circular bioeconomy. *Biotechnol. Rep.* 28: e00570.

Yang, G., Y. Li, S. Yang, J. Liao, X. Cai, Q. Gao, Y. Fang, F. Peng, and S. Zhang, 2021. Surface oxidized nanocobalt wrapped by nitrogen-doped carbon nanotubes for efficient purification of organic wastewater. *Sep. Purif. Technol.* 259: 118098.

Yang, G., Y. Sun, J.P. Zhang, and C. Wen, 2016. Fast carbonization using fluidized bed for biochar production from reed black liquor: optimization for H_2S removal. *Environ. Technol.* 37: 2447–2456.

Yang, W., G. Feng, D. Miles, L. Gao, Y. Jia, C. Li, and Z. Qu, 2020. Impact of biochar on greenhouse gas emissions and soil carbon sequestration in corn grown under drip irrigation with mulching. *Sci. Total Environ.* 729: 138752.

Yao, Y., B. Gao, J.J. Chen, M. Zhang, M. Inyang, Y.C. Li, A. Alva, and L.Y. Yang, 2013. Engineered carbon (biochar) prepared by direct pyrolysis of Mg-accumulated tomato tissues: characterization and phosphate removal potential. *Bioresour. Technol.* 138: 8–13.

Yao, Y., B. Gao, M. Inyang, A.R. Zimmerman, X. Cao, P. Pullammanappallil, and L. Yang, 2011. Biochar derived from anaerobically digested sugar beet tailings: characterization and phosphate removal potential. *Bioresour. Technol.* 102: 6273–6278.

Yao, Y., Y. Zhang, B. Gao, R.J. Chen, and F. Wu, 2018. Removal of sulfamethoxazole (SMX) and sulfapyridine (SPY) from aqueous solutions by biochars derived from anaerobically digested bagasse. *Environ. Sci. Pollut. Control. Ser.* 25: 25659–25667.

Yin, Q., M. Liu, and H. Ren, 2019. Biochar produced from the co-pyrolysis of sewage sludge and walnut shell for ammonium and phosphate adsorption from water. *J. Environ. Manag.* 249: 109410. http://doi.org/10.1016/j.jenvman.2019.109410.

Yorgun, S., and D. Yıldız, 2015. Slow pyrolysis of paulownia wood: effects of pyrolysis parameters on product yields and bio-oil characterization. *J. Anal. Appl. Pyrolysis.* 114: 68–78.

Yu, H., W. Zou, J. Chen, H. Chen, Z. Yu, J. Huang, H. Tang, X. Wei, and B. Gao, 2019. Biochar amendment improves crop production in problem soils: a review. *J. Environ. Manag.* 232: 8–21.

Zhang, Q.M., M. Saleem, and C.X. Wang, 2019. Effects of biochar on the earthworm (*Eisenia foetida*) in soil contaminated with and/or without pesticide mesotrione. *Sci. Total Environ.* 671: 52–58.

Zhang, S., L. Jin, J. Liu, Q. Wang, and L. Jiao, 2020 a. A label-free yellow-emissive carbon dot-based nanosensor for sensitive and selective ratiometric detection of chromium (VI) in environmental water samples. *Mater. Chem. Phys.* 248: 122912.

Zhang, S., X. Yang, L. Liu, M. Ju, and K. Zheng, 2018. Adsorption behavior of selective recognition functionalized biochar to Cd(II) in wastewater. *Materials* 11: 299.

Zhang, W., S. Mao, H. Chen, L. Huang, and R. Qiu, 2013. Pb (II) and Cr (VI) sorption by biochars pyrolyzed from the municipal wastewater sludge under different heating conditions. *Bioresour. Technol.* 147: 545–552.

Zhang, Y., G.M. Zhao, Y. Xuan, L. Gan, and M.Z. Pan, 2020b. Enhanced photocatalytic performance for phenol degradation using ZnO modified with nano-biochar derived from cellulose nanocrystals. *Cellulose.* 28: 991–1009.

Zhao, J., G. Liang, X. Zhang, X. Cai, R. Li, X. Xie, and Z. Wang, 2019. Coating magnetic biochar with humic acid for high efficient removal of fluoroquinolone antibiotics in water. *Sci. Total Environ.* 688: 1205–1215.

Zheng, W., M. Guo, T. Chow, D.N. Bennett, and N. Rajagopalan, 2010. Sorption properties of greenwaste biochar for two triazine pesticides. *J. Hazard Mater.* 181: 121–126.

Zheng, Y., B. Wang, A.E. Wester, J. Chen, F. He, H. Chen, and B. Gao, 2019. Reclaiming phosphorus from secondary treated municipal wastewater with engineered biochar. *Chem. Eng. J.* 362: 460–468.

Zhong, J., L. Li, Z. Zhong, Q. Yang, J. Zhang, and L. Wang, 2018. Advances on the research of the effect of biochar on the environmental behavior of antibiotics. *J. Safe. Environ.*18: 657–663.

Zhou, Y., X. Liu, Y. Xiang, P. Wang, J. Zhang, F. Zhang, J. Wei, L. Luo, M. Lei, and L. Tang, 2017. Modification of biochar derived from sawdust and its application in removal of tetracycline and copper from aqueous solution: adsorption mechanism and modelling. *Bioresour. Technol.* 245: 266–273.

Zhuang, Z.C., L. Wang, and J.C. Tang, 2020. Efficient removal of volatile organic compound by ball-milled biochars from different preparing conditions. *J. Hazard. Mater.* 406: 124676.

4 Biochar for the Remediation of Heavy-Metal-Contaminated Soil
Present Scenario and Future Challenges

Sai Shankar Sahu, Dipita Ghosh, and Subodh Kumar Maiti

CONTENTS

4.1 INTRODUCTION

Rapid industrialization, mining activities, and vehicular emissions lead to the increase in heavy metals pollution in soil. Contamination of heavy metals in soil is of great concern in countries like America, Australia, China, and India. It is estimated that worldwide about 20 million ha of land are potentially contaminated by heavy metals (Liu et al., 2018). The huge amount of industrial and hazardous waste generation and mining causes a significant increase in heavy metals contamination in soil (Wang et al., 2020). According to a report of the Central Pollution Control Board (CPCB), the Indian states of Gujarat, Maharashtra, and Andhra Pradesh generate 80% of hazardous wastes

DOI: 10.1201/9781003203438-4

containing high heavy metals (Kumar et al., 2019). The presence of heavy metals above the permissible limit can be detrimental to soil hydrology and biota, thus adversely affecting the food chain. Therefore, the development of effective means to remediate the contaminated soil needs attention.

Over a period of time, various in situ and ex situ techniques for remediation have been used to decontaminate soil contaminated with heavy metals. Examples are electrokinetic extraction, solidification, soil flushing, vitrification, phytoremediation, etc. These methods have different working mechanisms and limitations (Liu et al., 2018). Thus it is essential to find an effective and sustainable method for large-scale soil remediation. Some commonly used methods for remediation include solidification, stabilization, electrokinetic extraction, soil flushing, phytoremediation, bioremediation, and vitrification. *Solidification* refers to the addition of adsorptive materials like lime, fly ash, cement to immobilize contaminants. *Stabilization* is the technique that reduces the hazardousness of the contaminants by changing them into mobile and less soluble form. *Electrokinetic extraction* uses the principle of electrical adsorption to remove heavy metals from the contaminated soil. When a direct current is applied via electrodes, cations and anions of the contaminated soil move to the cathode and anode, respectively. Thereby the contaminants are removed by electroplating or ion exchange process (Wang et al., 2020). *Soil flushing* is an in situ technique to remove contaminant by passing a fluid for extraction through the soil. *Phytoremediation* is the process of remediation where plants are used to clean up contaminants from soil. Plants are grown in contaminated soil, which further immobilizes the heavy metals through roots and shoots. *Bioremediation* is the method in which microorganisms are used for detoxification and remediation of the contaminated soil. *Vitrification* is the method in which heat is used to transform contaminated soil into a solid form, where the heavy metals are captured. However, some of these techniques are costly and can even cause secondary pollution. Thus, a remediation technique which is both economical and ecologically sound is required for effective remediation of heavy-metal-contaminated soils. Biochar-based remediation is one such emerging technique which has the potential to address the aforementioned issues.

There is an increasing trend of using biochar for heavy metals remediation of heavy-metal-contaminated sites, metal mine spoils, and mine tailings. Biochar is a carbon-rich material obtained by the pyrolysis of biomass at a temperature around 250–900 °C in an oxygen-limited condition. Biochar is effective in improving soil properties by increasing the pH of acidic soil, increasing water retention capacity, and immobilizing heavy metals in the contaminated soils, which eventually favors plant growth (Ghosh et al., 2020). The current chapter focuses on the production conditions suited for biochar for heavy metals remediation, properties, and mechanisms of biochar, soil, and heavy metal interaction and its application for heavy-metals-contaminated soil. The future aspects and challenges associated with biochar-based heavy metal remediation are also discussed briefly.

4.2 BIOCHAR PRODUCTION

Biochar is produced by heating different biomass feedstock such as leaves, straw, wood, manure, sewage sludge, etc. in an anaerobic condition (or with limited air), in a process called pyrolysis. The production of biochar may also lead to the generation of syngas and bio-oil, which are used as a source of clean and green energy (Figure 4.1). Biochar production process can be divided into slow pyrolysis and fast pyrolysis depending upon the temperatures and heating rates (Table 4.1). In slow pyrolysis, thermal decomposition of the biomass is performed at relatively low heat and long residence time, which can range from some hours to even a few days. Slow pyrolysis is advantageous due to its high yielding capacity. The limitations include its processing at low heating rates and low biochar yield. Fast pyrolysis is associated with a short residence time and comparatively higher heating rate, which can be even less than 10 s. The pyrolysis temperature is controlled carefully, which is generally kept in the range of 300–900 °C. Fast pyrolysis enhances the yield of bio-oil, whereas slow pyrolysis increases biochar yield (Borchard et al., 2012). An absolute reactor with fluidized bed is used at a high mass conversion rate and with a higher temperature for the production of

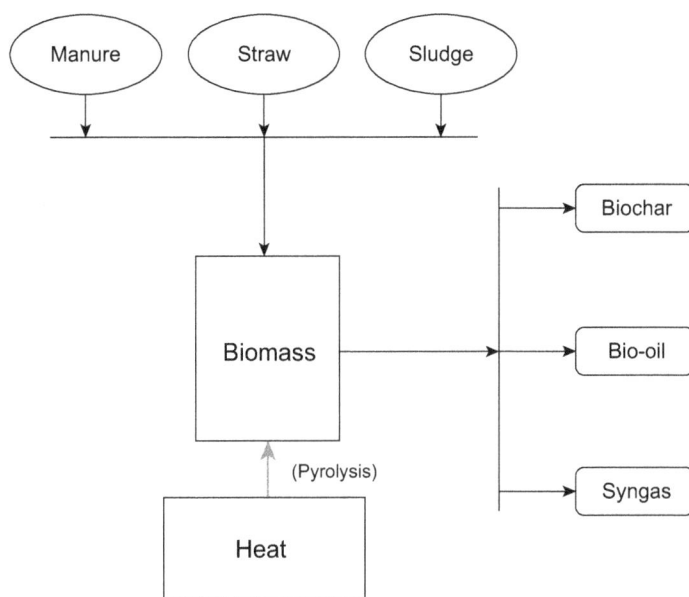

FIGURE 4.1 Schematic diagram of production of biochar.

TABLE 4.1
Biochar Production Process, Temperature, Heating Rate, Residence Time, and the End Products Obtained

Process of production	Temperature (°C)	Heating rate (°C/min)	Residence time	End products	References
Slow pyrolysis	300–900	5–7	>1 h	Biochar	Qian et al. (2015)
Fast pyrolysis	400–600	1000	~1 s	Bio-oil, biochar	Şensöz and Angin (2008)
Gasification	750–1500	100–200	10–20 s	Syngas, biochar	Qian et al. (2015)
Hydrothermal carbonization	150–350	–	1–12 h	Biochar, chemical products	Lehmann and Joseph (2012)

bio-oil. Pyrolysis temperature, gas flow rate, and particle size of the biomass influences the product quality and distribution of the reactor. Apart from the pyrolysis process, gasification and hydrothermal carbonization are two other methods that can be used for biochar production.

Gasification is the combustion of biomass partially under anaerobic condition and at a high temperature to produce gases like methane, carbon monoxide, carbon dioxide, and hydrogen, which are called syngas. Syngas can be converted into liquid fuel by liquefaction. Gasification enhances the yield of gases but reduces the production of biochar (Kumar et al., 2019).

The hydrothermal carbonization process is an environment-friendly technology. In this process, different biomass feedstocks are converted thermally into smokeless solid fuels that have high carbon content (Nizamuddin et al., 2017). The hydrothermal condition refers to the usage of aqueous medium in the process. Organic substances like saccharides or simpler compounds are mixed with water and thermally treated at a temperature range of 150–350 °C, resulting in organic substances that are soluble in water and a solid product that is rich in carbon, known as hydrothermal carbonization (Wang et al., 2020). Depending on the mechanisms of reaction and experimental conditions, the hydrothermal carbonization (HTC) process can be divided into two types: HTC with high temperature (300–800 °C) and HTC with low temperature (<300 °C) (Hu et al., 2010).

4.3   PROPERTIES OF BIOCHAR

Biochar is considered to be effective in soil remediation, mitigation of climate change, and waste management. The feedstock material, pyrolysis temperature, residence time, and its modification techniques determine its properties. Biochar characteristics such as pore distribution, pH, elemental composition, specific surface area, cation exchange capacity, and other physicochemical properties are influenced by these production conditions. This section focuses on how biochar physiochemical properties are influenced and the mechanism involved.

4.3.1   ELEMENTAL COMPOSITION OF BIOCHAR

Biochar is a carbon-rich substance mixed with other elements like hydrogen, oxygen, and nitrogen. These elements form alkyl groups with an aromatized structure, which are the primary components of biochar (Ye et al., 2015). Biochar can be categorized into two types according to the source of biomass materials such as plant biochar (bamboo, rice husk, straw, etc.) and animal biochar (animal manure). Carbon content in animal biochar is low, whereas other elements content is high due to the high protein content. Also, there is higher content of phosphorus and ash in animal biochar than that of plant, which further enhances the quality of soil fertility when used as an amendment (Taylor et al., 2013).

During biochar production, pyrolysis temperature plays an important role in elemental compositions. Higher the pyrolysis temperature, lower the hydrogen, oxygen, and nitrogen content. Also, with an increase in pyrolysis temperature, other elements like phosphorus, carbon, calcium, potassium, and magnesium content increase (Sun et al., 2014). Pyrolysis leads to deoxidation, dehydration, and decomposition of the volatile material present in biomass with the increase in pyrolysis temperature and residence time. A decrease in hydrogen and oxygen content or an increase in carbon content affects the hydrogen/carbon and oxygen/carbon ratios (Wang et al., 2020). Moreover, animal biochar and plant biochar respond similarly to an increase in pyrolysis temperature.

4.3.2   pH

The oxides and carbonates in biochar are generally alkaline in solution. The pH of biochar depends on pyrolysis temperature. It is observed that the pH of biochar increases with an increase in pyrolysis temperature. There is a high ash content in animal biochar, which makes it more alkaline. In the case of agricultural waste biochar, it has comparatively low ash content. However, with a rise in pyrolysis temperature, the functional groups that are acidic in nature get decomposed, and the organic acids get volatilized, ultimately increasing the pH. The raw materials used as biomass for the production of biochar also affect the pH. Inorganic carbonates generated at high temperatures are the major alkaline components of biochar, whereas organic anions are the ones that contribute to the alkalinity of biochar at relatively low temperatures (Wang et al., 2020).

4.3.3   SPECIFIC SURFACE AREA

Specific surface area (SSA) is an important factor that affects the adsorption ability of biochar. Pore structure and pore size distribution are more complex when they have a larger specific surface area due to the microporous structure of the biochar. The surface area of biochar is mainly influenced by the feedstock and pyrolysis temperature. SSA of dairy manure biochar increases with an increase in pyrolysis temperature, but plant biochar has more SSA than biochar produced from dairy manure (Cao and Harris, 2010). Plant-based biomass contains more organic carbon, which leads to high porosity when heated. It has been observed that SSA of wood biochar produced at 600 °C (373.5–401 m^2/g) was 70 times higher than that of biochar produced at 300°C (Sun et al., 2014). Thus, the specific surface area and porosity of soil and also the physical adsorption capacity of soil typically increase with the application of biochar (Liang et al., 2006).

4.3.4 Cation Exchange Capacity

The amount of exchangeable cations, such as Ca^{2+}, Mg^{2+}, K^+, Na^+, NH_4^+, that biochar is capable of holding is known as cation exchange capacity (CEC) (Weber and Quicker, 2018). As most nutrients used by plants are taken up in their ionic form, the CEC's negative surface charges attract cations and are used for describing the fertility of soil (Liang et al., 2017). CEC is dependent on the surface structure, functional groups, and the surface area that make the surface charges accessible (Liang et al., 2006). According to Rajkovich et al. (2012), CEC of oak and corn stover biochar has higher CEC than paper mill waste, food waste, and nut shells biochar. A possible explanation is that the decomposition of the acidic functional group such as cellulose and lignin in plant biomass can lead to a reduction in CEC with an increase in pyrolysis temperature. Cow manure biochar has relatively higher ash content and lower carbon/nitrogen as compared to plant biochar, which results in a higher electrical conductivity and CEC (Lu et al., 2017). The atomic ratios of oxygen/carbon, nitrogen/carbon, and hydrogen/carbon also decrease with the increase in pyrolysis temperature. Thus it can be concluded that the decrease in numbers of hydroxyl, carboxylic, and amino groups is due to the increase in temperature (Majumder et al., 2019). Functional groups with more oxygen content and oxygen/carbon ratio on the surface promote a rise in cation exchange capacity. Higher CEC values of biochar may help increase the CEC of the soil, which ultimately supports heavy metals immobilization (Wang et al., 2020).

4.3.5 Enhancement of Biochar Properties by Different Treatments

Modification techniques are often used to improve the adsorption capacity of biochar. Steam activation of biochar accelerates its positive effects on nutrient retention and uptake by plants as compared to nonactivated biochar (Borchard et al., 2012). It also has greater positive effects on soil than the nonactivated ones (Tang et al., 2013). The specific surface area and pH of biochar has been reported to increase with modification by steam activation (Hass et al., 2012). Magnetization of biochar can also improve sorption capacity. A strong ferromagnetic capacity is associated with the magnetic biochar, so further retained biochar is isolated and collected by magnetic separation.

The oxidation of biochar is also done to improve the sorption potential. It is achieved by the addition of different oxidants such as potassium permanganate, nitric acid, hydrogen peroxide, etc. Nitric acid is found to be more effective than potassium permanganate. The surface is seen to have more acidic functional groups after oxidation (Li et al., 2014).

Digestion of biochar can also enhance adsorption capacity. The anaerobic digestion of biochar produced from beet root, dairy manure, etc. have higher adsorption capacity than that of biochar without digestion treatment. The digestion of biochar can enhance adsorption capacity in such a way that it can be comparable to activated carbon (Yao et al., 2011). It has also been reported that the digested biochar has more specific surface area, cation exchange capacity, and a high pH value with more negative surface charges, which favors the adsorption capacity of biochar (Inyang et al., 2012).

Each modification method has different effects on the specific surface area, cation exchange capacity, functional groups, pH, and other physicochemical properties of biochar (Wang et al., 2020). The surface oxygen functional groups and the alkaline, porous structure of biochar are two main factors that are used for evaluating its heavy metals remediation potential (Ghosh and Maiti, 2020).

4.4 IMPACT OF BIOCHAR APPLICATION ON HEAVY-METAL-CONTAMINATED SOIL

Heavy metals are non-degradable in nature, and they pose a serious risk to the environment. According to Wang et al. (2020), the interaction between the heavy metals and organisms can be explained as the source-pathway-receptor model, which suggests that heavy-metal-contaminated sites can be remediated by removing sources, eliminating the pathway, or modifying the exposure to the receiver. Biochar has the property to immobilize heavy metals in the soil, so that mobility of the heavy metals is checked and bioavailability is increased (Table 4.2). Thus the effects of biochar application in heavy-metal-contaminated soil in terms of mobility, bioavailability, and strength are discussed next.

4.4.1 MOBILITY OF HEAVY METALS IN SOIL

Biochar has high adsorption ability, which can immobilize heavy metals and thereby reduce environmental risks. The mobility of heavy metals in soil is affected by biomass feedstock and the pyrolysis condition. The heavy metals present in soil particles can also be mobilized in a soil solution by biochar application (Wang et al., 2020). For example, there is a competitive adsorption on the surface of biochar between phosphorus and arsenic. When the biochar is applied, phosphorus increases in the soil, and hence more arsenic is forcibly leached out due to competitive adsorption (O'Connor et al., 2018).

Heavy metals can be immobilized by reduction of metal ions by the application of biochar. There are two forms of chromium: hexavalent chromium (VI) and trivalent chromium (III). Trivalent chromium is not toxic in nature, and it is bound to soil particles firmly; on the other hand, hexavalent chromium is very toxic and highly mobile. So the toxicity of hexavalent chromium can be significantly decreased by its reduction to trivalent chromium. The addition of biochar enhances the reduction of chromium (VI) and simultaneously improves the immobility of chromium (III) (Choppala and Bolan, 2013). In the aerobic condition, the toxicity and mobility of arsenic (III) is higher than that of arsenic (V). Reduction of arsenic (V) to arsenic (III) is done by biochar, which ultimately enhances the mobility of arsenic (Ahmad et al., 2014). So there is a need for modification of biochar to reduce these negative effects. It has been observed that the addition of iron oxide can magnetize biochar and hence reduce the mobility of arsenic through anion exchange (Warren et al., 2003). The effect of biochar on the immobilization of metals also depends on soil types. Different types of soil behave differently in the case of metal immobilization. In a study conducted by Shen et al. (2016), it was reported that hardwood biochar application significantly reduced the amount of leached zinc and nickel in a contaminated sandy soil after three years of application. However, the mobility of lead was not influenced by its application. Thus, biochar behaves differently with various ions depending upon the surface charges.

4.4.2 BIOAVAILABILITY OF HEAVY METALS

Plants absorb heavy metals from soil through their roots and transport them and accumulate them in different body parts. Thus, the objective of soil remediation by biochar is to limit the mobility and transportation of heavy metals in soil and to reduce their bioavailability (Table 4.2).

The addition of other materials with biochar has greater effectiveness in limiting the bioavailability of heavy metals in soil. It has been observed that the pH of soil increases and chromium immobilization is increased by biochar mixed with lime and lime-mixed curing agents. A study reveals that 5% chicken-manure-derived biochar can decrease the uptake of copper by *Oenothera picencis* plants in copper-mine-polluted soil (Meier et al., 2015). Hence various factors that influence the bioavailability of heavy metals by an amendment are the immobilization effect of soil heavy metals, the limitation of heavy metal accumulation and uptake by plants, the increase of plant biomass, and the dilution effect of heavy metals in plant tissues.

4.4.3 STRENGTH OF SOIL

There are studies that observe that the use of biochar in soil reduces its unconfined compressive strength and shear strength. With the increase in the use of biochar, there is an increase in

TABLE 4.2
Application of Biochar Produced from Various Feedstocks and Their Effect on Heavy Metal Remediation

Biomass feedstock	pH	Heavy metals incurred	Effects	References
Fruit biochar, compost	7.5–8	Cadmium, lead, arsenic, copper, zinc	Use of biochar reduces heavy metals concentration. High organic carbon content of compost has a notable effect on mobility of heavy metals, which efficiently decreases heavy metals bioavailability.	Beesley et al. (2014)
Orchard pruning	8.5–9.5	Arsenic	There is a decrease of 68% and 80% arsenic concentration in the roots and other parts of tomato seedlings, respectively.	Beesley et al. (2014)
Soyabean-straw-derived biochar	7.91	Arsenic, cadmium	Arsenic concentration is reduced by 88% in rice plants. Besides that, there is a decrease of 3.18 mg/kg and 565 µg/kg in As (III) and As (V) concentration, respectively.	Li et al. (2018)
Rice straw	8.2	Cadmium	Reduces the cadmium accumulation in rice grains.	Zheng et al. (2012)
Rice-husk-derived biochar	9.18	Mercury	Reduces the mercury concentration in rice grains.	Cheng et al. (2020)
Oak	7.86	Lead	Reduces bioavailability of lead in soil.	Ahmad et al. (2012)
Beet	9	Copper, nickel, cadmium, lead	Brings down the concentration of heavy metals in soil.	Inyang et al. (2012)
Hardwood	9.9	Zinc, cadmium	pH of soil increases, concentration of cadmium and zinc leachate.	Beesley and Marmiroli (2011)
Sludge	9.54	Cadmium, lead, zinc, nickel, copper	Immobilization of heavy metals is achieved when there is a 4% biochar addition.	Méndez et al. (2012)
Sugarcane	9.6	Lead	Reduces the concentration of lead.	Abdelhafez et al. (2014)
Stinging nettle	–	Arsenic and copper	Copper leaching is reduced, and the mobility of arsenic is affected a little.	Guo et al. (2020)
Sewage sludge	7.2	Cadmium	Cadmium concentration is reduced in the soil.	Zhang et al. (2016)
Dairy manure	7.1	Lead	Lead concentration is reduced in soil.	Cao and Harris (2010)
Miscanthus straw	8.7	Zinc, copper, lead, cadmium	Reduces the concentration of cadmium and zinc in soil, but the concentration of copper and lead increased in leachate.	Schweiker et al. (2014)
Rape straw	8	Cadmium	Reduces the cadmium concentration in soil efficiently.	Li et al. (2017)
Sawdust and swine manure	–	Lead	Lead concentration is reduced in soil efficiently.	Liang et al. (2017)

compressive strength, while there is no influence on internal friction angle with its application. The reduction in soil strength favors crop productivity in agriculture, but decreased soil strength is not acceptable in the case of soil solidification. Thus to improve the mechanical properties of soil, the biochar is generally used with solidifying materials.

4.5 MECHANISMS FOR INTERACTION OF BIOCHAR AND HEAVY METALS IN SOIL

Biochar has a high pH value, active surface functional groups, and porous and aromatized structure. These properties influence the remediation process and the mechanisms that include physical adsorption, electrostatic interaction, ion exchange, precipitation, and complexation (Figure 4.2).

4.5.1 PHYSICAL ADSORPTION

Physical adsorption is also designated as van der Waals adsorption. As the adsorbate and adsorbent molecules interact with each other, the phenomenon of physical adsorption occurs (Wang et al., 2020). The intermolecular force between them is reversible in nature. The pore volume of biochar, surface energy, and the specific surface area generally affect the physical adsorption of heavy metals on biochar (Zheng et al., 2012). The increase in pyrolysis temperature causes the increase in SSA and pore volume. The area of contact between biochar and heavy metal ions influences the adsorption process. For example, copper and uranium are immobilized significantly by biochar produced by switch grass (300 °C) and pine wood (700 °C) through the mechanism of physical adsorption (Liu et al., 2010).

4.5.2 ION EXCHANGE

The interchange of ions like sodium (Na^+), magnesium (Mg^{2+}), potassium (K^+), and calcium (Ca^{2+}) with the heavy metal ions on the surface of biochar is called ion exchange. The surface chemical properties of biochar influence the efficiency of ion exchange. High cation exchange capacity

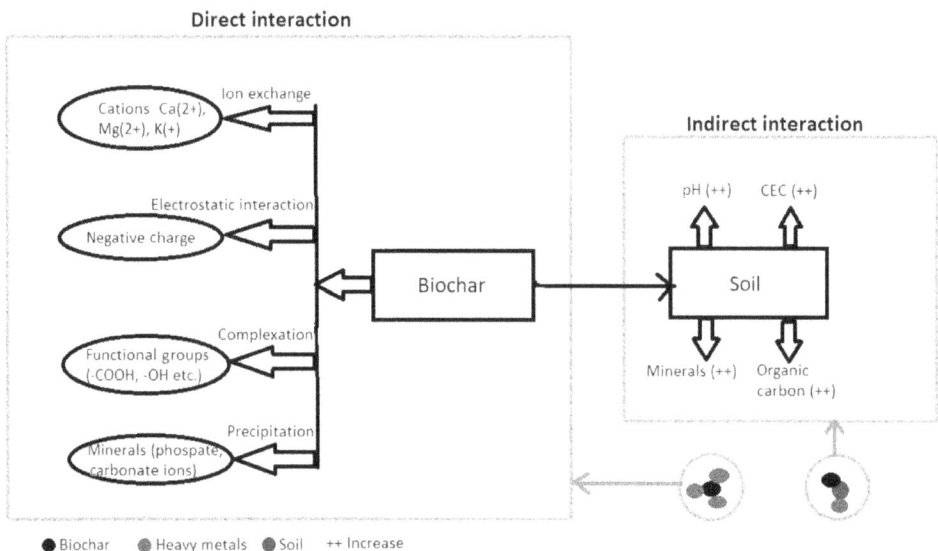

FIGURE 4.2 Schematic diagram of mechanism for biochar–heavy metal interaction in soil.

indicates that the ion exchange between biochar and metal ions is efficient enough to immobilize heavy metals in a contaminated soil (Ghosh and Maiti, 2021). The highest CEC of biochar occurs at a lower pyrolysis temperature, i.e., around 250–300 °C because at a higher pyrolysis temperature, there is a reduction of oxygen/carbon ratio and acidic functional groups, which leads to a reduction in the cation exchange capacity of biochar (Lee, Lawson, et al., 2010). El-Shafey (2010) reported that mercury and zinc remediation by husk biochar procured at 175–180 °C. Acidic oxygen functional groups like carboxyl (-COOH) and hydroxyl (-OH) can exchange with zinc (Zn^{2+}) and mercury (II) (Hg^{2+}) ions on the biochar surface to reduce protons. The equations of ion exchange are as follows:

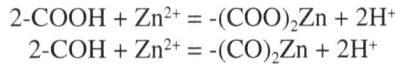

$$2\text{-COOH} + Zn^{2+} = \text{-(COO)}_2Zn + 2H^+$$
$$2\text{-COH} + Zn^{2+} = \text{-(CO)}_2Zn + 2H^+$$

The ion exchange potential of biochar is influenced by pH of the soil. When pH of the biochar at zero charge point is higher than that of pH of soil solution, the ion exchange process takes place, which causes a greater number of heavy metal ions to be attracted to the biochar surface. The adsorption capacity of hydrothermal biochar is found to be greater than that of pyrolyzed biochar, which shows that ion exchange is the most effective mechanism of heavy metal remediation by biochar application (Liu et al., 2010).

4.5.3 ELECTROSTATIC INTERACTION

Highly negatively charged biochar improves the electrostatic interaction between soil and metal ions to prevent mobility of heavy metals through electrostatic attraction. The π–π electron donor–acceptor interaction between biochar's π-electron-rich-graphene surface and π-electron-deficient-positively-charged heavy metal cations causes their immobilization (Ghosh and Maiti, 2020).

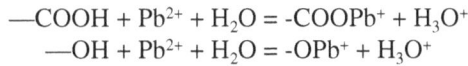

$$\text{—COOH} + Pb^{2+} + H_2O = \text{-COOPb}^+ + H_3O^+$$
$$\text{—OH} + Pb^{2+} + H_2O = \text{-OPb}^+ + H_3O^+$$

A study regarding the remediation effect of rice and wheat straw biochar depicts that biochar remediation is more effective than that of activated carbon due to the electrostatics interaction between the lead ion and the negatively charged biochar (Qiu et al., 2008). Thus it is an important mechanism of biochar remediation that immobilizes the heavy metals in soil.

4.5.4 COMPLEXATION

Biochar contains oxygen-containing functional groups like -COH, -COOH, and -OH on its surface that form complexes with heavy metal ions which are stable in nature (Wang et al., 2020). Complexation can occur between the carbon–oxygen ligand of the oxygen functional group and the positively charged metal ion. Cao and Harris (2010) suggested that the C=O ligand of the oxygen functional group forms the surface complexation of lead (II) ions with free carboxyl and hydroxyl functional groups and inner-sphere complexation of lead (II) ions with free hydroxyl functional groups of mineral oxides, as mentioned in the following equations:

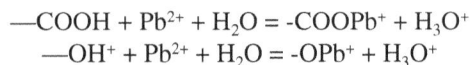

$$\text{—COOH} + Pb^{2+} + H_2O = \text{-COOPb}^+ + H_3O^+$$
$$\text{—OH}^+ + Pb^{2+} + H_2O = \text{-OPb}^+ + H_3O^+$$

4.5.5 PRECIPITATION

Biochar can get precipitated with cations present in the metals and form insoluble carbonates and phosphates, which prevent mobility of the heavy metals in soil. Plant biomass contains cellulose and hemicellulose, which, when pyrolyzed at high temperature, produce alkaline biochar and form metal precipitates. There is a higher ash content in animal biochar that reacts with heavy metals and leads to the formation of insoluble minerals. Yao et al. (2011) reported that cow manure biochar has a higher phosphate content and that it can prevent the mobility of lead in soil by the formation of insoluble pyromorphite [$Pb_5(PO_4)_3(OH)$]. Another study by Cao and Harris (2010) reported that, dairy-manure-derived biochar adsorb 84–87% heavy metals through precipitation, whereas surface sorption can only absorb 13–16%. All these results show that precipitation can effectively reduce the mobilization of heavy metals in soil.

4.6   POTENTIAL RISKS ASSOCIATED WITH APPLICATION OF BIOCHAR

The possibility of ecological risks and the factor of stability in the long run should be taken into account while applying biochar for soil remediation for immobilization/mobilization. It is observed that biochar may be the carrier of heavy metals (Xu et al., 2019), volatile organic compounds (VOCs) (Buss, 2016), dioxins, polycyclic aromatic hydrocarbons (PAHs) (Oleszczuk and Zieli, 2016), and other toxic substances. Biochar plays a vital role in greenhouse gases emission reduction and carbon utilization. However, a few studies report that, under certain conditions, the application of biochar can promote the emission of CO_2, N_2O, CH_4 (Ribas et al., 2019). Though biochar can reduce methane emissions, it may also aid in nitrous oxide emissions reduction depending on the condition in which it is applied. Therefore, the effect of reduction in emission due to biochar cannot be noticed in the case of all greenhouse gases; thus the use of biochar can also have negative effects (Cheng et al., 2020).

To mitigate the potential risks of biochar, different measures can be taken, including choosing appropriate pyrolysis temperature and biochar feedstock. This may help reduce the heavy metals concentration in biochar or convert them to less bioavailable forms (Zhou et al., 2017).

4.7   FUTURE OPPORTUNITIES AND CHALLENGES

Thus, in order to meet the demand and the application-based results, certain future opportunities and challenges need to be addressed.

1. Biochar produced from different feedstocks and pyrolysis conditions have different chemical and physical properties, and so does the remediation performance. Thus there is a need for a unique classification system and standard for biochar application, which will be helpful in the large-scale application of biochar.
2. Secondary ecological risks that include the formation of heavy metals, VOCs, and PAHs in the process of production of biochar. These contaminants are generally derived from the biomass, so they should be screened and pretreated to avoid the aforementioned issues. In addition, there is a need for optimization of factors such as residence time, pyrolysis temperature, etc., along with other obligatory criteria to reduce the inception of possible pollutants.
3. The study on soil heavy metal immobilization by biochar is generally based on pot experiment and short-term field trials. However, the focus needs to be shifted to long-term, large-scale applications.
4. Soil is a complex system, and various factors influence the immobilization of heavy metals in soil. Most of the studies are focused on the adsorption mechanism of biochar immobilization. However, other mechanisms such as ion exchange, electrostatic interaction, complexation, and precipitation also play important roles in immobilizing the heavy metals. Thus, it is crucial to study the interaction between heavy metals and biochar, which is helpful for soil remediation in the case of practical implementation.

5. Multifunctional biochar material, such as biochar with chemical modification and micro-organisms, should be used for effective engineering applications.
6. In order to achieve economic availability for large-scale application, cheaper raw materials and mass-scale in situ production processes need to be developed.

4.8 CONCLUSION

Biochar is a suitable approach to clean up heavy metals from soil as it is a sustainable, environment-friendly, and economically viable technology. Various thermal and chemical conversion methods are used for the preparation of biochar. The performance attribute of biochar is greatly influenced by the pyrolysis temperature, feedstock materials, and the residence time that jointly regulate the remediation of heavy-metals-contaminated soil. The mechanisms of biochar remediation include physical adsorption, ion exchange, electrostatic attraction, complexation, and precipitation. Modification techniques such as activation, magnetization, digestion, and oxidation are often used for improving the adsorption of heavy metals on the biochar surface. Biochar reduces the mobility and bioavailability of heavy metals. However, research needs to be done in field application in order to optimize methods, materials, and conditions for effective heavy metal remediation.

REFERENCES

Abdelhafez, A.A., Li, J., Abbas, M.H.H., 2014. Chemosphere Feasibility of biochar manufactured from organic wastes on the stabilization of heavy metals in a metal smelter contaminated soil. *Chemosphere* 117, 66–71. https://doi.org/10.1016/j.chemosphere.2014.05.086

Ahmad, M., Rajapaksha, A.U., Lim, J.E., Zhang, M., Bolan, N., Mohan, D., Vithanage, M., Lee, S.S., Ok, Y.S., 2014. Biochar as a sorbent for contaminant management in soil and water: A review. *Chemosphere* 99, 19–33. https://doi.org/10.1016/j.chemosphere.2013.10.071

Ahmad, M., Soo, S., Yang, J.E., Ro, H., Han, Y., Sik, Y., 2012. Ecotoxicology and Environmental safety effects of soil dilution and amendments (mussel shell, cow bone, and biochar) on Pb availability and phyto-toxicity in military shooting range soil. *Ecotoxicol. Environ. Saf.* 79, 225–231. https://doi.org/10.1016/j.ecoenv.2012.01.003

Beesley, L., Inneh, O.S., Norton, G.J., Moreno-Jimenez, E., Pardo, T., Clemente, R., Dawson, J.J.C., 2014. Assessing the influence of compost and biochar amendments on the mobility and toxicity of metals and arsenic in a naturally contaminated mine soil. *Environ. Pollut.* 186, 195–202. https://doi.org/10.1016/j.envpol.2013.11.026

Beesley, L., Marmiroli, M., 2011. The immobilisation and retention of soluble arsenic, cadmium and zinc by biochar. *Environ. Pollut.* 159, 474–480. https://doi.org/10.1016/j.envpol.2010.10.016

Borchard, N., Wolf, A., Laabs, V., Aeckersberg, R., Scherer, H.W., Moeller, A., Amelung, W., 2012. Physical activation of biochar and its meaning for soil fertility and nutrient leaching—a greenhouse experiment. *Soil Use Manag.* 28, 177–184. https://doi.org/10.1111/j.1475-2743.2012.00407.x

Buss, W., 2016. High-VOC biochar—effectiveness of post-treatment measures and potential health risks related to handling and storage. *Environ. Sci. Pollut. Res. Int.* 23(19), 19580–19589. https://doi.org/10.1007/s11356-016-7112-4

Cao, X., Harris, W., 2010. Properties of dairy-manure-derived biochar pertinent to its potential use in remediation. *Bioresour. Technol.* 101, 5222–5228. https://doi.org/10.1016/j.biortech.2010.02.052

Cheng, S., Chen, T., Xu, W., Huang, J., Jiang, S., Yan, B., 2020. Application research of biochar for the remediation of soil heavy metals contamination: A review. *Molecules.* 25, 1–21.

Choppala, G., Bolan, N., 2013. Concomitant reduction and immobilization of chromium in relation to its bioavailability in soils. *Environ. Sci. Pollut. Res. Int.* 22, 8969–8978. https://doi.org/10.1007/s11356-013-1653-6

El-Shafey, E.I., 2010. Removal of Zn (II) and Hg (II) from aqueous solution on a carbonaceous sorbent chemically prepared from rice husk. *J. Hazard. Mater.* 175, 319–327. https://doi.org/10.1016/j.jhazmat.2009.10.006

Ghosh, D., Maiti, S.K., 2020. Can biochar reclaim coal mine spoil? *J. Environ. Manage.* 272, 111097. https://doi.org/10.1016/j.jenvman.2020.111097

Ghosh, D., Maiti, S.K., 2021. Biochar assisted phytoremediation and biomass disposal in heavy metal contaminated mine soils: A review. *Int. J. Phytoremediation* 23, 559–576. https://doi.org/10.1080/1522651 4.2020.1840510

Guo, M., Song, W., Tian, J., 2020. Biochar-facilitated soil remediation: Mechanisms and efficacy variations. *Front. Environ. Sci.* 8. https://doi.org/10.3389/fenvs.2020.521512

Hass, A., Gonzalez, J.M., Lima, I.M., Godwin, H.W., Halvorson, J.J., Boyer, D.G., 2012. Chicken manure biochar as liming and nutrient source for acid appalachian soil. *J. Environ. Qual.* 41, 1096–1106. https://doi.org/10.2134/jeq2011.0124

Hu, B., Wang, K., Wu, L., Yu, S.H., Antonietti, M., Titirici, M.M., 2010. Engineering carbon materials from the hydrothermal carbonization process of biomass. *Adv. Mater.* 22, 813–828. https://doi.org/10.1002/adma.200902812

Inyang, M., Gao, B., Yao, Y., Xue, Y., Zimmerman, A.R., 2012. Bioresource technology removal of heavy metals from aqueous solution by biochars derived from anaerobically digested biomass. *Bioresour. Technol.* 110, 50–56. https://doi.org/10.1016/j.biortech.2012.01.072

Kumar, V., Sharma, A., Kaur, P., Singh Sidhu, G.P., Bali, A.S., Bhardwaj, R., Thukral, A.K., Cerda, A., 2019. Pollution assessment of heavy metals in soils of India and ecological risk assessment: A state-of-the-art. *Chemosphere* 216, 449–462. https://doi.org/10.1016/j.chemosphere.2018.10.066

Lee, K., Lawson, R.J., Olenchock, S.A., Vallyathan, V., Southard, R.J., Thorne, P.S., Saiki, C., Schenker, M.B., Lee, K., Lawson, R.J., Olenchock, S.A., Vallyathan, V., Southard, R.J., Thorne, P.S., Saiki, C., Schenker, M.B., Exposures, P., Lee, K., Lawson, R.J., Olenchock, S.A., Vallyathan, V., Southard, R.J., Thorne, P.S., Saiki, C., Schenker, M.B., 2010. Personal exposures to inorganic and organic dust grapes in manual harvest of california citrus and table grapes. *J. Occup. Environ. Hyg.* 1, 505–514. https://doi.org/10.1080/15459620490471616

Lehmann, J., Joseph, S. (Ed.), 2012. *Biochar for environmental management*, second ed., Biochar for Environmental Management. Taylor & Francis.

Li, G., Khan, S., Ibrahim, M., Sun, T., Tang, J., Cotner, J.B., Xu, Y., 2018. Biochars induced modi fi cation of dissolved organic matter (DOM) in soil and its impact on mobility and bioaccumulation of arsenic and cadmium. *J. Hazard. Mater.* 348, 100–108. https://doi.org/10.1016/j.jhazmat.2018.01.031

Li, H., Dong, X., da Silva, E.B., de Oliveira, L.M., Chen, Y., Ma, L.Q., 2017. Mechanisms of metal sorption by biochars: Biochar characteristics and modifications. *Chemosphere* 178, 466–478. https://doi.org/10.1016/j.chemosphere.2017.03.072

Li, Y., Shao, J., Wang, X., Deng, Y., Yang, H., Chen, H., 2014. Characterization of modified biochars derived from bamboo pyrolysis and their utilization for target component (furfural) adsorption. *Energy Fuels* 28, 5119–5127. https://doi.org/10.1021/ef500725c

Liang, B., Lehmann, J., Solomon, D., Kinyangi, J., Grossman, J., O'Neill, B., Skjemstad, J.O., Thies, J., Luizão, F.J., Petersen, J., Neves, E.G., 2006. Black carbon increases cation exchange capacity in soils. *Soil Sci. Soc. Am. J.* 70, 1719–1730. https://doi.org/10.2136/sssaj2005.0383

Liang, J., Yang, Z., Tang, L., Zeng, G., Yu, M., Li, Xiaodong, Wu, H., Qian, Y., Li, Xuemei, Luo, Y., 2017. Changes in heavy metal mobility and availability from contaminated wetland soil remediated with combined biochar-compost. *Chemosphere* 181, 281–288. https://doi.org/10.1016/j.chemosphere.2017.04.081

Liu, L., Li, W., Song, W., Guo, M., 2018. Remediation techniques for heavy metal-contaminated soils: Principles and applicability. *Sci. Total Environ.* 633, 206–219. https://doi.org/10.1016/j.scitotenv.2018.03.161

Liu, Z., Zhang, F., Wu, J., 2010. Characterization and application of chars produced from pinewood pyrolysis and hydrothermal treatment. *Fuel* 89, 510–514. https://doi.org/10.1016/j.fuel.2009.08.042

Lu, K., Yang, X., Gielen, G., Bolan, N., Ok, Y.S., Niazi, N.K., Xu, S., Yuan, G., Chen, X., Zhang, X., Liu, D., Song, Z., Liu, X., Wang, H., 2017. Effect of bamboo and rice straw biochars on the mobility and redistribution of heavy metals (Cd, Cu, Pb and Zn) in contaminated soil. *J. Environ. Manage.* 186, 285–292. https://doi.org/10.1016/j.jenvman.2016.05.068

Majumder, S., Neogi, S., Dutta, T., Powel, M.A., Banik, P., 2019. The impact of biochar on soil carbon sequestration: Meta-analytical approach to evaluating environmental and economic advantages. *J. Environ. Manage.* 250, 109466. https://doi.org/10.1016/j.jenvman.2019.109466

Meier, S., Curaqueo, G., Khan, N., Bolan, N., Cea, M., Eugenia, G.M., Cornejo, P., Ok, Y.S., Borie, F., 2015. Chicken-manure-derived biochar reduced bioavailability of copper in a contaminated soil. *J. Soils Sediments*. 17. https://doi.org/10.1007/s11368-015-1256-6

Méndez, A., Gómez, A., Paz-ferreiro, J., Gascó, G., 2012. Effects of sewage sludge biochar on plant metal availability after application to a Mediterranean soil. *Chemosphere* 89, 1354–1359. https://doi.org/10.1016/j.chemosphere.2012.05.092

Nizamuddin, S., Baloch, H.A., Griffin, G.J., Mubarak, N.M., Bhutto, A.W., Abro, R., Mazari, S.A., Ali, B.S., 2017. An overview of effect of process parameters on hydrothermal carbonization of biomass. *Renew. Sustain. Energy Rev.* 73, 1289–1299. https://doi.org/10.1016/j.rser.2016.12.122

O'Connor, D., Peng, T., Zhang, J., Tsang, D.C.W., Alessi, D.S., Shen, Z., Bolan, N.S., Hou, D., 2018. Biochar application for the remediation of heavy metal polluted land: A review of in situ field trials. *Sci. Total Environ.* 619–620, 815–826. https://doi.org/10.1016/j.scitotenv.2017.11.132

Oleszczuk, P., Zieli, A., 2016. Effect of pyrolysis temperatures on freely dissolved polycyclic aromatic hydrocarbon (PAH) concentrations in sewage sludge-derived biochars. *Chemosphere* 153, 68–74. https://doi.org/10.1016/j.chemosphere.2016.02.118

Qian, K., Kumar, A., Zhang, H., Bellmer, D., Huhnke, R., 2015. Recent advances in utilization of biochar. *Renew. Sustain. Energy Rev.* 42, 1055–1064. https://doi.org/10.1016/j.rser.2014.10.074

Qiu, Y., Cheng, H., Xu, C., Sheng, G.D., 2008. Surface characteristics of crop-residue-derived black carbon and lead (II) adsorption. *Water Res.* 42, 567–574. https://doi.org/10.1016/j.watres.2007.07.051

Rajkovich, S., Enders, A., Hanley, K., Hyland, C., Zimmerman, A.R., Lehmann, J., 2012. Corn growth and nitrogen nutrition after additions of biochars with varying properties to a temperate soil. *Biol. Fertil. Soils* 48, 271–284. https://doi.org/10.1007/s00374-011-0624-7

Ribas, A., Mattana, S., Llurba, R., Debouk, H., Sebastià, M.T., Domene, X., 2019. Science of the total environment biochar application and summer temperatures reduce N 2 O and enhance CH 4 emissions in a Mediterranean agroecosystem: Role of biologically-induced anoxic microsites. *Sci. Total Environ.* 685, 1075–1086. https://doi.org/10.1016/j.scitotenv.2019.06.277

Schweiker, C., Wagner, A., Peters, A., Bischoff, W.-A., Kaupenjohann, M., 2014. Biochar reduces zinc and cadmium but not copper and lead leaching on a former sewage field. *J. Environ. Qual.* 43, 1886–1893. https://doi.org/10.2134/jeq2014.02.0084

Şensöz, S., Angin, D., 2008. Pyrolysis of safflower (Charthamus tinctorius L.) seed press cake: Part 1. The effects of pyrolysis parameters on the product yields. *Bioresour. Technol.* 99, 5492–5497. https://doi.org/10.1016/j.biortech.2007.10.046

Shen, Z., Mcmillan, O., Jin, F., Al-tabbaa, A., 2016. Salisbury biochar did not affect the mobility or speciation of lead in kaolin in a short-term laboratory study. *J. Hazard. Mater.* 316, 214–220. https://doi.org/10.1016/j.jhazmat.2016.05.042

Sun, Y., Gao, B., Yao, Y., Fang, J., Zhang, M., Zhou, Y., Chen, H., Yang, L., 2014. Effects of feedstock type, production method, and pyrolysis temperature on biochar and hydrochar properties. *Chem. Eng. J.* 240, 574–578. https://doi.org/10.1016/j.cej.2013.10.081

Tang, J., Zhu, W., Kookana, R., Katayama, A., 2013. Characteristics of biochar and its application in remediation of contaminated soil. *J. Biosci. Bioeng.* 116, 653–659. https://doi.org/10.1016/j.jbiosc.2013.05.035

Taylor, P., Kelly, B.C.O., Pichan, S.P., 2013. Effects of decomposition on the compressibility of fibrous peat—a review. *Geomech. Geoengin.* 8, 37–41. https://doi.org/10.1080/17486025.2013.804210

Wang, Q., Huang, Q., Guo, G., Qin, J., Luo, J., Zhu, Z., Hong, Y., Xu, Y., Hu, S., Hu, W., Yang, C., Wang, J., 2020. Reducing bioavailability of heavy metals in contaminated soil and uptake by maize using organic-inorganic mixed fertilizer. *Chemosphere* 261, 128122. https://doi.org/10.1016/j.chemosphere.2020.128122

Wang, Y., Liu, Y., Zhan, W., Zheng, K., Wang, J., Zhang, C., Chen, R., 2020. Stabilization of heavy metal-contaminated soils by biochar: Challenges and recommendations. *Sci. Total Environ.* 729, 139060. https://doi.org/10.1016/j.scitotenv.2020.139060

Warren, G.P., Alloway, B.J., Lepp, N.W., Singh, B., Bochereau, F.J.M., Penny, C., 2003. Field trials to assess the uptake of arsenic by vegetables from contaminated soils and soil remediation with iron oxides. *Sci. Total Environ.* 311, 19–33. https://doi.org/10.1016/S0048-9697(03)00096-2

Weber, K., Quicker, P., 2018. Properties of biochar. *Fuel* 217, 240–261. https://doi.org/10.1016/j.fuel.2017.12.054

Xu, Y., Qi, F., Bai, T., Yan, Y., Wu, C., An, Z., Luo, S., 2019. A further inquiry into co-pyrolysis of straws with manures for heavy metal immobilization in manure-derived biochars. *J. Hazard. Mater.* 380, 120870. https://doi.org/10.1016/j.jhazmat.2019.120870

Yao, Y., Gao, B., Inyang, M., Zimmerman, A.R., Cao, X., Pullammanappallil, P., Yang, L., 2011. Bioresource Technology Biochar derived from anaerobically digested sugar beet tailings: Characterization and phosphate removal potential. *Bioresour. Technol.* 102, 6273–6278. https://doi.org/10.1016/j. biortech.2011.03.006

Ye, L., Zhang, J., Zhao, J., Luo, Z., Tu, S., Yin, Y., 2015. Properties of biochar obtained from pyrolysis of bamboo shoot shell. *J. Anal. Appl. Pyrolysis.* 114, 172–178. Elsevier B.V. https://doi.org/10.1016/j.jaap.2015.05.016

Zhang, G., Guo, X., Zhao, Z., He, Q., Wang, S., Zhu, Y., Yan, Y., Liu, X., Sun, K., Zhao, Y., Qian, T., 2016. Effects of biochars on the availability of heavy metals to ryegrass in an alkaline contaminated soil. *Environ. Pollut.* 218, 513–522. https://doi.org/10.1016/j.envpol.2016.07.031

Zheng, R., Cai, C., Liang, J., Huang, Q., Chen, Z., Huang, Y., Peter, H., Arp, H., Sun, G., 2012. Chemosphere the effects of biochars from rice residue on the formation of iron plaque and the accumulation of Cd, Zn, Pb, As in rice (Oryza sativa L.) seedlings. *Chemosphere* 89, 856–862. https://doi.org/10.1016/j. chemosphere.2012.05.008

Zhou, J., Ma, H., Gao, M., Sun, W., Zhu, C., 2017. Changes of chromium speciation and organic matter during low-temperature pyrolysis of tannery sludge. *Environ. Sci. Pollut. Res. Int.* 25, 2495–2505. https://doi. org/10.1007/s11356-017-0271-0

5 Review on Biochar Production from Biomass and Its Applications in Removing Environmental Pollutants

Veerupaksh Galyan, Sidhika Joshi, Vandana Singh,
Mohit Sahni, S. Shankara Narayanan, Sanjay Kumar,
Kanupriya, Ankit Kumar, and Soumya Pandit

CONTNETS

5.1 INTRODUCTION

The rapid growth of industry and increasing dependence on agrochemical-based crop production practices after the green revolution have tremendously shot up the levels of harmful heavy metals in the food chain and the neighboring environment. The expanding urge of energy production with the necessity to cut down the emissions of greenhouse gases, protection of human health [Spokas et al. 2009], and the threat of declining oil reserves have put forward our attention towards, the potential use of biomass to be used as a sustainable energy source [Limousy et al. 2015]. Conventional methods of fossil fuel extraction usually include ion exchange, chemical precipitation, phase separation, and various other techniques. These methods are costly, and, in addition, they produce a substantial amount of harmful chemical precipitates that are detrimental to the surrounding environment. Therefore, the majority of the developed countries have started working on this absolute conviction of biomass use and its relentless potential for development [Ajoku et al. 2012]. The energy produced from biomass is widely used in rural areas as the primary source for energy production for household purposes [Jeguirium et al. 2014]. For their fuel requirements, most developing countries like

DOI: 10.1201/9781003203438-5

Biomass Sources

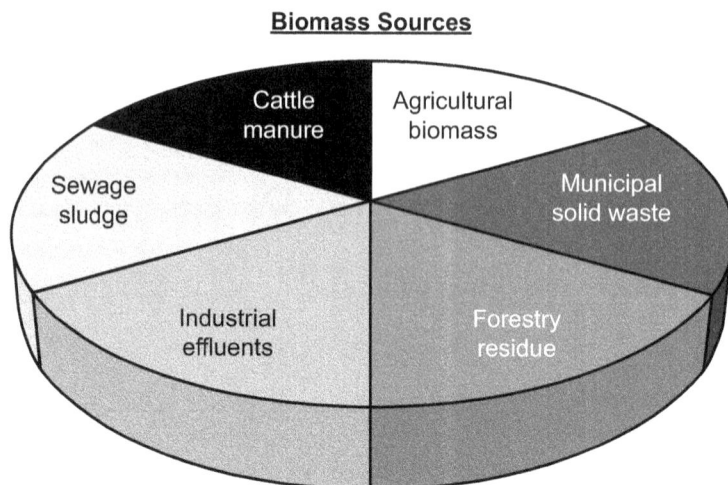

FIGURE 5.1 Pie chart of available sources for the production of biochar.

India or Bangladesh are still dependent on the accessibility of natural resources such as coal, natural gas, etc. (Figure 5.1). Renewable energy resources are often called alternative energy resources since they can provide sustainable energy with reduced emissions of air pollutants and various greenhouse gases [Demirbas et al. 2005].

Biomass can be transformed into solid, liquid, or gaseous fuels through various chemical, physical, and biological processes [Campbell et al. 1983]. Pyrolysis is a remarkable technique through which biomass can be converted into biochar [Bridgewater et al. 2003]. Pyrolysis is a thermal conversion method under limited or no supply of oxygen with temperatures varying from 300 to 700°C. It transforms organic materials into various useful products including carbonaceous solid biochar, liquid biofuel, and gases like methane for combustion.

Biochar is of high interest as it has several useful applications. It is a value-added product as it is rich in carbon and hence can produce high energy content as compared to different types of coals [Y. Lee et al. 2013]. It is used for carbon adsorption and has various soil applications that lead to a circular economy as biomass. The release of moisture, decomposition of volatile matter, and changes in the structure of carbon during the process of pyrolysis produce biochar equipped with unique physicochemical properties and functionality [Tan et al. 2017]. These properties allow biochar to be used for different applications such as adsorption of heavy metals [J. Tang et al. 2020] and herbicides [S. Yavari et al. 2017], crop improvement by pH buffering [L. Zhao et al. 2013] and nutrient retention [P. Pariyar et al. 2020], refused derived fuel [Y.E. Lee et al. 2017], and carbon sequestration in soil [R. K. Liew et al. 2018]. There are more than 55 potential uses of biochar in which the selection of the final application is based on the various properties of biochar that are textural, such as chemical composition and the presence of distinct surface functional groups [Guizani et al. 2019].

The quality and quantity of the biochar that is generated depending on the source material and the operating criteria followed, such as particle size, heating temperature, retention time [Tripathi M. et al. 2016], and many others. Different feedstocks produce biochar with different qualities that are more suitable for certain applications. The main focus of biochar production is on the properties of biochar under different process parameters and single feedstock streams such as lignocellulosic biomass, food waste, including crops and animal manure. This chapter focuses on analyzing the distinct properties of biochar derived from lignocellulosic biomass, manure and food waste with different compositions and process temperatures toward their optimal utilization for the final desired application and its environmental implications.

5.2 SELECTION OF FEEDSTOCKS FOR THE DESIRED APPLICATIONS OF BIOCHAR

The three main components of lignocellulosic feedstock utilized for the production of biochar, including agricultural and arboraceous biomass, are mainly 40–60 wt.% of cellulose, 10–25 wt.% of lignin, and 15–30 wt.% of hemicellulose [S.Y. Foong et al. 2020]. Different concentrations of these three components determine the presence of various elements like carbon, hydrogen, oxygen, and nitrogen in the biochar and affect its properties [Zhu et al. 2019]. Food waste predominantly comprises carbohydrate lipids and proteins, which vary depending on the kind of food waste taken, such as rice, vegetable waste, bread, fish waste, and mixed food waste. When the pyrolysis of three types of oil palm biomass (palm kernel shell, empty fruit brunch, and palm oil sludge) was performed at 500°C [X.J. Lee et al. 2017], the solid residues, which were PKS and EFB, produced biochar with low ash content C < 20 wt.% and high heating value of 27.50 and 26.18 MJ/Kg, respectively.

The biochar derived can be utilized as refused derived fuel due to its considerable heating value comparable to coal. The difference in the properties was observed because of the higher content of lignin and volatile matter in the solid residue. Biochar derived from food waste such as cooked rice, bread crumbs, and fish residue had an ash content lower than 10 wt.% [M. Fu et al. 2019]. The surface assimilation magnitude of the biochar is greatly affected by the feedstock composition.

Biochar with higher lignocellulosic content and lower ash content, which are tea leaves and nut husks, showed higher efficiency for the adsorption of ammonia than meat and starchy staples because of its higher surface area and several pores (Figure 5.2) [S. Xue et al. 2019]. The carbon embedded in the biochar varies from 13 to 80 wt.% depending on the type of food waste from which it has been derived [M. Fu et al. 2019]. One distinct characteristic is that all the biochar has a relatively lower surface area of less than 2 m^2/g, except for mixed food waste.

Biochar produced from pig manure has 8.28 m^2/g of surface area [X. Shen et al. 2020], whereas the lignocellulosic biomass-derived biochar has a larger pore size and a surface area ranging from 10 to 300 m^2/g [Meschewski et al. 2019]. Biochar having a larger surface area and pore volume of 25.17 m^3/g at 700°C, along with higher pH, is useful in eliminating the heavy metals embedded in manure based-biochar [J. Wang et al. 2019]. Biochar derived from wastewater sludge is rich in phosphorous, potassium, and other minerals, with phosphorous ranging from 1.7 to 4.7 g/kg and potassium ranging from 0.5 to 3.8 g/kg and an ash content greater than 50 wt.% showing high agronomic value [L. Zhao et al. 2014].

The biochar derived from potassium- and phosphorous-rich feedstock might also be suitable for carbon sequestration as soluble phosphorous resulted in cross-linking of carbon bonds during pyrolysis [Oldfield et al. 2018]. On comparing the cation exchange capacity of biochar obtained from sawdust, food waste, rice husk, paper sludge, it is noted that the biochar derived from manure has the highest cation exchange capacity of 67.23 $cmol_c$/kg, and food waste-derived biochar has the lowest with 6.69 $cmol_c$/kg. The huge variation was observed in cation exchange capacity ranging from 41.7 to 562 $cmol_c$/kg, following different feedstocks, with sawdust biochar having the lowest CEC and peanut shell biochar having the highest CEC. Higher cation exchange capacity indicates greater ability to absorb nutrients and be able to retain nutrients like ammonia and potassium, thereby preventing nutrient leaching, and suitability for soil and crop amendment for agricultural applications. The immobilization of the heavy metals is attributed to the higher pH and larger surface area, cation exchange capacity, and porous structure of biochar [W. Jia et al. 2017].

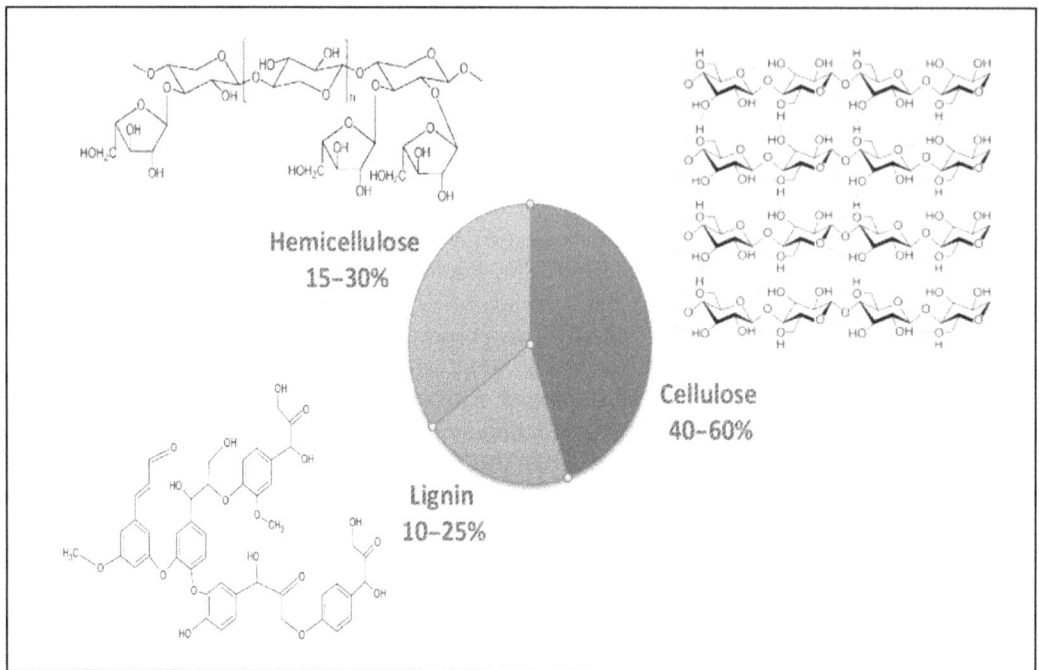

FIGURE 5.2 Percentages of cellulose, hemicellulose, and lignin in lignocellulosic biomass.

Biochar is well-known for carbon sequestration and is beneficial in improving soil productivity. The suitability for the use of biochar to be used in crop improvement programs depends on its predetermined carbon content, the O/C and molar H/C ratios that indicate its stability and resistance, and the carbon species retained in biochar. During pyrolysis, 50% of the carbon is retained in the biochar, whereas 10% of carbon gets mineralized when biochar is added to the soil. Disintegration by the chemical breakdown in presence of oxygen and biological mineralization reflects the recalcitrance and stability of the biochar [H. Nan et al. 2020]. Biochar generated at temperatures above 300°C has a lower H/C ratio and lower O/C ratio because of the release of oxygen and hydrogen upon the cracking of the residual biochar. A declining hydrogen-to-carbon ratio indicates higher aromaticity for higher resistance, whereas a declining oxygen-to-carbon ratio indicates lower polarity with lower absorption, having lower reactivity.

Biochar generated at high temperature has a high pH for buffering capacity, high surface area, a high number of pores, and higher aromaticity but has lower polarity due to thermal disruption of the surface functional groups due to high temperature. Based on the results obtained, the carbon char and the carbonization degree increased with pyrolysis temperature and is dependent on the carbon present in the feedstock from which the biochar is produced. Soil amendment and adsorption materials are more widely suited for biochar produced from the various feedstocks. The beneficial effect of soil amendment can be in the form of structural support in terms of C supply and porosity, whereas adsorption can be of nutrient retention and pollutant removal. The textural properties such as pore volume, pore size, and surface area increase with increasing temperature at which biochar is produced.

5.3 THERMOCHEMICAL CONVERSION AND TECHNIQUES USED FOR THE SYNTHESIS OF BIOCHAR

The use of heat and chemical reactions to release hydrogen from biomass is a crucial method of biomass transformation to energy and various value-added products. It is a process that comprises

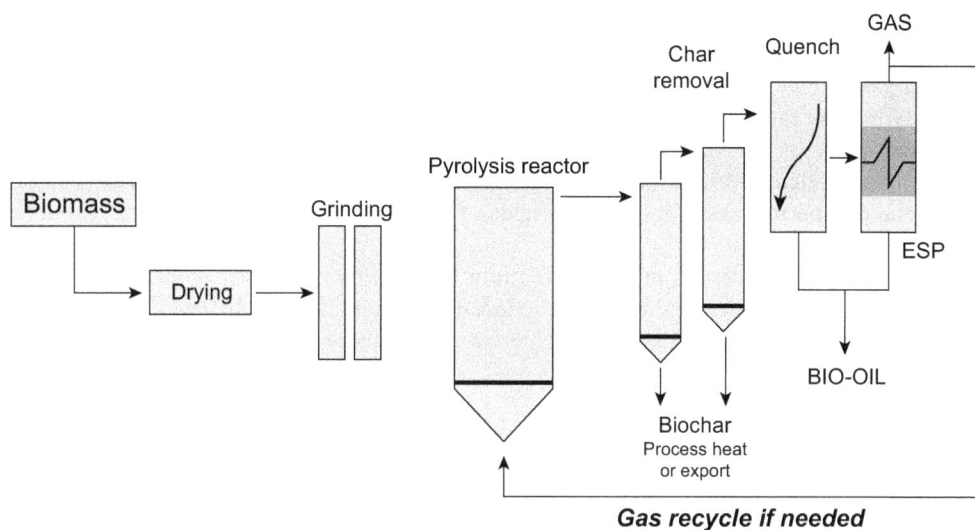

FIGURE 5.3   Process of pyrolysis by which generation of biochar, bio-oil, and biogas takes place

heating feedstock in the presence of oxygen to generate various energy outputs (Figure 5.3). Thermochemical conversion includes various techniques, including pyrolysis, liquefaction, gasification, and combustion. These techniques are different from one another in their spans of operating atmosphere and degree of control of oxidation and heating temperatures. Pyrolysis is the primary step for the process of gasification as well as for combustion.

Pyrolysis is described as a process of thermal breakdown of biomass carried out first in the absence of oxygen and involves the utilization of static atmosphere created by the gases, i.e., through nitrogen and argon. This is an endothermic process in which the products generated have high energy content. The mixture of products that are formed at the end of the process is highly heterogeneous, consisting of a variety of different compounds that include liquid mixtures, volatile gaseous products, and solid residues having high carbon content. The noncondensable gases that are generated during the process are mainly CO, CH_4, CO_2, H_2, C_2H_6, and H_2O. The liquid phase obtained is usually referred to as pyrolysis oil, which comprises water and oxygenated aliphatic chemicals [W. Jia et al. 2017].

Pyrolysis oils differ in their composition, even though they are likely to have corresponding properties; the water content in pyrolysis oil is normally in the range of 15–30 wt.% and have an oxygen composition of 35–40% diffused over all the products obtained from the process. They tend to have low pH due to the presence of carboxylic acids such as formic acid and acetic acid. Acids, esters, ketones, aldehydes, sugars, phenols, furans, syringols are some of the common components of pyrolysis oils. They have highly oxygen-containing functionalities.

Oils obtained from pyrolysis might be employed as fuels for power and heat production, but, due to their destructive nature and low pH, they cause many challenges in energy generation. Further upgrading is required before they can be used as fuels for various purposes like transportation. The change could be made by the incorporation of catalysts and various additives that can enhance the quality of these oils. For instance, the percentage of oxygen can be reduced by employing the process of hydrotreating, i.e., the removal of oxygen in the form of water by the addition of hydrogen gas or by using shape-selective catalysts like zeolites.

Various types of pyrolysis processes have been modified to generate a particular type of product. Slow pyrolysis is the process that uses a long retention time along with a low heating rate.

This helps in the formation of biochar and is also known as carbonization. The product obtained in slow pyrolysis is in the ratio of 35 wt.% of biochar with 30 wt.% of pyrolysis oils, along with 35 wt.% of gases.

A steady heating rate along with a long retention time favors different types of chemical reactions to take place at varying degrees of temperatures while maintaining a balance with one another. Long residence time allows reactions such as condensation and repolymerization to occur over a long period, resulting in high yields of biochar and generating a wider product distribution (Figure 5.4).

The carbonization process is modified to attain high yields of biochar without much control over particle size. Large particles result in a reduced consistent heating diffusion, as they heat up slowly leading to a fall in overall average temperatures. Liquid products that are formed alongside biochar are usually viscous and have poor dissolution in water, which reduce their relevancy to be used as fuels.

On the other hand, the fast pyrolysis method uses considerably high temperatures of heating ranging from 10 to 200 Ks and have a shorter residence time, usually less than 2 s. The fast-heating rate decreases the number of reactions taking place during the heating process due to which disintegration happens quickly, reducing the production of biochar. Thermochemical conversions that use very high heating rates (up to 10^4 K/s) and very short residence times are labeled as flash pyrolysis. In this process of flash pyrolysis, heating must be accompanied by speedy quenching along with cooling so that no more conversions take place and to accelerate the production of liquid products, which later condenses upon cooling. It is necessary to separate the biochar from the liquid components of the mixture as fast as possible so that biochar does not act as a cracking catalyst [M. Van et al. 2010]. The temperature has a very important role during the process of pyrolysis as heating above 650°C favors the production of gases, whereas slow heating at lower temperatures favors the generation of biochar.

The composition of the products obtained can be adjusted by various factors such as heating rate, temperature, particle size, residence time. Different components of biomass have distinct thermal properties and go through pyrolysis at different temperatures. Hemicellulose undergoes pyrolysis at lower temperatures, ranging from 210 to 310°C. Cellulose undergoes the process of pyrolysis around 300 to 315°C, depending on the source from which it is produced.

The process of pyrolysis is significantly affected by the presence of catalysts; in general, catalysts decrease the oxygen content in the oils obtained from pyrolysis. Torrefaction is a low-temperature

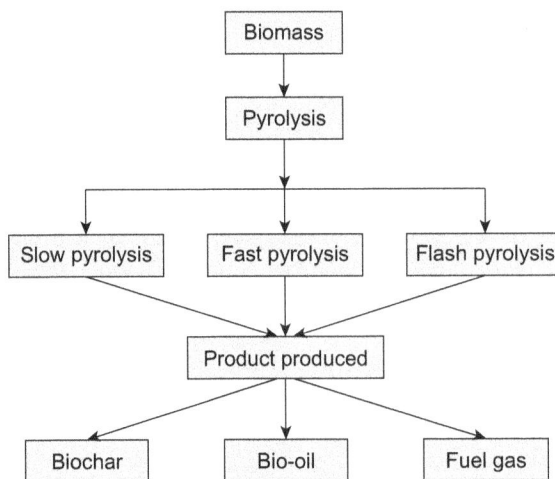

FIGURE 5.4 Pyrolysis process divided into three types: slow, fast, and flash pyrolysis.

thermal pretreatment designed to eliminate water and other volatiles producing a solid product with low oxygen content and higher energy density than the original biomass feedstock. Torrefaction takes place at 300°C and ends up when the temperature is dropped down to 200°C. Pyrolysis, when performed under anaerobic conditions, is referred to as thermal decomposition, and when this same process is executed in the open air at high temperatures, it is called gasification. Pyrolysis products that are produced through gasification require less pretreatment. Gases involved in gasification are CO, H_2, CH_4, and CO_2. Gasification is divided into three stages: (1) drying (up to 120°C), in which evaporation of water vapors takes place; (2) devitalization, in which evaporation of volatile matter takes place, and biochar is produced; (3) gasification of biochar (above 350°C), in which complete gasification is achieved above 500°C. The presence of oxygen is necessary during the process of gasification (the inert conditions used for pyrolysis instead favor devolatization, generating higher yields of biochar).

5.4 APPLICATIONS OF BIOCHAR

5.4.1 USE IN BIOETHANOL PRODUCTION

Bioethanol, which is a very economically effective and environmentally friendly way of energy production, has some complications during its manufacturing, and therefore biochar can be a good alternative to fixing these issues. Biochar has a great tool in its physiochemical property, i.e., adsorption, which can help in getting rid of fermentation inhibitors as well as in transfixing fermentative microbes, which go on to create a highly enhanced environment for bioethanol production. Therefore, if we keep biochar and bioethanol usage hand in hand with each other, it will help greatly in creating a better future for humankind.

5.4.2 REMOVAL OF FERMENTATIVE INHIBITORS

The removal of fermentative inhibitors from the pretreated hydrolysate is very important due to their negative effects on the fermentative microorganisms to achieve a high yield of bioethanol [J. Hou et al. 2017; Q. Zhang et al. 2017]. This can be achieved by a proper detoxification technique. Most of the impurities present in the environment have been known to be effectively removed by biochar, whether they are organic, inorganic, or heavy metals [M. Brodin et al. 2017]. Biochar is successful in adsorbing most of these pollutants through its physiochemical properties [K. T. Klasson et al. 2011]. Even though there have been very few studies on the usage of biochar in the removal of fermentative inhibitors, it has been noticed that the pyrolysis process of biochar can remove inhibitors present in bioethanol production such as is true for furan compounds. As it was noticed that pyrolysis requires greater energy input for the removal of inhibitory elements in bioethanol production, the use torrefaction was proposed for the production of biochar to reduce this energy input, which would have created an issue regarding the economic expenditure to carry out the process.

5.4.3 IMMOBILIZATION OF FERMENTATIVE INHIBITORS

The fermentation ability that is inherent to microorganisms in the process of bioethanol production can be fixed with the help of biochar [J. Mongokolkajit et al. 2011]. There have been many procedures to help fix these microorganisms for the production of bioethanol, but too few of avail due to issues arising from their instability in mechanical properties [M. Kyriakou et al. 2019]. As soil-dwelling plants and animals are affected negatively by the heavy metals in the environment, biochar has been applied as an economically friendly way to transfix heavy metals and deplete their negative effects [M. Kyriakou et al. 2019]. It is also known to help in the

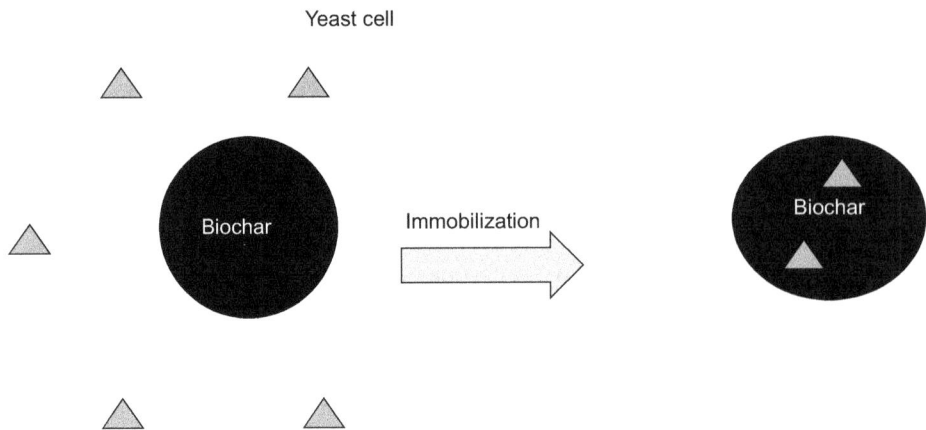

FIGURE 5.5 Immobilization of biochar.

production of methane gas [D. Mohan et al. 2006]. Even then, very little research has been done on the utilization of biochar for the fixation of fermentative microbes. In addition, the fixation of yeast cells was also very successful in using biochar for the production of bioethanol (Figure 5.5) [M. Kyriakou et al. 2019].

5.5 ENVIRONMENTAL APPLICATIONS OF BIOCHAR

Biochar has a huge usage in mitigating environmental pollution (Figure 5.6). By manipulating the production conditions of biochar, such as the biomass used and the pyrolysis temperature, it was noticed we can create a very efficient mechanism to fight against pollution. For example [D. Mohan et al. 2006; Ahmad et al. 2014], in the case where we carried out pyrolysis at a higher temperature, we noticed that it had a great affinity to organic pollutants due to its physical properties such as surface area, etc., high carbon-to-nitrogen ratio, high pH [Ahmad et al. 2014; Keiluweit 2010], and low dissolved organic carbon. On the other hand, it was noticed that the biochar produced at a lower pyrolysis temperature had a greater affinity toward the inorganic pollutants in the environment [Ahmad et al. 2014; Keiluweit et al. 2010; Ronsse et al. 2013]. It was further shown with the help of studies that wood or crop residues were a more efficient source of biomass for pollutant removal, whereas manure and industrial wastes were less effective [Ahmad et al. 2014]. The pyrolysis temperatures also showed a variation regarding the pH and such properties that affected the removal process along with the contaminant type [Ahmad et al. 2014]. Therefore, it is concluded that biochar variant is a key factor in contamination removal [Ahmad et al. 2014].

5.5.1 REMOVAL OF ORGANIC POLLUTANTS

It has been noticed through research that biochar has a significant application in the removal of organic pollutants [S.Y. Foong et al. 2020]. Some of these organic pollutants include agricultural chemicals, drugs, chemicals used in industries, unstable organic compounds, and cationic dyes [Kong et al. 2011; Zhang et al. 2013; Cao et al. 2011; Mondal et al. 2016; Jung et al. 2015]. Moreover, several pollutants are present in water sewage systems, animal manure and sewage, toxic organic compounds in landfill leachate, etc. [Xu et al. 2011; Yu et al. 2009; Adeel et al. 2017]. The removal of these organic pollutants is examined by their interactions with different properties of biochar.

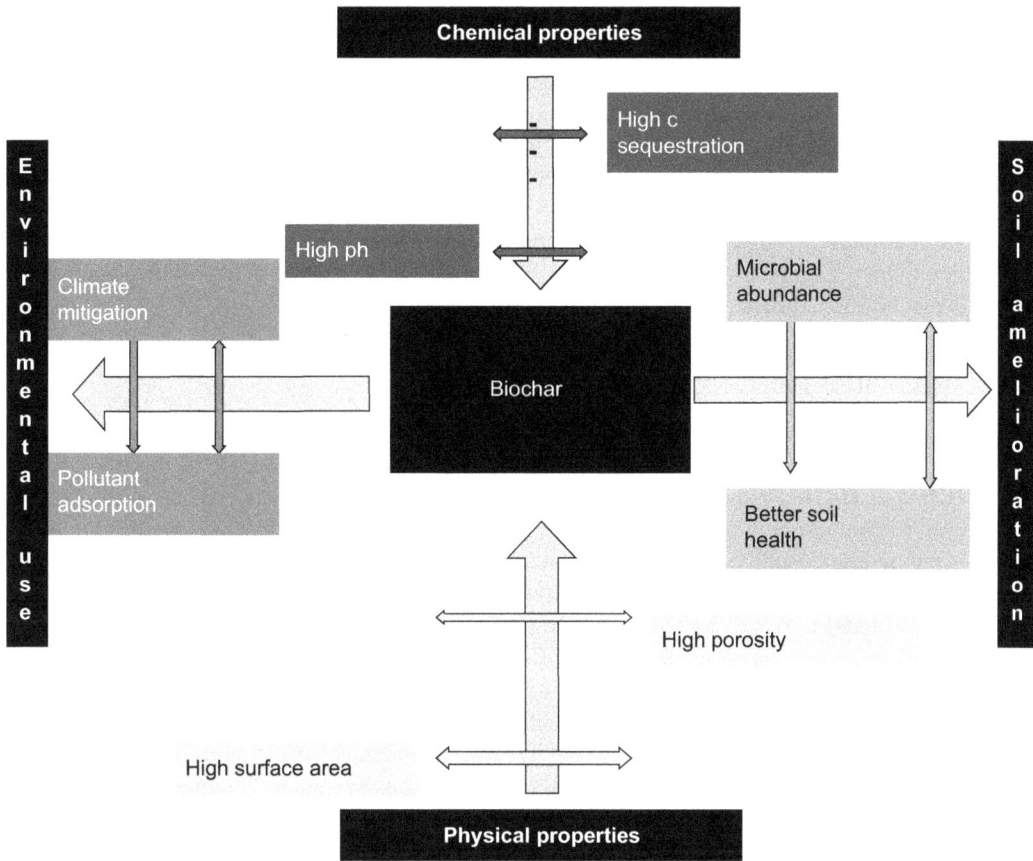

FIGURE 5.6 Applications of biochar according to its physical and chemical properties.

For organic pollutants, removal is mostly through chemical and physical adsorption [Brebu et al. 2010; Li et al. 2014] and partitioning, chemical transformation, and finally mineralization through biodegradation (by different types of microorganisms present on biochar) [Teixido et al. 2011; Xie et al. 2015; Beesely et al. 2010; Xu et al. 2013a]. Properties such as hydrolysis temperature, biomass source, and others have a significant effect on the adsorption quality of biochar. Even though the higher surface area and microporosity are very desirable characteristics for the extraction of nonpolar pollutants, they are often lacking in biochar produced at lower temperatures [Ahmad et al. 2012; Kookana et al. 2010]. While high pyrolysis temperatures result in low O-bearing functional groups [Zhang et al. 2005], which in turn decreases affinity for polar organic compounds, low pyrolysis temperature constitutes more O-bearing functional groups wherein increasing affinity for polar organic compounds [Jones et al. 2011].

The presence of pollutants in soil and their intake by organisms living and growing in the soil have been seen to be reduced by biochar [Zheng et al. 2010]. As opposed to wood-derived biochar produced at a low pyrolysis temperature, that produced at a significantly high pyrolysis temperature had more high surface area and microporosity, which contributed to the reduction in the availability of pesticides in soil. Even though the pollutant removal efficiency was greater with biochar-unamended soil, the fixation capacity of these pesticides was still higher with biochar-amended soil and hence was less toxic for plants as the uptake was lower [Zhang et al. 2013; Vithanage et al.

2016]. However, the depletion of pesticides did increase with the subsequent increase in biochar in the amended soil. Also, it has been noted that the leaching effect of simazine and atrazine on groundwater was significantly depleted by green-waste-derived biochar [Chen et al. 2011; Lima et al. 2010]. It has also been observed that cationic dyes, which are deficient in electrons, can be electrostatically adsorbed onto anionic biochar, which is electron rich [Komkiene et al. 2016]. It has also been observed that polar antibiotic sulfamethazine is adsorbed better at a higher pH as there is a hydrogen bond formed between it and -COOH and -OH groups of woody biochar, whereas at a lower or neutral pH, π–π electron donor–acceptor interactions and cation exchange are the dominant interactions between biochar and SMZ [Nelissen et al. 2014; Hartley et al. 2009]. Therefore, it shows us just how dependent the adsorption is on pH levels.

5.5.2 REMOVAL OF INORGANIC POLLUTANTS

Inorganic compounds are highly toxic and therefore act as a huge liability to public health conditions and the environment [Cao et al. 2009]. The most toxic and carcinogenic of these are heavy metals [Khan et al. 2013]; most of the inorganic compounds present in industrial waste and the like are varied from them [Choppala et al. 2012; Bolan et al. 2013]. The biochar produced at a low pyrolysis temperature usually has a high content of carbon, microporosity, and numerous functional groups. But the most valued attribute for the removal of heavy metals remains the ion exchange mechanism [Mandal et al. 2017a].

The metal reduction capacity to a stable form is highly dependent on the physiochemical attributes of biochar [Shang et al. 2013]. The interaction therefore of heavy metals is very dependent on either the high or the low end of pyrolysis temperature, the biomass it has been derived from, the pH and rate of application. There have been several research studies regarding the efficiency of extraction of heavy metals in an aqueous system by biochar produced from sewage sludge, agricultural wastes, and animal manure [Shang et al. 2013; Xu et al. 2014a; Jo et al. 2010]. Cu^{+2}, among the heavy metals, had a significant attraction toward functional groups like -COOH and -OH on the surface of woody biochar and was heavily pH dependent on the biochar. The lower the pH was, more the mechanism of adsorption is inclined toward cationic exchange, whereas higher the pH, the more it is inclined toward functional group interaction [Yang et al. 2011; Smith et al. 2016]. pH was involved in many attributes of biochar like C and O concentrations, polarity index, etc., which was significant for high adsorption [Lehmann et al. 2015, 2011].

On the other hand, agricultural-waste- and animal-manure-derived biochar was used in the removal of highly toxic Hg^{+2} from surface and groundwater. Animal manure was seen to be very effective for the removal of Hg^{+2} due to high sulfur by precipitation at a significant pyrolysis temperature. The weightage of biochar added was subsequently responsible for the extraction of heavy metals due to higher pH and higher surface capacity [Cao et al. 2011; S. Kirdponpattara et al. 2013]. The feedstock type was very important for the extraction of heavy metals as hardwood-, green-waste-, sewage-sludge-, coconut-shell-, etc. derived biochar was very efficient for the extraction of heavy metals [Lal et al. 2007; Lehmann 2015, 2011].

Woody biochar was the most valued for the depletion of Cu, Cd^{+2}, and As^{+3}. In regard to Cd^{+2} and Zn^{+2}, there was a sign of reduced mobility in the soil when there was an increase in pH due to the dosage increase [Lehmann et al. 2011]. It has been noted that the physiochemical properties of woody biochar are also very different at different temperature levels. Lead produced from gasoline combustion has been observed to be highly toxic, but it can be removed with the help of alkaline biochar from dairy manure and sewage sludge [K.T. Klasson et al. 2011; Keiluweit et al. 2010]. The removal process of Pb^{+2} has been due to the increase in pH of soil by biochar, which precipitates it in the form of hydroxides, oxides, etc. The removal of Ni^{+2}, Co^{+2}, Pb^{+2}, Cu^{+2}, and Cd^{+2} has been attributed to cottonseed hull and sewage sludge biochar, which has the properties of basic pH, etc., which does not allow the accumulation of pollutants in the plant through acidic

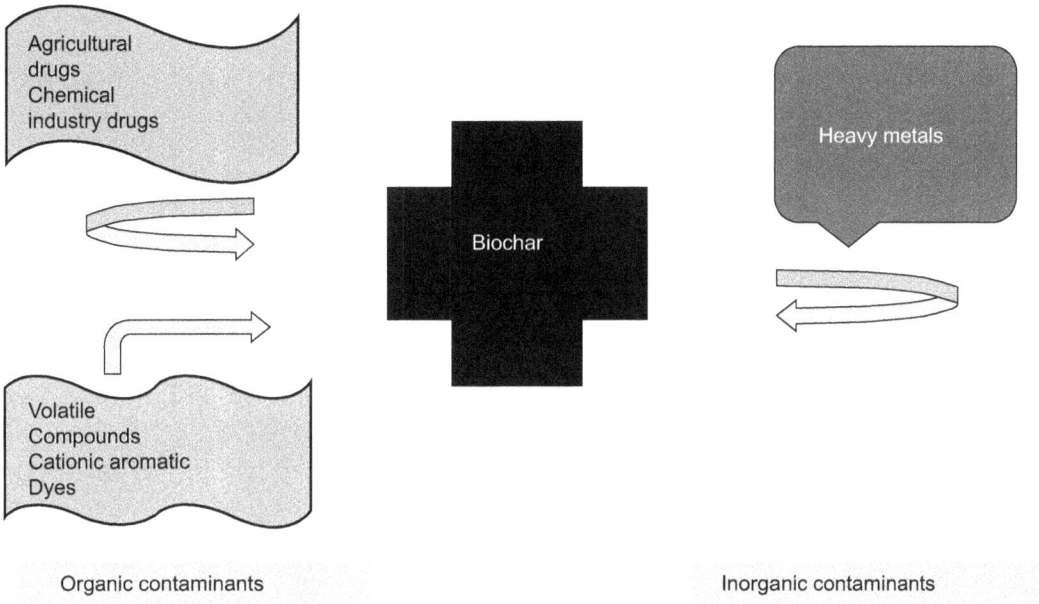

FIGURE 5.7 Adsorption of different pollutants such as heavy metals, dyes, pharmaceutical compounds by biochar.

soil [Ronsse et al. 2013; M. Brodin 2017]. The acidic as well as the basic functional groups present on biochar are responsible for their ability to remove mainly Cr^{+2} and Cr^{+3} [Uchimiya et al. 2013; Kong et al. 2011].

Biochar that has been modified prepyrolysis has shown to have a better removal capacity of Cr^{+6} over manure-derived biochar [Zhang et al. 2013]. It can therefore be summarized that the removal of inorganic pollutants is heavily dependent on the process, feedstock type, and many of the physiochemical attributes of biochar (Figure 5.7).

5.5.3 BIOCHAR USAGE FOR REMOVAL OF TOXIC GASES

It has been observed that the extraction of gases high in toxicity from gaseous conditions is highly efficient (90% for H_2S) in the case of biomass-derived biochar [Cao et al. 2011; Mondal et al. 2016]. The removal of H_2S has been highly efficient due to physiochemical attributes such as pH level, moisture content, surface area, and functional groups that are present on the surface of biochar [Gupta et al. 2022]. Biochar converted H_2S into $(K, Na)_2SO^{-4}$ through ionic surface interaction with -COOH and -OH functional groups in the presence of moisture and O_2, which could be available for plants as SO_4^{-2} [Mondal et al. 2016]. The toxic volatile organic compound trichloroethylene from industries and urban areas has a dire effect on groundwater nearby [Jung et al. 2015; Xu et al. 2011]. Soya-bean-straw- and peanut-shell-derived biochar pyrolyzed at a high temperature as opposed to a low temperature was seen to be effective for the removal of trichloroethylene. The better removal of trichloroethylene was attributed to its physicochemical properties such as hydrophobicity and surface area [R. French et al. 2010].

5.6 CLIMATE MITIGATION

The emission of greenhouse gases has increased by 35% in the past decades due to the release of these gases into the surrounding atmosphere through the process of burning of fossil energy, removal of forests, depletion of soil quality, which have, in turn, affected the climate drastically [Yu et al. 2009]. Hence the removal of CO_2 gas has become a key factor in the reduction of greenhouse gases for which biochar has shown efficiency due to its nature of sequestering carbon through its recalcitration property [Adeel et al. 2017]. The ability of biochar to sequester carbon in itself to a more stable form in the soil has proven to be of great importance for climate control and mitigation [Brebu et al. 2010]. If we were to convert the biomass that has absorbed CO_2 during photosynthesis into biochar, then it would be fixed into it and hence there would be a low emission of carbon into the environment [Li et al. 2014; Lehmann et al. 2011; Liu et al. 2015]. It has also been noted that biochar usage would remove N_2O emissions to a totality of 80%. The mechanisms proposed include N_2O entrapment in water-saturated soil pores and microbial activity associated with biological denitrification [Wang et al. 2013]. Due to nitrogen fixation in agriculture, there is a high increase of N_2O emissions by two thirds [Harter et al. 2016; Cayuela et al. 2014]. Hence with the usage of biochar in these conditions, the emissions of N_2O can be made unavailable to crops and therefore increase crop yield [Laird et al. 2011]. Therefore, not only can the use of biochar be very beneficial to the crop yield in many agricultural processes can, but it can also help in climate mitigation strategies, making the application of biochar multidiverse.

5.7 FUTURE PROSPECTS

Biochar has emerged as a new and highly efficient sustainable form of a renewable resource, with its applications ranging from pollutant control for better produce conditions to its valued properties in mitigating climate change conditions. It is a highly environment-friendly resource that can have huge contributions to a circular economy designed to save resources for the future. It therefore has not only soil and water enhancement capabilities but also a capacity to mitigate greenhouse gas emissions. Since the applications of biochar are very reliant on the methods and biomass type used for its production, to get the best results, it is therefore very important to study these specific conditions to arrange and characterize them so that the required end product is not only highly efficient and specific for the application required but also economically plausible for use on a larger scale

5.8 CONCLUSION

Biochar is therefore a viable resource for pollutant removal, climate mitigation, soil and water enhancement, and bioethanol production improvement. Its application is dependent on methods of production and their temperature levels, as well as on the type of biomass it is derived from. Its efficiency levels and economic value can also be manipulated by the arrangement of these conditions. With the help of its physiochemical properties like pH, microporosity, recalcitrance, etc., it can be converted into a very valued resource for the environment.

REFERENCES

Adeel, M., Song, X., Wang, Y., Francis, D., Yang, Y., 2017. Environmental impact of estrogens on human, animal and plant life: a critical review. *Environ. Int.* 99, 107–119.
Ahmad, M., Lee, S.S., Dou, X., Mohan, D., Sung, J.K., Yang, J.E., Ok, Y.S., 2012. Effects of pyrolysis temperature on soybean stover and peanut shell-derived biochar properties and TCE adsorption in water. *Bioresour. Technol.* 118, 536–544.
Ahmad, M., Rajapaksha, A.U., Lim, J.E., Zhang, M., Bolan, N., Mohan, D., Ok, Y.S., 2014. Biochar as a sorbent for contaminant management in soil and water: a review. *Chemosphere* 99, 19–33.
Ajoku, K.B., 2012. Modern use of solid biomass in Africa: prospect for utilization of agro-waste resources in Nigeria. In: Janssen R., Rutz D., editors. *Bioenergy for sustainable development in Africa*. Netherland: Springer; 2012, pp. 131–146. http://doi.org/10.1007/978-94-007-2181-4_12, ISBN 978-94-007-2180-7
Beesley, L., Jiménez, E.M., Eyles, J.L.G., 2010. Effects of biochar and green waste compost amendments on mobility, bioavailability and toxicity of inorganic and organic contaminants in a multi-element polluted soil. *Environ. Poll.* 158, 2282–2287.

Bolan, N.S., Choppala, G., Kunhikrishnan, A., Park, J.H., Naidu, R., 2013. Microbial transformation of trace elements in soils in relation to bioavailability and remediation. *Rev. Environ. Contam. Toxicol.* 225, 1–56.

Brebu, M., Vasile, C., 2010. Thermal degradation of lignin—a review. *Cellul. Chem. Technol.* 44 (9), 353.

Bridgewater, AV., 2003. Renewable fuels and chemicals by thermal processing of biomass. *Chem. Eng. J.* 91, 87–102. http://doi.org/10.1016/S1385-8947(02)00142-0

Brodin, M., Vallejos, M., Opedal, M.T., Area, M.C., Chinga-Carrasco, G., 2017. Lignocellulosics as sustainable resources for production of bioplastics – A review. *J. Clean. Prod.* 162, 646.

Campbell, I., 1983. *Biomass catalysts and liquid fuels.* Lancaster: Technomic Publishing Co. Inc.

Cao, X., Ma, L., Gao, B., Harris, W., 2009. Dairy-manure derived biochar effectively sorbs lead and atrazine. *Environ. Sci. Technol.* 43, 3285–3291.

Cao, X., Ma, L., Liang, Y., Gao, B., Harris, W., 2011. Simultaneous immobilization of lead and atrazine in contaminated soils using dairy-manure biochar. *Environ. Sci. Technol.* 45, 4884–4889.

Cayuela, M.L., Van Zwieten, L., Singh, B.P., Jeffery, S., Roig, A., Sánchez-Monedero, M.A., 2014. Biochar's role in mitigating soil nitrous oxide emissions: a review and metaanalysis. *Agr. Ecosyst. Environ.* 191, 5–16.

Chen, B., Chen, Z., Lv, S., 2011. A novel magnetic biochar efficiently sorbs organic pollutants and phosphate. *Bioresour. Technol.* 102 (2), 716–723.

Choppala, G.K., Bolan, N.S., Mallavarapu, M., Chen, Z., Naidu, R., 2012. The influence of biochar and black carbon on reduction and bioavailability of chromate in soils. *J. Environ. Qual.* 41, 1–10.

Demirbas, A., 2005. Potential applications of renewable energy sources, biomass combustion problems in boiler power systems and combustion related environmental issues. *Prog. Energy Combust. Sci.* 31, 171–192. http://doi.org/10.1016/j.pecs.2005.02.002

Foong, S.Y., Liew, R.K., Yang, Y., Cheng, Y.W., Yek, P.N.Y., Wan Mahari, W.A., Lee, X.Y., Han, C.S., Vo, D.V.N., Le, Q.V., Aghbashlo, M., Tabatabaei, M., Sonne, C., Peng, W., Lam, S.S., 2020. Valorisation of biomass waste to engineered activated biochar by microwave pyrolysis: progress, challenges and future directions. *Chem. Eng. J.* 389, 124401.

French, R., Czernik, S., 2010. Catalytic pyrolysis of biomass for biofuel production. *Fuel Process. Technol.* 91, 25–32.

Fu, M.M., Mo, C.H., Li, H., Zhang, Y.N., Huang, W.X., Wong, M.H., 2019. Comparison of physico-chemical properties of biochars and hydrochars produced from food wastes. *J. Clean. Prod.* 236, 117637.

Guizani, C., Jeguirim, M., Valin, S., Peyrot, M., Salvador, S., 2019. The heat treatment severity index: a new metric correlated to the properties of biochars obtained from entrained flow pyrolysis of biomass. *Fuel.* 244, 61–68.

Gupta, R., Pandit, C., Pandit, S., et al., 2022. Potential and future prospects of biochar-based materials and their applications in removal of organic contaminants from industrial waste water. *J. Mater. Cycles Waste Manag.* 24, 852–876. https://doi.org/10.1007/s10163-022-01391-z

Harter, J., Guzman-Bustamante, I., Kuehfuss, S., Ruser, R., Well, R., Spott, O., Kappler, A., Behrens, S., 2016. Gas entrapment and microbial N2O reduction reduce N2O emissions from a biochar-amended sandy clay loam soil. *Sci. Rep.* 6.

Hartley, W., Dickinson, N.M., Riby, P., Lepp, N.W., 2009. Arsenic mobility in brownfield soils amended with green waste compost or biochar and planted with Miscanthus. *Environ. Poll.* 157, 2654–2662.

Hou, J., Zhang, S., Qiu, Z., Han, H., Zhang, Q., 2017. Layer-tunable phosphorene modulated by the cation insertion rate as a sodium-storage anode. *Bioresour. Technol.* 224, 465.

Jeguirim, M., Bikai, J., Elmay, Y., Limousy, L., Njeugna, E., et al., 2014. Thermal characterization and pyrolysis kinetics of tropical biomass feedstocks for energy recovery. *Energy Sustain. Dev.* 23, 188–193. http://doi.org/10.1016/j.esd.2014.09.009

Jia, W., Wang, B., Wang, C., Sun, H., 2017. Tourmaline and biochar for the remediation of acid soil polluted with heavy metals. *J. Environ. Eng.* 5, 2107–2114.

Jo, Y.J., Lee, J.Y., Yi, M.J., Kim, H.S., Lee, K.K., 2010. Soil contamination with TCE in an industrial complex: contamination levels and implication for groundwater contamination. *Geosci. J.* 14, 313–320.

Jones, D.L., Jones, G.E., Murphy, D.V., 2011. Biochar mediated alterations in herbicide breakdown and leaching in soil. *Soil Biol. Biochem.* 43, 804–813.

Jung, C., Oh, J., Yoon, Y., 2015. Removal of acetaminophen and naproxen by combined coagulation and adsorption using biochar: influence of combined sewer overflow components. *Environ. Sci. Poll. Res.* 22 (13), 10058–10069.

Keiluweit, M., Nico, P.S., Johnson, M.G., Kleber, M., 2010. Dynamic molecular structure of plant biomass-derived black carbon (biochar). *Environ. Sci. Technol.* 44, 1247–1253.

Khan, S., Chao, C., Waqas, M., Arp, H.P.H., Zhu, Y.G., 2013. Sewage sludge biochar influence upon rice (Oryza sativa L.) yield, metal bioaccumulation and greenhouse gas emissions from acidic paddy soil. *Environ. Sci. Technol.* 47, 8624–8632.

Kirdponpattara, S., Phisalaphong, M., 2013. Bacterial cellulose–alginate composite sponge as a yeast cell carrier for ethanol production. *Biochem. Eng. J.* 77, 103.

Klasson, K.T., Uchimiya, M., Lima, I., Boihem Jr., L., 2011. Influence of pyrolysis temperature on biochar property and function as a heavy metal sorbent in soil. *J. Agric. Food Chem.* 6, 3242.

Komkiene, J., Baltrenaite, E., 2016. Biochar as adsorbent for removal of heavy metal ions [cadmium(II), copper(II), lead(II), zinc(II)] from aqueous phase. *Int. J. Environ. Sci. Technol.* 13, 471.

Kong, H., He, J., Gao, Y., Wu, H., Zhu, X., 2011. Cosorption of phenanthrene and mercury (II) from aqueous solution by soybean stalk-based biochar. *J. Agric. Food Chem.* 59, 12116–12123.

Kookana, R.S., 2010. The role of biochar in modifying the environmental fate, bioavailability, and efficacy of pesticides in soils: a review. *Soil Res.* 48 (7), 627–637.

Kyriakou, M., Chatziiona, V.K., Costa, C.N., Kallis, M., Koutsokeras, L., Constantinides, G., Koutinas, M., 2019. Biowaste-based biochar: a new strategy for fermentative bioethanol overproduction via whole-cell immobilization. *Appl. Energy* 242, 480.

Laird, D.A., Rogovska, N.P., Garcia-Perez, M., Collins, H.P., Streubel, J.D., Smith, M., 2011. Pyrolysis and biochar—opportunities for distributed production and soil quality enhancement. In: Braun R., Karlen D., Johnson D., editors. *Sustainable alternative fuel feedstock opportunities, challenges and roadmaps for six US regions, Sustainable feedstocks for advanced biofuels.* Atlanta, GA: SWCS Publisher.

Lal, R., Follett, R.F., Stewart, B.A., Kimble, J.M., 2007. Soil carbon sequestration to mitigate climate change and advance food security. *Soil Sci.* 172 (12), 943–956.

Lee, X.J., Lee, L.Y., Gan, S., Thangalazhy-Gopakumar, S., Ng, H.K., 2017. Biochar potential evaluation of palm oil wastes through slow pyrolysis: thermochemical characterisation and pyrolytic kinetic studies. *Bioresour. Technol.* 236, 155–163.

Lee, Y.E., Jo, J.H., Kim, S.M., Yoo, Y.S., 2017. Recycling possibility of the salty food waste by pyrolysis and water scrubbing. *Energies.* 10, 210.

Lee, Y.E., Park, J., Ryu, C., Gang, K.S., Yang, W., Park, Y.K., Jung, J., Hyun, S., et al., 2013. Comparison of biochar properties from biomass residues produced by slow pyrolysis at 500C. *Bioresour. Technol.* 148, 196–201. http://doi.org/10.1016/j.biortech.2013.08.135

Lehmann, J., Joseph, S., 2015. *Biochar for environmental management: Science, technology and implementation.* United Kingdom: Routledge, Taylor and Francis Group.

Lehmann, J., Rillig, M.C., Thies, J., Masiello, C.A., Hockaday, W.C., Crowley, D., 2011. Biochar effects on soil biota: a review. *Soil Biol. Biochem.* 43, 1812–1836.

Li, Y., Shao, J., Wang, X., Deng, Y., Yang, H., Chen, H., 2014. Characterization of modified biochars derived from bamboo pyrolysis and their utilization for target component (furfural) adsorption. *Energy Fuel.* 28 (8), 5119–5127.

Liew, R.K., Nam, W.L., Chong, M.Y., Phang, X.Y., Su, M.H., Yek, P.N.Y., Ma, N.L., Cheng, C.K., Chong, C.T., Lam, S.S., 2018. Oil palm waste: an abundant and promising feedstock for microwave pyrolysis conversion into good quality biochar with potential multi-application. *Process Saf. Environ. Prot.* 115, 57–69.

Lima, I.M., Boateng, A.A., Klasson, K.T., 2010. Physicochemical and adsorptive properties of fast-pyrolysis bio-chars and their steam activated counterparts. *J. Chem. Technol. Biotechnol.* 85 (11), 1515–1521.

Limousy, L., Jeguirim, M., Labbe, S., Balay, F., Fossard, E., et al., 2015. Performance and emissions characteristics of compressed spent coffee ground/wood chip logs in a residential stove. *Energy Sustain. Dev.* 28, 52–59. http://doi.org/10.1136/10.1016/j.esd.2015.07.002

Liu, W.J., Jiang, H., Yu, H.Q., 2015. Development of biochar-based functional materials: toward a sustainable platform carbon material. *Chem. Rev.* 115 (22), 12251–12285.

Mandal, A., Singh, N., Purakayastha, T.J., 2017a. Characterization of pesticide sorption behaviour of slow pyrolysis biochars as low cost adsorbent for atrazine and imidacloprid removal. *Sci. Total Environ.* 577, 376–385.

Meschewski, E., Holm, N., Sharma, B.K., Spokas, K., Minalt, N., Kelly, J.J., 2019. Pyrolysis biochar has negligible effects on soil greenhouse gas production, microbial communities, plant germination and initial seedling growth. *Chemosphere* 228, 565–576.

Mohan, D., Pittman, C.U., Steele, P.H., 2006. Pyrolysis of wood/biomass for bio-oil: a critical review. *Energy Fuel.* 20, 848.

Mondal, S., Bobde, K., Aikat, K., Halder, G., 2016. Biosorptive uptake of ibuprofen by steam activated biochar derived from mung bean husk: equilibrium, kinetics, thermodynamics, modeling and eco-toxicological studies. *J. Environ. Manage.* 182, 581–594.

Mongkolkajit, J., Pullsirisombat, J., Limtong, S., Phisalaphong, M., 2011. Biotechnol. *Bioprocess Eng.* 16, 505.

Nan, H., Zhao, L., Yang, F., Liu, Y., Xiao, Z., Cao, X., Qiu, H., 2020. Different alkaline minerals interacted with biomass carbon during pyrolysis: which one improved biochar carbon sequestration? *J. Clean. Prod.* 255, 120162.

Nelissen, V., Ruysschaert, G., Müller-Stöver, D., Bodé, S., Cook, J., Ronsse, F., Shackley, S., Boeckx, P., Haugard-Nielsen, H., 2014. Short-term effect of feedstock and pyrolysis temperature on biochar characteristics, soil and crop response in temperate soils. *Agronomy* 4, 52–73.

Oldfield, T.L., Sikirica, N., Mondini, C., López, G., Kuikman, P.J., Holden, N.M., 2018. Biochar, compost and biocharcompost blend as options to recover nutrients and sequester carbon. *J. Environ. Manage.* 218, 465–476.

Pariyar, P., Kumari, K., Jain, M.K., Jadhao, P.S., 2020. Evaluation of change in biochar properties derived from different feedstock and pyrolysis temperature for environmental and agricultural application. *Sci. Total Environ.* 713, 136433.

Ravindranath, N.H., Hall, D.O., 1995. *Biomass, energy, and environment: A developing country perspective from India.* Oxford: Oxford University Press.

Ronsse, F., Hecke, S., Dickinson, D., Prins, W., 2013. Production and characterization of slow pyrolysis biochar: influence of feedstock type and pyrolysis conditions. *GCB. Bioenerg.* 104–115.

Shang, G., Shen, G., Liu, L., Chen, Q., Xu, Z., 2013. Kinetics and mechanisms of hydrogen sulfide adsorption by biochars. *Bioresour. Technol.* 133, 495–499.

Shen, X., Zeng, J., Zhang, D., Wang, F., Li, Y., Yi, W., 2020. Effect of pyrolysis temperature on characteristics, chemical speciation and environmental risk of Cr, Mn, Cu and Zn in biochars derived from pig manure. *Sci. Total Environ.* 704, 135283.

Smith, P., 2016. Soil carbon sequestration and biochar as negative emission technologies. *Global Change Biol.* 22 (3), 1315–1324.

Spokas, K.A., Koskinen, W.C., Baker, J.M., Reicosky, D.C., 2009. Impacts of woodchip biochar additions on greenhouse gas production and sorption/degradation of two herbicides in a Minnesota soil. *Chemosphere* 77, 574–581.

Tan, X.F., Liu, S.B., Liu, Y.G., Gu, Y.L., Zeng, G.M., Hu, X.J., Wang, X., Liu, S.H., Jiang, L.H., 2017. Biochar as potential sustainable precursors for activated carbon production: multiple applications in environmental protection and energy storage. *Bioresour. Technol.* 227, 359–372.

Tang, J., Zhang, L., Zhang, J., Ren, L., Zhou, Y., Zheng, Y., Luo, L., Yang, Y., Huang, H., Chen, A., 2020. Physicochemical features, metal availability and enzyme activity in heavy metal-polluted soil remediated by biochar and compost. *Sci. Total Environ.* 701, 134751.

Teixidó, M., Pignatello, J.J., Beltrán, J.L., Granados, M., Peccia, J., 2011. Speciation of the ionizable antibiotic sulfamethazine on black carbon (biochar). *Environ. Sci. Technol.* 45, 10020–10027.

Tripathi, M., Sahu, J.N., Ganesan, P., Jewaratnam, J., 2016. Thermophysical characterisation of oil palm shell (OPS) char synthesised by the microwave pyrolysis of OPS. *Appl. Therm. Eng.* 105, 605–612.

Uchimiya, M., Ohno, T., He, Z., 2013. Pyrolysis temperature-dependent release of dissolved organic carbon from plant, manure, and biorefinery wastes. *J. Anal. Appl. Pyrol.* 104, 84–94.

Van de Velden, M., Baeyens, J., Brems, A., Janssens, B., Dewil, R., 2010. Fundamentals, kinetics and endothermicity of the biomass pyrolysis reaction. *Renew. Energ.* 35, 232–242.

Vithanage, M., 2016. Mechanistic modeling of glyphosate interaction with rice husk derived engineered biochar. *Micropor. Mesopor. Mat.* 225, 280–288.

Wang, J., Wang, S., 2019. Preparation, modification and environmental application of biocar: a review. *J. Clean. Prod.* 227, 1002–1022.

Wang, Z., Zheng, H., Luo, Y., Deng, X., Herbert, S., Xing, B., 2013. Characterization and influence of biochar on nitrous oxide emission from agricultural soil. *Environ. Poll.* 174, 289–296.

Xie, T., Reddy, K.R., Wang, C., Yargicoglu, E., Spokas, K., 2015. Characteristics and applications of biochar for environmental remediation: a review. *Crit. Rev. Environ. Sci. Technol.* 45 (9), 939–969.

Xu, G., Sun, J., Shao, H., Chang, S.X., 2014a. Biochar had effects on phosphorus sorption and desorption in three soils with differing acidity. *Ecol. Eng.* 62, 54–60.

Xu, R.K., Xiao, S.C., Yuan, J.H., Zhao, A.Z., 2011. Adsorption of methyl violet from aqueous solutions by the biochar derived from crop residues. *Bioresour. Technol.* 102, 10293–10298.

Xu, X., Cao, X., Zhao, L., 2013a. Comparison of rice husk-and dairy manure-derived biochars for simultaneously removing heavy metals from aqueous solutions: role of mineral components in biochars. *Chemosphere* 92 (8), 955–961.

Xue, S., Zhang, X., Ngo, H.H., Guo, W., Wen, H., Li, C., Zhang, Y., Ma, C., 2019. Food waste based biochars for ammonia nitrogen removal from aqueous solutions. *Bioresour. Technol.* 292, 121927.

Yang, J.E., Skogley, E.O., Ok, Y.S., 2011. Carbonaceous resin capsule for vapor-phase monitoring of volatile monoaromatic hydrocarbons in soil. *Soil Sediment Contam. An Int. J.* 20, 205–220.

Yavari, S., Malakahmad, A., Sapari, N.B., Yavari, S., 2017. Synthesis optimisation of oil palm empty fruit bunch and rice husk biochars for removal of imazapic and imazapyr herbicides. *J. Environ. Manage.* 193, 201–210.

Yu, X.Y., Ying, G.G., Kookana, R.S., 2009. Reduced plant uptake of pesticides with biochar additions to soil. *Chemosphere* 76 (5), 665–671.

Zhang, K., Cheng, X., Dang, H., Ye, C., Zhang, Y., Zhang, Q., 2013. Linking litter production, quality and decomposition to vegetation succession following agricultural abandonment. *Soil Biol. Biochem.* 57, 803–813.

Zhang, P., Sheng, G., Feng, Y., Miller, D.M., 2005. Role of wheat-residue-derived char in the biodegradation of benzonitrile in soil: nutritional stimulation versus adsorptive inhibition. *Environ. Sci. Technol.* 39 (14), 5442–5448.

Zhang, Q., Huang, H., Han, H., Qiu, Z., Achal, V., 2017. Lithium-sulfur batteries: review on high-loading and high-energy lithium–sulfur batteries. *Energy* 135, 32.

Zhao, L., Cao, X., Mašek, O., Zimmerman, A., 2013. Heterogeneity of biochar properties as a function of feedstock sources and production temperatures. *J. Hazard. Mater.* 256–257, 1–9.

Zhao, L., Cao, X., Zheng, W., Kan, Y., 2014. Phosphorus-assisted biomass thermal conversion: reducing carbon loss and improving biochar stability. *PLoS One* 9, 115373.

Zheng, W., Guo, M.X., Chow, T., Bennett, D.N., Rajagopalan, N., 2010. Sorption properties of greenwaste biochar for two triazine pesticides. *J. Hazard. Mater.* 181, 121–126. https://doi.org/10.1016/j.jhazmat.2010.04.103

Zhu, X., Li, Y., Wang, X., 2019. Machine learning prediction of biochar yield and carbon contents in biochar based on biomass characteristics and pyrolysis conditions. *Bioresour. Technol.* 288, 121527.

6 Modified Biochar for Wastewater Treatment and Reuse

Bikram Mishra and Neelancherry Remya

CONTENTS

6.1 INTRODUCTION

Urbanization, industrialization, and population rise have led to an increase in water demand. As a result, the amount of wastewater generated has also increased significantly. To ensure the safety of the environment, the wastewater generated must be treated before disposal. In the past, various physical, chemical, and biological methods have been employed for the treatment of wastewater. In recent years, research on the use of biochar for wastewater treatment has increased. Biochar is a carbonaceous material obtained from the thermochemical conversion of biomass feedstock in the presence of limited or no oxygen (Dai et al. 2019). Biomass feedstock generally used as an organic waste material, which includes forest and crop residues, algae, wood chips, sewage sludge, and organic municipal waste (Chao et al. 2018; Huang et al. 2017; Vause, Heaney, and Lin 2018). The biomass majorly contains lignin, cellulose, and hemicelluloses, which can add up to more than 65% carbon on a dry weight basis. Methods employed for thermochemical conversion include pyrolysis, torrefaction, gasification, and hydrothermal carbonization. The physical and chemical properties of biochar are affected by factors such as feedstock type, thermochemical conversion method, temperature, and duration. Pretreatment of feedstock and post-treatment of biochar also affect biochar properties. Biochar is an effective, low-cost, environment-friendly adsorbent, which has a relatively large surface area and surface functional groups (Cha et al. 2016). Biochar can be used as an adsorbent for the removal of various pollutants like metals, nutrients, and emerging contaminants.

Although raw biochars are good adsorbents, modifications to raw biochar are made to improve the pore structure, surface area, and functionality. The functional groups, such as hydroxyl, carbonyl, ether, amide, amine, alkyl, alkyne, alkene, and carboxyl, on the biochar surface are responsible for the adsorption of pollutants on biochar. Improvements in those functional groups can be

DOI: 10.1201/9781003203438-6

done via steam activation, chemical modification, impregnation, and heat treatment (Shen, Li, and Liu 2010). Modified biochars can be used as alternative sorbents compared to activated carbon, graphene, and carbon nanotubes for pollutant removal from wastewater. In this chapter, we discuss the biochar modification methods, properties of modified biochar, use of modified biochar for wastewater treatment, and reusability of the modified biochar.

6.2   BIOCHAR PRODUCTION

Biomass feedstock is converted to biochar by the process of carbonization. The carbonization methods generally used are pyrolysis, microwave-assisted pyrolysis, hydrothermal carbonization, and gasification. Out of all these processes, pyrolysis is the most commonly adopted process used for biochar production. Table 6.1 summarizes and compares the biochar production methods.

Pyrolysis is a thermochemical process for the conversion of biomass into biochar in limited or no oxygen conditions (Cha et al. 2016). Operating temperature, heating rate, and residence time of the pyrolysis process are the factors that can affect the physicochemical properties and composition of biochar. Prolonged residence time provides more complete feedstock decomposition with a decrease in biochar yield (Mohamed et al. 2016). The heating rate influences biochar properties and determines pyrolysis speed (Cho et al. 2017). With the increase in pyrolysis temperature, biochar yield decreases, whereas ash and carbon content increases.

Microwave-assisted pyrolysis (MAP) is one type of pyrolysis technique that uses microwave irradiation for heating the feedstock. MAP offers a shorter processing time, lower energy requirement, more effective heat transfer, and better selective heating as compared to conventional techniques (Zhang et al. 2017; Durán-Jiménez et al. 2018). In the MAP technique, microwave power and irradiation time are important factors that affect biochar properties (Durán-Jiménez et al. 2018; Nhuchhen et al. 2018). Biochar yield decreases with an increase in microwave power, which can be attributed to high heating rates at high microwave power levels (Nhuchhen et al. 2018).

Hydrothermal carbonization (HTC) is the conversion of wet feedstock into biochar without predrying. Biochar produced by HTC is usually called hydrochar. The wet biomass is heated in the temperature range of 120–260°C in a confined system under pressure (2–10 MPa) for 5–240 min (Kambo and Dutta 2015; Fang et al. 2018). The reaction temperature is the key parameter. With an increase in temperature, abundant acidic functional groups are formed on the hydrochar surface, which can benefit in pollutant adsorption capabilities of hydrochar (Saha, Saba, and Reza 2019). The porosity of hydrochar increases with an increase in residence time and temperature.

TABLE 6.1
Common Biochar Production Methods (Xiang et al. 2020)

Carbonization methods	Temperature/ power range	Residence time	Key parameters	Advantages
Pyrolysis	300–850°C	1–3 h	Temperature, heating rate, residence time	Robust, simple, cost-effective, small scale
MAP	400–500 W	1–10 min	Microwave power, irradiation time	Fast, selective, volumetric, and efficient heating
HTC	120–260°C	1–16 h	Temperature, residence time, pressure, water-to-biomass ratio	Suitable for biomass feedstock with high moisture content
Gasification	>700°C	10–20 s	Temperature, particle size, residence time, pressure, gasification-agent-to-biomass ratio	Biochar contains a high level of alkali salts (Ca, K, Mg, etc.), but biochar yield is less as the targeted product is syngas.

Gasification is the process of converting biomass to gas fuel using gasification agents (air, oxygen, steam, etc.). Gasification requires a higher temperature (greater than 700°C) than pyrolysis. Solid, liquid, and gaseous products are formed during the gasification process. However, biochar yield via the gasification process is lower than that via the pyrolysis process because gaseous products are the targeted products in this process. Biochar produced by gasification usually contains high levels of alkali salts and alkaline earth minerals (Kambo and Dutta 2015), which can precipitate many heavy metal pollutants.

6.3   BIOCHAR MODIFICATIONS

To improve the properties of the biochar for pollutant removal, many methods are adopted to modify the biochar. Modifications methods can be classified as physical modification and chemical modification. Physical modification mainly includes heat treatment, steam, and gas purging. Chemical modifications mainly include acid modification, alkaline modification, oxidizing agent modification, metal salts or metal oxide modification, and carbonaceous material modification.

6.3.1   ACID MODIFICATION

Acid modification is done to introduce the acid functional groups on the surface of biochar and to remove impurities such as metals. Generally, acids such as nitric acid, phosphoric acid, sulfuric acid, oxalic acid, and citric acid are used in the process (Rajapaksha et al. 2016). It is typically done by soaking or suspending the biochar in the acid solution in the ratio of up to 1:10 (biochar/acid) (Zhang et al. 2015; Jin et al. 2014; Zhou et al. 2013). The soaking and stirring time may last hours or days, followed by washing, drying, and pyrolysis to obtain the desired modified biochar. The acid modification changes the surface area of biochar, and the effect on the surface area varies with the types and concentrations of acids, feedstock, and preparation conditions. To investigate the effect of the concentration and types of acids on the surface area of biochar, other conditions should stay the same. The nitric acid (2 M) treatment of biochar introduced carboxylic, lactonic, phenolic, and carbonyl groups on the surface of bamboo-derived biochar (Li et al. 2014). The surface area of reed-derived biochar increased from 58.75 to 88.35 m^2/g after hydrochloric acid (1 M) treatment (Peng, Lang, and Wang 2016). However, the surface area of rice-straw-derived biochar decreased from 71.35 to 56.9 m^2/g after 2% treatment of sulfuric acid (Yakout, Daifullah, and El-Reefy 2018). Weak acids such as oxalic acid and citric acid can also introduce the carboxyl group on the biochar surface through esterification. The modification using oxalic acid and citric acid slightly decreased the surface area of biochar from 1.57 to 0.69 and 1.21 m^2/g, respectively (Sun et al. 2015).

6.3.2   ALKALINE MODIFICATION

Alkaline modification increases the surface area and the oxygen-containing functional groups. The alkaline agents generally used are potassium hydroxide and sodium hydroxide. Alkaline modification is conducted by soaking or suspending biochar in different basic concentrations from room temperature to 100°C. Soaking and stirring time may vary last hours or days (mostly 6–24 h) depending on the feedstock. Potassium hydroxide modification increased the surface area (from 14.4 to 49.1 m^2/g) and the oxygen-containing functional groups of municipal solid-waste-derived-biochar that was prepared by pyrolysis, resulting in the enhanced removal of As(V) (Jin et al. 2014). Sodium hydroxide modification significantly increased the surface area of coconut-derived biochar up to 2885 m^2/g, which was higher than that of potassium hydroxide modification of 1940 m^2/g (Tan, Ahmad, and Hameed 2008; Cazetta et al. 2011). In addition to the different types of bases, the ratio between base and biochar can significantly affect the properties of biochar.

6.3.3 Metal Salts or Metal Oxide Modification

Metal salts or metal oxides modification can be done in two ways: (1) metal salts or metal oxides are mixed with feedstock first, and then pyrolysis is done to synthesize biochar; (2) pyrolysis of feedstock is done first to prepare the biochar, then the biochar is soaked with metal salts or metal oxides under certain conditions. Both ways are widely used for modifying biochar. The common metals include iron, magnesium, aluminum, titanium, and manganese.

The main reasons leading to the use of metal salts or metal oxides to modify biochar are:

1. To enhance the adsorption of pollutants: For instance, biochar has a very low adsorption capacity for anionic dyes due to the negative charge on the surface of biochar. Metal modification can change the surface properties of biochar, further improving the adsorption capacity to anionic dyes.
2. *To remove biochar*: The small size of biochar makes its removal from the water very difficult while applying biochar for the removal of pollutants in water and wastewater. Iron salts or iron metal oxides modification can increase the magnetic characteristic of biochar, which can contribute to the removal of biochar.
3. *To enhance the catalytic characteristic of biochar*: For instance, to enhance the catalytic characteristic of biochar during the persulfate activation process, metal salts or metal oxides can be employed to synthesize the metal-biochar composite for removal of emerging contaminants (Wang and Wang 2018).

Liu et al. (2015) modified corn straw biochar using iron and found that the surface of the modified biochar was covered by small granules of Fe_3O_4, which improved the removal efficiency of phosphorus from aqueous solutions. The modification of rice-husk-derived biochar by Fe(III) enhanced the removal of As(III) and As(V) (Samsuri, Sadegh-Zadeh, and Seh-Bardan 2013). Similarly, the modification of swine-manure-derived biochar by MnO_2 could enhance the removal of Pb(II) and Cd(II) (Liang et al. 2017). To enhance the catalytic characteristic of biochar, nanoscale zero-valent iron was used to modify the rice-hull-derived biochar (Yan et al. 2015).

6.3.4 Carbonaceous Material Modification

Biochar can also be modified using carbonaceous materials like carbon nanotubes or graphene to enhance the properties of pristine biochar. Carbon nanotubes and graphene can adsorb heavy metals and organic pollutants. Generally, carbon nanotubes and graphene are expensive, but they can be used with biochar to produce nanocomposite, which decreases the production cost. This modification leads to an increase in the surface area of biochar.

Carbon nanotube-biochar nanocomposite was synthesized by Inyang et al. (Inyang et al. 2014) for the removal of dyes. The composite was prepared using a 1% carbon nanotube, which had a higher surface area of 390 m^2/g as compared to 9 m^2/g of unmodified biochar. Similarly, Du et al. (Du et al. 2020) prepared graphene-biochar nanocomposite for the removal of crystal violet dye. The surface area of the nanocomposite was 145.96 m^2/g compared to 78 m^2/g of unmodified biochar.

6.3.5 Steam and Gas Purging

In steam modification, the pyrolysis of feedstock is done first, and then gasification of biochar using steam is carried out. During the process, the oxygen from water molecules transfers to the free active sites of the carbon surface, and the hydrogen gas produced due to the loss of oxygen in water molecules reacts with carbon on the biochar surface to form surface hydrogen complexes. Steam modification facilitates the devolatilization and formation of crystalline carbon in the biochar. It also removes the trapped products of incomplete combustion produced during the first pyrolysis process. Therefore, steam modification increases the surface area and improves the structure of biochar. Rajapaksha et al. (Rajapaksha et al. 2014) modified tea-waste-derived biochar using steam for the removal of antibiotics in water. The surface area of modified biochar (576.09 m^2/g) significantly increased in comparison with unmodified biochar (342.22 m^2/g).

Similar to steam modification, the gas-purging modification also involves two processes, first is the pyrolysis of feedstock. The second is to modify the biochar by purging carbon dioxide or ammonia gas. Gas purging modification can increase the surface area of biochar and improve the structure of biochar. Carbon dioxide modification can promote the formation of pores and improve the structure of micropores (Xiong et al. 2013). The modification using ammonia gas can introduce the nitrogen-containing groups on the biochar (Xiong et al. 2013).

6.4 APPLICATION OF MODIFIED BIOCHAR FOR WASTEWATER TREATMENT

Biochar is an effective adsorbent for the removal of various pollutants due to its properties like large surface area and abundant surface functional groups. Modifications to biochar are done to improve the properties to target specific pollutants. Adsorption of pollutants on biochar in water depends on the properties of the pollutant and the type of biochar. Wastewater has been an environmental issue, which is a by-product of domestic, industrial, commercial, and agricultural activities. Biochar has great potential to be used for wastewater treatment. Table 6.2 summarizes the use of modified biochar for the removal of different pollutants.

Industrial wastewater comes from various sources like the chemical industry, tanneries, mining, smelting, etc. Mainly heavy metals, dyes, and organic pollutants are present in industrial wastewater. Biochar has been applied in industrial wastewater treatment. Zero-valent magnetic biochar derived from peanut shells (Liu et al. 2019) is a promising material for the removal of Cr(VI) from wastewater. Biochar produced at higher temperature (800°C) shows higher Cr(VI) adsorption due to large surface area, total pore volume, and more reductive zero-valent iron. Cr(VI) removal was due to electrostatic attraction and ion exchange. The removal of Cr(VI) was also pH dependent. Similarly, biochar derived from peanut shells and impregnated with hydrated manganese oxide was used for the removal of Pb^{2+} (Wan et al. 2018). The embedded hydrated manganese oxide nanoparticles provide the preferable capture of target cations through specific inner-sphere complexation, while the nondiffusive negatively charged oxygen-containing groups bound to BC facilitate the pre-enrichment and permeation of Pb^{2+} cations into the pore channels. Pb^{2+} removal was effective in a wide pH range (3–7).

Zhang et al. (2020) produced a green biochar/iron oxide composite for the removal of methylene blue dye from wastewater. The modified biochar adsorption capacity for methylene blue was considerably enhanced (862 mg/g) compared to unmodified biochar, and it exhibited good performance for a wide range of pH values (2.05–9.2). Enhanced methylene blue adsorption in the biochar/iron oxide composite is attributed to increased electrostatic attraction. Similarly, Du et al. (Du et al. 2020) developed novel Fe_3O_4-graphene biochar from rice straw for crystal violet adsorption. The Fe_3O_4 nanoparticles were impregnated on graphene-modified biochar, which leads to an increase in surface area, pore volume, thermal stability, and crystal violet adsorption, but also improved the magnetic properties. The removal of crystal violet was due to electrostatic attraction and $\pi-\pi$ interaction. The maximum adsorption capacity of crystal violet was 436.68 mg/g at pH of 6 and temperature of 40°C.

Biochar can be used in municipal wastewater treatment either directly or in combination with biofilter and other technologies, which result in nutrient recovery (nitrogen and phosphorous). Xue et al. (Xue et al. 2016) synthesized Mg-Fe-layered double hydroxide biochar composite by a liquid-phase deposition method using wheat straw biochar. The biochar composite was used for nitrate removal from aqueous solutions. The maximum adsorption capacity for nitrate was 24.8 mg/g. Surface adsorption and interlayer anion exchange were the main mechanisms of nitrate removal. The biochar composite can be effectively used for the treatment of nitrate-contaminated water as

TABLE 6.2
Use of Modified Biochar for Removal of Different Pollutants

Biochar feedstock	Treatment/ modification	Pyrolysis temperature (°C)	Biochar dose (g/L)	Pollutant	Initial concentration (mg/L)	Adsorption capacity (mg/g)	Removal mechanism	Ref.
Peanut shell	Magnetized	800	1	Cr(VI)	300	223.21	Electrostatic attraction and ion exchange	Liu et al. (2019)
Peanut shell	Hydrated manganese oxide impregnated	400	0.2	Pb^{2+}	20	36	Outer sphere complexation and inner-sphere complexation	Wan et al. (2018)
Banana peel	Iron oxide modified	600	0.5	Methylene blue	500	862	Electrostatic attraction	Zhang et al. (2020)
Rice straw	Graphene-Fe_3O_4 modified	600	0.3	Crystal violet	600	436.68	Electrostatic attraction and π–π interaction	Du et al. (2020)
Wheat straw	Mg-Fe double-layered hydroxides	600	2	Nitrate	45	24.8	Surface adsorption and interlayer anion exchange	Xue et al. (2016)
Hickory wood chips	Aluminum salt modified	600	2.5	Phosphorous	6.4	8.346	Electrostatic attraction	Zheng et al. (2019)
Paper mill sludge	Zero-valent iron modified	700	0.2	Pentachlorophenol	10	–	Adsorption and reductive dechlorination	Devi and Saroha (2014)
Bur cucumber plants	Steam activation	700	1	Sulfamethazine	2.5–50	37.7	Electrostatic attraction and π–π interaction	Rajapaksha et al. (2015)
Peanut shell	Magnetized	800	2	Tricholoroethylene	9.2	4.6	Hydrophobic partitioning, pore filling, and reductive degradation	Liu et al. (2019)

well as a soil amendment for the reduction of soil nitrate leaching. Engineered biochar loaded with aluminum oxyhydroxide was developed by Zheng et al. (Zheng et al. 2019) for reclaiming phosphorous from secondary treated municipal wastewater. The electrostatic attraction was the main mechanism of phosphorous removal wastewater. The phosphorus-loaded biochar can be used as a slow-release fertilizer for crop production.

Emerging contaminants are now a major environmental issue. Emerging contaminants enter into the aquatic environment from different sources like industrial wastewater, domestic wastewater, runoff from agricultural land, etc. Pharmaceutical compounds, pesticides, cleaning solvents, and personal care products, etc. come under the category of emerging contaminants. Zero-valent iron magnetic biochar composite was synthesized from paper mill sludge biochar and used for the removal of pentachlorophenol from real and synthetic effluent (Devi and Saroha 2014). The removal of pentachlorophenol from solution occurs due to the adsorption of pentachlorophenol on the biochar composite and the subsequent reductive dechlorination of the adsorbed pentachlorophenol by forming chloride ions. Leaching tests confirmed the stability of zero-valent iron on the biochar matrix. Steam-activated biochar derived from bur cucumber plants can effectively remove sulfamethazine (Rajapaksha et al. 2015). Steam-activated biochar produced at 700°C showed the highest sorption capacity (37.7 mg/g) at pH 3, with a 55% increase in sorption capacity compared to nonactivated biochar produced at the same temperature. The main removal mechanism of sulfamethazine was an electrostatic attraction and π-π interaction. Magnetic zero-valent iron biochar synthesized using peanut shells can effectively remove trichloroethylene from aqueous solutions (Liu et al. 2019). Modified biochar showed a maximum adsorption capacity of 4.6 mg/g for trichloroethylene. The removal of trichloroethylene was due to pore filling, hydrophobic partitioning, and reductive degradation.

Modified biochar show good performance in the removal of targeted pollutants. However, in the practical situation, many pollutants coexist in the water and wastewater. Competitive adsorption could occur resulting in inconsistent results with the results of the laboratory. In addition, actual flow conditions could also affect the adsorption of pollutants by biochar. Therefore, further studies should simulate the actual situation in the laboratory and investigate the performance of biochar in the removal of pollutants.

6.5 POST-TREATMENT REUSABILITY/DISPOSAL OF BIOCHAR

Modified biochars have better adsorption capacity for targeted pollutants than unmodified biochar. The adsorption capacity of any adsorbent is finite; thus an extended contact with sorbate molecules leads to the creation of thermodynamic equilibrium between sorbents and sorbates. Hence it is necessary to regenerate the adsorbent for reuse or simply disposed of. The desorption method to be used should be economical and effective so that it can reduce the cost of regeneration and reusability of biochar adsorbents, as adsorption–desorption can be repeated a few times (Wang et al. 2015). Biochar saturated with metal contaminants can be regenerated using $NaNO_3$, HNO_3, or KNO_3 solutions at different concentrations, as these desorbents provide a sufficient amount of cations that can replace adsorbed metal ions. Acid solutions are particularly effective in desorbing metal ions due to low pH (Sounthararajah et al. 2015). Other techniques such as heat treatment, reduction of partial pressure, purging with inert fluid or gas, and changing of the chemical state such as pH can be employed for the regeneration of spent biochar. HCl, NaOH, acetic acid, EDTA, and NaCl solutions are the most widely used desorbents for modified biochar. Metal ion desorption is favored by using acids as desorbent, while desorption of organic compounds using NaOH as a desorbent was favored. Ethanol was used as a desorbent to regenerate the iron oxide/biochar composite for the desorption of methylene blue (Zhang et al. 2020). The adsorption capacity of the biochar composite slightly decreased after five adsorption–desorption cycles, but the performance remained the same throughout. The biochar composite is stable and has reuse potential. The exhausted hydrated manganese oxide biochar composite could be efficiently regenerated using 0.2 M HCl and 4 wt%

$CaCl_2$ binary solutions in situ at 298 K (Wan et al. 2018). The preloaded Pb(II) and Cd(II) were almost completely desorbed. To investigate the reusability of hydrated manganese oxide biochar composite, batch adsorption–desorption cycles were performed. The results suggest that hydrated manganese oxide biochar composite can be repeatedly used to remove Pb(II) and Cd(II) for at least 5 runs without any observable sorption capacity loss. However, Zheng et al. (Zheng et al. 2019) in their study did not regenerate the engineered biochar to reuse; instead, they proposed to use the exhausted phosphorous-laden biochar as slow-releasing fertilizer. The experimental results showed phosphorous-laden biochar increased the seed germination rate from 30% in untreated sand to 46.7 % in the sand with phosphorous-laden biochar. So modified biochar can either be regenerated for reuse as an adsorbent or used as slow-releasing fertilizer for agricultural production (in the case of nutrient recovery using biochar).

6.6 CONCLUSION

Biochar is an efficient and low-cost adsorbent, which can be produced from a variety of biomass materials by different thermochemical conversion processes like pyrolysis, hydrothermal carbonization, gasification, etc. Biochar properties are related to the type of feedstock and conditions of the thermal conversion process. Modifications of biochar are done to increase surface area, improve pore structure, enhance reaction activity, or introduce functional groups on the biochar surface. Steam activation, acid modification, alkali modification, metal salt or metal oxide impregnation, and carbonaceous material impregnation are some common modification methods. Modified biochar can be used for wastewater treatment and helps in targeting particular contaminants like heavy metals, nutrient recovery, and emerging contaminants. Adsorption of contaminants onto the surface of modified biochar is mainly due to electrostatic attraction, ion exchange, π–π interaction, surface complexation, etc. The spent biochar can be regenerated by using different techniques like changing pH, partial pressure, heat treatment, or using chemicals (acids, alkali). The spent biochar can also be used as slow-releasing fertilizer for crop production if nutrient recovery using biochar is the main target.

Although research has been done on wastewater treatment using modified biochar, knowledge gaps still need to be filled. Most of the research works are at the laboratory scale. Research to use biochar at a larger scale using real wastewater is required. Further research on regeneration techniques and reusability of biochar and on the disposal of spent biochar is needed.

REFERENCES

Cazetta, André L., Alexandro M. M. Vargas, Eurica M. Nogami, Marcos H. Kunita, Marcos R. Guilherme, Alessandro C. Martins, Tais L. Silva, Juliana C. G. Moraes, and Vitor C. Almeida. 2011. "NaOH-Activated Carbon of High Surface Area Produced from Coconut Shell: Kinetics and Equilibrium Studies from the Methylene Blue Adsorption." *Chemical Engineering Journal* 174 (1): 117–125. https://doi.org/10.1016/J.CEJ.2011.08.058.

Cha, Jin Sun, Sung Hoon Park, Sang-Chul Jung, Ryu Changkook, Jong-Ki Jeon, Min-Chul Shin, and Young-Kwon Park. 2016. "Production and Utilization of Biochar: A Review." *Journal of Industrial and Engineering Chemistry* 40: 1–15.

Chao, Xu, Xiang Qian, Zhu Han-hua, Wang Shuai, Zhu Qi-hong, Huang Dao-you, and Zhang Yang-zhu. 2018. "Effect of Biochar from Peanut Shell on Speciation and Availability of Lead and Zinc in an Acidic Paddy Soil." *Ecotoxicology and Environmental Safety* 164 (November): 554–561. https://doi.org/10.1016/j.ecoenv.2018.08.057.

Cho, Dong Wan, Gihoon Kwon, Kwangsuk Yoon, Yiu Fai Tsang, Yong Sik Ok, Eilhann E. Kwon, and Hocheol Song. 2017. "Simultaneous Production of Syngas and Magnetic Biochar via Pyrolysis of Paper Mill Sludge Using CO2 as Reaction Medium." *Energy Conversion and Management* 145: 1–9. https://doi.org/10.1016/j.enconman.2017.04.095.

Dai, Yingjie, Naixin Zhang, Chuanming Xing, Qingxia Cui, and Qiya Sun. 2019. "The Adsorption, Regeneration and Engineering Applications of Biochar for Removal Organic Pollutants: A Review." *Chemosphere* 223: 12–27. https://doi.org/10.1016/j.chemosphere.2019.01.161.

Devi, Parmila, and Anil K. Saroha. 2014. "Synthesis of the Magnetic Biochar Composites for Use as an Adsorbent for the Removal of Pentachlorophenol from the Effluent." *Bioresource Technology* 169: 525–531. https://doi.org/10.1016/j.biortech.2014.07.062.

Du, Cong, Yonghui Song, Shengnan Shi, Bei Jiang, Jiaqi Yang, and Shuhu Xiao. 2020. "Preparation and Characterization of a Novel Fe3O4-Graphene-Biochar Composite for Crystal Violet Adsorption." *Science of The Total Environment* 711 (April): 134662. https://doi.org/10.1016/J.SCITOTENV.2019.134662.

Durán-Jiménez, G., V. Hernández-Montoya, M. A. Montes-Morán, S. W. Kingman, T. Monti, and E. R. Binner. 2018. "Microwave Pyrolysis of Pecan Nut Shell and Thermogravimetric, Textural and Spectroscopic Characterization of Carbonaceous Products." *Journal of Analytical and Applied Pyrolysis* 135: 160–168. https://doi.org/10.1016/j.jaap.2018.09.007.

Fang, June, Lu Zhan, Yong Sik Ok, and Bin Gao. 2018. "Minireview of Potential Applications of Hydrochar Derived from Hydrothermal Carbonization of Biomass." *Journal of Industrial and Engineering Chemistry* 57: 15–21. https://doi.org/10.1016/j.jiec.2017.08.026.

Huang, Yu Fong, Pei Hsin Cheng, Pei Te Chiueh, and Shang Lien Lo. 2017. "Leucaena Biochar Produced by Microwave Torrefaction: Fuel Properties and Energy Efficiency." *Applied Energy* 204 (Complete): 1018–1025. https://doi.org/10.1016/j.apenergy.2017.03.007.

Inyang, Mandu, Bin Gao, Andrew Zimmerman, Ming Zhang, and Hao Chen. 2014. "Synthesis, Characterization, and Dye Sorption Ability of Carbon Nanotube—Biochar Nanocomposites." *Chemical Engineering Journal* 236 (January): 39–46. https://doi.org/10.1016/J.CEJ.2013.09.074.

Jin, Hongmei, Sergio Capareda, Zhizhou Chang, Jun Gao, Yueding Xu, and Jianying Zhang. 2014. "Biochar Pyrolytically Produced from Municipal Solid Wastes for Aqueous As(V) Removal: Adsorption Property and Its Improvement with KOH Activation." *Bioresource Technology* 169: 622–629.

Kambo, Harpreet Singh, and Animesh Dutta. 2015. "A Comparative Review of Biochar and Hydrochar in Terms of Production, Physico-Chemical Properties and Applications." *Renewable and Sustainable Energy Reviews* 45: 359–378. https://doi.org/10.1016/j.rser.2015.01.050.

Li, Yunchao, Jingai Shao, Xianhua Wang, Yong Deng, Haiping Yang, and Hanping Chen. 2014. "Characterization of Modified Biochars Derived from Bamboo Pyrolysis and Their Utilization for Target Component (Furfural) Adsorption." *Energy and Fuels* 28 (8): 5119–5127. https://doi.org/10.1021/ef500725c.

Liang, Jie, Xuemei Li, Zhigang Yu, Guangming Zeng, Yuan Luo, Longbo Jiang, Zhaoxue Yang, Yingying Qian, and Haipeng Wu. 2017. "Amorphous MnO2 Modified Biochar Derived from Aerobically Composted Swine Manure for Adsorption of Pb(II) and Cd(II)." *ACS Sustainable Chemistry and Engineering* 5(6): 5049–5058. https://doi.org/10.1021/acssuschemeng.7b00434.

Liu, Fenglin, Jiane Zuo, Tong Chi, Pei Wang, and Bo Yang. 2015. "Removing Phosphorus from Aqueous Solutions by Using Iron-Modified Corn Straw Biochar." *Frontiers of Environmental Science and Engineering* 9 (6): 1066–1075. https://doi.org/10.1007/s11783-015-0769-y.

Liu, Yuyan, Saran P. Sohi, Siyuan Liu, Junjie Guan, Jingyao Zhou, and Jiawei Chen. 2019. "Adsorption and Reductive Degradation of Cr(VI) and TCE by a Simply Synthesized Zero Valent Iron Magnetic Biochar." *Journal of Environmental Management* 235: 276–281. https://doi.org/10.1016/j.jenvman.2019.01.045.

Mohamed, Badr A., Chang Soo Kim, Naoko Ellis, and Xiaotao Bi. 2016. "Microwave-Assisted Catalytic Pyrolysis of Switchgrass for Improving Bio-Oil and Biochar Properties." *Bioresource Technology* 201: 121–132. https://doi.org/10.1016/j.biortech.2015.10.096.

Nhuchhen, D. R., M. T. Afzal, T. Dreise, and A. A. Salema. 2018. "Characteristics of Biochar and Bio-Oil Produced from Wood Pellets Pyrolysis Using a Bench Scale Fixed Bed, Microwave Reactor." *Biomass and Bioenergy* 119: 293–303. https://doi.org/10.1016/j.biombioe.2018.09.035.

Peng, Peng, Yin Hai Lang, and Xiao Mei Wang. 2016. "Adsorption Behavior and Mechanism of Pentachlorophenol on Reed Biochars: PH Effect, Pyrolysis Temperature, Hydrochloric Acid Treatment and Isotherms." *Ecological Engineering* 90 (May): 225–233. https://doi.org/10.1016/J.ECOLENG.2016.01.039.

Rajapaksha, Anushka Upamali, Season S. Chen, Daniel C. W. Tsang, Ming Zhang, Meththika Vithanage, Sanchita Mandal, Bin Gao, Nanthi S. Bolan, and Yong Sik Ok. 2016. "Engineered/Designer Biochar for Contaminant Removal/Immobilization from Soil and Water: Potential and Implication of Biochar Modification." *Chemosphere* 148 (April): 276–291. https://doi.org/10.1016/J.CHEMOSPHERE.2016.01.043.

Rajapaksha, Anushka Upamali, Meththika Vithanage, Mahtab Ahmad, Dong Cheol Seo, Ju Sik Cho, Sung Eun Lee, Sang Soo Lee, and Yong Sik Ok. 2015. "Enhanced Sulfamethazine Removal by Steam-Activated Invasive Plant-Derived Biochar." *Journal of Hazardous Materials* 290: 43–50. https://doi.org/10.1016/j.jhazmat.2015.02.046.

Rajapaksha, Anushka Upamali, Meththika Vithanage, Ming Zhang, Mahtab Ahmad, Dinesh Mohan, Scott X. Chang, and Yong Sik Ok. 2014. "Pyrolysis Condition Affected Sulfamethazine Sorption by Tea Waste Biochars." *Bioresource Technology* 166 (August): 303–308. https://doi.org/10.1016/J.BIORTECH.2014.05.029.

Saha, Nepu, Akbar Saba, and M. Toufiq Reza. 2019. "Effect of Hydrothermal Carbonization Temperature on PH, Dissociation Constants, and Acidic Functional Groups on Hydrochar from Cellulose and Wood." *Journal of Analytical and Applied Pyrolysis* 137: 138–145. https://doi.org/10.1016/j.jaap.2018.11.018.

Samsuri, Abd Wahid, Fardin Sadegh-Zadeh, and Bahi Jalili Seh-Bardan. 2013. "Adsorption of As(III) and As(V) by Fe Coated Biochars and Biochars Produced from Empty Fruit Bunch and Rice Husk." *Journal of Environmental Chemical Engineering* 1 (4): 981–988. https://doi.org/10.1016/J.JECE.2013.08.009.

Shen, Wenzhong, Zhijie Li, and Yihong Liu. 2010. "Surface Chemical Functional Groups Modification of Porous Carbon." *Recent Patents on Chemical Engineering* 1 (1): 27–40. https://doi.org/10.2174/1874478810801010027.

Sounthararajah, D. P., P. Loganathan, J. Kandasamy, and S. Vigneswaran. 2015. "Adsorptive Removal of Heavy Metals from Water Using Sodium Titanate Nanofibres Loaded onto GAC in Fixed-Bed Columns." *Journal of Hazardous Materials* 287 (April): 306–316. https://doi.org/10.1016/J.JHAZMAT.2015.01.067.

Sun, Lei, Dongmei Chen, Shungang Wan, and Zebin Yu. 2015. "Performance, Kinetics, and Equilibrium of Methylene Blue Adsorption on Biochar Derived from Eucalyptus Saw Dust Modified with Citric, Tartaric, and Acetic Acids." *Bioresource Technology* 198 (December): 300–308. https://doi.org/10.1016/J.BIORTECH.2015.09.026.

Tan, I. A. W., A. L. Ahmad, and B. H. Hameed. 2008. "Adsorption of Basic Dye on High-Surface-Area Activated Carbon Prepared from Coconut Husk: Equilibrium, Kinetic and Thermodynamic Studies." *Journal of Hazardous Materials* 154 (1–3): 337–346. https://doi.org/10.1016/J.JHAZMAT.2007.10.031.

Vause, Danielle, Natalie Heaney, and Chuxia Lin. 2018. "Differential Release of Sewage Sludge Biochar-Borne Elements by Common Low-Molecular-Weight Organic Acids." *Ecotoxicology and Environmental Safety* 165 (December): 219–223. https://doi.org/10.1016/j.ecoenv.2018.09.005.

Wan, Shunli, Jiayu Wu, Shanshan Zhou, Rui Wang, Bin Gao, and Feng He. 2018. "Enhanced Lead and Cadmium Removal Using Biochar-Supported Hydrated Manganese Oxide (HMO) Nanoparticles: Behavior and Mechanism." *Science of The Total Environment* 616–617 (March): 1298–1306. https://doi.org/10.1016/J.SCITOTENV.2017.10.188.

Wang, Jianlong, and Shizong Wang. 2018. "Activation of Persulfate (PS) and Peroxymonosulfate (PMS) and Application for the Degradation of Emerging Contaminants." *Chemical Engineering Journal* 334 (February): 1502–1517. https://doi.org/10.1016/J.CEJ.2017.11.059.

Wang, Sheng ye, Yan kui Tang, Cheng Chen, Jin tao Wu, Zhining Huang, Ya yuan Mo, Kai xuan Zhang, and Ji bo Chen. 2015. "Regeneration of Magnetic Biochar Derived from Eucalyptus Leaf Residue for Lead(II) Removal." *Bioresource Technology* 186 (June): 360–364. https://doi.org/10.1016/J.BIORTECH.2015.03.139.

Xiang, Wei, Xueyang Zhang, Jianjun Chen, Weixin Zou, Feng He, Xin Hu, Daniel C. W. Tsang, Yong Sik Ok, and Bin Gao. 2020. "Biochar Technology in Wastewater Treatment: A Critical Review." *Chemosphere* 252 (August): 126539. https://doi.org/10.1016/J.CHEMOSPHERE.2020.126539.

Xiong, Zhang, Zhang Shihong, Yang Haiping, Shi Tao, Chen Yingquan, and Chen Hanping. 2013. "Influence of NH3/CO2 Modification on the Characteristic of Biochar and the CO2 Capture." *Bioenergy Research* 6 (4): 1147–1153. https://doi.org/10.1007/s12155-013-9304-9.

Xue, Lihong, Bin Gao, Yongshan Wan, June Fang, Shengsen Wang, Yuncong Li, Rafael Muñoz-Carpena, and Linzhang Yang. 2016. "High Efficiency and Selectivity of MgFe-LDH Modified Wheat-Straw Biochar in the Removal of Nitrate from Aqueous Solutions." *Journal of the Taiwan Institute of Chemical Engineers* 63 (June): 312–317. https://doi.org/10.1016/J.JTICE.2016.03.021.

Yakout, Sobhy M., Abd El Hakim M. Daifullah, and Sohair A. El-Reefy. 2018. "Pore Structure Characterization of Chemically Modified Biochar Derived From Rice Straw." *Environmental Engineering and Management Journal* 14 (2): 473–480. https://doi.org/10.30638/eemj.2015.049.

Yan, Jingchun, Lu Han, Weiguo Gao, Song Xue, and Mengfang Chen. 2015. "Biochar Supported Nanoscale Zerovalent Iron Composite Used as Persulfate Activator for Removing Trichloroethylene." *Bioresource Technology* 175 (January): 269–274. https://doi.org/10.1016/J.BIORTECH.2014.10.103.

Zhang, Ming Ming, Yun Guo Liu, Ting Ting Li, Wei Hua Xu, Bo Hong Zheng, Xiao Fei Tan, Hui Wang, Yi Ming Guo, Fang Ying Guo, and Shu Fan Wang. 2015. "Chitosan Modification of Magnetic Biochar Produced from Eichhornia Crassipes for Enhanced Sorption of Cr(vi) from Aqueous Solution." *RSC Advances* 5 (58): 46955–46964. https://doi.org/10.1039/C5RA02388B.

Zhang, Ping, David O'Connor, Yinan Wang, Lin Jiang, Tianxiang Xia, Liuwei Wang, Daniel C. W. Tsang, Yong Sik Ok, and Deyi Hou. 2020. "A Green Biochar/Iron Oxide Composite for Methylene Blue Removal." *Journal of Hazardous Materials* 384 (February): 121286. https://doi.org/10.1016/J.JHAZMAT.2019.121286.

Zhang, Yaning, Paul Chen, Shiyu Liu, Peng Peng, Min Min, Yanling Cheng, Erik Anderson, et al. 2017. "Effects of Feedstock Characteristics on Microwave-Assisted Pyrolysis—A Review." *Bioresource Technology* 230: 143–151. https://doi.org/10.1016/j.biortech.2017.01.046.

Zheng, Yulin, Bing Wang, Anne Elise Wester, Jianjun Chen, Feng He, Hao Chen, and Bin Gao. 2019. "Reclaiming Phosphorus from Secondary Treated Municipal Wastewater with Engineered Biochar." *Chemical Engineering Journal* 362: 460–468. https://doi.org/10.1016/j.cej.2019.01.036.

Zhou, Yanmei, Bin Gao, Andrew R. Zimmerman, June Fang, Yining Sun, and Xinde Cao. 2013. "Sorption of Heavy Metals on Chitosan-Modified Biochars and Its Biological Effects." *Chemical Engineering Journal* 231 (September): 512–518.

7 Utilization of Sustainable Biochar Approach to Improve Contaminated Agricultural Soils for Plant Growth Promotion

Kamal Prasad and Ajay Kumar Singh

CONTENTS

7.1 INTRODUCTION

Today's scenario, agricultural productivity is severely affected by insects, plant diseases, weeds, drought, high salinity, high temperature, and soil quality (Thalmann and Santelia 2017). Generally, chemical pesticides (CP) are used to control and improve the yield and quality of plants. The excessive use of CPs and their appearance in the environment have caused serious problems, especially the pollution of soil, water, and air, which has adversely affected the ecosystem and the entire food chain (Moon et al. 2013). The area of the soil is directly related to the availability of nutrients. When CPs are used, a part of them remains in the soil, and the accumulation affects the microorganisms living in it. Therefore, the concentration of CP residues in the soil must be reduced and effective. A recovery method is needed. Plant growth requires a variety of soil nutrients such as nitrogen (N), phosphorus (P), and potassium (K) to grow, but the level of nutrients in the soil is decreasing in India. The soil in many countries and regions not only lacks macronutrients in addition to N, P, and K, micronutrients (such as sulfur, calcium, and magnesium) and trace elements (such as sulfur, calcium, magnesium, boron, zinc, copper, and iron) (Pathak 2010) are also needed. Of the large amount of chemical fertilizers (CF) used in soil for this purpose, only a small percentage of water-soluble nutrients are absorbed by plants, and the rest remain in the soil. This not only reduces the nutrient content of crops but also affects soil fertility in the long term (Hariprasad and Dayananda 2013;

Yargholi and Azarneshan 2014). In addition to CFs, CPs are also a major scourge of agriculture. The negative impact of CPs on the environment affect the microbial properties of the soil. Large amounts of CFs and CPs and their long-term storage in the soil have a negative impact on soil quality, health, and soil microflora. This destroys soil health and significantly reduces the overall biomass of bacteria and fungi (Prashar and Shah 2016). Long-term application of CFs (N and NPK) and/or organic manures leads to changes in the structural diversity of agricultural soils and the dominant bacterial community (Wu et al. 2012).

This approach is an environment-friendly, cost-effective, biocompatible, and safe substitute for green techniques. It is needed as an alternative to pricey and poisonous approaches, along with chemical and physical methods. Biofertilizers, on the other hand, can reenergize the soil by enhancing the soil fertility and therefore may be used as an effective device for sustainable agriculture, rendering agroecosystems stress free. Additionally, the utility of natural amendments to soils, from a remedial point of view, has commonly been justified based on their distinctly low cost, which generally calls for different styles of disposal (burial in a landfill, incineration, etc.). Soil amendments should have excessive binding potential, offer environmental protection, and not have any terrible impact on soil structure, soil fertility, or the environment at the whole (Paz-Ferreiro et al. 2013). The use of biochar has become common as a sustainable method and a promising technique to enhance soil in a highly satisfactory manner and to do away with heavy metal pollution of the soil (Lahori et al. 2017). This 2000-year-old exercise converts agricultural waste right into a soil enhancer that can maintain carbon, improve food security, encourage soil biodiversity, and discourage deforestation. The system creates a fine-grained, tremendously porous charcoal that enables soils keep vitamins and water. Biochar is located in soils around the world due to plant life fires and ancient soil control practices. Intensive observation of biochar-wealthy darkish earths within the Amazon (terrapreta) has created a much wider appreciation of biochar's precise role as a soil enhancer.

Biochar is created from biomass via a shift under hot temperature associated with low oxygen conditions. It's a carbon-rich organic material, an organic amendment. Biomass, akin to wood, manure, or leaves, is heated in a closed instrumentation with very little or no access to air. In other words, biochar is made by the thermal decomposition of organic material with restricted availability of oxygen and at varying temperatures (about 550°C). However, the number of the materials produced depends on process conditions. Recently, biochar obtained from the destructive distillation of organic wastes has become a substitute that not only influences the sequestration of soil carbon but additionally modifies its physicochemical and biological properties (Garcia et al. 2016; Zhang et al. 2017). However, biochar is distinguished from charcoal and similar materials in that it's made with the intent of being applied to soil as a method of rising soil productivity, carbon (C) storage, and probably filtration of percolating soil water (to cut pollution of surface and groundwater bodies).

7.2 CURRENT SCENARIO: BIOCHAR AS A POWERFUL TOOL TO COMBAT CLIMATE CHANGE

The carbon in biochar resists degradation and might keep carbon in soils for many years. Biochar is produced via pyrolysis or gasification, strategies that warms biomass in the absence or reduced levels of oxygen. In addition to growing a soil enhancer, sustainable biochar practices can produce oil and fuel line by-products that may be used as gasoline, supplying clean, renewable energy. When the biochar is buried as a soil enhancer, it can become "carbon negative." Biochar and bioenergy comanufacturing can assist fight international weather changes by offering an alternate means of displacing fossil gasoline use and of sequestering carbon in stable soil carbon pools. The approach may additionally lessen emissions of nitrous oxide. People can use this easy and powerful tool to lessen carbon emissions.

Biochar has the potential to generate renewable energy on the basis of agriculture in an environmentally friendly way. Specifically, the quality of biochar depends on several factors, such as the type of soil, the metal and the raw material used for carbonation, the pyrolysis conditions, and the amount of biochar and biochar applied to the soil (Debela et al. 2012). In addition, supplementing the soil with biochar has been shown to be beneficial in improving soil quality and maintaining nutrients, thereby promoting plant growth (Bonanomi et al. 2017). Since biochar contains organic matter and nutrients, its addition increases soil pH, electrical conductivity (EC), organic carbon (C), total nitrogen (TN), available phosphorus (P), and cation exchange capacity (CEC) (Dume et al. 2016). Previously, Verheijen et al. (2009) reported that the use of biochar affects the toxicity, transport, and fate of various heavy metals in the soil due to the improvement in the soil's absorption capacity. The presence of plant nutrients and ash in the biochar and its large surface area, porous nature, and the ability to act as a breeding ground for microorganisms have been identified as the main reasons for the improvement in soil properties and the increased nutrient uptake by plants in biochar-treated soils (Nigussie et al. 2012). Chan et al. (2008) reported that the use of biochar reduced the tensile strength of the soil cores, suggesting that the use of biochar may reduce the risk of soil compaction. The advantages of inoculating rhizobacteria in the soil have already been much discussed, but the addition of biochar can also add more nutrients to the soil and thus benefit agricultural crops. The mixture of microorganisms that promote plant growth with biochar is known to be the best combination for French bean growth and yield, as proposed by Saxena et al. (2013). Adding biochar to the soil is extremely beneficial to improving soil quality and stimulating plant growth. Therefore, biochar can play an important role in the development of sustainable agricultural systems. Today, it is considered an effective method for remediating contaminated soil (Placek et al. 2016) and for achieving high yields without polluting the environment. The positive effects of biochar on plant growth and soil quality indicate that the use of biochar is a good way to overcome nutrient deficiency and is therefore a suitable method to improve nutrient cycling on an agricultural scale. Therefore, a comprehensive approach was used to study the positive effects of biochar correction on soil stability and plant growth stimulation.

7.3 BIOCHAR PREPARATION AND CHARACTERISTICS

Biochar is composed of elements such as carbon, hydrogen, sulfur, oxygen, and nitrogen, as well as minerals in ashes. In the pyrolysis process, the thermal decomposition of biomass occurs in an environment with limited oxygen. Biochar is black, porous and fine-grained, lightweight, bulky, and has a high pH, all of which has a positive effect when it is applied to soil to increase its biological cleanliness. It is a stable biomass that can be mixed with soil to specifically change the characteristics of the soil atmosphere in order to increase crop yields and reduce environmental pollution. The raw materials used (the biomass) and process parameters determine the characteristics of biochar.

7.4 BIOMASS UTILIZATION FOR PRODUCTION OF BIOCHAR

A variety of organic materials are suitable as raw materials for the production of biochar, which can be made from agricultural wastes such as grass, cow dung, wood chips, rice husk, wheat straw, and cassava roots (Ronsse et al. 2013; Kiran et al. 2017). According to reports, the production of high-nutrient biochar depends on the type of raw materials used and pyrolysis conditions (Chan et al. 2007). Biochar is produced from surplus biomass such as plant residues, manure liquid, wood residue, forest and green waste using modern pyrolysis technology. Agricultural waste (bark, straw, chaff, seeds, chaff, cake, sawdust, walnut husk, wood chips, animal beds, corn cobs, straw, etc.), industrial waste (pulp, vinasse), and municipal waste/urban garbage (Novotny et al. 2015; Kameyama et al. 2016) are widespread, which also ensures the management of garbage through its production and use (Woolf et al. 2010). The raw materials currently used for commercial purposes include bark, wood shavings, plant waste (husk, straw, and rice husk), grass and organic waste, including distiller's grains, industrial sugarcane cake, factory waste, chicken manure, dairy products. Manure, sewage sludge, and paper sludge (Tumuluru et al. 2011; Sohi et al. 2009; Reddy 2015).

A 40 wt.% yield of biochar from maize stover was obtained by Peterson et al. (2012). The biomass used for the assembly of biochar is especially composed of polysaccharide, hemicellulose, and lignin polymers (Sullivan and Ball 2012). Among these, cellulose has been found to be the biggest component of maximum plant-derived biomasses, but lignin is additionally vital in woody biomass.

7.5 PRODUCTION OF BIOCHAR

Biochar is often factory-made on a very small scale in inexpensive modified stoves or kilns or through large-scale, cost-intensive production methods in larger pyrolysis plants and using greater amounts of feedstocks. Biochar is created from many biomass feedstocks through pyrolysis, as mentioned, generating oil and gases as by-products (Zhu et al. 2018). The dry waste obtained is reduced to small items under 3 cm before use. The feedstock is heated, either with no oxygen or with little oxygen, at temperatures of 350–700°C. The shift is mostly classified according to the temperature and the length of time for heating; quick pyrolysis takes place at temperatures of >500°C and usually happens in the order of seconds (heating rates ≥ 1000°C/min). This condition maximizes the generation of bio-oil. Slow pyrolysis, on the other hand, typically takes additional time, from 30 min to many hours for the feedstock to totally pyrolyze (heating rates ≤ 100°C/min) and at constant time, and it yields more biochar. In slow pyrolosis, the temperature remains 250–500°C (Brown et al. 2011). The type of biochar produced depends on three variables: the biomass used, temperature, and heating rate. High temperature or low temperature has a significant impact on charcoal production. It has been observed that low-temperature biochar (<550°C) has an amorphous carbon structure, and its aromaticity is lower than that of biochar obtained at high temperature (Joseph et al. 2010). The high temperature leads to a reduction in the carbon yield in all pyrolysis reactions (Antal and Gronli 2003). Peng et al. (2011) reported the impact of carbonization time on biochar yield; at the same temperature, the yield decreased with the increase in time. The pyrolysis process seriously affects the quality of biochar and its potential agricultural value in terms of agronomic performance or carbon sequestration. The range is from 24 to 77% (Dutta 2010; Stoyle 2011). The pyrolysis process can be expressed as follows:

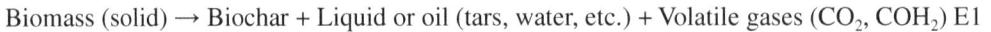

Biomass (solid) → Biochar + Liquid or oil (tars, water, etc.) + Volatile gases (CO_2, COH_2) E1

7.6 CHARACTERISTICS OF BIOCHAR

- Biochar stores recalcitrant form of carbon in soil.
- Enhances plant growth. Raises and sustain crop yields. Helps improve good and problematic nutrient-poor soils, including acidic tropical humid and drier environment soils.
- Compensates for greenhouse gas emissions associated with agricultural development.
- May improve soil moisture retention, increasing agricultural resilience, and provides support to intensive sustainable agriculture, which could help to reduce pressure for new forest clearances and enhance biodiversity conservation benefits (Table 7.1).
- Enables production of useful materials from uncropped land making use of unused waste with increased adaptability to environmental change by making production more resilient.
- Reduces the need for fertilizer/manure/compost. Reduces costs of sewage and animal waste treatment and cuts emissions that they would otherwise cause if held in lagoons or heaps.
- Offers a more environmentally friendly way of processing plastics and refuse if biochar is too contaminated for agricultural use for growing nonfood crops or is sent to landfill to sequester carbon.
- Nutrient affinity, i.e., the retention of plant nutrients, notably N, on permeable soils under rainy conditions is found to be higher with biochar application. Biochar may play a role

in bioremediation by binding agrochemicals and help reduce phosphate and nitrate and agrochemicals pollution of streams and groundwater. Thus it helps resolve major problems hindering sustained and improved agriculture. Reduces plant uptake of pesticides from contaminated soils.

- Reduces soil acidity/raises pH. Reduces aluminum toxicity and increases cation exchange capacity.

7.7 IMPACT OF BIOCHAR ON SOIL PROPERTIES

- By improving moisture retention, biochar may reduce the demand for irrigation and make cropping more secure.
- Supports biofuel production, reduces its carbon footprint, and even enables it to move toward being carbon neutral.
- Increases soil microbial biomass and supports other beneficial organism like earthworms. Supports nitrogen fixation. Increases arbuscular mycorrhizal fungi in the soil (Table 7.1).
- Offers opportunities for the poor to benefit from the carbon offset market and also reduces the dependency of farmers on input suppliers.
- In periurban/urban agriculture, biochar may be a useful input to counter harmful compounds like heavy metals, dioxins, PAHs (polycyclic aromatic hydrocarbons) present in sewage or refuse inputs.

7.8 PHYSICAL, CHEMICAL, AND BIOLOGICAL PROPERTIES OF BIOCHAR

Biochar is a stable form of carbon that can be stored in the soil for thousands of years (Shenbagavalli and Mahimairaja 2012) and that is used to add carbon to the soil and improve soil quality. The pyrolysis conditions and materials used can significantly affect the properties of biochar, and the physical properties of biochar contribute to its function as an environmental management tool. According to reports, when biochar is used as a soil amendment, it can stimulate soil fertility and improve soil quality and health. The pH value increases the ability to retain water, attracts more beneficial fungi and other microorganisms, improves the cation exchange capacity, and maintains nutrients in the soil (Ajema 2018). Biochar reduces soil density and hardening, increases soil aeration and cation exchange capacity, and changes soil structure and consistency by changing its physical and chemical properties. It also helps rebuild degraded soil. Compared with other soil organic matter, due to its higher surface area, negative surface charge, and charge density (Liang et al. 2006), it can

TABLE 7.1
Utilization of Biochar in Soil and Their Impact on Soil Properties

Soil properties	Findings
Cation exchange capacity	50% increase
Fertilizer use efficiency	10–30% increase
Liming agent	1 unit pH increase
Crop productivity	20–120% increase
Biological nitrogen fixation	50–72% increase
Soil moisture retention	Up to 18% increase
Mycorrhizal fungi	40% increase
Bulk density	Soil dependent
Methane emission	100% decrease
Nitrous oxide emissions	50% decrease

adsorb cations per carbon unit (Lehmann et al. 2006), thereby increasing yield (Lehmann 2007). Carbon can be added to the soil for binding; surfaces with high biochar absorption rates can be characterized as soil additives, which can prevent dangerous elements in the soil. The physical traits of biochar are immediately and in a roundabout way associated with how they have an effect on soil systems. Soils have their individual physical residences depending on the character of local mineral and natural components, their relative amounts, and the way minerals and other components are related. When biochar is present within the soil mixture, its contribution to the physical nature of the systems is significant, affecting intensity, texture, structure, porosity, and consistency by converting the surface area, pore and particle-length distribution, density, and packing (Blanco-Canqui 2017). The impact of biochar on soil residences immediately influences the vegetation because the intensity of penetration and the accessibility of air and water within the root region depend in particular on the physical composition of the soil horizons. This influences the soil's physical, chemical, and biological properties through the enhancement of soil nutrients and water-holding capacity, pH, bulk density, and stimulation of soil microbial activities by improving aggregation, porosity, surface area, and habitat for soil microbes in biochar-amended soils, especially to a degraded soil for a large and long-term carbon sink restoration. The smaller the pores on biochar, the longer they are able to maintain capillary soil water. The addition of biochar can lessen the outcomes of drought on crop productiveness in drought-affected regions with its moisture-retention capacity. It has been proven that biochar gets rid of soil constraints that restrict the growth of vegetation and neutralizes acidic soil due to its simple nature (Hammes and Schmidt 2009). Carbon dioxide and oxygen occupy air-crammed areas at the pores of biochar, or they may be chemosorbed at the surface. As biochar can include nutrients, microorganisms, and syngases, it could additionally maintain fertilizers within the soil longer than in untreated soil and prevent the fertilizers from leaching into the water of rivers and lakes. In terms of chemical properties, biochar reduces the acidity of the soil by increasing the pH value (also known as the lime effect) and by helping the soil store nutrients and fertilizers (Lehmann et al. 2003). The use of biochar improves soil fertility through two mechanisms: by adding nutrients into the soil (such as K, part of P, and many trace elements) or by retaining nutrients from other sources, including nutrients from the soil itself. However, the main benefit is the preservation of nutrients from other sources. The addition of biochar has a positive effect on plant growth only when nutrients from other sources (such as inorganic or organic fertilizers) are used. Biochar increases the availability of C, N, Ca, Mg, K, and P for plants. Because biochar slowly absorbs and releases fertilizer (DeLuca et al. 2015), it also helps prevent fertilizer dehydration and leaching, thereby reducing hazardous CF use and reducing agricultural pollution (Cao et al. 2018). Exposure to CP and to complex nitrogen fertilizers in the soil also thereby reduces environmental pollution.

Healthy soil should contain broad and balanced life forms, including bacteria, fungi, protozoa, nematodes, arthropods, and earthworms. It has recently been reported that biochar increases soil microbial respiration and creates space for soil microbes (Slapakova et al. 2018), thereby increasing soil biodiversity and density. In the competition from saprophytes, biochar therefore serves as an inoculum for arbuscular mycorrhizal fungi (Saito and Marumoto 2002). The average residence time of biochar in the soil is 0.01–1 million years, with an average of 0.5 million years (Skjemstad et al. 1998; Swift 2001; Krull et al. 2003); however, their rebellious and physical properties are major obstacles to assessing long-term stability. Soil microorganisms that are available for marketing include *Azospirillum*, *Azotobacter*, *Bacillus thuringiensis*, *Bacillus megaterium*, *Glomus fasciculatum*, *G. mosseae*, *Pseudomonas fluorescens*, *Rhizobium*, and *Trichoderma viride* (Hazarika and Ansari 2007).

7.9 EFFICACY OF BIOCHAR FOR SOIL AMENDMENT

The problems of food insecurity, declining soil fertility, climate change, and productivity are the driving forces behind the introduction of recent technologies or new systems in farming. The modification of soils for his correction aims at reducing the chance of waste transfer to waters or receptor organisms in proximity. The organic material reminiscent of biochar may function as a preferred alternative for this purpose as a result of its supply being biological, and it should be directly applied to soils with very little pretreatment (Beesley et al. 2011). Two aspects make biochar amendment superior to other organic materials: the primary aspect is the high stability against decay, so it stays in the soil for extended times providing long-run advantages to soil; the second aspect is therefore the great capability to retain the nutrients. Biochar modification improves soil quality by increasing soil pH, its moisture-holding capacity, its cation exchange capacity, and its microorganism flora (Mensah and Frimpong 2018). The addition of biochar to the soil has shown the increase in convenience for basic cations in concentrations of phosphorus and total N (Glaser et al. 2002; Lehmann et al. 2003). Typically, base-forming hydrogen ion concentrations and mineral constituents of biochar (ash content, together with N, P, K, and trace elements) offer the necessary agronomical edges to several soils, a minimum of them within the short to medium term. Once biochar with a better pH value was applied to the soil, the amended soil typically became less acidic (Yuan et al. 2011). Acidic biochar may also increase soil pH when employed in soil with a lower pH value. The pH of biochar, like the opposite properties, is influenced by the kind of feedstock, production temperature, and production duration. Additional precious assets of biochar are suppression of emissions of greenhouse gases in soil. It has additionally been proven (Zhang et al. 2010) of decreasing the emissions of methane and nitrous oxide from agricultural soils, which may also have extra weather mitigation effects because those are powerful greenhouse gases. Spokas et al. (2009) mentioned decreased carbon dioxide manufacturing as a means of adding various concentrations of biochar from 2 to 60% (w/w), suppressing nitrous oxide manufacturing in degrees by better than 20% (w/w), and reducing ambient methane oxidation in any respect in degrees over unamended soil.

Some research has proven the control of pathogens with the aid of using biochar in agricultural soil. Bonanomi et al. (2015) pronounced that biochar is powerful in opposition to airborne (*Botrytis cinerea* and a one-of-a-kind species of powdery mildew) and soil-borne pathogens (*Rhizoctonia solani* and species of *Fusarium* and *Phytophthora*). The utilization of the biochar derived from citrus wood was able to control airborne gray mold, *Botrytis cinerea* on *Lycopersicon esculentum*, *Capsicum annuum*, and *Fragaria × ananassa*. The addition of biochar in 0.32, 1.60, and 3.20% (w/w) to *Asparagus* soils infested with *Fusarium* augmented the biomass of *Asparagus flowers* and decreased *Fusarium* root rot sickness (Elmer et al. 2010). Similarly, *Fusarium* root rot sickness in *Asparagus* was additionally decreased with the aid of using biochar inoculated with mycorrhizal fungi (Thies and Rillig 2009). An observation of suppression of bacterial wilt in tomatoes confirmed that biochar received from municipal natural waste decreased the occurrence of the sickness in *Ralstonia-solanacearum*-infested soil (Nerome et al. 2005). Ogawa (2009) recommended using biochars and biochar-amended composts for controlling the sicknesses because of microorganisms and fungi in soil. The sickness suppression mechanism has been attributed to the presence of calcium compounds, in addition to enhancements in the physical, chemical, and organic traits of the soil. The prevention of "dispersed water pollution" by absorbing ammonia or adjusting the kinetics of soil solutions containing nitrates, phosphorus, and other nutrients has been studied in detail. The introduction of biochar into the soil affects various soil limitations, such as high availability of organic (Yu et al. 2009) and inorganic (Hua et al. 2009) pollutants, cation exchange capacity (CEC), and nutrient retention (Major et al. 2010a; Singh et al. 2010). Pesticides, nutrients, and minerals in the soil are absorbed to prevent these chemicals from being transported to the surface or groundwater and then being decomposed by agricultural activities. Xie et al. (2013) reported that biochar amendment enhanced soil fertility and crop production, particularly in soils with low nutrients. However, in soils with high fertility, no noticeable increase in production was noticed, and some studies even reported inhibition of plant growth. The observations of Taghizadeh-Toosi et al. (2012) indicated that ammonia adsorbed by biochar could be later released to the soil. Saarnio et al. (2013) showed that biochar application along with fertilizers can lead to better plant growth, but sometimes a negative effect was also observed without fertilization due to reduced bioavailability through sorption of nitrogen. It has been shown that application of biochar in the soil has a positive to neutral and even negative impact on crop production. Hence, it is crucial that the mechanisms for action of biochar in the soil be understood before its application. The effect of adding biochar on plant productivity depends on the added amount. The recommended application rate for any soil improvement should be based on extensive field trials. There is not enough data to make general recommendations. The properties of biochar materials can vary greatly, so the properties of the corresponding biochar materials (such as pH and ash content) also affect the application rate. In the pot experiment with biochar-modified nutrient-poor soil, Rondon et al. (2007) found that the yield per hectare was reduced by 165 tons. A test in the United States showed that peanut shell and pine chip biochar applied at 11 tons and 22 tons per hectare, respectively, can make corn yield lower than standard fertilized control plots (Gaskin et al. 2010). Therefore, it is very important to control the consumption of biochar to prevent the negative effects of biochar.

7.10 BIOCHAR STIMULATION FOR SOIL MICROFLORA AND PLANT GROWTH

Several reports indicate that biochar has the ability to stimulate soil microbial communities, thereby increasing soil carbon storage. In addition to adsorbing organic matter, nutrients, and gases, biochar is likely to become a habitat for bacteria, actinomycetes, and fungi (Thies and Rillig 2009). It is assumed that faster heating of biomass (rapid pyrolysis) leads to the formation of biochar, with fewer microorganisms, smaller pores, and more liquid and gas components (Nartey and Zhao 2014). It is well-known that the water retention of biochar is improved when applied to the soil (Busscher et al. 2010) and that it affects the soil microbial population. Although the interaction between biochar and soil microorganisms is complex, biochar provides a suitable habitat for a large number of diverse soil microbes. The addition of biochar and phosphate solubilizing fungal strains promoted the growth and productivity of cowpea and glycine plants, and the results were better than those observed with the control or when the strains and biochar were used alone (Saxena et al. 2016; Saxena et al. 2017). The use of biochar multiplied mycorrhizal growth in clover bioassay flowers by offering the right situations for colonization of plant roots (Solaiman et al. 2010). Warnock et al. (2007) summarized four mechanisms as a means by which biochar can have an effect on the functioning of mycorrhizal fungi: (1) modifications within the bodily and chemical properties of soil, (2) oblique consequences on mycorrhizae via propagation to different soil microbes, (3) plant–fungus signaling interference and detoxing of poisonous chemical compounds on biochar, and (iv) offering a safe haven from mushroom browsers. Carrots and legumes grown on steep slopes and in soils with much lower than 5.2 pH confirmed the considerably advanced bloom due to the addition of biochar (Rondon et al. 2004). It was observed that biochar multiplied the organic N_2 fixation (BNF) of *Phaseolus vulgaris* (Rondon et al. 2007), especially because of greater availability of micronutrients after the addition of biochar. Lehmann et al. (2003) said that biochar decreased leaching of NH^{4+} as a means of helping it in the floor soil, wherein it became available for plant uptake. Mycorrhizal fungi have been regularly covered in crop control techniques as they have been broadly used as dietary supplements for soil inoculum (Schwartz et al. 2006). When each biochar and the mycorrhizal fungi are used accordingly with control practices, it made capability synergism, which may undoubtedly have an effect on soil quality, glaringly clear. The fungal hyphae and microorganisms that colonize the biochar particles (or different porous materials) can be protected from soil predators, along with mites, *Collembola*, and larger (>16 μm in diameter) protozoans, and nematodes (Saito 1990; Pietikainen et al. 2000; Ezawa et al. 2002). Biochar can enhance the quality of unharvested agricultural products (Major et al. 2005) and promote plant growth (Oguntunde et al. 2004). Compared with a control that did not change in the second or fourth year after application, a single application of 20 tons/ha of biochar in the Colombian savanna soil can increase corn yield by 28–140% (Major et al. 2010). When biochar was utilized in the soil, regions with low phosphorus content (Asai et al. 2009; Silber 2010) in northern Laos achieved higher grain yields of upland rice (Oryza sativa). However, in some cases, growth retardation has been miserable (Mikan and Abrams 1995) with regard to the infectious soil pathogens chiefly involved with the impact of AM fungal inoculations on asparagus tolerance to the soil-borne root rot pathogen *Fusarium*. Matsubara et al. (2002) confirmed that charcoal amendments had a restrictive effect on pathogens. An additional study that supported these earlier findings made explicit that biochar made up of ground hardwood added to asparagus field soil diode led to a decrease in root lesions caused by *Fusarium oxysporum, F. asparagi*, and *F. proliferatum* compared to the nonamended control (Elmer and Pignatello 2011). Biochar reduces the necessity for CF, which ends up in the reduction in emissions from fertilizer production, and turning agricultural waste into biochar additionally reduces methane levels caused by the natural decomposition of waste.

7.11   CONSORTIUM BIOCHAR

Biochar mixed with different soil amendments corresponding to manure, compost, or lime before soil application can improve potency by reducing the amount of field operations required. Since biochar has been shown to sorb nutrients and shield them from leaching (Novak et al. 2009), a mixture of biochar could improve the efficiency of manure and other amendments. However, Kammann et al. (2016) acknowledged in their recent review that only a few studies that directly combined organic amendments with biochars were available. They found that cocomposted biochars had an interesting plant-growth-promoting impact as compared to biochars once used pure; however, no systematic studies have been done to understand the interactive effects of biochars with nonpyrogenic organic amendments (NPOAs). Biochar can even be mixed with liquid manures and used as slurry. Additionally, combined biochar and compost applications have varied benefits over mix of biochar or compost with soil separately. These benefits, consistent with Liu et al. (2012), embody additional economical use of nutrients, biological activation of biochar, an increased provision of plant-available nutrients by biological nitrogen fixation, reduction of nutrient leaching, and therefore the contribution of combined nutrients as compared to one application of compost and biochar. Diminutive biochar is possibly best fitted to this kind of application. Biochar was conjointly mixed with manure in ponds, and doubtlessly reduced losses of N gas were recorded once it had been applied to the soil (Yanai et al. 2007; Spokas and Reikosky 2009).

7.12   CONCLUSION

The depletion of agricultural land due to the pressure of a fast increasing population requires sustainable crop production. The use of biochar has been suggested to restore contaminated farmland, to improve soil fertility by reducing acidity, and to increase the availability of nutrients for plant growth and productivity. Therefore, adding biochar to the soil may be one of the best ways to overcome biological stress in the soil and increase crop yield and to make more efficient use of nitrogen and water. Therefore, from this comprehensive overview, it can be concluded that biochar has the potential to improve soil properties, microbial population, biological nitrogen fixation, and plant growth. Therefore, it is recommended to use biochar as a soil stimulant for the long-term recovery of carbon sinks.

REFERENCES

Ajema L. (2018) Effects of biochar application on beneficial soil organism review. *International Journal of Research Studies in Science, Engineering and Technology* 5(5):9–18

Antal MJ, Grønli M. (2003) The art, science, and technology of charcoal production. *Industrial and Engineering Chemistry Research* 42(8):1619–1640

Asai H, Samson BK, Stephan HM, Songyikhangsuthor K, Homma K, Kiyono Y. (2009) Biochar amendment techniques for upland rice production in Northern Laos: 1. Soil physical properties, leaf SPAD and grain yield. *Field Crops Research* 111(1–2):81–84

Beesley L, Moreno-Jimenez E, Gomez-Eyles JL, Harris E, Robinson B, Sizmur T. (2011) A review of biochars' potential role in the remediation, revegetation and restoration of contaminated soils. *Environmental Pollution* 159:3269–3282

Blanco-Canqui H. (2017) Biochar and soil physical properties. *Soil Science Society of America Journal* 81:687–711

Bonanomi G, Ippolito F, Cesarano G, Nanni B, Lombardi N, Rita A. (2017) Biochar as plant growth promoter: Better off alone or mixed with organic amendments? *Frontiers in Plant Science* 8:1570.

Bonanomi G, Ippolito F, Scala F. (2015) A "black" future for plant pathology? Biochar as a new soil amendment for controlling plant diseases. *Journal of Plant Pathology* 97(2):223–234

Brown TR, Wright MM, Brown RC. (2011) Estimating profitability of two biochar production scenarios: Slow pyrolysis vs fast pyrolysis. *Biofuels, Bioproducts and Biorefining* 5(1):54–68

Busscher WJ, Novak JM, Evans DE, Watts DW, Niandou MAS, Ahmedna M. (2010) Influence of pecan biochar on physical properties of a Norfolk loamy sand. *Soil Science* 175:10–14

Cao Y, Gao Y, Qi Y, Li J. (2018) Biochar-enhanced composts reduce the potential leaching of nutrients and heavy metals and suppress plant-parasitic nematodes in excessively fertilized cucumber soils. *Environmental Science and Pollution Research International* 25(8):7589–7599

Chan KY, Van Zwieten L, Meszaros I, Downie A, Joseph S. (2007) Agronomic values of greenwaste biochar as a soil amendment. *Australian Journal of Soil Research* 45:629–634

Chan KY, Van Zwieten L, Meszaros I, Downie A, Joseph S. (2008) Using poultry litter biochars as soil amendments. *Australian Journal of Soil Research* 46(5):437–444

Debela F, Thring RW, Arocena JM. (2012) Immobilization of heavy metals by co-pyrolysis of contaminated soil with woody biomass. *Water, Air, and Soil Pollution* 223:1161–1170

DeLuca TH, Gundale MJ, MacKenzie MD, Jones DL. (2015) Biochar effects on soil nutrient transformation. In: Lehmann J, Joseph S, editors. *Biochar for Environmental Management: Science and Technology.* 2nd ed. London, UK: Routledge, pp. 421–454

Dume B, Mosissa T, Nebiyu A. (2016) Effect of biochar on soil properties and lead (Pb) availability in a military camp in Southwest Ethiopia. *African Journal of Environmental Science and Technology* 10(3):77–85

Dutta B. (2010) *Assessment of pyrolysis techniques of lignocellulosic biomass for biochar production* [dissertation—master's thesis]. McGill University

Elmer WH, Pignatello JJ. (2011) Effect of biochar amendments on mycorrhizal associations and Fusarium crown and root rot of asparagus in replant soils. *Plant Disease* 95:960–966

Elmer WH, White JC, Pignatello JJ. (2010) *Impact of Biochar Addition to Soil on the Bioavailability of Chemicals Important in Agriculture. Report.* New Haven: University of Connecticut

Ezawa T, Yamamoto K, Yoshida S. (2002) Enhancement of the effectiveness of indigenous arbuscular mycorrhizal fungi by inorganic soil amendments. *Soil Science and Plant Nutrition* 48:897–900

García AC, de Souza LGA, Pereira MG, Castro RN, García-Mina JM, Zonta E. (2016) Structure-property-function relationship in humic substances to explain the biological activity in plants. *Scientific Reports* 6:20798

Gaskin JW, Speir RA, Harris K, Das KC, Lee RD, Morris LA. (2010). Effect of peanut hull and pine chip biochar on soil nutrients, corn nutrient status, and yield. *Agronomy Journal* 102:623–633

Glaser B, Lehmann J, Zech W. (2002) Ameliorating physical and chemical properties of highly weathered soils in the tropics with charcoal: A review. *Biology and Fertility of Soils* 35:219–230

Hammes K, Schmidt MWI. (2009) Changes in biochar in soils. In: Lehmann M, Joseph S, editors. *Biochar for Environmental Management Science and Technology.* London: Earthscan, pp. 169–182

Hariprasad NV, Dayananda HS. (2013) Environmental impact due to agricultural runoff containing heavy metals—A review. *International Journal of Scientific and Research Publications* 3(5):1–6

Hazarika BN, Ansari S. (2007) Biofertilizers in fruit crops—A review. *Agricultural Reviews* 28(1):69–74

Hua L, Wu WX, Liu YX, McBride M, Chen YX. (2009) Reduction of nitrogen loss and Cu and Zn mobility during sludge composting with bamboo charcoal amendment. *Environmental Science and Pollution Research International* 16:1–9

Joseph SD, Camps-Arbestain M, Lin Y, Munroe P, Chia CH, Hook J. (2010) An investigation into the reactions of biochar in soil. *Australian Journal of Soil Research* 48:501–515

Kameyama K, Miyamoto T, Iwata Y, Shiono T. (2016) Influences of feedstock and pyrolysis temperature on the nitrate adsorption of biochar. *Soil Science and Plant Nutrition* 62(2):180–184

Kammann C, Glaser B, Schmidt HP. (2016) Combining biochar and organic amendments. In: Shackley S, Ruysschaert G, Zwart K, Glaser B, editors. *Biochar in European Soils and Agriculture: Science and Practice.* New York: Routledge, pp. 136–164

Kiran YK, Barkat A, Xiao-qiang CUI, Ying F, Feng-shan P, Lin T. (2017) Cow manure and cow manure-derived biochar application as a soil amendment for reducing cadmium availability and accumulation by *Brassica chinensis* L. in acidic red soil. *Journal of Integrative Agriculture* 16(3):725–734

Krull ES, Skjemstad J, Graetz D, Grice K, Dunning W, Cook G. (2003) 13C-depleted charcoal from C4 grasses and the role of occluded carbon in phytoliths. *Organic Geochemistry* 34:1337–1352

Lahori AH, Zhanyu G, Zhang Z, Li R, Mahar A, Awasthi M. (2017) Use of biochar as an amendment for remediation of heavy metal-contaminated soils: Prospects and challenges. *Pedosphere* 27:991–1014

Lehmann J. (2007) Bioenergy in the black. *Frontiers in Ecology and the Environment* 5(7):381–387

Lehmann J, da Silva JP Jr, Steiner C, Nehls T, Zech W, Glaser B. (2003) Nutrient availability and leaching in an archaeological anthrosol and a ferrasol of the Central Amazon basin: Fertilizer, manure, and charcoal amendments. *Plant and Soil* 249:343–357

Lehmann J, Gaunt J, Rondon M. (2006) Biochar sequestration in the terrestrial ecosystem—A review. *Mitigation and Adaptation Strategies for Global Change* 11:403–427

Liang B, Lehmann J, Solomon D, Kinyangi J, Grossman J, O'Neill B. (2006) Black carbon increases cation exchange capacity in soils. *Soil Science Society of America Journal* 70:1719–1730

Liu J, Schulz H, Brandl S, Miehtke H, Huwe B, Glaser B. (2012) Short-term effect of biochar and compost on soil fertility and water status of a Dystric Cambisol in NE Germany under field conditions. *Journal of Plant Nutrition and Soil Science* 175(5):1–10

Major J, Lehmann J, Rondon M, Goodale C. (2010) Fate of soil-applied black carbon: Downward migration, leaching and soil respiration. *Global Change Biology* 16(4):1366–1379

Major J, Rondon M, Molina D, Riha SJ, Lehmann J. (2010a) Maize yield and nutrition during four years after biochar application to a Colombian savanna oxisol. *Plant and Soil* 333:117–128

Major J, Steiner C, Ditommaso A, Falcao NP, Lehmann J. (2005) Weed composition and cover after three years of soil fertility management in the central Brazilian Amazon: Compost, fertilizer, manure and charcoal applications. *Weed Biology and Management* 5:69–76

Matsubara Y, Hasegawa N, Fukui H. (2002) Incidence of Fusarium root rot in asparagus seedlings infected with arbuscular mycorrhizal fungus as affected by several soil amendments. *Journal of the Japanese Society for Horticultural Science* 71:370–374

Mensah AK, Frimpong KA. (2018) Biochar and/or compost applications improve soil properties, growth, and yield of maize grown in acidic rainforest and coastal savannah soils in Ghana. *International Journal of Agronomy* 1–8. http://doi.org/10.1155/2018/6837404

Mikan CJ, Abrams MD. (1995) Altered forest composition and soil properties of historic charcoal hearths in southeastern Pennsylvania. *Canadian Journal of Forest Research* 25:687–696

Moon DH, Park JW, Chang YY, Ok YS, Lee SS, Ahmad M. (2013) Immobilization of lead in contaminated firing range soil using biochar. *Environmental Science and Pollution Research* 20:8464–8471

Nartey OD, Zhao B. (2014) Biochar preparation, characterization, and adsorptive capacity and its effect on bioavailability of contaminants: An overview. *Advances in Materials Science and Engineering* 2014:1–12

Nerome M, Toyota K, Islam TM, Nishimima T, Matsuoka T, Sato K. (2005) Suppression of bacterial wilt of tomato by incorporation of municipal biowaste charcoal into soil. *Soil Microorganisms* 59(1):9–14

Nigussie A, Kissi E, Misganaw M, Ambaw G. (2012) Effect of biochar application on soil properties and nutrient uptake of Lettuces (Lactuca sativa) grown in chromium polluted soils. *American-Eurasian Journal of Agriculture and Environmental Science* 12(3):369–376

Novak JM, Lima I, Gaskin JW, Steiner C, Das KC, Ahmedna M. (2009) Characterization of designer biochar produced at different temperatures and their effects on a loamy sand. *Annals of Environmental Science* 3:195–206

Novotny EH, Maia CMB de F, Carvalho MT de M, Madari BE. (2015) Biochar: Pyrogenic carbon for agricultural use—A critical review. *Revista Brasileira de Ciência do Solo* 39(2):321–344

Ogawa M. (2009) Charcoal use in agriculture in Japan. *Keynote address, 1st Asia Pacific Biochar Conference*, May 17–20, 2009, Gold Coast, Australia

Oguntunde PG, Fosu M, Ajayi AE, van de Giesen N. (2004) Effects of charcoal production on maize yield, chemical properties and texture of soil. *Biology and Fertility of Soils* 39:295–299

Pathak H. (2010) Trend of fertility status of Indian soils. *Current Advances in Agricultural Sciences* 2(1):10–12

Paz-Ferreiro J, Lu H, Fu S, Mendez A, Gasco G. (2013) Use of phytoremediation and biochar to remediate heavy metal polluted soils: A review. *Solid Earth Discussions* 5:2155–2179

Peng X, Ye LL, Wang CH, Bo S. (2011) Temperature and duration dependent rice straw derived biochar: Characteristics and its effects on soil properties of an Ultisol in southern China. *Soil and Tillage Research* 112(2):159–166

Peterson SC, Jackson MA, Kim S, Palmquist DE. (2012) Increasing biochar surface area: Optimization of ball milling parameters. *Powder Technology* 228:115–120

Pietikainen J, Kiikkila O, Fritze H. (2000) Charcoal as a habitat for microbes and its effect on the microbial community of the underlying humus. *Oikos* 89:231–242

Placek A, Grobelak A, Kacprzak M. (2016) Improving the phytoremediation of heavy metals contaminated soil by use of sewage sludge. *International Journal of Phytoremediation* 18(6):605–618

Prashar P, Shah S. (2016) Impact of fertilizers and pesticides on soil microflora in agriculture sustainable agriculture reviews. *Sustainable Agriculture Reviews* 19:331–362

Reddy KR. (2015) Characteristics and applications of biochar for environmental remediation: A review. *Critical Reviews in Environmental Science and Technology* 45:939–969

Rondon M, Lehmann J, Ramírez J, Hurtado M. (2007) Biological nitrogen fixation by common beans (Phaseolus vulgaris L.) increases with bio-char additions. *Biology and Fertility of Soils* 43:699–708

Rondon M, Ramirez A, Hurtado M. (2004) *Charcoal additions to high fertility ditches enhance yields and quality of cash crops in Andean hillsides of Columbia CIAT annual report.* Cali, Colombia

Ronsse F, van Hecke S, Dickinson D, Prins W. (2013) Production and characterization of slow pyrolysis biochar: Influence of feedstock type and pyrolysis conditions. *GCB Bioenergy* 5:104–115

Saarnio S, Heimonen K, Kettunen R. (2013) Biochar addition indirectly affects N2O emissions via soil moisture and plant N uptake. *Soil Biology and Biochemistry* 58:99–106

Saito M. (1990) Charcoal as a micro habitat for VA mycorrhizal fungi, and its practical application. *Agriculture, Ecosystems and Environment* 29:341–344

Saito M, Marumoto T. (2002) Inoculation with arbuscular mycorrhizal fungi: The status quo in Japan and the future prospects. *Plant and Soil* 24(4):273–279

Saxena J, Rana G, Pandey M. (2013) Impact of addition of biochar along with Bacillus sp. on growth and yield of French beans. *Scientia Horticulturae* 162:351–356

Saxena J, Rawat J, Kumar R. (2017) Conversion of biomass waste into biochar and the effect on mung bean crop production. *Clean Soil, Air, Water* 45(7):1501020 (1–9)

Saxena J, Rawat J, Sanwal P. (2016) Enhancement of growth and yield of glycine max plants with inoculation of phosphate solubilizing fungus Aspergillus niger K7 and biochar amendment in soil. *Communications in Soil Science and Plant Analysis* 47(20):2334–2347

Schwartz MW, Hoeksema JD, Gehring CA, Johnson NC, Klironomos JN, Abbott LK. (2006). The promise and the potential consequences of the global transport of mycorrhizal fungal inoculum. *Ecology Letters* 9:501–515

Shenbagavalli S, Mahimairaja S. (2012) Production and characterization of biochar from different biological wastes. *International Journal of Plant, Animal and Environmental Sciences* 2(1):197–201

Silber A, Levkovitch I, Graber ER. (2010) pH-dependent mineral release and surface properties of cornstraw biochar: Agronomic implications. *Environmental Science and Technology* 44:9318–9323

Singh PB, Hatton JB, Singh B, Cowie LA, Kathuria A. (2010) Influence of biochars on nitrous oxide emission and nitrogen leaching from two contrasting soils. *Journal of Environmental Quality* 39:1224–1235

Skjemstad JO, Janik LJ, Taylor JA (1998). Non-living soil organic matter: What do we know about it? *Australian Journal of Experimental Agriculture* 38:667–680

Slapakova B, Jerabkova V, Tejnecky DO. (2018) The biochar effect on soil respiration and nitrification. *Plant, Soil and Environment* 64(3):114–119

Sohi S, Loez-Capel S, Krull E, Bol R. (2009) Biochar's roles in soil and climate change: A review of research needs. *CSIRO Land and Water Science Report* 5(09):17–31

Solaiman ZM, Blackwell P, Abbott LK, Storer P. (2010) Direct and residual effect of biochar application on mycorrhizal root colonisation, growth and nutrition of wheat. *Australian Journal of Soil Research* 48:546–554

Spokas KA, Koskinen WC, Baker JM, Reicosky DC. (2009) Impacts of woodchip biochar additions on greenhouse gas production and sorption/degradation of two herbicides in a Minnesota soil. *Chemosphere* 77:574–581

Spokas KA, Reikosky DC. (2009) Impacts of sixteen different biochars on soil greenhouse gas production. *Annals of Environmental Science* 3:179–193

Stoyle A. (2011) *Biochar production for carbon sequestration* [master's thesis]. Shanghai Jiao Tong University

Sullivan AL, Ball R. (2012) Thermal decomposition and combustion chemistry of cellulosic biomass. *Atmospheric Environment* 47:133–141

Swift RS. (2001). Sequestration of carbon by soil. *Soil Science* 166:858–871

Taghizadeh-Toosi A, Clough TJ, Sherlock RR, Condron LM. (2012) Biochar adsorbed ammonia is bioavailable. *Plant and Soil* 350:57–69

Thalmann M, Santelia D. (2017) Starch as a determinant of plant fitness under abiotic stress. *The New Phytologist* 214(3):943–951

Thies JE, Rillig M. (2009) Characteristics of biochar: Biological properties. In: Lehmann M, Joseph S, editors. *Biochar for Environmental Management Science and Technology*. London: Earthscan, pp. 85–105

Tumuluru JS, Sokhansanj S, Hess JR, Wright CT, Boardman RD. (2011) A review on biomass torrefaction process and product properties for energy applications. *Industrial Biotechnology* 7:384–402

Verheijen F, Jeffery S, Bastos AC, van der Velde M, Diafas I. (2009) *Biochar Application to Soils—A Critical Scientific Review of Effects on Soil Properties, Processes and Functions*. Luxemburg: Office for the Official Publications of the European Communities, p. 149

Warnock DD, Lehmann J, Kuyper TW, Rillig MC. (2007). Mycorrhizal responses to biochar in soil—Concepts and mechanisms. *Plant and Soil* 300:9–20

Woolf D, Amonette J, Street-Perrott F, Lehmann J, Joseph S. (2010) Sustainable biochar to mitigate global climate change. *Nature Communications* 1:1–9

Wu F, Gai Y, Jiao Z, Liu Y, Ma X, An L. (2012) The community structure of microbial in arable soil under different long-term fertilization regimes in the Loess Plateau of China. *African Journal of Microbiology Research* 6:6152–6164

Xie Z, Xu Y, Liu G, Liu Q, Zhu J, Tu C. (2013) Impact of biochar application on nitrogen nutrition of rice, greenhouse-gas emissions and soil organic carbon dynamics in two paddy soils of China. *Plant and Soil* 370(1–2):527–540

Yanai Y, Toyota K, Okazaki M. (2007) Effects of charcoal addition on N2O emissions from soil resulting from rewetting air-dried soil in short-term laboratory experiments. *Soil Science and Plant Nutrition* 53:181–188

Yargholi B, Azarneshan S. (2014) Long-term effects of pesticides and chemical fertilizers usage on some soil properties and accumulation of heavy metals in the soil (case study of Moghan plain's (Iran) irrigation and drainage network). *International Journal of Agriculture and Crop Sciences* 7:518–523

Yu XY, Ying GG, Kookana RS. (2009) Reduced plant uptake of pesticides with biochar additions to soil. *Chemosphere* 76:665–671

Yuan J, Xu R, Zhang H. (2011) The forms of alkalis in the biochar produced from crop residues at different temperatures. *Bioresource Technology* 102:3488–3497

Zhang A, Cui L, Pan G, Li L, Hussain Q, Zhang X. (2010) Effect of biochar amendment on yield and methane and nitrous oxide emissions from a rice paddy from Tai Lake plain, China. *Agriculture, Ecosystems and Environment* 139(4):469–475

Zhang R, Zhang Y, Song L, Song X, Hanninen H, Wu J. (2017) Biochar enhances nut quality of Torreya grandis and soil fertility under simulated nitrogen deposition. *Forest Ecology and Management* 391:321–329

Zhu L, Lei H, Zhang Y, Zhang X, Bu Q, Wei Y. (2018) A review of biochar derived from pyrolysis and its application in biofuel production. *SF Journal of Materials Chemistry Engineering* 1(1):1007

8 Prospects of Biochar Technology for Clean Water Provision

*Piyush Gupta, Namrata Gupta, Neha Rana,
Sapna Salar, and Subhakanta Dash*

CONTENTS

DOI: 10.1201/9781003203438-8

8.1 INTRODUCTION

Soap or detergent manufacturing or agricultural processes release enormous amounts of inorganic and organic pollutants, like surfactants, colors (dyes and pigments), heavy metals, medicines, personal care items, and pesticides into our surface water bodies. The majority of these contaminants exist in recalcitrant forms and show heightened persistency in nature. Uncontrolled emissions of these toxic pollutants are a matter of concern due to their adverse impacts on our ecosystems (Bogusz et al. 2015). Different conventional technologies have been used for the removal of pollutants from wastewaters: coagulation-flocculation, chemical precipitation, membrane filtration, ion exchange, adsorption, electrochemical treatment, reverse osmosis, flotation, and solvent extraction for the removal of pollutants. However, the limitations of these technologies range from inefficiency in removing pollutants at lower concentrations and in completely converting lethal contaminants into biodegradable and less toxic by-products; to greater consumption of energy and chemicals, higher installation, operation, and maintenance costs; to process complexity; etc. (Cheng et al. 2016; Ejraei et al. 2019; Fu and Wang 2011).

An effective and feasible treatment process must be compatible in terms of the economical and environmental aspects for commercialization and field-scale applications. The incorporation of inexpensive and readily accessible materials into various treatment methods has the potential to lower the overall treatment cost along with an increase in efficiency of the process. It is becoming increasingly popular to use biochar, which is a low-cost and environmentally friendly material that is usually prepared from organic wastes, for instance, forest residues, agricultural wastes, and municipal wastes, in a variety of environmental applications. Organic wastes are generally transformed into char using a variety of techniques (Meyer et al. 2011), comprising pyrolysis, microwave-assisted pyrolysis, gasification, hydrothermal carbonization, and torrefaction. Pyrolysis is the most common carbonization process for preparing biochar, whereas chars produced through gasification, torrefaction, and HTC do not generally meet the definition of biochar specified in the guidelines for the European Biochar Certificate (EBC). Biochar and its activated derivatives have been reported to be extremely effective materials for the removal of a wide range of contaminants (Gai et al. 2014; Mohan et al. 2014), including pathogenic organisms, inorganic contaminants such as heavy metals (Yang et al. 2019), and organic contaminants such as dyes (Park et al. 2019) due to enhanced properties such as a high carbon content, increased surface area, a high cation/anion exchange capacity, and a stable structure. Physicochemical characteristics of biochar, such as their elemental composition, surface area, distribution of pore size, surface functional groups, and cation–anion exchange capacity, all play a role in their ability to remove organic and inorganic pollutants from wastewater. These physicochemical characteristics vary depending on the feedstock and the preparation method used to prepare biochar. In general, the most widely used biochar modification methods can be divided into two categories: chemical modification methods (which primarily consist of acidification, alkalinization, and oxidizing agent modification), and physical modification methods (which are generally performed by steam activation and gas purging) (Rizwan et al. 2016).

Despite the fact that biochar has a broad range of potential applications in wastewater treatment, the negative impacts of biochar application should be taken into consideration. Biochar may contain a variety of heavy metals along with other contaminants, depending on feedstock type and the conversion technique employed in its production. As a result, more research is needed to determine the stability of biochar and the relationship between it and the experimental conditions used during biochar production (Kim et al. 2015). In this chapter, recent studies on the preparation, modification, characterization, regeneration, and application of biochar from pyrolysis as well as chars from other thermal conversion processes for the treatment of various types of wastewaters are discussed.

8.2 PRODUCTION METHODS OF BIOCHAR

Biochar synthesis has been studied using a variety of methods and factors, including heating temperature and rate, as well as the residence time of the biochar. The characteristics of biochar may vary depending on the method and parameters used in its production. In response to the growing interest in employing biochar in a variety of applications, there has been a rise in biomass conversion into biochar. For biochar manufacturing, thermochemical conversion processes such as pyrolysis, hydrothermal carbonization, torrefaction, gasification, etc. are used (Pang 2019; Lin et al. 2016). In order to get higher rates of biochar output, the preparation technique and process parameters, such as heating rate, temperature, residence time, and so on, should be appropriate and optimized for the biomass type being used. These variables are critical because they have the potential to influence the physical and chemical phases of biochar during the manufacturing process. In the beginning, weight loss occurred as a result of water loss around 100°C, which was followed by cellulose, hemicellulose, and lignin decomposition that occurred around 220°C. Finally, weight loss happens as a result of the combustion of carbonaceous wastes in the atmosphere. Process conditions, including thermochemical conversion pathways, are presented in Table 8.1.

8.2.1 PYROLYSIS

Pyrolysis is the term used to describe the thermal breakdown of organic compounds in an oxygen-free atmosphere at temperatures ranging from 250 to 900 °C (Osayi et al. 2014). Waste biomass may be converted into value-added products such as biochar, syngas, and bio-oil using this technique, which is an alternative strategy to traditional methods. At particular temperatures, various cellulosic components such as cellulose, hemicellulose, and lignin undergo depolymerization, fragmentation, and cross-linking, resulting in a variety of product states such as solid, liquid, and gas. The solid and liquid products are char and bio-oil, whereas the gaseous products, e.g., carbon dioxide, carbon monoxide, and hydrogen, are the components of syngas. How much biochar is produced is dependent on the kind and nature of biomass utilized in the pyrolysis process. Among the operational process conditions that influence product efficiency, temperature is the most important (Wei et al. 2019). It is generally accepted that increasing the temperature of the

TABLE 8.1
Process Conditions of Thermochemical Conversion Methods

Technique	Temperature (°C)	Residence time	Biochar Yield (%)	Bio-oil yield (%)	Syngas production (%)	References
Pyrolysis	300–700 (slow)	< 2 s (slow)	35 (slow)	30 (slow)	35 (slow)	Cantrell et al. (2012)
	500–1000 (fast)	Hour-day (fast)	12 (fast)	75 (fast)	13 (fast)	
Hydrothermal carbonization	180–300	1–16 h	50–80	5–20	2–5	Funke and Ziegler (2010)
Gasification	750–900	10–20 s	10	5	85	Klinghoffer et al. (2015)
Torrefaction	290	10–60 min	80	0	20	(Bergman et al. 2005)
Flash carbonization	300–600	< 30 min	37	–	–	(Nunoura et al. 2006)

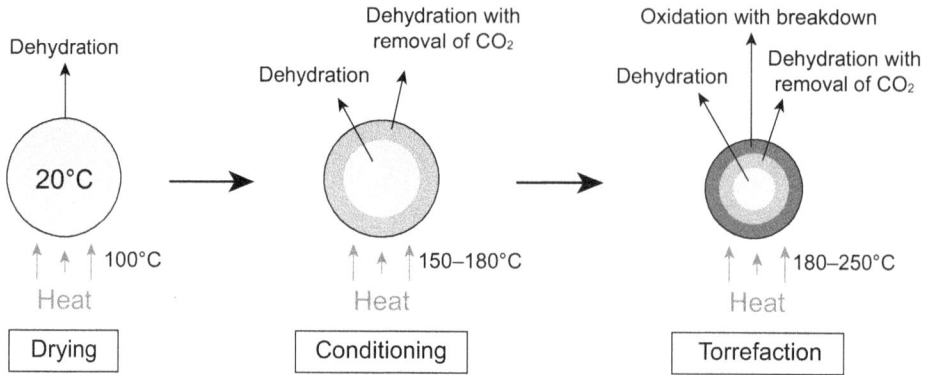

FIGURE 8.1 Stages involved in biomass conversion into biochar.

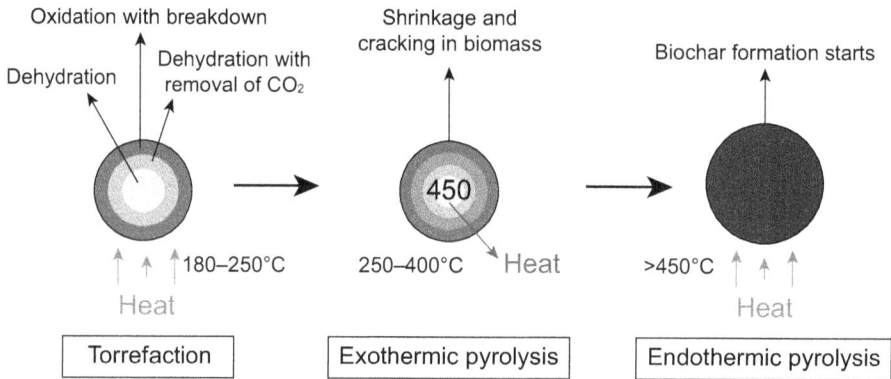

FIGURE 8.2 Stages involved in biomass conversion into biochar.

pyrolysis process reduces the output of biochar while simultaneously increasing the generation of syngas.

Pyrolysis may be divided into two categories: fast pyrolysis and slow pyrolysis, which are determined by the heating rate, temperature, residence time, and pressure.

Slow pyrolysis is carried out in an oxygen-starved environment in kilns or retorts. In slow pyrolysis, the rate of heating is very low, around 5–7°C/min and possesses a longer residence time of more than 1 h (Liu et al. 2015b). The slow pyrolysis method has a better yield of char contrasted with other pyrolysis and carbonization strategies. The biochar can be utilized as a dirt enhancer to improve soil quality (Al Arni 2018).

Fast pyrolysis is a direct thermochemical procedure that can liquefy solid biomass into liquid bio-oil with a high potential for energy application and that converts finely ground feedstock into bio-oils, gas and char, in seconds. Fast pyrolysis tends to be used by commercial biochar/bio-oil producers (Wang et al. 2014).

In the progression of pyrolysis, the several stages are shown in Figures 8.1 and 8.2.

Stage 1: Drying and Conditioning

*   The majority of biomass is composed of five major constituents: cellulose, hemicellulose, lignin, water, and minerals (ash), in varying quantities depending on the source of the biomass material.

- The water adsorbed on the cellulose/lignin structure of "seasoned" wood ranges from 12 to 19% by volume. Freshly cut wood or agricultural wastes can have a water content of 40–60% by weight when they are first harvested (wet basis, i.e., expressed as a percentage of the wet weight of the biomass).
- As the biomass is heated over 100°C, the majority of the water is eliminated.
- When the temperature reaches 150°C, the biomass begins to decompose.
- When the biomass reaches a temperature of around 150°C, it begins to decompose and soften (referred to as conditioning). In addition to tiny quantities of carbon dioxide and volatile organic chemicals, chemically bonded water (derived from the structure of the molecules in the biomass) is emitted by the biomass.

Stage 2: Torrefaction

- A thermochemical process that is applied to an increase in the biomass's temperature until the temperature range of 200–280°C is reached results in breaking of the chemical bonds within the constituents of the biomass.
- The temperature of the dry biomass needs to be increased in order to break the molecular bonds and induce an endothermic reaction.
- CO_2 and CO are emitted, as well as volatile organic compounds and CO_2, from the breakdown of hemicellulose and cellulose during this stage.
- Biomass that has been through the torrefaction process is far more brittle than fresh biomass, and grinding (e.g., for boiler fuel) is therefore much easier. Storability is increased because of resistance to biological decomposition and water absorption.
- Smoke water and wood vinegar are terms for the vapors that condense out of low-temperature pyrolysis. The term "pyroligneous acid" refers to "liquid smoke." The degree of concentration and temperature of production influence the way it is utilized. For example, it can be used as a fungicide, plant growth stimulant, an accelerant for seed germination, or a boost for composting and biochar efficacy.

Stage 3: Exothermic Pyrolysis

- At 250–300°C, depending on the nature of the feedstock, thermal breakdown of the biomass becomes more severe, releasing a flammable combination of H_2, CO, CH_4, CO_2, other hydrocarbons, and tars, as well as other combustible gases.
- Pyrolysis becomes exothermic as a result of the breakdown of large polymers in the biomass, which releases energy. Some of the oxygen trapped within the biomass structure is released and enters into energy-releasing oxidation reactions with the gases and char. The energy released causes the heat necessary to break down more chemical bonds in the biomass to be produced. After reaching a temperature of around 400°C, the process becomes self-sustaining and may continue on its own, leaving an oxygen-depleted, carbon-enriched, charcoal-like residue.
- In practice, heat is lost from the pyrolysis area, necessitating the use of external heat sources to raise and then maintain the temperature throughout pyrolysis.
- The maximum yield is reached prior to the completion of exothermic pyrolysis, although the stable carbon content is rather low. The ash component of a wood biochar is typically 1.5–5 wt%, the volatiles 25–35 wt%, and the remaining 60–70 wt% is fixed carbon.

Stage 4: Endothermic Pyrolysis

- The biochar that remains after exothermic pyrolysis contains a significant quantity of volatile chemicals.
- Additional heating is necessary to enhance the fixed carbon content, surface area, and porosity of the material by driving out and decomposing a greater proportion of the volatile compounds.

- At 550–600ºC, wood biochar typically has a fixed carbon content of approximately 80–85% and a volatile carbon content of about 12%, with a yield of about 25–30 wt% of the oven-dry feedstock.

Stage 5: Activation and Gasification

- Once the temperature has risen over 600 °C, introducing a small quantity of air and steam can boost the surface temperature of the biochar to 700–800ºC, triggering two processes:
  - *Activation*: Air, steam, and heat may all activate the surface of biochar, causing it to release additional volatiles. It is possible to improve the surface area of the biochar as well as the cation exchange capacity of biochar by adding acidic functional groups. The yield is decreased.
  - *Gasification*: Gasification is a process that occurs when a large amount of air and/or steam is introduced. This can result in the production of a gas that is reasonably pure and can be used to generate power. The production of biochar is poor (typically less than 20%), and the ash concentration is considerable.

8.2.2 HYDROTHERMAL CARBONIZATION

Biochar manufacturing using hydrothermal carbonization is believed to be an affordable and efficient option, as the process may be carried out at low temperatures ranging from 180 to 250°C (Lee et al. 2018). It is referred to as a hydrochar because the end product produced of the hydrothermal process is distinctly different from dry procedures such as pyrolysis and gasification (Fang et al. 2018). Once blended, the wet biomass is placed in a closed reactor where it undergoes an anaerobic process. Stability is increased by allowing the temperature to gently climb. According to their respective temperatures, the following items are produced: biochar when heated to a temperature between 250–400°C is referred to as hydrothermal carbonization (Zhang et al. 2017b), while syngas is generated when the temperature is greater than 400°C, such as CO, CO_2, H_2, and CH_4 (Khorram et al. 2016). Hydrolysis is performed first, which separates and partitions off the starting compound. Hydrolysis also produces intermediate products such as 5-hydroxymethylfurfural and its derivatives. The hydrochar is produced through condensation, polymerization, and intramolecular dehydration, which follow the reaction path that starts with condensation, polymerization, and intramolecular dehydration. Lignin has a high molecular weight and highly complex structure, therefore the mechanism is more difficult to understand. The dealkylation and hydrolysis reactions to form phenolic compounds, including phenols, catechols, syringols, etc., lead to the lignin breakdown (Jain et al. 2016). Finally, the final polymer is created by the cross-linking and repolymerization of intermediates. The dissolved lignin and unconverted lignin are turned into hydrochar by pyrolysis.

8.2.3 GASIFICATION

The technique of decomposition is heating the carbonaceous material to produce syngas (comprising carbon monoxide, carbon dioxide, methane, hydrogen, and hydrocarbons, which are present in trace amounts) and using gasification agents such as oxygen, air, steam, etc. The temperature is the most essential component in determining the production of syngas. At high temperatures, carbon monoxide (CO) levels increase, but the amount of hydrogen produced is offset by an increase in the generation of methane, carbon dioxide, and hydrocarbons (Prabakar et al. 2018). Syngas is the primary product, with the char regarded as a by-product with lower yield. The stages involved in gasification mechanism are discussed next.

8.2.3.1 Drying

Without energy recovery, the moisture content of the biomass evaporates. Depending on the biomass used, the moisture content fluctuates. Drying is an independent procedure performed when the biomass contains a high moisture content during the gasification process.

8.2.3.2 Combustion

The main energy sources for a gasification process are the oxidation and combustion reactions of the gasification agents. The combustible species contained in the gasifier react with the gasification agents to produce CO_2, CO, and water.

8.2.4 TORREFACTION AND FLASH CARBONIZATION

The biochar generation process that's currently the most widely used is torrefaction. The method is described as "light pyrolysis" due to its modest heating rate. A variety of decomposition mechanisms are used to strip biomass of its oxygen, moisture, and carbon dioxide when it is exposed to inert, pressurized air with no oxygen at a temperature of 300°C. By torrefaction, the biomass attributes, such as particle size, moisture content, surface area, heating rate, energy density, etc., are modified. This torrefaction procedure requires the biomass to be processed under steam in temperatures no higher than 260°C with a residence time of 10 min.

When biomass is put under high pressure, a flash fire is set off that converts the biomass into various solid- and gas-phase products. This whole procedure is carried out at temperatures ranging from 300 to 600 °C with reaction times ranging from 30 min to 2 h. Biomass in its original form comprises around 40% of biomass, which is then used as solid matter, while the remaining 60% is reduced to molecules by the pressure-driven process. However, for literature, the procedure of flash carbonization is only utilized on a very restricted scale.

Incomplete pyrolysis is the process mechanism of torrefaction, and it progresses as follows: the reaction requirements are in a temperature range of 200–300°C, a residence duration of less than 30 min, and a heating rate that is less than 50°C per minute. Dry torrefaction is divided into multiple parts, with the first ones being the heating, drying, torrefaction, and cooling stages. Another way to think of drying is to classify it as predrying and postdrying.

8.2.4.1 Heating

The biomass is heated to the desired drying temperature, and moisture is released from the biomass until the desired drying temperature is maintained.

8.2.4.2 Preheating

In this method, the biomass must be heated to 100°C, and then the moisture inside is removed.

8.2.4.3 Postdrying

The water content of the solution is totally evaporated at a temperature of 200°C. Temperature increases cause mass loss.

8.2.4.4 Torrefaction

In order to maintain a consistent temperature during the process, the ingredients are heated to 200°C.

8.2.4.5 Cooling

In this stage, the product is permitted to cool down to a certain temperature.

Biomass Preprocessing: Increases Biochar Properties

The preprocessing of the biomass can affect the pyrolysis rate and the characteristics of the biochar. The techniques might include:

- Treating the biomass with phosphoric acid prior to processing, improving functional groups, decreasing pH, and making a phosphate fertilizer with a long release time

- Using alkali, such potassium hydroxide, to soften the biomass, resulting in biomass decomposition
- To create a magnetic biochar, treating the biomass with Fe salts prior to the pyrolysis process (e.g., to remove heavy metals from water)
- Reducing the pace of pyrolysis, boosting uptake of N, and raising nutrient-rich nanoparticle concentration on the surface using clay, ferrous sulfate, or other minerals such as rock phosphate mixed with the biomass
- Using lump biomass, or briquettes of briquette-low-density biomass, to make handling easier and to enhance charcoal production

Biomass Postprocessing: Increases Biochar Properties

The efficacy of biochar can be changed by postprocessing the biochar. Techniques that can be used include:

- Treating with phosphoric acid to enhance functional groups, lowering pH, and producing a slow-release phosphate fertilizer
- Treating the biomass with an alkali (e.g., potassium hydroxide) to raise the pH and to increase the potassium content; infusing the biomass with organic or inorganic nitrogen compounds (e.g., urine) to raise the nitrogen content; combining the biomass with a nutrient-rich organic material (e.g., manure) (The combination may require heating, sterilization, and drying in order to remove biohazards and make handling easier.)
- Adding minerals, such as rock phosphate, gypsum, dolomite, iron oxides, and lime to meet particular soil limitations
- Adding chemicals, such as urea and diammonium phosphate, to create a complex fertilizer
- Granulating or pelletizing to make it easier to handle and apply biochar
- Adding steam or oxygen to help it decompose faster

8.3 MODIFICATION OF BIOCHAR

Biochar is frequently modified in order to enhance its remediation capabilities. Chemical and physical modification are the two types of modifications that can be used to change the characteristics of biochar. Metal oxides modification, alkaline modification, oxidizing agent modification, acidic modification, and carbonaceous materials modification are all examples of chemical alteration in use today. Physical alteration procedures such as steam and gas purging are further examples. Figure 8.3 portrays the different approaches for the production and modification of biochar (Fdez-Sanroman et al. 2020).

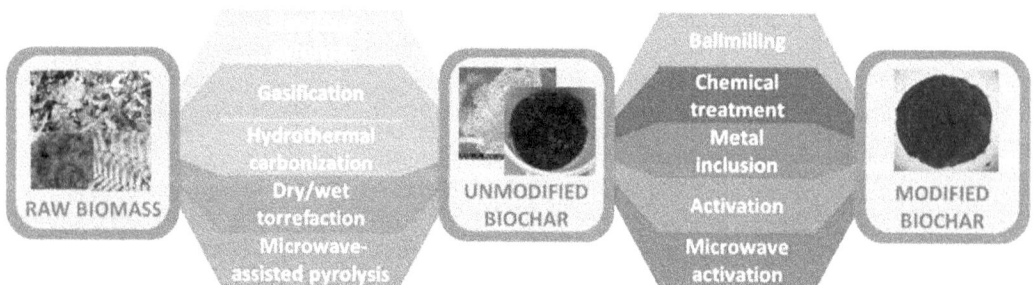

FIGURE 8.3   Production and modification methods for biochar.

Source: Fdez-Sanroman et al. (2020).

Chemical modification is the process of oxidizing the surface of biochar to enhance the oxygen-containing functional groups such as -OH, -COOH, and so on, increasing its hydrophilicity. At the same time, the biochar's pore size and structure is altered, and its adsorption capacity for polar adsorbates is improved. Acidic modification is a method in which metal contaminants are removed from the surface of biochar and are replaced with acid functional groups. Hydrochloric acid, sulfuric acid, nitric acid, phosphoric acid, and citric acid are some of the acids that have been utilized in this procedure. Peng et al. (2016) treated reed-derived biochar with 1 M hydrochloric acid resulting in a modified product. Li et al. (2014) modified bamboo-based biochar with nitric acid. Wang and Wang (2019) increased the amount of oxygen-containing functional groups on biochar through the use of oxidizing agents. Typically, hydrogen peroxide is employed in this process. The employment of oxidizing agents for modification results in the introduction of carboxyl groups (Tan et al. 2015). Jing et al. (2014) used methanol to transform a municipal solid-waste-derived biochar, revealing esterification between the carbonyl groups and the biochar, resulting in a significant increase in tetracycline adsorption capacity.

Alkaline modification is accomplished by using potassium hydroxide and sodium hydroxide (Wang and Wang 2019). Alkali activation of biochar employing potassium hydroxide (KOH) and sodium hydroxide (NaOH) can increase O content and surface basicity while dissolving ash and condensed organic matter (e.g., lignin and celluloses) to facilitate subsequent activation. Increased surface area and oxygen-containing functional groups in biochar are achieved through alkaline alteration. Jing et al. (2014) treated municipal-solid-waste-derived biochar with potassium hydroxide to get a modified biochar with better characteristics. Fan et al. (2010) used sodium hydroxide to modify bamboo-derived biochar to enhance its performance.

Biochars have a greater surface area, negative surface charge, and a high pH. Due to specific adsorption on oxygenated functional groups, electrostatic attraction to aromatic groups, and precipitation on the mineral ash components of biochar, biochars are considered effective sorbents for metal cations but poor sorbents for oxyanion contaminants. Biochar-based composites are capable of removing negatively charged oxyanions from aqueous solutions by utilizing the greater surface area of biochars as a platform to embed a metal oxide with contrasting chemical properties (and usually a positive charge). The goal of most methods to make metal-oxide-biochar-based composites is to guarantee that the metal is uniformly spread across the surface of the biochar. Biochar is essentially a porous carbon scaffold upon which metal oxides precipitate to increase the metal oxide's surface area. Adsorption, catalysis, and magnetic characteristics of biochar are improved by the addition of metal oxides. This technique of modification is most commonly used because targeted pollutants are rapidly adsorbed, biochar may be recycled, and the catalytic properties of biochar are enhanced. When carbonaceous elements are included into the biochar, the surface area of the biochar increases. Adsorption of heavy metals and organic contaminants by carbon nanotubes and graphene has been demonstrated, but they are, however, limited in their applicability due to the high expense of material preparation (Wang and Wang 2019).

Different researchers proposed a number of modification methods to get a carbon-rich version of biochar with enhanced properties, e.g., $KMnO_4$-modified hickory-wood-derived biochar (Wang et al. 2015), metal-oxide-modified biochar (Yan et al. 2015), nanoscale zero-valent-iron-modified rice-hull-derived biochar (Jung et al. 2016), methanol-modified biochar obtained from municipal solid waste (Jing et al. 2014), polydopamine (PDA), and nanoscale ZVI-modified biochar (Wang et al. 2018).

Steam activation is a typical approach for improving the structural porosity of biochar and removing contaminants such as incomplete combustion products. The purpose of this treatment is to enhance the surface area available for sorption. Lima and Marshall (2005) used pyrolysis at 700°C, followed by steam activation at 800°C with a range of water flow rates and durations to make activated biochars from poultry manure feedstocks. Although steam activation enhances the surface area and porosity of biochar, some researchers discovered that activation with steam at 800°C had no effect on the Cu^{2+} sorption capacity of biochar generated by slow pyrolysis of *Miscanthus* at 500°C. They discovered that whereas steam activation increased the surface area of the biochar, the number of functional groups was reduced, while aromaticity increased. Similarly, Lou et al. (2016) found that steam activation of pine sawdust biochar increased surface area but had little effect on surface functional group attributes. As a result, steam activation appears to be more successful in the context of inorganic pollutant sorption when used before a second activation/modification phase that generates functional groups, as the steam just increases the surface area of biochar.

8.4 CHARACTERIZATION OF BIOCHAR

The chemical composition and shape of biochar following pyrolysis are determined by characterization. Characterization techniques used include scanning electron microscopy (SEM), transmission electron microscopy (TEM), X-ray diffraction (XRD), X-ray photoelectron spectroscopy (XPS), Fourier transform infrared spectrometry (FTIR), energy dispersive X-ray spectroscopy (EDS), scanning transmission electron microscopy (STEM), and extended X-ray absorption fine structure (EXAFS). The pore volume, surface area, and diameter of biochar can be determined using the nitrogen adsorption isotherm and the Brunauer–Emmett–Teller (BET) adsorption model under various nitrogen partial pressures (Kumi et al. 2020). A technique known as XRD is used to identify carbon crystallites by measuring the angle and intensity of diffracted biochar beams. Carbon crystallites can be either graphitized or nongraphitized in nature. Sharp and narrow peaks are indicative of graphitized carbon, whereas broad peaks are indicative of nongraphitized carbon (Yoo et al. 2018). This function uses waves in the frequency domain of light intensity to identify functional groupings in the biochar using Fourier transforms. A functional group can be identified by the different vibration positions found in infrared spectroscopy (Wang et al. 2017). By studying the distribution of photoelectrons in relation to their energy, XPS may also be used to detect the valence state of chemical elements. Because of the interaction between the electron beam and the biochar, SEM exposes the surface morphologies of the substance. In the same way as SEM, TEM analyses sample surface morphologies but with a better level of resolution, and they can be used to measure the biochar's lattice constant (K). SEM and TEM operate on the same concept as STEM but with a better resolution. STEM enables atomic-scale characterization of biochar. EDS is used to determine the X-ray photon characteristic energy of each element to estimate the molar proportion of elements on the surface of biochar. Biochar's chemical composition can be analyzed using electron microscopy and energy dispersive spectroscopy (EDS). EXAFS may examine the polyaromatic structure of biochar via a multiple scattering resonance mechanism (Kumi et al. 2020). The pH of biochar can be determined using pH probe by mixing biochar with distilled water, whereas elements like carbon and nitrogen are measured using element analyzer.

8.5 TREATMENT OF WASTEWATERS USING BIOCHAR

The high specific surface area and availability of surface functional groups in biochar are just two of the many unique characteristics of this material that have already been discussed. Because of these characteristics, biochar is a highly effective adsorbent, eliminate a wide range of contaminants. Thus, the effort to improve environmental quality, the use of biochar to eliminate pollutants from industrial and agricultural sectors has been significantly important in recent years (Wang et al. 2017). Wastewater, which is generated as a result of residential, industrial, commercial, and agricultural operations, has been a worrying point around the world. Especially when it comes to wastewater treatment, biochars hold a lot of potential. It is the primary focus of this section to investigate the application of biochar in the remediation of industrial wastewater, municipal wastewater, agricultural wastewater, and stormwater (Xiang et al. 2020).

8.5.1 Agricultural Wastewater

A growing amount of pesticides and hazardous heavy metals are being released into farmlands as the agricultural business develops at a quick pace. As a result of this rapid expansion, agricultural contamination is becoming dangerous (Wei et al. 2018). The use of biochar and its improved forms in the treatment of agricultural wastewater contamination has been studied extensively by researchers. Pesticides such as pentachlorophenol and atrazine are two of the most widely used in agricultural production worldwide. Mandal and Singh (2017) studied the adsorption of imidacloprid & atrazine from agriculture wastewater using biochar of rice straw and H_3PO_4 modified rice straw biochars. Atrazine removal capabilities of corn & soybean straw biochars are both high, in which pore volume and pH values play a significant role (Liu et al. 2015a; Zhao et al. 2013). Sulfamethazine may be successfully removed from water using steam-activated biochar, and the rate of removal is dependent on the pH value (Rajapaksha et al. 2015). In order to remove pentachlorophenol (PCP) from the effluent, zero-valent iron, magnetic paper mill sludge biochar (ZVI-MBC) was used (Devi and Saroha 2014). With the ZVI-MBC, PCP in the effluent can be adsorbed and dechlorinated concurrently, resulting in full PCP removal from the effluent. Biochar has also been researched for its ability to remove glyphosate, diuron, and carbaryl from agricultural effluent. Several factors influence the adsorption capacity of biochar for pesticides, including the feedstock used, the functional materials used, and the pollutants targeted (Wei et al. 2018). Agricultural wastewater contains harmful heavy metals, which are a major source of concern. Metals such as arsenic, cadmium, copper, and lead are some of the most commonly encountered hazardous metals. Biochar has been shown to increase the adsorption capacity of Cu^{2+} and As^{5+} in agricultural wastewater by up to 69.4 mg/g and 34.1 mg/g, respectively; the adsorption capacity of Cd^{2+} and Pb^{2+} by up to 0.4 mg/g to 12.3 mg/g and 36 mg/g to 35 mg/g, respectively; and the adsorption capacity of Cu^{2+} and As^{5+} in agricultural wastewater by up to 34.1 mg/g (Higashikawa et al. 2016; Son et al. 2018; Zhou et al. 2017). Heavy metals in agricultural wastewater can be removed using adsorbents via a number of different mechanisms, including electrostatic interactions, surface complexation, ion exchange, intermolecular contact, cation–pi bonding, and π–π interactions. There are significant differences in the adsorption behavior of biochars for different agricultural pollutants (Wei et al. 2018). For the most part, nanomaterial content, surface tension, surface free energy, and porous structure all have strong relationships with adsorption capabilities (Son et al. 2018; Wan et al. 2018). The adsorption mechanism of biochars is also influenced by other parameters such as inner-sphere complexes, π–π interaction, hydrophobic effect, precipitation, ion exchange, and a variety of other factors (Lefevre et al. 2018; Yao et al. 2018).

8.5.2 STORMWATER

Cities face concerns about stormwater runoff and its effect on water quality owing to the urbanization process. The increase in the concentration of organic matter, biological contaminants, and metals and the resulting elevated load on natural water quality due to stormwater runoff means that the water must be treated before it is released (Ashoori et al. 2019; Mohanty et al. 2014; Ulrich et al. 2017).

In the stormwater treatment realm, stormwater purification systems like bioretention and biofiltration are popular, but neither method is good at clearing out the pollutants that frequently come with rainwater (Ulrich et al. 2017). Biochar and its many kinds of modification, as an efficient medium, have been used to treat stormwater systems. The study has revealed that adding aluminum to biochar (charred organic material) has the capability to clean runoff water contaminated with several toxic heavy metals, including As^{5+}, Pb^{2+}, Zn^{2+}, Cu^{2+}, and PO_4^{3-} (Liu et al. 2019). A new technique of using biochar to extract metals from stormwater runoff has proven highly effective, with the amount of copper and zinc removed exceeding 85% and 95%, respectively. However, the filter medium must be carefully scrutinized and developed to satisfy stormwater treatment needs (Gray 2016).

BPA removal is being improved with the inclusion of biochar in biofilters. Wood dust biochar demonstrates a high adsorption capability with regard to BPA and is capable of improving *Phragmites australis* growth by boosting SSA and pore volume. Additionally, the increased SSA and pore volume help in BPA adsorption, as well as in promoting improved phosphorus and TOC, TSS, nitrogen, and *E. coli* removal rates (Ashoori et al. 2019). Biochar has resulted in increased pollutant removal in stormwater biofilters, with the most dramatic improvement seen in the removal of toxic trace organic contaminants (TOrCs), which had been inadequately eliminated in traditional systems. When it comes to TOrC removal, using the new biofilter columns with biochar makes the filters up to 99% more effective than if they were left without amendment. At the same time, it's possible for a biofilter that is altered with biochar to get rid of over 60% of TOC, TN, and TP (Ulrich et al. 2017).

8.5.3 INDUSTRIAL WASTEWATER

Industrial wastewater originates from a variety of origins, including mines, smelters, battery manufacturers, the chemical industry, leather manufacturers, dye manufacturers, and other processes and industries. In addition, heavy metals and organic contaminants are the most dominant pollutants in industrial wastewater. Biochars have been used in the treatment of industrial effluent, among other things. The entry of industrial contaminants such as poisonous elements, chemical substances, organic waste, and heavy metals into natural ecosystems has a significant impact on environmental contamination (both on land and in water bodies). These contaminants have a variety of adverse effects on human bodies and living organisms.

It is undesirable to discharge industrial wastewater effluents into the environment because the breakdown of industrial wastewater effluents, which include dyes, heavy metals, pesticides, herbicides, and other pollutants, has the potential to cause toxicity, carcinogenicity, teratogenicity, and mutation in humans. Therefore, the discharge of industrial wastewater effluents into the environment is prohibited. The discharge of such effluents into the environment also constitutes a persistent threat to groundwater and water bodies because they limit the reoxygenation capacity of aquatic plants as well as the ability of aquatic species to execute biological activities in their natural environments. The huge quantities of industrial effluents produced, on the other hand, might be mitigated through the use of sludge in the treatment process. Beyond having a negative impact on human health, excess sludge generated by wastewater treatment facilities can also have a negative impact on water bodies, aquatic vegetation, and the condition of the soil. Wastewater treatment plant biological sludge, which is commonly disposed of as soil amendment or by open drying, contains hazardous compounds and heavy metals that readily ionize in soil, and so plants' roots absorb these toxic materials and metals, which subsequently bioaccumulate in the plant's tissues. This sludge can, however, be managed properly in order to reap economic and environmental benefits by producing biochar through oxygen-limited pyrolysis, which can then be utilized to remediate industrial wastewater by adsorption.

Due to the environmental consequences of a wide range of industrial effluents (Xiang et al. 2020), as well as increased public concern about environmental issues, researchers have been searching for novel ways of treating these effluents and developing new materials that are capable of alleviating these environmental problems. Biochar adsorption is a cost-effective and environmentally friendly alternative to existing wastewater treatment methods. It is also simple to apply and efficient. Many liquids and gaseous molecules could benefit from the adsorptive properties of biochar, which could be used to purify and detoxify them, as well as to filter the concentration of a number of these substances. It is incredibly significant in a variety of industries, including apparel, food processing, chemical manufacturing, and extraction of oil and natural gas. Biochar is also used in mining and pharmaceutical manufacturing, among other fields.

It is possible to cast membranes, beads, and solutions out of a biochar mixture that has been crosslinked with chitosan. It has the potential to be used efficiently as an adsorbent for the adsorption of heavy metals in industrial wastewater. The amount of biochar and chitosan used in the adsorption of copper, lead, arsenic, cadmium, and other heavy metals in industrial effluent depends on the ratio of biochar to chitosan (Hussain et al. 2017). A biochar's pH value, surface area, and pore volume are all important factors to consider throughout the CV sorption process (Wathukarage et al. 2019). Biochar made from bagasse was used to absorb lead from the effluent of the battery manufacturing sector. Adsorption capacity can reach 12.7 mg/g, and the adsorptive process is dependent on the medium's pH value, contact time, and dose (Poonam et al. 2018). More study and implementation in real-world conditions are needed, as the majority of investigations on the use of biochar to remove contaminants from industrial wastewater have been conducted only in a laboratory setup

8.5.4 MUNICIPAL WASTEWATER

The use of biochar in municipal wastewater treatment can be advantageous since it can be used alone or in conjunction with other technologies to recover labile nitrogen and phosphorus (Cole et al. 2017). Zheng et al. (2019) used engineered biochar that contained aluminum oxyhydroxides (AlOOH) to recover and reuse phosphorus from secondary treated wastewater. In general, electrostatic attraction is the primary mechanism through which phosphorus is absorbed. It is feasible to employ phosphorus adsorbed on manufactured biochar as a slow-release fertilizer for crop production by incorporating it into the biochar.

Using biochar made from digested sludge, ammonium was removed from municipal wastewater using an adsorbent system. As a result of its greater surface area and functional group density, biochar generated at 450°C has the maximum ammonium removal capability, and the process is controlled by chemisorption (Tang et al. 2019). It has been demonstrated that biochar generated from waste sludge can be utilized as catalysts to ozonate refinery effluent and achieve a high removal rate of total organic carbon (TOC). Because biochar contains functional carbon groups, Si/O structures, and metallic oxides, it has the potential to induce oxidation through the generation of hydroxyl radicals and mineralized petroleum pollutants in soils and plants (Chen et al. 2019).

The use of biochar, which is created from municipal waste, can be used to treat municipal wastewater once it has gone through the biofiltration stage. Biochar is an excellent alternative for use as a biofilter in municipal wastewater treatment because of its huge porous surface area. Upon passing through the charcoal biofilter, the COD, TSS, TKN, and total phosphorus (TP) levels in wastewater are reduced by 90%, 89%, 64%, and 78%, respectively (Manyuchi et al. 2018).

8.6 MECHANISMS INVOLVED IN REMOVAL OF CONTAMINANTS

Adsorption takes place when an adsorbate forms a bond with the surface of an adsorbent and remains there until equilibrium is attained. Adsorption is a three-step process that includes the following steps:

1. Adsorption via physical means, where the adsorbate is aggregated on the surface of the adsorbent
2. Precipitation and complexation, which involve the deposition of the adsorbate on the adsorbent's surface
3. Pore filling, which occurs when the adsorbate is condensed in the pore of the adsorbent (Fagbohungbe et al. 2017)

The process has three phases/stages. The first stage is characterized by the absence of adsorption and is referred to as the clean zone. The second stage is characterized by the presence of adsorption and is referred to as the dirty zone, or the mass transfer zone, where adsorption is taking place. In the third and final phase, equilibrium has been attained; this is referred to as the inactive zone, or exhausted zone. During the process, the fully covered or exhausted zone expands; however, the clean zone shrinks in proportion to the process. It is possible that a rise in the concentration of the adsorbate impacts the mass transfer zone; otherwise, it is not affected. Such a pattern persists until the breaking point is reached. The many mechanisms involved in the remediation of both inorganic and organic pollutants as shown in Figure 8.4 (Enaime et al. 2020) are discussed next.

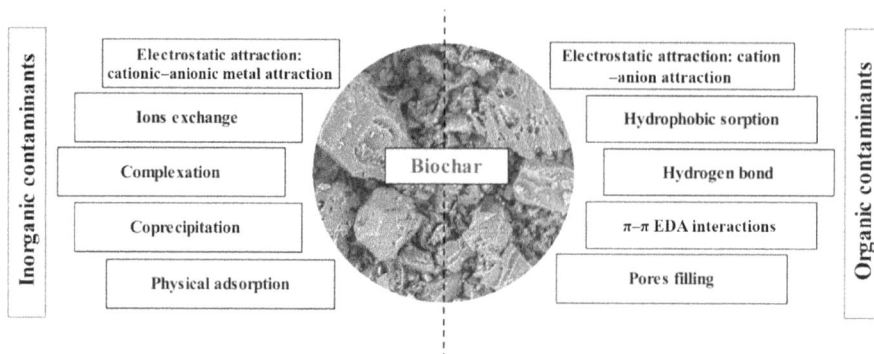

FIGURE 8.4   Mechanisms involved in adsorption of organic and inorganic pollutants on biochar
Source: Enaime et al. (2020).

8.6.1   ADSORPTION MECHANISM FOR INORGANIC POLLUTANTS

Surface sorption, electrostatic interaction, ion exchange, precipitation, and complexation are some of the mechanisms involved in the removal of harmful metals from the environment.

8.6.1.1   Surface Sorption

During the metal ion diffusion in the tiny pores of the sorbent a physical process takes place that involves the formation of chemical bonds. The carbonization temperature of the sorbent (biochar) has an effect on the volume of the pores and the surface area of the sorbent (biochar). Kumar et al. (2017) conducted an investigation into the adsorption of uranium onto pinewood biochar at temperatures ranging from 300 to 700°C. Their findings demonstrated that biochar prepared at a high temperature can completely remove uranium, whereas biochar prepared at a low temperature could only half remove it. That is because high carbonization increases the surface area of the biochar as well as the volume of pores in the material, resulting in increased efficacy. Another study, conducted by Wang et al. (2015), studied the adsorption of dangerous metals on biochar produced from hickory wood and treated with $KMnO_4$, which was published in *Nature Communications*. The biochar that they created had a Pb adsorption capacity of 153.1 mg/g, a Cu adsorption capacity of 34.2 mg/g, and a Cd adsorption capacity of 28.1 mg/g, according to the results. Because the valences of various metals differ, it is possible that the discrepancies in adsorption are due to variances in their affinity for the biochar.

8.6.1.2   Interaction of Metals with Biochar via Electrostatic Attraction

It is necessary to use electrostatic interaction between charged biochar and metal ions in order to minimize the mobilization of potentially dangerous metals from the biochar in order to achieve success with this strategy (Mukherjee et al. 2011). According to Qiu et al. (2009), the removal of lead from aqueous solution using biochar formed from rice and wheat suggests a high removal of lead due to the attraction of the positively charged Pb and negatively charged charcoal. Other researchers have claimed that increasing the pyrolysis temperature to greater than 400°C can increase carbonization of biochar, which leads in an increase in the electrostatic interaction of the biochar and thus increases its potential to adsorb pollutants.

8.6.1.3   Capacity for Ion Exchange

A major element of this mechanism is the exchange of protons and ionized cations with dissolved salts on the surface of biochar, which serves as the fundamental driving force behind the process. In order to effectively remove heavy metals from water, the adsorption ability of biochar must be matched to the size of the contaminated particles as well as the surface functional group of the biochar (Rizwan et al. 2016).

8.6.1.4 Precipitation

Using biochar to eliminate inorganic pollutants from the environment is one of the most effective methods of accomplishing this goal. It is also one of the most cost-effective. When using biochar, the formation of mineral precipitates into the solution or on the surface of the sorbing material is important because it has an alkaline property. Biochar is produced by decomposing cellulose and hemicellulose materials at temperatures greater than $300^{\circ}C$ and is particularly important for this reason (Cao and Harris 2010). Using sugarcane and straw dust as a source of biochar, Puga et al. (2015) discovered that the precipitation of Cd and Zn was enhanced when using the material. However, they asserted that the capacity of surface precipitation of biochar is dependent on the pyrolysis temperature, and the further research into the optimization of the pyrolysis temperature is required in the future.

8.6.1.5 Complexation

Complexation is characterized by the formation of multiatom configurations as a result of the inter-action of certain metal ligands in order to form complexes; it is a mechanism of metal complex-ation. Using biochar that has been produced at a low temperature, heavy metals can be bonded to the material. When compared to biochar prepared from animals such as dairy manure and poultry litter, it has been demonstrated that biochar prepared from vegetal biomass is more efficient in the binding of potentially toxic metals such as Cu, Cd, Ni, and Pb to form metal complexes with car-boxylic and phenolic functional groups (Cao et al. 2009; Zhang et al. 2017a).

8.6.2 Adsorption Mechanism for Organic Pollutants

Partitioning, pore filling, electrostatic interaction, and electron donor–acceptor (EDA) interaction are some of the various mechanisms involved in the adsorption of organic contaminants.

8.6.2.1 Through Partitioning

Biochar's noncarbonized component contains pores, which allow for the diffusion of adsorbate material. As a result, this section can easily interact with the organic adsorbate, resulting in the organic adsorbate becoming bound to the surface. But it is important to note that the characteristics of noncarbonized biochar (crystalline or amorphous carbon) as well as the characteristics of the carbonized crystalline and graphene fractions of the biochar are important in terms of adsorption of organic pollutant compounds. Sun et al. (2011) studied whether organic components of biochar generated from wood and grass may improve the adsorption of norfurazon and furidone through partitioning in their research. As a general rule, when the biochar contains a high concentration of volatile organic matter and a high concentration of organic contaminants, the partitioning process is more visible and highly efficient (Keiluweit et al. 2010).

8.6.2.2 Through Filling of Pores

Biochar is used in this procedure because it has mesopores (2–50 nm) and micropores (2 nm) that allow organic pollutants to accumulate on the surface. It is dependent on the nature, type of bio-char used, and polarity of the organic contamination to determine how the pore filling mechanism works. For this pore filling procedure to be effective, the biochar must have a limited amount of volatile matter and be present in a low concentration of organic pollutants, in general.

8.6.2.3 Through Electrostatic Interactions

In this mechanism, ionizable organic molecules are attracted to the positively charged surface of biochar as a result of electrostatic contact between the organic compounds and the biochar surface.

This is the most important mechanism because it involves the adsorption of organic molecules to the positively charged surface of biochar. The pH and ionic strength of water have an effect on the ability of an aqueous solution to attract or repel impurities (Ahmad et al. 2014; Zheng et al. 2013). Researchers found that the electrostatic interaction between organic contaminants and biochar surface was more positive at lower pH levels and more negative at higher pH levels, according to Mukherjee et al. (2011), who investigated the electrostatic interaction between organic contaminants and biochar surface. In accordance with this finding, the pH of the solution/effluent has an effect on the net charge of the biochar's surface. Additional research has found an association between the ionic strength of the aqueous solution and the electrostatic interaction of organic contaminants with biochar.

8.6.2.4 Through Interaction between Electron Donors and Acceptors

When aromatic compounds adsorb on biochar, which has a graphene-like structure, it is commonly utilized to better understand the electron donor and acceptor interaction mechanism between the two molecules. In order to accomplish complete graphitization of the carbon, the biochar preparation procedure should be carried out at a temperature greater than 1100°C (Spokas 2010).

8.7 FACTORS INFLUENCING THE ADSORPTION OF CONTAMINANTS ON BIOCHAR

Adsorption capacity of biochar is regulated by a variety of parameters, including the material's characteristics, pH, temperature, and amount of adsorbent used. The following are a few of the factors that influence the adsorption of contaminants onto the surface of biochar (Yaashikaa et al. 2019; Enaime et al. 2020).

8.7.1 FEATURES OF BIOCHAR

Temperature, feedstock, and residence time are all pyrolytic factors that have a substantial effect on the adsorption efficacy of biochar. The biochar's features were found to be significantly influenced by temperature, as demonstrated in the study (Zhou et al. 2010). As a result of the increased surface area formed by the carbonization of the organic content in the biomass, the adsorption rate increased as a result of the higher temperature employed in the process of carbonization (Ahmad et al. 2013). According to the findings of the study, when the pyrolytic temperature was elevated over 500°C, the surface area and pH increased, resulting in greater adsorption of cadmium and chromium ions. The hydrothermal treatment approach produced biochar that had more oxygen-containing functional groups than the pyrolysis method, according to the results of the study. A further benefit of hydrothermal carbonization treatment is that it is more efficient in producing biochar when compared to pyrolysis treatment. The process by which biochar is produced has an impact on the effectiveness of the biochar product, in addition to its features such as feedstock composition, pH, surface area, and temperature (Bernardo et al. 2013; Ding et al. 2014).

8.7.2 pH

Throughout the adsorption process, the pH of the solution is the most important variable to consider. The effect of pH on the adsorption process is dependent on the qualities of biochar as well as the pollutant to be removed (Regmi et al. 2012). The pH of the adsorbent material, as well as the speciation and ionization of the adsorbate, all have substantial effects on the surface charge of the adsorbent material. As a result, most of the study focused on pH. Increasing the pH of biochar causes changes in the functional groups found on its surface, such as hydroxyl, carboxylate, and other similar groups (Kołodyńska et al. 2017; Parshetti et al. 2014). When the pH of a solution is low, these groups become protonated and gain a positive charge, which promotes the adsorption of anions. Electrostatic repulsion occurs between positively charged biochar and negatively charged cation pollution when the two materials are in contact. As a result, when the pH is low, adsorption occurs at a slower rate. As the pH value rises, the rivalry between protons and metal ions diminishes. Functional groups are deprotonated to make binding sites more readily available, which is primarily responsible for this phenomenon. When the pH of the solution is greater than the point of zero charges on the surface of the biochar, a negative charge is present on the biochar. In conclusion, cations are adsorbed by the biochar surface when the pH is high, and anions are absorbed by the biochar surface when the pH is low (Kumar et al. 2011a, 2011b).

8.7.3 ADSORBENT DOSAGE

The adsorption capacity is highly dependent on the dosage of adsorbent utilized. To make the method cost-effective, the adsorbent dosage must be adjusted accordingly. A significant amount of biochar slowed the adsorption process greatly. An increase in adsorbent dose resulted in a higher elimination of pollutants due to the availability of more active sites. When heavy metals and dyes were removed from the water, these results were observed (Chen et al. 2011; Tsai and Chen 2013).

8.7.4 TEMPERATURE

The majority of research indicates that the adsorption process is endothermic, meaning that its efficiency increases as the temperature rises. When adequate energy was provided in conjunction with a greater temperature, the removal efficiency of metals and dyes was increased (Meng et al. 2014).

8.8   REGENERATION OF BIOCHAR

In order to determine the possibility for reuse of biochar, the desorption process used for its recovery is crucial. There are several ways for recovering metal from solutions (Rosales et al. 2015), including precipitation, complexation, and ion exchange. It is possible to achieve pollutant desorption from biochar through the process of solvent regeneration, which takes into account the equilibrium relationship between the biochar, the solvent, and the pollutant in order to break the adsorption equilibrium by adjusting the temperature and pH value of the solvent.

A number of research studies have been conducted on the effectiveness of recycling biochar using various organic solvents in diverse applications. Thermal desorption or microwave irradiation are two other approaches for producing pollutant desorption from biochar. In this procedure, the adsorbed pollutant is carbonized and degraded until the molecule is smaller than the pore size of the biochar and escapes (Dai et al. 2019).

8.9   CHALLENGES AND FUTURE PERSPECTIVES

Biochar is used as adsorbent with a good adsorption capability for a variety of applications, including environmental remediation. As a result of the fact that biochar research began in the later years of the twentieth century, we currently lack thorough knowledge about the process of mechanism for pollutant adsorption by biochar (Varjani et al. 2019). Another crucial element is the recovery and regeneration of biochar following the adsorption of contaminants. This is due to the fact that it is critical to convert the hazardous pollutants that have been bonded to the biochar into a nontoxic condition before using it (Chen et al. 2011; Li et al. 2018; Yaashikaa et al. 2019). Cost, the raw feedstock utilized in biochar production, methods of recovering and regenerating biochar, and biochar features such as functional groups and surface area are all regarded as essential parameters for the optimal exploitation of biochar in various applications. It is reasonable to expect that future breakthroughs in the valorization of lignocellulosic biomass will be accompanied by improvements in the fields of synthetic and microbiological chemicals. The majority of the research has been devoted to the removal of metals and dyes from contaminated water. The study on adsorption capability is still not available. However, the use of biochar has a number of advantages when it comes to solving environmental problems and maintaining ecosystems. Furthermore, investigation is required to gain a thorough understanding of biochar as well as the process by which it can be applied in the field.

Additionally, research is required to (1) develop a new low-cost, high-efficiency biochar modification technology, (2) expand the practical application of biochar in wastewater treatment, particularly industrial and municipal wastewater treatment, and (3) further enhance biochar's adsorption capacity for heavy metals, organic pollutants, nitrogen, and phosphorus, among other thin-layer contaminants.

8.10 CONCLUSION

Biochar has the potential to be utilized in industrial wastewater treatment; however, it must first be enhanced in terms of adsorptive capacity and stability before it can be employed in this environment. Due to its great efficacy and low cost, biochar is gaining traction in wastewater treatment. Biochar can be made from a range of biomass materials, including agricultural crop leftovers, forestry residues, sewage sludge, manure, and solid organic municipal trash. Biochar has been effectively used in a number of applications, including wastewater treatment. In many applications, biochar can be a feasible alternative to activated carbon due to the large surface area, high porosity, high inorganic mineral content, and surface functional groups present in the material. When the preparation techniques for other carbon compounds were compared to the preparation procedures for biochar, it was discovered that pyrolysis is less expensive, easier, and more environmentally friendly (activated carbon and graphene). Additionally, to serve as an environmentally friendly treatment approach, the conversion of wastewater sludge to biochar has the potential to improve the physiochemical and adsorption properties of the resulting material. Chemicals that oxidize oxygen-containing groups in biochar are employed to modify them, whereas acid and alkali modification are utilized to modify the functional groups and surface area. When metal ions or metal oxides are introduced to a mixture, the adsorption, magnetic, and catalytic capacities are all enhanced. In addition to carbonaceous components, the surface area of biochar can be modified. Hence this chapter summarized an overview of current biochar manufacturing technologies, with an emphasis on feedstock pretreatment, thermal conversion, and post-treatment, biochar adsorption mechanisms, biochar modification techniques, biochar characterization techniques, and biochar's application in wastewater treatment.

REFERENCES

Ahmad, Mahtab, Sang Soo Lee, Anushka Upamali Rajapaksha, Meththika Vithanage, Ming Zhang, Ju Sik Cho, Sung-Eun Lee, and Yong Sik Ok. "Trichloroethylene adsorption by pine needle biochars produced at various pyrolysis temperatures." *Bioresource Technology* 143 (2013): 615–622.
Ahmad, Mahtab, Anushka Upamali Rajapaksha, Jung Eun Lim, Ming Zhang, Nanthi Bolan, Dinesh Mohan, Meththika Vithanage, Sang Soo Lee, and Yong Sik Ok. "Biochar as a sorbent for contaminant management in soil and water: A review." *Chemosphere* 99 (2014): 19–33.
Al Arni, Saleh. "Comparison of slow and fast pyrolysis for converting biomass into fuel." *Renewable Energy* 124 (2018): 197–201.
Ashoori, Negin, Marc Teixido, Stephanie Spahr, Gregory H. LeFevre, David L. Sedlak, and Richard G. Luthy. "Evaluation of pilot-scale biochar-amended woodchip bioreactors to remove nitrate, metals, and trace organic contaminants from urban stormwater runoff." *Water Research* 154 (2019): 1–11.
Bergman, Patrick C. A., A. R. Boersma, R. W. R. Zwart, and J. H. A. Kiel. "Torrefaction for biomass co-firing in existing coal-fired power stations." *Report No. ECNC05013 Energy Research Centre of The Netherlands (ECN)*, Petten, The Netherlands (2005), p. 71.

Bernardo, Maria, Sandra Mendes, Nuno Lapa, Margarida Gonçalves, Benilde Mendes, Filomena Pinto, Helena Lopes, and Isabel Fonseca. "Removal of lead (Pb2+) from aqueous medium by using chars from co-pyrolysis." *Journal of Colloid and Interface Science* 409 (2013): 158–165.

Bogusz, Aleksandra, Patryk Oleszczuk, and Ryszard Dobrowolski. "Application of laboratory prepared and commercially available biochars to adsorption of cadmium, copper and zinc ions from water." *Bioresource Technology* 196 (2015): 540–549.

Cantrell, Keri B., Patrick G. Hunt, Minori Uchimiya, Jeffrey M. Novak, and Kyoung S. Ro. "Impact of pyrolysis temperature and manure source on physicochemical characteristics of biochar." *Bioresource Technology* 107 (2012): 419–428.

Cao, Xinde, and Willie Harris. "Properties of dairy-manure-derived biochar pertinent to its potential use in remediation." *Bioresource Technology* 101, no. 14 (2010): 5222–5228.

Cao, Xinde, Lena Ma, Bin Gao, and Willie Harris. "Dairy-manure derived biochar effectively sorbs lead and atrazine." *Environmental Science & Technology* 43, no. 9 (2009): 3285–3291.

Chen, Chunmao, Xin Yan, YingYing Xu, Brandon A. Yoza, Xin Wang, Yue Kou, Huangfan Ye, Qinghong Wang, and Qing X. Li. "Activated petroleum waste sludge biochar for efficient catalytic ozonation of refinery wastewater." *Science of the Total Environment* 651 (2019): 2631–2640.

Chen, Xincai, Guangcun Chen, Linggui Chen, Yingxu Chen, Johannes Lehmann, Murray B. McBride, and Anthony G. Hay. "Adsorption of copper and zinc by biochars produced from pyrolysis of hardwood and corn straw in aqueous solution." *Bioresource Technology* 102, no. 19 (2011): 8877–8884.

Cheng, Zihang, Fenglian Fu, Dionysios D. Dionysiou, and Bing Tang. "Adsorption, oxidation, and reduction behavior of arsenic in the removal of aqueous as (III) by mesoporous Fe/Al bimetallic particles." *Water Research* 96 (2016): 22–31.

Cole, Andrew J., Nicholas A. Paul, Rocky De Nys, and David A. Roberts. "Good for sewage treatment and good for agriculture: Algal based compost and biochar." *Journal of Environmental Management* 200 (2017): 105–113.

Dai, Yingjie, Naixin Zhang, Chuanming Xing, Qingxia Cui, and Qiya Sun. "The adsorption, regeneration and engineering applications of biochar for removal organic pollutants: A review." *Chemosphere* 223 (2019): 12–27.

Devi, Parmila, and Anil K. Saroha. "Synthesis of the magnetic biochar composites for use as an adsorbent for the removal of pentachlorophenol from the effluent." *Bioresource Technology* 169 (2014): 525–531.

Ding, Wenchuan, Xiaoling Dong, Inyang Mandu Ime, Bin Gao, and Lena Q. Ma. "Pyrolytic temperatures impact lead sorption mechanisms by bagasse biochars." *Chemosphere* 105 (2014): 68–74.

Ejraei, Ayoub, Mohammad Ali Aroon, and Afshin Ziarati Saravani. "Wastewater treatment using a hybrid system combining adsorption, photocatalytic degradation and membrane filtration processes." *Journal of Water Process Engineering* 28 (2019): 45–53.

Enaime, Ghizlane, Abdelaziz Baçaoui, Abdelrani Yaacoubi, and Manfred Lübken. "Biochar for wastewater treatment—conversion technologies and applications." *Applied Sciences* 10, no. 10 (2020): 3492.

Fagbohungbe, Michael O., Ben M. J. Herbert, Lois Hurst, Cynthia N. Ibeto, Hong Li, Shams Q. Usmani, and Kirk T. Semple. "The challenges of anaerobic digestion and the role of biochar in optimizing anaerobic digestion." *Waste Management* 61 (2017): 236–249.

Fan, Ye, Bin Wang, Songhu Yuan, Xiaohui Wu, Jing Chen, and Linling Wang. "Adsorptive removal of chloramphenicol from wastewater by NaOH modified bamboo charcoal." *Bioresource Technology* 101, no. 19 (2010): 7661–7664.

Fang, June, Lu Zhan, Yong Sik Ok, and Bin Gao. "Mini review of potential applications of hydrochar derived from hydrothermal carbonization of biomass." *Journal of Industrial and Engineering Chemistry* 57 (2018): 15–21.

Fdez-Sanroman, Antía, Marta Pazos, Emilio Rosales, and María Angeles Sanromán. "Unravelling the environmental application of biochar as low-cost biosorbent: A review." *Applied Sciences* 10, no. 21 (2020): 7810.

Fu, Fenglian, and Qi Wang. "Removal of heavy metal ions from wastewaters: A review." *Journal of Environmental Management* 92, no. 3 (2011): 407–418.

Funke, Axel, and Felix Ziegler. "Hydrothermal carbonization of biomass: A summary and discussion of chemical mechanisms for process engineering." *Biofuels, Bioproducts and Biorefining* 4, no. 2 (2010): 160–177.

Gai, Xiapu, Hongyuan Wang, Jian Liu, Limei Zhai, Shen Liu, Tianzhi Ren, and Hongbin Liu. "Effects of feedstock and pyrolysis temperature on biochar adsorption of ammonium and nitrate." *PLoS ONE* 9, no. 12 (2014): e113888.

Gray, Myles. "Black is green: Biochar for stormwater management." *Proceedings of the Water Environment Federation* 2016, no. 6 (2016): 2108–2123.

Higashikawa, Fábio Satoshi, Rafaela Feola Conz, Marina Colzato, Carlos Eduardo Pellegrino Cerri, and Luís Reynaldo Ferracciú Alleoni. "Effects of feedstock type and slow pyrolysis temperature in the production of biochars on the removal of cadmium and nickel from water." *Journal of Cleaner Production* 137 (2016): 965–972.

Hussain, Athar, Jaya Maitra, and Kashif Ali Khan. "Development of biochar and chitosan blend for heavy metals uptake from synthetic and industrial wastewater." *Applied Water Science* 7, no. 8 (2017): 4525–4537.

Jain, Akshay, Rajasekhar Balasubramanian, and M. P. Srinivasan. "Hydrothermal conversion of biomass waste to activated carbon with high porosity: A review." *Chemical Engineering Journal* 283 (2016): 789–805.

Jing, Xiang-Rong, Yuan-Ying Wang, Wu-Jun Liu, Yun-Kun Wang, and Hong Jiang. "Enhanced adsorption performance of tetracycline in aqueous solutions by methanol-modified biochar." *Chemical Engineering Journal* 248 (2014): 168–174.

Jung, Kyung-Won, Brian Hyun Choi, Tae-Un Jeong, and Kyu-Hong Ahn. "Facile synthesis of magnetic biochar/Fe_3O_4 nanocomposites using electro-magnetization technique and its application on the removal of acid orange 7 from aqueous media." *Bioresource Technology* 220 (2016): 672–676.

Keiluweit, Marco, Peter S. Nico, Mark G. Johnson, and Markus Kleber. "Dynamic molecular structure of plant biomass-derived black carbon (biochar)." *Environmental Science & Technology* 44, no. 4 (2010): 1247–1253.

Khorram, Mahdi Safaei, Qian Zhang, Dunli Lin, Yuan Zheng, Hua Fang, and Yunlong Yu. "Biochar: A review of its impact on pesticide behavior in soil environments and its potential applications." *Journal of Environmental Sciences* 44 (2016): 269–279.

Kim, Jin Hyo, Yong Sik Ok, Geun-Hyoung Choi, and Byung-Jun Park. "Residual perfluorochemicals in the biochar from sewage sludge." *Chemosphere* 134 (2015): 435–437.

Klinghoffer, Naomi B., Marco J. Castaldi, and Ange Nzihou. "Influence of char composition and inorganics on catalytic activity of char from biomass gasification." *Fuel* 157 (2015): 37–47.

Kołodyńska, D., J. A. Krukowska, and P. Thomas. "Comparison of sorption and desorption studies of heavy metal ions from biochar and commercial active carbon." *Chemical Engineering Journal* 307 (2017): 353–363.

Kumar, Amit, Ajay Kumar, Gaurav Sharma, Mu Naushad, Florian J. Stadler, Ayman A. Ghfar, Pooja Dhiman, and Reena V. Saini. "Sustainable nano-hybrids of magnetic biochar supported g-C_3N_4/$FeVO_4$ for solar powered degradation of noxious pollutants-Synergism of adsorption, photocatalysis & photo-ozonation." *Journal of Cleaner Production* 165 (2017): 431–451.

Kumar, Ponnusamy Senthil, Subramaniam Ramalingam, and Kannaiyan Sathishkumar. "Removal of methylene blue dye from aqueous solution by activated carbon prepared from cashew nut shell as a new low-cost adsorbent." *Korean Journal of Chemical Engineering* 28, no. 1 (2011a): 149–155.

Kumar, Sandeep, Vijay A. Loganathan, Ram B. Gupta, and Mark O. Barnett. "An assessment of U (VI) removal from groundwater using biochar produced from hydrothermal carbonization." *Journal of Environmental Management* 92, no. 10 (2011b): 2504–2512.

Kumi, Andy G., Mona G. Ibrahim, Mahmoud Nasr, and Manabu Fujii. "Biochar synthesis for industrial wastewater treatment: A critical review." In *Materials Science Forum*, vol. 1008, pp. 202–212. Switzerland: Trans Tech Publications Ltd., 2020.

Lee, Jongkeun, Kwanyong Lee, Donghwan Sohn, Young Mo Kim, and Ki Young Park. "Hydrothermal carbonization of lipid extracted algae for hydrochar production and feasibility of using hydrochar as a solid fuel." *Energy* 153 (2018): 913–920.

Lefèvre, Emilie, Nathan Bossa, Courtney M. Gardner, Gretchen E. Gehrke, Ellen M. Cooper, Heather M. Stapleton, Heileen Hsu-Kim, and Claudia K. Gunsch. "Biochar and activated carbon act as promising amendments for promoting the microbial debromination of tetrabromobisphenol A." *Water Research* 128 (2018): 102–110.

Li, Xue, Jiwei Luo, Hui Deng, Peng Huang, Chengjun Ge, Huamei Yu, and Wen Xu. "Effect of cassava waste biochar on sorption and release behavior of atrazine in soil." *Science of the Total Environment* 644 (2018): 1617–1624.

Li, Yunchao, Jingai Shao, Xianhua Wang, Yong Deng, Haiping Yang, and Hanping Chen. "Characterization of modified biochars derived from bamboo pyrolysis and their utilization for target component (furfural) adsorption." *Energy & Fuels* 28, no. 8 (2014): 5119–5127.

Lima, Isabel M., and Wayne E. Marshall. "Granular activated carbons from broiler manure: Physical, chemical and adsorptive properties." *Bioresource Technology* 96, no. 6 (2005): 699–706.

Lin, Yousheng, Xiaoqian Ma, Xiaowei Peng, Zhaosheng Yu, Shiwen Fang, Yan Lin, and Yunlong Fan. "Combustion, pyrolysis and char CO_2-gasification characteristics of hydrothermal carbonization solid fuel from municipal solid wastes." *Fuel* 181 (2016): 905–915.

Liu, Na, Alberto Bento Charrua, Chih-Huang Weng, Xiaoling Yuan, and Feng Ding. "Characterization of biochars derived from agriculture wastes and their adsorptive removal of atrazine from aqueous solution: A comparative study." *Bioresource Technology* 198 (2015a): 55–62.

Liu, Qingsong, Laying Wu, Matthew Gorring, and Yang Deng. "Aluminum-impregnated biochar for adsorption of arsenic (V) in urban stormwater runoff." *Journal of Environmental Engineering* 145, no. 4 (2019): 04019008.

Liu, Wu-Jun, Hong Jiang, and Han-Qing Yu. "Development of biochar-based functional materials: Toward a sustainable platform carbon material." *Chemical Reviews* 115, no. 22 (2015b): 12251–12285.

Lou, Kangyi, Anushka Upamali Rajapaksha, Yong Sik Ok, and Scott X. Chang. "Pyrolysis temperature and steam activation effects on sorption of phosphate on pine sawdust biochars in aqueous solutions." *Chemical Speciation & Bioavailability* 28, no. 1–4 (2016): 42–50.

Mandal, Abhishek, and Neera Singh. "Optimization of atrazine and imidacloprid removal from water using biochars: Designing single or multi-staged batch adsorption systems." *International Journal of Hygiene and Environmental Health* 220, no. 3 (2017): 637–645.

Manyuchi, M. M., C. Mbohwa, and E. Muzenda. "Potential to use municipal waste bio char in wastewater treatment for nutrients recovery." *Physics and Chemistry of the Earth*, Parts A/B/C 107 (2018): 92–95.

Meng, Jun, Xiaoli Feng, Zhongmin Dai, Xingmei Liu, Jianjun Wu, and Jianming Xu. "Adsorption characteristics of Cu (II) from aqueous solution onto biochar derived from swine manure." *Environmental Science and Pollution Research* 21, no. 11 (2014): 7035–7046.

Meyer, Sebastian, Bruno Glaser, and Peter Quicker. "Technical, economical, and climate-related aspects of biochar production technologies: A literature review." *Environmental Science & Technology* 45, no. 22 (2011): 9473–9483.

Mohan, Dinesh, Ankur Sarswat, Yong Sik Ok, and Charles U. Pittman Jr. "Organic and inorganic contaminants removal from water with biochar, a renewable, low cost and sustainable adsorbent—a critical review." *Bioresource Technology* 160 (2014): 191–202.

Mohanty, Sanjay K., Keri B. Cantrell, Kara L. Nelson, and Alexandria B. Boehm. "Efficacy of biochar to remove Escherichia coli from stormwater under steady and intermittent flow." *Water Research* 61 (2014): 288–296.

Mukherjee, A., A. R. Zimmerman, and W. Harris. "Surface chemistry variations among a series of laboratory-produced biochars." *Geoderma* 163, no. 3–4 (2011): 247–255.

Nunoura, Teppei, Samuel R. Wade, Jared P. Bourke, and Michael Jerry Antal. "Studies of the flash carbonization process. 1. Propagation of the flaming pyrolysis reaction and performance of a catalytic afterburner." *Industrial & Engineering Chemistry Research* 45, no. 2 (2006): 585–599.

Osayi, Julius I., Sunny Iyuke, and Samuel E. Ogbeide. "Biocrude production through pyrolysis of used tyres." *Journal of Catalysts* 2014 (2014).

Pang, Shusheng. "Advances in thermochemical conversion of woody biomass to energy, fuels and chemicals." *Biotechnology Advances* 37, no. 4 (2019): 589–597.

Park, Jong-Hwan, Jim J. Wang, Yili Meng, Zhuo Wei, Ronald D. DeLaune, and Dong-Cheol Seo. "Adsorption/desorption behavior of cationic and anionic dyes by biochars prepared at normal and high pyrolysis temperatures." *Colloids and Surfaces A: Physicochemical and Engineering Aspects* 572 (2019): 274–282.

Parshetti, Ganesh K., Shamik Chowdhury, and Rajasekhar Balasubramanian. "Hydrothermal conversion of urban food waste to chars for removal of textile dyes from contaminated waters." *Bioresource Technology* 161 (2014): 310–319.

Peng, Peng, Yin-Hai Lang, and Xiao-Mei Wang. "Adsorption behavior and mechanism of pentachlorophenol on reed biochars: PH effect, pyrolysis temperature, hydrochloric acid treatment and isotherms." *Ecological Engineering* 90 (2016): 225–233.

Poonam, Bharti, Sushil Kumar, and Narendra Kumar. "Kinetic study of lead (Pb^{2+}) removal from battery manufacturing wastewater using bagasse biochar as biosorbent." *Applied Water Science* 8, no. 4 (2018): 1–13.

Prabakar, Desika, Varshini T. Manimudi, Swetha Sampath, Durga Madhab Mahapatra, Karthik Rajendran, and Arivalagan Pugazhendhi. "Advanced biohydrogen production using pretreated industrial waste: Outlook and prospects." *Renewable and Sustainable Energy Reviews* 96 (2018): 306–324.

Puga, A. P., C. A. Z. Abreu, L. C. A. Melo, and L. Beesley. "Biochar application to a contaminated soil reduces the availability and plant uptake of zinc, lead and cadmium." *Journal of Environmental Management* 159 (2015): 86–93.

Qiu, Yuping, Zhenzhi Zheng, Zunlong Zhou, and G. Daniel Sheng. "Effectiveness and mechanisms of dye adsorption on a straw-based biochar." *Bioresource Technology* 100, no. 21 (2009): 5348–5351.

Rajapaksha, Anushka Upamali, Meththika Vithanage, Mahtab Ahmad, Dong-Cheol Seo, Ju-Sik Cho, Sung-Eun Lee, Sang Soo Lee, and Yong Sik Ok. "Enhanced sulfamethazine removal by steam-activated invasive plant-derived biochar." *Journal of Hazardous Materials* 290 (2015): 43–50.

Regmi, Pusker, Jose Luis Garcia Moscoso, Sandeep Kumar, Xiaoyan Cao, Jingdong Mao, and Gary Schafran. "Removal of copper and cadmium from aqueous solution using switchgrass biochar produced via hydrothermal carbonization process." *Journal of Environmental Management* 109 (2012): 61–69.

Rizwan, Muhammad, Shafaqat Ali, Muhammad Farooq Qayyum, Muhammad Ibrahim, Muhammad Zia-ur-Rehman, Tahir Abbas, and Yong Sik Ok. "Mechanisms of biochar-mediated alleviation of toxicity of trace elements in plants: A critical review." *Environmental Science and Pollution Research* 23, no. 3 (2016): 2230–2248.

Rosales, Emilio, Laura Ferreira, M. Ángeles Sanromán, Teresa Tavares, and Marta Pazos. "Enhanced selective metal adsorption on optimised agroforestry waste mixtures." *Bioresource Technology* 182 (2015): 41–49.

Son, Eun-Bi, Kyung-Min Poo, Jae-Soo Chang, and Kyu-Jung Chae. "Heavy metal removal from aqueous solutions using engineered magnetic biochars derived from waste marine macro-algal biomass." *Science of the Total Environment* 615 (2018): 161–168.

Spokas, Kurt A. "Review of the stability of biochar in soils: Predictability of O:C molar ratios." *Carbon Management* 1, no. 2 (2010): 289–303.

Sun, Ke, Kyoung Ro, Mingxin Guo, Jeff Novak, Hamid Mashayekhi, and Baoshan Xing. "Sorption of bisphenol A, 17α-ethinyl estradiol and phenanthrene on thermally and hydrothermally produced biochars." *Bioresource Technology* 102, no. 10 (2011): 5757–5763.

Tan, Xiaofei, Yunguo Liu, Guangming Zeng, Xin Wang, Xinjiang Hu, Yanling Gu, and Zhongzhu Yang. "Application of biochar for the removal of pollutants from aqueous solutions." *Chemosphere* 125 (2015): 70–85.

Tang, Yao, Md Samrat Alam, Kurt O. Konhauser, Daniel S. Alessi, Shengnan Xu, WeiJun Tian, and Yang Liu. "Influence of pyrolysis temperature on production of digested sludge biochar and its application for ammonium removal from municipal wastewater." *Journal of Cleaner Production* 209 (2019): 927–936.

Tsai, W.-T., and H.-R. Chen. "Adsorption kinetics of herbicide paraquat in aqueous solution onto a low-cost adsorbent, swine-manure-derived biochar." *International Journal of Environmental Science and Technology* 10, no. 6 (2013): 1349–1356.

Ulrich, Bridget A., Megan Loehnert, and Christopher P. Higgins. "Improved contaminant removal in vegetated stormwater biofilters amended with biochar." *Environmental Science: Water Research & Technology* 3, no. 4 (2017): 726–734.

Varjani, Sunita, Gopalakrishnan Kumar, and Eldon R. Rene. "Developments in biochar application for pesticide remediation: Current knowledge and future research directions." *Journal of Environmental Management* 232 (2019): 505–513.

Wan, Shunli, Jiayu Wu, Shanshan Zhou, Rui Wang, Bin Gao, and Feng He. "Enhanced lead and cadmium removal using biochar-supported hydrated manganese oxide (HMO) nanoparticles: Behavior and mechanism." *Science of the Total Environment* 616 (2018): 1298–1306.

Wang, Bing, Bin Gao, and June Fang. "Recent advances in engineered biochar productions and applications." *Critical Reviews in Environmental Science and Technology* 47, no. 22 (2017): 2158–2207.

Wang, Hongyu, Bin Gao, Shenseng Wang, June Fang, Yingwen Xue, and Kai Yang. "Removal of Pb (II), Cu (II), and Cd (II) from aqueous solutions by biochar derived from $KMnO_4$ treated hickory wood." *Bioresource Technology* 197 (2015): 356–362.

Wang, Jianlong, and Shizong Wang. "Preparation, modification and environmental application of biochar: A review." *Journal of Cleaner Production* 227 (2019): 1002–1022.

Wang, Xiangyu, Weitao Lian, Xin Sun, Jun Ma, and Ping Ning. "Immobilization of NZVI in polydopamine surface-modified biochar for adsorption and degradation of tetracycline in aqueous solution." *Frontiers of Environmental Science & Engineering* 12, no. 4 (2018): 1–11.

Wang, Yan, Renzhan Yin, and Ronghou Liu. "Characterization of biochar from fast pyrolysis and its effect on chemical properties of the tea garden soil." *Journal of Analytical and Applied Pyrolysis* 110 (2014): 375–381.

Wathukarage, Awanthi, Indika Herath, M. C. M. Iqbal, and Meththika Vithanage. "Mechanistic understanding of crystal violet dye sorption by woody biochar: Implications for wastewater treatment." *Environmental Geochemistry and Health* 41, no. 4 (2019): 1647–1661.

Wei, Dongning, Bingyu Li, Hongli Huang, Lin Luo, Jiachao Zhang, Yuan Yang, Jiajun Guo, Lin Tang, Guangming Zeng, and Yaoyu Zhou. "Biochar-based functional materials in the purification of agricultural wastewater: Fabrication, application and future research needs." *Chemosphere* 197 (2018): 165–180.

Wei, Jing, Chen Tu, Guodong Yuan, Ying Liu, Dongxue Bi, Liang Xiao, Jian Lu et al. "Assessing the effect of pyrolysis temperature on the molecular properties and copper sorption capacity of a halophyte biochar." *Environmental Pollution* 251 (2019): 56–65.

Xiang, Wei, Xueyang Zhang, Jianjun Chen, Weixin Zou, Feng He, Xin Hu, Daniel C. W. Tsang, Yong Sik Ok, and Bin Gao. "Biochar technology in wastewater treatment: A critical review." *Chemosphere* 252 (2020): 126539.

Yaashikaa, P. R., P. Senthil Kumar, Sunita J. Varjani, and A. Saravanan. "Advances in production and application of biochar from lignocellulosic feedstocks for remediation of environmental pollutants." *Bioresource Technology* 292 (2019): 122030.

Yan, Jingchun, Lu Han, Weiguo Gao, Song Xue, and Mengfang Chen. "Biochar supported nanoscale zerovalent iron composite used as persulfate activator for removing trichloroethylene." *Bioresource Technology* 175 (2015): 269–274.

Yang, Wenchao, Zhaowei Wang, Shuang Song, Jianbo Han, Hong Chen, Xiaomeng Wang, Ruijun Sun, and Jiayi Cheng. "Adsorption of copper (II) and lead (II) from seawater using hydrothermal biochar derived from Enteromorpha." *Marine Pollution Bulletin* 149 (2019): 110586.

Yao, Ying, Yan Zhang, Bin Gao, Renjie Chen, and Feng Wu. "Removal of sulfamethoxazole (SMX) and sulfapyridine (SPY) from aqueous solutions by biochars derived from anaerobically digested bagasse." *Environmental Science and Pollution Research* 25, no. 26 (2018): 25659–25667.

Yoo, Seunghyun, Stephen S. Kelley, David C. Tilotta, and Sunkyu Park. "Structural characterization of loblolly pine derived biochar by X-ray diffraction and electron energy loss spectroscopy." *ACS Sustainable Chemistry & Engineering* 6, no. 2 (2018): 2621–2629.

Zhang, Hanyu, Zhaowei Wang, Ruining Li, Jialei Guo, Yan Li, Junmin Zhu, and Xiaoyun Xie. "TiO_2 supported on reed straw biochar as an adsorptive and photocatalytic composite for the efficient degradation of sulfamethoxazole in aqueous matrices." *Chemosphere* 185 (2017a): 351–360.

Zhang, Qian, Qingfeng Li, Linxian Zhang, Zhongliang Yu, Xuliang Jing, Zhiqing Wang, Yitian Fang, and Wei Huang. "Experimental study on co-pyrolysis and gasification of biomass with deoiled asphalt." *Energy* 134 (2017b): 301–310.

Zhao, Xuchen, Wei Ouyang, Fanghua Hao, Chunye Lin, Fangli Wang, Sheng Han, and Xiaojun Geng. "Properties comparison of biochars from corn straw with different pretreatment and sorption behaviour of atrazine." *Bioresource Technology* 147 (2013): 338–344.

Zheng, Hao, Zhenyu Wang, Jian Zhao, Stephen Herbert, and Baoshan Xing. "Sorption of antibiotic sulfamethoxazole varies with biochars produced at different temperatures." *Environmental Pollution* 181 (2013): 60–67.

Zheng, Yulin, Bing Wang, Anne Elise Wester, Jianjun Chen, Feng He, Hao Chen, and Bin Gao. "Reclaiming phosphorus from secondary treated municipal wastewater with engineered biochar." *Chemical Engineering Journal* 362 (2019): 460–468.

Zhou, Nan, Honggang Chen, Junting Xi, Denghui Yao, Zhi Zhou, Yun Tian, and Xiangyang Lu. "Biochars with excellent Pb (II) adsorption property produced from fresh and dehydrated banana peels via hydrothermal carbonization." *Bioresource Technology* 232 (2017): 204–210.

Zhou, Zunlong, Dongjin Shi, Yuping Qiu, and G. Daniel Sheng. "Sorptive domains of pine chars as probed by benzene and nitrobenzene." *Environmental Pollution* 158, no. 1 (2010): 201–206.

9 Biochar for Removal of Dyes from Industrial Effluents

Neha Rana, Sapna Salar, and Piyush Gupta

CONTENTS

9.1 INTRODUCTION

Every day, a large amount of wastewater is generated by industry (coal and steel, nonmetallic minerals, and metal surface processing industries such as iron picking and electroplating), which has a significant impact on the environment (Inyang et al. 2012). As a result, advanced oxidation processes (AOPs), reverse osmosis, adsorption, ion exchange, ozonation, precipitation, filtration with coagulation, and the coagulation process have all been used to treat industrial wastewater (Park et al. 2011). However, the majority of these processes come at a high cost of operation and capital. This has been identified as the primary impediment to their application in both developed and developing countries for the abatement/removal of potentially toxic contaminants from polluted waters (Ambaye et al. 2020; Giannakis et al. 2017; Lou et al. 2017; Villegas-Guzman et al. 2017).

Biochar is a solid made by pyrolyzing biomass at temperatures below 700°C in the absence or low presence of oxygen (Park et al. 2011). The resultant solid is high in carbon and has good adsorption properties; thus it can remove organic and inorganic pollutants from wastewater. Various waste items, such as straw, feces, and sludge, have been evaluated as biochar source materials. Park et al. (2016) demonstrated that biochar can be produced by pyrolysis, hydrothermal carbonization, and gasification. One can conclude that the performance of biochar can be affected by (1) the type of the organic materials feedstock, (2) the preparation temperature, and (3) the molecular weight of the biochar. Biochar's adsorption properties have been used to manage waste resources (Park et al. 2011), improve soil performance (Liu and Zhang 2009), reduce climate change, and

DOI: 10.1201/9781003203438-9

produce renewable biofuels (Kołodyńska et al. 2012). Many scientists investigated the feasibility of adsorbing pesticides, medicines, hormones, and potentially harmful metals using biochar made from animal manure, plant residues, and biosolids (Downie et al. 2009; Joseph et al. 2010; Sun et al. 2014). When compared to activated carbons, they found that biochar was more efficient in adsorbing pollutants. They discovered that biochar can remove up to 60% of organic contaminants such as triazine herbicide and ethinyl estradiol, atrazine and bisphenol. They came to the conclusion that biochar's potential to absorb pollutants is dependent on the preparation feedstock as well as the physiochemical features of both the biochar and the pollutant. Two dyes, safranin and methylene blue, were removed using rice husk (Inyang et al. 2014). Sawdust is a type of agricultural waste that can be used to remove colors, harmful metals, and salts from wastewater (Peng et al. 2011). Many substances (such as cellulose, lignin, and hemicellulose) and polyphenolic groups are found in this material, and they play an important role in attaching to organic compounds by various mechanisms such as ion exchange, complexation, and hydrogen bonding (Thies and Rilling, 2009). Cao and Harris (2010) found that adsorbents made from blast furnace sludge, dust, and slag have very poor dye adsorption ability. These adsorbents were discovered to have a limited surface area and poor porosity. Chemical activation of sludge-based adsorbents with metal hydroxide reagents (e.g., KOH) is, nevertheless, the most successful method for obtaining high (>1800 m^2/g) surface area of sludge-based adsorbents (Van Zwieten et al. 2010).

Biochar can be employed as a release control agent in fertilizers due to its high phosphorus and nitrogen adsorption capacity, as previously described (Chen et al. 2011a, b; Yao et al. 2011a, b; Zhang et al. 2013a, b). Because biochar is a nontoxic absorbent, studying its physicochemical properties and a comprehensive understanding of its adsorption mechanisms is of current interest. Electrostatic interaction, ion exchange, pore filling, and precipitation can all be used to explain how biochar removes organic and inorganic contaminants. This is dependent on the biochar's physiochemical properties, such as dosage, pyrolysis temperature, and pH of the medium/effluent (Ahmad et al. 2014; Lam et al. 2016; Mubarak et al. 2016; Pellera et al. 2012; Qayyum et al. 2017; Rehman et al. 2017; Vithanage et al. 2016; Younis et al. 2016). Water contamination has become a serious global environmental issue. Water contamination is caused by a variety of toxins, most of which are released as a result of industrial and agricultural operations. The discharge of untreated dye effluents is a major source of water contamination. Not only are these colors hazardous to plants and aquatic life, but they are also dangerous to humans. Cancer, allergies, and skin illnesses are just a few of the health problems that can result from humans ingesting and absorbing dye-contaminated water (Lin et al. 2020). Biochar has been discovered as a possible wastewater treatment candidate. Reactive dye wastewater pollutes the environment, putting human life at jeopardy. Ion exchange, oxidation, flocculation, precipitation reverse osmosis, and adsorption are some of the processes used to remove reactive dye pollution (Oliveira et al. 2019). Adsorption is regarded as an effective method when compared to all other methods because of its basic equipment requirements, cheap budget operating costs, and few secondary pollutants (Esan et al. 2019). This chapter throws some light on various aspects of biochar usage for dyes removal from industrial effluents.

9.2 COLORANTS AND TOXICITY

Color-giving organic molecules that are water or oil soluble are referred to as dyes, as opposed to pigments, which are insoluble. Natural dyes are those that come from natural sources like animals, flowers, roots, mollusks, minerals, and so on. Natural dyes can be divided into two categories: adjectival and substantive (Bhuiyan et al. 2017). Substantive dyes, on the other hand, contain a natural mordant called tannin and can offer fast color without the use of extra mordants (Zerin et al. 2020). Adjective dyes only give lasting color when combined with a mordant to attach them to the fabric. Mordants are substances that act as a link between the dye molecules and the fabric molecules, keeping the dye in place (Nambela et al. 2020). Natural mordants include weak acids such as tannic acid and acetic acid (Degano et al. 2019; Saxena and Raja 2014), as well as metal salts such as copper sulfate and ferrous sulfate. Natural dyes are favorable since they are largely nontoxic and renewable in nature, but their application in the industrial environment is not cost-effective, and they do not provide an even consistent hue when compared to synthetic dyes. Natural dyes can still be employed in small-scale and home enterprises. The majority of synthetic colors are unsaturated chemical compounds. Mauve, a reddish-purple dye that swiftly fades in water or bright sunlight to a pale purple tint, was the first synthetic dye developed. The first synthetic color was made from coal tar, which is a by-product of coal carbonization. Synthetic dyes are currently more cost-effective and have superior color fastness than natural dyes; hence they have a monopoly on the market. Synthetic dyes, while more cost-effective, are far more poisonous and polluting than natural dyes, resulting in environmental contamination and negative health consequences for living species (Zerin et al. 2020).

9.3 TREATMENT METHODS EMPLOYED FOR REMOVAL OF DYES FROM WATER SYSTEMS

The different treatment technologies for removal of dyes from water systems are summarized next. The advantages and drawbacks of the different processes are tabulated in Table 9.1.

9.3.1 COAGULATION

Since the early twentieth century, coagulation has been one of the most widely utilized wastewater treatment methods. The act of adding chemical substances to bind particles together until they reach

TABLE 9.1
Treatment Technologies for Removal of Dyes from Water

Treatment technology	Advantages	Disadvantages
Coagulation	• Reduced time for suspended solids settling	• Excessive price for frequent monitoring and proper dosing
Advanced oxidation processes (AOP)	• Easier removal of small particles	• Significant sludge production
Membrane processes	• Effective for removing bacteria, protozoa, and virus	• Fenton's reagent AOP produces iron sludge
Biological processes	• ·OH radicals can cure a wide variety of organic materials.	• High capital and operating costs
Adsorption processes	• No sludge production	• High flow rate can cause damage to the membrane
	• Minor, valuable products can be recovered from the feed stream	• Results in membrane fouling effects and requires extensive cleaning
	• Process can be easily scaled up	• Expensive equipment
	• No-phase shifts involved between the feed and product stream	• Slow process
	• Eco-friendly as uses simple and nontoxic materials	• Favorable environment is essential
	• Effective removal of almost all biodegradable organic matter	• Biological sludge generation
	• Efficient color attenuation	• Dye molecule remediation is difficult
	• Eco-friendly and common wastewater treatment technique	• Adsorbent performance degrades after repeated operational cycles
	• Highly efficient process	• Spent adsorbent is likely to be a hazardous waste
	• Applicable for a wide range of target pollutants	• Adsorbent material regeneration is costly
	• Simple treatment technology involved	

a huge mass and eventually settle down is known as coagulation. Coagulants are positively charged compounds that are quickly blended with the effluent to ensure uniform dispersion. The majority of dissolved/suspended particles in wastewater have a negative charge that is neutralized by these coagulants, allowing them to cling together. The coagulation process is typically used as a first step in the wastewater treatment process. Iron or aluminum salts are the most often utilized coagulants. However, wastewater containing dyes has a lot of color and a lot of chemical oxygen demand (COD). As a result, traditional procedures such as coagulation are ineffective and can cause a sludge disposal problem (Issa Hamoud H. et al. 2017). To boost its efficiency, a combination of the traditional coagulation technique and other treatment approaches must be developed. For example, a combination of coagulation and adsorption approaches was found to be a viable method for removing reactive colors from water (Furlan et al. 2010). The adsorbent was activated carbon made from coconut shells, and the coagulation was done with aluminum chloride. Orange 16 and black 5 reactive dyes were shown to have removal efficiencies of 84 and 90%, respectively (Furlan et al. 2010).

9.3.2 ADVANCED OXIDATION PROCESSES (AOPs)

AOPs are chemical treatment procedures that exploit the oxidizing action generated by the in situ generation of hydroxyl radicals (•OH) to remove organic and inorganic pollutants from wastewater (Babu et al. 2019). Oxidizing agents (H_2O_2, O_3, $KMnO_4$), catalysts, and UV radiation all create these radicals. Photocatalysis, ozonation, and other AOPs are examples (Nidheesh et al. 2018). The sections that follow provide a brief overview of some of the AOPs utilized in dye removal from wastewater.

9.3.2.1 Ozonation

Ozonation refers to the process of removing the majority of dyes found in wastewater having polycyclic aromatic structures containing components such as nitrogen, metals, and sulfur, making physical, chemical, and biological treatment of the wastewater problematic. Ozonation, an AOP process that involves the chemical treatment of wastewater by dissolving ozone in water, actively destroys conjugated chains contained in dye compounds, which are responsible for color imparting. The use of ozonation as a treatment method is favorable because no sludge is produced. The dye is degraded in a single phase, and ozone decomposes into stable oxygen in the process (Muniyasamy et al. 2020). However, because ozonation seldom achieves complete oxidation as a treatment, byproducts are frequently generated in the effluent.

9.3.2.2 Fenton's Reagent and Fenton-like Processes

Fenton's reagent is a ferrous iron and hydrogen peroxide solution that is used as a catalyst in the oxidation of different pollutants found in wastewater. In the presence of light irradiation and hydrogen peroxide, a composite of La-Fe-O (Xu et al. 2020a) was utilized as a photo-Fenton catalyst, and rhodamine B dye was oxidized to 98% in 25 min.

9.3.2.3 Membrane Processes

Membranes for wastewater treatment have been used since around the 1960s. Because membrane techniques were prohibitively expensive at the time, this method was reserved for specific applications. Membranes have become more cost-effective during the 2000s, and they are now utilized in conjunction with other traditional water treatment methods. When a driving force is given to a membrane, it is a thin, semipermeable material that is attached to a porous support and is used to remove dissolved chemicals based on attributes such as size or charge. Reverse osmosis (RO), forward osmosis (FO), nanofiltration, and ultrafiltration use membrane processes.

9.3.2.4 Nanofiltration

Nanofiltration membranes have pore sizes ranging from 0.1 to 10 nm. Nanofiltration membranes have the benefit of separating dyes with a high molecular weight and have a dye removal rejection effectiveness of >90%, making them a potential technique (Jin et al. 2020). Semi-xylenol orange, tropaeolin O, and Victoria blue B dyes were removed with 99, 98.3, and 99.2% efficiency, respectively, using a positively charged polyethylenimine-modified nanofiltration membrane (Qi et al. 2019).

9.3.2.5 Forward Osmosis (FO)

FO is a water treatment method that separates water from dissolved solutes by using an osmotic pressure gradient. In comparison to reverse osmosis and other membrane separation methods, the FO process is more energy efficient because it does not require any driving power (Meng et al. 2020). Lin et al. (2020) employed a thin-film composite membrane in forward osmosis and discovered that it had a dye rejection rate of 96% for regularly used dyes in the textile industry. FO is a water treatment method that separates water from dissolved solutes by using an osmotic pressure gradient. Because the FO process does not require the use of a driving force, it is more energy efficient.

9.3.2.6 Biological Process

Bacteria, germs, and other microorganisms are used in biological processes to treat wastewater. These biological processes are environmentally favorable, are energy efficient, produce little sludge, and utilize no or very little chemicals. Increasing the efficiency of biological processes could be done by altering environmental conditions to favor microbe development. Because algae contain proteins, lipids, and functional groups such as amino, carboxylate, sulfate, and others, they are often explored by researchers as a potential choice for biosorption (Marzbali et al. 2020). Because algae have a large surface area and a high binding affinity in their cell structure, they have a high biosorption capacity (Ihsanullah et al. 2020). Because algae are readily available (very common in salty oceans and freshwater lakes), they could be used to remove color from textile effluent in the future. Chemically (sulfuric acid) modified defatted *Laminaria japonica* biomass (renewable brown algae) had a methylene blue adsorption capacity of 549.45 mg/g, and quasi-equilibrium was reached in 60 min under ideal circumstances of 0.6 g/L biosorbent, pH 6, and temperature of 308 K (Shao et al. 2017). Bioreactors are also becoming more popular for removing colors. For the treatment of azo dyes, a combination of aerobic and anaerobic methods is usually used. For example, the combination process of an upflow bioelectrocatalyzed electrolysis reactor and aerobic biocontact oxidation reactor was used to remove alizarin yellow R dye in just 6 h of hydraulic retention time (Cui et al. 2014)

9.3.2.7 Adsorption Process

Adsorption has been found to be one of the best treatment processes for dye removal from water among other traditional water treatment methods due to its cheap cost, affordability, increased efficiency, and the fact that it requires minimal maintenance (Sirajudheen et al. 2020). Adsorption also has the advantage of leaving no harmful residues and treating huge volumes of water (Saxena et al. 2020). An adsorbent could potentially be reused in subsequent treatment operations (Morais da Silva et al. 2020). Water treatment has traditionally relied on common materials such as activated carbon, zeolites, activated alumina, silica gel, and polymeric adsorbents. Many researchers are interested in using biomaterials instead of traditional materials for adsorption procedures since biomaterials have a low commercial value and are abundant (Deniz et al. 2013). Naturally derived biopolymers that are hyperreactive and chemically stable and that have good physicochemical qualities have gotten a lot of attention, and they're worth investigating for use as green adsorbents (Fan et al. 2020)

9.4 BIOCHAR AS AN ADSORBENT

Biochar has been suggested as a unique material for wastewater treatment due to its environmental friendliness and versatility. Biochar's adsorption characteristics have made it a feasible option for adsorbing potentially hazardous and environmentally detrimental colorants.

9.4.1 Biochar for Adsorption of Dyes

Effluent released from the textile manufacturing, dye manufacturing, paper-pulp mills, and tannery industries carry dye contaminants into the water streams. Several classes of dyes like basic, acidic, reactive, direct, azo, mordent, sulfur, and vat dyes are the major classes of dyes used extensively by the textile, paper, and leather industries. Biochar was found to be a more economical and environmentally safer option. Table 9.2 elaborates the description of different biochars for dyes removal.

9.4.2 Influence of Operational Parameters on Adsorption Dyes Using Biochar

Biochar works as an adsorbent only in very restricted circumstances. The efficiency of biochar is influenced by the dye/biochar concentration, temperature, and solution pH (Chen et al. 2020; Chu et al. 2020; Mahmoud et al. 2020; Park et al. 2019; Yek et al. 2020; Zhang et al. 2020). At 20°C, nickel-modified biochar was capable of adsorbing 479.49 mg/g of methylene blue from wastewater (Yao et al. 2020). The adsorption by biochar was initially high when the concentration of methylene blue was low due to the vast number of active sites still available on the biochar for adsorption. The efficacy of methylene blue adsorption by biochar was reported to be hampered by competitive adsorption (Yao et al. 2020). The rate of decolorization increased as the temperature of the solution increased, according to research on the adsorption and degradation of acid red dye Rubeena et al. 2018). When the pH was changed from 3 to 11, the dye removal effectiveness only fell by about

---

TABLE 9.2
Biochar Materials for Removal of Dyes

Biochar	Dye	References
Steam activated spent mushroom substrate (SMS)	Congo red and crystal violet	Damertey et al. (2019)
Nickel aluminium layered double oxides modified magnetic biochar	Acridine orange	Wang et al. (2020)
Chemically modified lychee seed biochar	Methylene blue	Ang et al. (2020)
Triethylenetetramine biochar	Sunset yellow dye	Mahmoud et al. (2020)
Orange peel waste microwave activated biochar	Congo red dye	Yek et al. (2020)
Fe_2O_3/TiO_2 functionalized biochar	Methylene blue	Chen et al. (2020)
Switchgrass biochar Orange G, Congo red	Methylene blue	Park et al. (2019)
Biochar derived from Opuntia ficus-indica (OFI) cactus	Malachite green	Choudhary et al. (2020)
N-doped biochar	Acid red 18	Wang et al. (2018)
Ag-TiO_2 biochar	Methyl orange	Shan et al. (2020)
Biochar derived from mixed municipal discarded material	Methylene blue	Hoslett et al. (2020)
Mesoporous nano-zerovalent manganese (nZVMn) and Phoenix dactylifera leaves biochar (PBC) composite	Congo red	Iqbal et al. (2021)
Cetyl trimethyl ammonium bromide modified magnetic biochar derived from pine nutshells	Acid chrome blue K	Wang et al. (2020)
Pristine and ball milled biochar	Reactive red 120	Xu et al. (2020)
Wood based biochar	Acid orange 7	Zhu et al. (2019)
Pyrolyzed rice husk biochar	Malachite green	Ganguly et al. (2020)
Animal waste biochar	Basic red 9	Côrtes et al. (2019)
Chitosan based material	Basic Blue 7	Morais et al. (2020)
Surfactant modified chitosan beads	Tartrazine	Pal et al. (2019)
Ziziphus Lotus stones	Basic Yellow 28	Boudechiche et al. (2019)
Biochar derived from Caulerpa scalpelliformis	Remazol brilliant blue	Gokulan et al. (2019)
Dried, powdered weed.	Safranine	Shrivastava et al. (2010)

5%, according to the scientists (Chu et al. 2020). However, a closer examination revealed that the decline in removal efficiency was not uniform across the board but was consistent between pH 3–9, with a dramatic drop when pH neared 11. The authors explained that as pH increased, the decomposition of hydrogen peroxide decreased, as did the conversion of carbonate and bicarbonates to their corresponding acids, resulting in a low reaction with OH radicals and, as a result, a decrease in the solubility of iron present in the bicarbonates to their corresponding acids. The ideal pH for adsorption has been determined in part by the nature of dyes. When a dye is anionic, greater pH increases the presence of OH ions, which makes the surface of biochar negative, promoting electrostatic interaction between the positively charged cationic dye and the negatively charged biochar surface (Damertey et al. 2019; Wang et al. 2020). The amount of adsorbent and adsorbate used has an impact on the amount of dye adsorption and is a crucial element to consider. The removal percentage and capacity of adsorption was strongly impacted by the dosage of adsorbent in a research of dye adsorption using biochar (Yu et al. 2020). Even a slight increase in adsorbent dosage resulted in a considerable rise in removal percentage (almost an 80% increase) that remained constant once a particular maximum dosage was reached. According to the authors (Yu et al. 2020), this event occurred because the number of active sites and adsorption surface area increased with increasing adsorbent dosages, resulting in a significant rise in the percentages of dye removed from the water.

9.5 MECHANISMS INVOLVED IN DYES ADSORPTION ONTO BIOCHAR

Many complicated interactions (both chemisorption and physisorption) between the adsorbate (dye) and adsorbent are involved in the removal of colors from wastewater using biochar (biochar). According to current research, adsorption involves numerous mechanisms functioning together, some of which are dominant and others that are dependent on the system's current state. Mechanisms such as pore filling effect, van der Waals interaction, electrostatic interaction, chemical action, ion exchange, surface complexation-interactions, and cation- interactions could all play a role in the adsorption process depending on the dyes, biochar, and solvent (Kah et al. 2017). Fan et al. (2016) studied the adsorption of positively charged methylene blue dye (MBD) utilizing municipal sewage-sludge- and tea-waste-produced biochar. The effect of pH on adsorption revealed that adsorption was impacted by pH; for pH lower than 2, MBD removal efficiency was only 80%, while it was over 100% at pH 11 (Fan et al. 2016). This strongly shows that electrostatic contact had a role in the adsorption process. The numerous H^+ ions occupied the limited bonding sites on biochar and resisted the positively charged MBD molecules at low pH, but as pH was increased, the bonded sites were deprotonated and the MBD molecules were free to interact, dramatically boosting the adsorption capacity. The concentration of released metal cations (Ca^{2+}, Na^+, K^+, Mg^{2+}) by the biochar in the equilibrium solution was measured, and it was discovered that the concentration of these ions, especially Na^+ and K^+ ions, was greatly increased after MBD adsorption, indicating that these ions were involved in the ion exchange process. The pore filling effect was the primary mechanism in a study of malachite green dye (MGD) on activated biochar generated from *Opuntia ficus-indica* with a mesoporous structure, which suggested that the pore filling effect was the main mechanism (Choudhary et al. 2020). The adsorption process could have been aided by hydrogen bonding between the MGD functional groups and the biochar. According to XRD and FTIR measurements, the activated biochar also contained electron-rich amine and hydroxyl groups, but the MGD had electron-deficient functional groups as part of the aromatic ring. Electron donor–acceptor (EDA) interactions and cation–cation interactions would have been feasible depending on the pH of the solution.

So far, only single colors have been used in studies, which were adsorbed in batches. While studies like Del Bubba et al. (2020) and Santra et al. (2020) employed textile effluents that contained more than one color, they failed to list them; thus the impact of various dyes and biochar interactions is unknown. In an industrial wastewater treatment context, it's critical to test biochar's efficacy while adsorbing various dyes at the same time. Multiple treatment technologies could be combined

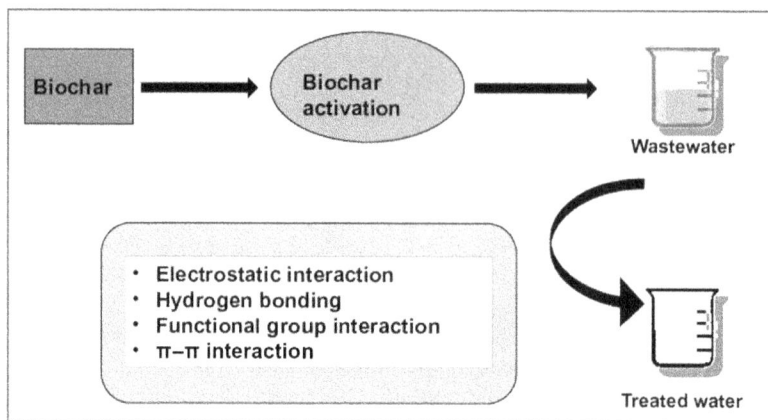

FIGURE 9.1   Mechanism involved in adsorption of dyes from wastewater using biochar.

to treat wastewater, as De Gisi et al. (2017) indicates, depending on the kind of effluent. The mechanism for dyes removal using biochar from wastewater is shown in Figure 9.1.

9.6   CONCLUSION AND FUTURE PROSPECTS

As evidenced by the literature reviewed here, biochar has been found to be an excellent choice for the removal of dyes from wastewater. Advanced oxidation processes (AOPs), such as ozonation and Fenton oxidation, remove dyes and colorants effectively, with removal rates ranging from 85 to 95% in 30 min to 1 h. These techniques, however, have drawbacks, such as being costly and requiring a high level of competence. Membrane-based procedures, such as nanofiltration and forward osmosis, on the other hand, are highly successful and achieve removal rates of more than 99% in just a few minutes. These membranes are extremely fragile and must be replaced frequently. After surface modification, biochar formed through the hydrothermal processing of biomass would operate as an efficient adsorbent capable of removing colors from synthetic and actual industrial wastewater. For processing various types of feedstocks (agricultural waste, algae biomass, sludge, plant residue, etc.) for biochar synthesis, pyrolysis and hydrothermal carbonization have been the most popular processes. Chemisorption is present in the majority of biochar-based adsorption methods. The potential for employing biochar as an adsorbent in the future is enormous. Engineered biochar and selective alternations will boost biochar's efficiency. Embedding metal frameworks produces catalyst-quality biochar, which can be employed as an inexpensive catalyst in a variety of sophisticated oxidation processes. Furthermore, the regeneration and reuse of used biochar adsorbents are hot topics in biochar adsorbent research. There have been few studies on the use of waste biochar as a soil amendment and additive in biological degradation processes. After being employed as adsorbents, biochar can be used as biocarriers. The problem of dye contamination in water is becoming more prevalent in most nations throughout the world; hence studies on biochar–dye interaction will be significant.

REFERENCES

Ahmad M, Rajapaksha AU, Lim JE, Zhang M, Bolan N, Mohan D, Vithanage M, Lee SS, Ok YS. 2014. Biochar as a sorbent for contaminant management in soil and water: a review. Chemosphere 99: 19–33.
Ambaye TG, Vaccari M, Castro FD, Prasad S, Rtimi S. 2020. Emerging technologies for the recovery of rare earth elements (REEs) from the end-of-life electronic wastes: a review on progress, challenges, and perspectives. Environ Sci Poll Res 27(29): 36052–36074.
Ang TN, Young BR, Taylor M, Burrell R, Aroua MK, Chen W.-H, Baroutian S. 2020. Enrichment of surface oxygen functionalities on activated carbon for adsorptive removal of sevoflurane. Chemosphere 260: 127496.

Babu DS, Srivastava V, Nidheesh PV, Kumar MS. 2019. Detoxification of water and wastewater by advanced oxidation processes. Sci. Total Environ 696: 133961.

Bhuiyan MAR, Islam A, Ali A, Islam MN. 2017. Colour and chemical constitution of natural dye henna (Lawsonia inermis L) and its application in the colouration of textiles. J. Clean. Prod. 167: 14–22.

Boudechiche N, Fares M, Ouyahia S, Yazid H, Trari M, Sadaoui Z. 2019. Comparative study on removal of two basic dyes in aqueous medium by adsorption using activated carbon from Ziziphus lotus stones. Microchem J 146: 1010–1018.

Cao X, Harris W. 2010. Properties of dairy-manure-derived biochar pertinent to its potential use in remediation. Bioresour Technol 101: 5222–5228.

Chen B, Chen Z, Lv S. 2011a. Novel magnetic biochar efficiently sorbs organic pollutants and phosphate. Bioresour Technol 102: 716–723.

Chen XL, Li F, Chen HY, Wang HJ, Li GG. 2020. Fe_2O_3/TiO_2 functionalized biochar as a heterogeneous catalyst for dyes degradation in water under Fenton processes. J. Environ. Chem. Eng. 8: 103905.

Choudhary M, Kumar R, Neogi S. 2020. Activated biochar derived from Opuntia ficus-indica for the efficient adsorption of malachite green dye, Cu^{+2} and Ni^{+2} from water. J Hazard Mater 392: 122441.

Chu JH, Kang JK, Park SJ, Lee CG. 2020. Application of magnetic biochar derived from food waste in heterogeneous sono-Fenton-like process for removal of organic dyes from aqueous solution. J Water Process Eng 37: 101455.

Côrtes LN, Druzian SP, Streit AFM, Godinho M, Perondi D, Collazzo GC, Oliveira MLS, Cadaval TRS, Dotto GL. 2019. Biochars from animal wastes as alternative materials to treat coloured effluents containing basic red 9. J Environ Chem Eng 7.

Cui D, Guo YQ, Lee HS, Cheng HY, Liang B, Kong FY, Wang YZ, Huang LP, Xu MY, Wang AJ. 2014. Efficient azo dye removal in bioelectrochemical system and post-aerobic bioreactor: Optimization and characterization. Chem Eng J 243: 355–363.

Damertey D, Jung H, Soo S, Sung D, Han S. 2019. Decolourization of cationic and anionic dye-laden wastewater by steam- activated biochar produced at an industrial-scale from spent mushroom substrate. Bioresour Technol 277: 77–86.

Degano I, Mattonai M, Sabatini F, Colombini MP. 2019. A mass spectrometric study on tannin degradation within dyed woolen yarns. Molecules 24: 2318.

De Gisi S, Notarnicola M. 2017. Industrial wastewater treatment; Abraham, M, Ed, Elsevier: Oxford: 23–42.

Del Bubba M, Anichini B, Bakari Z, Bruzzoniti MC, Camisa R, Caprini C, Checchini L, Fibbi D, El Ghadraoui A, Liguori F, et al. 2020. Physicochemical properties and sorption capacities of sawdust-based biochars and commercial activated carbons towards ethoxylated alkylphenols and their phenolic metabolites in effluent wastewater from a textile district. Sci Total Environ 708: 135217.

Deniz F. 2013. Adsorption properties of low-cost biomaterial derived from *Prunus amygdalus L.* for dye removal from water. Sci World J: 961671.

Downie A, Crosky A, Munroe P. 2009. Physical properties of biochar. Biochar for environmental management. Sci Technol: 13–32.

Esan OS, Kolawole AO, Olumuyiwa AC. 2019. The removal of single and binary basic dyes from synthetic wastewater using bentonite clay adsorbent. Am J Polym Sci Technol 5(1): 16.

Fan H, Ma Y, Wan J, Wang Y, Li Z, Chen Y. 2020. Adsorption properties and mechanisms of novel biomaterials from banyan aerial roots via simple modification for ciprofloxacin removal. Sci Total Environ 708: 134630.

Fan S, Tang J, Wang Y, Li H, Zhang H, Tang J, Wang Z, Li X. 2016. Biochar prepared from co-pyrolysis of municipal sewage sludge and tea waste for the adsorption of methylene blue from aqueous solutions: kinetics, isotherm, thermodynamic and mechanism. J Mol Liq 220: 432–441.

Furlan FR, de Melo da Silva LG, Morgado AF, de Souza AAU, Guelli Ulson de Souza SMA. 2010. Removal of reactive dyes from aqueous solutions using combined coagulation/flocculation and adsorption on activated carbon. Resour Conserv Recycl 54: 283–290.

Ganguly P, Sarkhel R, Das P. 2020. Synthesis of pyrolyzed biochar and its application for dye removal: batch, kinetic and isotherm with linear and non-linear mathematical analysis. Surf Interfaces 20: 100616.

Giannakis S, Rtimi S, Pulgarin C. 2017. Light-assisted advanced oxidation processes for the elimination of chemical and microbiological pollution of wastewaters in developed and developing countries. Molecules 22: 1070.

Gokulan R, Avinash A, Prabhu GG, Jegan J. 2019. Remediation of remazol dyes by biochar derived from Caulerpa scalpelliformis—an eco-friendly approach. J Environ Chem Eng 7: 103297.

Hoslett J, Ghazal H, Mohamad N, Jouhara H. 2020. Removal of methylene blue from aqueous solutions by biochar prepared from the pyrolysis of mixed municipal discarded material. Sci Total Environ 714: 136832.

Ihsanullah I, Jamal A, Ilyas M, Zubair M, Khan G, Atieh MA. 2020. Bioremediation of dyes: current status and prospects. J Water Process Eng 38: 101680.

Inyang M, Gao B, Yao Y, Xue Y, Zimmerman AR, Pullammanappallil P, Cao X. 2012. Removal of heavy metals from aqueous solution by biochars derived from anaerobically digested biomass. Bioresour Technol 110: 50–56.

Inyang M, Gao B, Zimmerman A, Zhang M, Chen H. 2014. Synthesis, characterization, and dye sorption ability of carbon nanotube—biochar nanocomposites. Chem Eng J 236: 39–46.

Iqbal J, Shah NS, Sayed M, Niazi NK, Imran M, Khan JA, Khan ZUH, Hussien AGS, Polychronopoulou K, Howari F. 2021. Nano-zerovalent manganese/biochar composite for the adsorptive and oxidative removal of Congo-red dye from aqueous solutions. J Hazard Mater 403: 123854.

Issa Hamoud H, Finqueneisel G, Azambre B. 2017. Removal of binary dyes mixtures with opposite and similar charges by adsorption, coagulation/flocculation and catalytic oxidation in the presence of CeO_2/H_2O_2 Fenton-like system. J Environ Manag 195: 195–207.

Jin J, Du X, Yu J, Qin S, He M, Zhang K, Chen G. 2020. High performance nanofiltration membrane based on SMA-PEI cross-linked coating for dye/salt separation. J Memb Sci 611: 118307.

Joseph SD, Camps-Arbestain M, Lin Y, Munroe P, Chia CH, Hook J, Van Zwieten L, Kimber S, Cowie A, Singh BP, Lehmann J. 2010. An investigation into the reactions of biochar in soil. Soil Res 48: 501–515.

Kah M, Sigmund G, Xiao F, Hofmann T. 2017. Sorption of ionizable and ionic organic compounds to biochar, activated carbon and other carbonaceous materials. Water Res 124: 673–692.

Kołodyńska D, Wnętrzak R, Leahy JJ, Hayes MH, Kwapiński W, Hubicki ZJ. 2012. Kinetic and adsorptive characterization of biochar in metal ions removal. Chem Eng J 197: 295–305.

Lam YF, Lee LY, Chua SJ, Lim SS, Gan S. 2016. Insights into the equilibrium, kinetic and thermodynamics of nickel removal by environmentally friendly Lansium domesticum peel biosorbent. Ecotoxicol Environ Saf 127: 61–70.

Lin YT, Kao FY, Chen SH, Wey MY, Tseng HH. 2020. A facile approach from waste to resource: Reclaimed rubber-derived membrane for dye removal. J Taiwan Inst Chem Eng 112: 286–295.

Liu Z, Zhang FS. 2009. Removal of lead from water using biochars prepared from hydrothermal liquefaction of biomass. J Hazard Mater. 167: 933–939.

Lou W, Kane A, Wolbert D, Rtimi S, Assadi AA. 2017. Study of a photocatalytic process for removal of antibiotics from wastewater in a falling film photoreactor: Scavenger study and process intensification feasibility. Chem Eng Proc Proc Intensif 122: 213–221.

Mahmoud ME, Abdelfattah AM, Tharwat RM, Nabil GM. 2020. Adsorption of negatively charged food tartrazine and sunset yellow dyes onto positively charged triethylenetetramine biochar: Optimization, kinetics and thermodynamic study. J Mol Liq 318: 114297.

Marzbali MH, Mir AA, Pazoki M, Pourjamshidian R, Tabeshnia M. 2020. Removal of direct yellow 12 from aqueous solution by adsorption onto spirulina algae as a high-efficiency adsorbent. J Environ Chem Eng 5: 1946–1956.

Meng L, Wu M, Chen H, Xi Y, Huang M, Luo X. 2020. Rejection of antimony in dyeing and printing wastewater by forward osmosis. Sci Total Environ 745.

Morais da Silva PM, Camparotto NG, Grego Lira KT, Franco Picone CS, Prediger P. 2020. Adsorptive removal of basic dye onto sustainable chitosan beads: Equilibrium, kinetics, stability, continuous-mode adsorption and mechanism. Sustain. Chem Pharm 18.

Mubarak NM, Sahu JN, Abdullah EC, Jayakumar NS. 2016. Plam oil empty fruit bunch based magnetic biochar composite comparison for synthesis by microwave-assisted and conventional heating. J Analy Appl Pyrol 120: 521–528.

Muniyasamy A, Sivaporul G, Gopinath A, Lakshmanan R, Altaee A, Achary A, Velayudhaperumal Chellam P. 2020. Process development for the degradation of textile azo dyes (mono-, di-, poly-) by advanced oxidation process—Ozonation: Experimental & partial derivative modelling approach. J Environ Manag: 265.

Nambela L, Haule LV, Mgani Q. 2020. A review on source, chemistry, green synthesis and application of textile colourants. J Clean Prod 246: 119036.

Nidheesh PV, Zhou M, Oturan MA. 2018. An overview on the removal of synthetic dyes from water by electrochemical advanced oxidation processes. Chemosphere 197: 210–227.

Oliveira JA, Cunha FA, Ruotolo LAM. 2019. Synthesis of zeolite from sugarcane bagasse fly ash and its application as a low-cost adsorbent to remove heavy metals. J Clean Prod 229: 956–963.

Pal P, Pal A. 2019. Dye removal using waste beads: Efficient utilization of surface-modified chitosan beads generated after lead adsorption process. J Water Process Eng 31.

Park JH, Choppala GK, Bolan NS, Chung JW, Chuasavathi T. 2011. Biochar reduces the bioavailability and phytotoxicity of heavy metals. Plant Soil: 348–439.

Park JH, Ok YS, Kim SH, Cho JS, Heo JS, Delaune RD, Seo DC. 2016. Competitive adsorption of heavy metals onto sesame straw biochar in aqueous solutions. Chemosphere 142: 77–83.

Park JH, Wang JJ, Meng Y, Wei Z, DeLaune RD, Seo DC. 2019. Adsorption/desorption behavior of cationic and anionic dyes by biochars prepared at normal and high pyrolysis temperatures. Colloids Surf. A Physicochem. Eng Asp 572: 274–282.

Pellera FM, Giannis A, Kalderis D, Anastasiadou K, Stegmann R, Wang JY, Gidarakos E. 2012. Adsorption of Cu (II) ions from aqueous solutions on biochars prepared from agricultural byproducts. J Environ Manag 96: 35–42.

Peng XY, Ye LL, Wang CH, Zhou H, Sun B. 2011. Temperature-and duration-dependent rice straw-derived biochar: characteristics and its effects on soil properties of a Ultisol in southern China. Soil Tillage Res 112: 159–166.

Qayyum MF, Liaqat F, Rehman RA, Gul M, Hye MZ, Rizwan M, Rehman MZ. 2017. Effects of co-composting of farm manure and biochar on plant growth and carbon mineralization in an alkaline soil. Environ Sci Pollut Res 24: 26060–26068.

Qi Y, Zhu L, Shen X, Sotto A, Gao C, Shen J. 2019. Polythyleneimine-modified original positive charged nano filtration membrane: Removal of heavy metal ions and dyes. Sep Purif Technol 222: 117–124.

Rehman MZ, Khalid H, Akmal F, Ali S, Rizwan M, Qayyum MF, Iqbal M, Khalid MU, Azhar M. 2017. Effect of limestone, lignite and biochar applied alone and combined on cadmium uptake in wheat and rice under rotation in an effluent irrigated the field. Environ Pollut 227: 560–568.

Rubeena KK, Prasad PH, Laiju AR, Nidheesh P. V. 2018. Iron impregnated biochars as heterogeneous Fenton catalyst for the degradation of acid red 1 dye. J Environ Manag 226: 320–328.

Santra B, Ramrakhiani L, Kar S, Ghosh S, Majumdar S. 2020. Ceramic membrane-based ultrafiltration combined with adsorption by waste derived biochar for textile effluent treatment and management of spent biochar. J Environ Health Sci Eng. 18(2): 973–992.

Saxena M, Sharma N, Saxena R. 2020. Highly efficient and rapid removal of a toxic dye: adsorption kinetics, isotherm, and mechanism studies on functionalized multiwalled carbon nanotubes. Surf Interfaces 21: 100639.

Saxena S, Raja ASM. 2014. Natural dyes: sources, chemistry, application and sustainability issues BT—roadmap to sustainable textiles and clothing: eco-friendly raw materials, technologies, and processing methods; Muthu, S.S., Ed, Springer Singapore: Singapore: 37–80.

Shan R, Lu L, Gu J, Zhang Y, Yuan H, Chen Y, Luo B. 2020. Photocatalytic degradation of methyl orange by Ag/TiO2/biochar composite catalysts in aqueous solutions. Mater Sci Semicond Process 114: 105088.

Shao H, Li Y, Zheng L, Chen T, Liu J. 2017. Removal of methylene blue by chemically modified defatted brown algae Laminaria japonica. J. Taiwan Inst. Chem Eng 80: 525–532.

Shrivastava VS. 2010. The biosorption of safranine onto parthenium hysterophorus L: equilibrium and kinetics investigation. Desalination and Water Treatment 22: 146–155.

Sirajudheen P, Karthikeyan P, Vigneshwaran S, Meenakshi S. 2020. Synthesis and characterization of La(III) supported carboxymethylcellulose-clay composite for toxic dyes removal: evaluation of adsorption kinetics, isotherms and thermodynamics. Int J Biol Macromol 161: 1117–1126.

Sun Y, Gao B, Yao Y, Fang J, Zhang M, Zhou Y, Chen H, Yang L. 2014. Effects of feedstock type, production method, and pyrolysis temperature on biochar and hydrochar properties. Chem Eng J 240: 574–578.

Thies JE, Rillig MC. 2009. Characteristics of biochar: biological properties. Biochar for environmental management. Sci Technol 8: 5–105.

Van Zwieten L, Kimber S, Morris S, Chan KY, Downie A, Rust J, Joseph S, Cowie A. 2010. Effects of biochar from slow pyrolysis of papermill waste on agronomic performance and soil fertility. Plant Soil 327: 235–246.

Villegas-Guzman P, Giannakis S, Rtimi S, Grandjean D, Bensimon M, de Alencastro F, Torres-Palma R, Pulgarin C. 2017. A green solar photo-Fenton process for the elimination of bacteria and micropollutants in municipal wastewater treatment using mineral iron and natural organic acids. Appl Catal B Environ 219: 538–549.

Vithanage M, Mayakaduwa S, Herath I, Ok YS, Mohan D. 2016. Kinetics, thermodynamics and mechanistic studies of carbofuran removal using biochars from tea waste and rice husks. Chemosphere 150: 781–789.

Wang H, Wang S, Gao Y. 2020. Cetyl trimethyl ammonium bromide modified magnetic biochar from pine nut shells for efficient removal of acid chrome blue K. Bioresour Technol 312: 123564.

Wang H, Zhao W, Chen Y, Li Y. 2020. Nickel aluminum layered double oxides modified magnetic biochar from waste corncob for efficient removal of acridine orange. Bioresour Technol 315: 123834.

Wang L, Yan W, He C, Wen H, Cai Z, Wang Z, Chen Z, Liu W. 2018. Microwave-assisted preparation of nitrogen-doped biochars by ammonium acetate activation for adsorption of acid red 18. Appl Surf Sci 433: 222–231.

Xu X, Geng A, Yang C, Carabineiro SAC, Lv K, Zhu J. 2020a. One-pot synthesis of La—Fe—O @ CN composites as photo-Fenton catalysts for highly efficient removal of organic dyes in wastewater. Ceram Int 46: 10740–10747.

Xu X, Xu Z, Huang J, Gao B, Zhao L, Qiu H, Cao X. 2020b. Sorption of reactive red by biochars ball milled in different atmospheres: co-effect of surface morphology and functional groups. Chem Eng J: 127468.

Yao Y, Gao B, Inyang M, Zimmerman AR, Cao X, Pullammanappallil P, Yang L. 2011a. Biochar derived from anaerobically digested sugar beet tailings: characterization and phosphate removal potential. Bioresour Technol 102: 6273–6278.

Yao Y, Gao B, Inyang M, Zimmerman AR, Cao X, Pullammanappallil P, Yang L. 2011b. Removal of phosphate from aqueous solution by biochar derived from anaerobically digested sugar beet tailings. J Hazard Mater 190: 501–507.

Yao X, Ji L, Guo J, Ge S, Lu W, Chen Y, Cai L, Wang Y, Song W. 2020. An abundant porous biochar material derived from wakame (Undaria pinnatifida) with high adsorption performance for three organic dyes. Bioresour Technol 2020 318: 124082.

Yek PNY, PengW, Wong CC, Liew RK, Ho YL, Wan MahariWA, Azwar E, Yuan TQ, Tabatabaei M, Aghbashlo M, et al. 2020. Engineered biochar via microwave CO2 and steam pyrolysis to treat carcinogenic Congo red dye. J Hazard Mater: 395.

Younis U, Malik SA, Rizwan M, Qayyum MF, Ok YS, Shah MHR, Rehman RA, Ahmad N. 2016. Biochar enhances the cadmium tolerance in spinach (Spinacia oleracea) through modification of Cd uptake and physiological and biochemical attributes. Environ Sci Poll Res 23: 21385–21394.

Yu KL, Lee XJ, Ong HC, Chen WH, Chang JS, Lin CS, Show PL, Ling TC. 2020. Adsorptive removal of canic methylene blue and anionic Congo red dyes using wet-torrefied microalgal biochar: Equilibrium, kinetic and mechanism modeling. Environ Pollut: 115986.

Zerin I, Farzana N, Sayem ASM, Anang DM, Haider J. 2020. Potentials of natural dyes for textile applications; Hashmi, S., Choudhury, IA, Eds, Elsevier: Oxford: 873–883.

Zhang H, Lu T, Wang M, Jin R, Song Y, Zhou Y, Qi Z, Chen W. 2020. Inhibitory role of citric acid in the adsorption of tetracycline onto biochars: Effects of solution pH and Cu2+. Colloids Surf. A Physicochem Eng Asp: 595.

Zhang H, Wang L, He K, Lu A, Sarmah J, Li NS, Bolan J, Pei H, Huang. 2013a Using biochar for remediation of soils contaminated with heavy metals and organic pollutants. Environ Sci Pollut Res 20: 8472–8483.

Zhang W, Mao S, Chen H, Huang L, Qiu R. 2013b Pb (II) and Cr (VI) sorption by biochars pyrolyzed from the municipal wastewater sludge under different heating conditions. Bioresour Technol 147: 545–552.

Zhu K, Wang X, Chen D, Ren W, Lin H, Zhang H. 2019. Wood-based biochar as an excellent activator of peroxydisulfate for Acid Orange 7 decolourization. Chemosphere 231: 32–40.

10 Efficacy of Biochar in Decontamination of Pesticides

Sapna Salar, Neha Rana, and Piyush Gupta

CONTENTS

10.1 INTRODUCTION

Excessive use of chemical pesticides in today's world is seen to be a major contributor to a variety of ecological issues and unfavorable environmental repercussions. Agricultural activities, which use agrochemicals including synthetic fertilizers, chemical growth agents, and pesticides, are among the major sources of environmental contamination. Despite many advantages, agrochemicals may have a variety of

DOI: 10.1201/9781003203438-10

unintended harmful effects that endanger the ecosystem (Zanella et al. 2011). The use of pesticides to control pests (insects, weeds, and diseases) is an important aspect of crop management in contemporary agriculture that can help with the provision of safe, high-quality, and low-cost food, but many pesticides, such as dichlorodiphenyltrichloroethane (DDT) and lindane, can persist in soil and water for years. These have a negative impact on the environment as a whole and can build up in the food chain (Khorram et al. 2016). Several methods for removing pesticides from liquids and purifying mixtures of substances have been developed and successfully implemented. Biological, chemical, and physical remediation are the three main approaches used. Chemical methods, in addition to being toxic, enter the soil and disrupt its enzymatic activity and fertility. Pesticide residues in water are a major source of concern since they pose a great threat to aquatic life and also to humans. Accidental contamination, industrial effluent, surface water contamination, and residues from pesticide-treated soils, washing of spray equipment after spraying processes, drainage into ponds, lakes, streams, and river water, and aerial sprays are all ways pesticides might end up in water (Carter and Heather 1995). Biomagnification refers to the increase in pesticide concentrations caused by their persistent and nonbiodegradable nature in the tissues of organisms at each successive level of the food chain. As a result, creatures at higher levels of the food chain suffer more harm than those at lower levels. Farmers should emphasize using pesticides that are more poisonous to target species than capable of polluting the environment and adversely affecting livestock and human health. Various techniques such as biological treatment, membrane filtering, improved oxidation treatment, photocatalyst, incineration, dumping in the ground, and ion exchange therapy are some of the approaches that have been introduced and used to remove pesticides. Every method has its own set of advantages and disadvantages. However, most of these treatment methods have limitations such as high maintenance and operational expenses, as well as generation of nondegradable by-products following treatment (Al Hattab et al. 2012). Thus adsorption is a preferred method due to simplicity of design and operation toward a sustainable environment and effective removal of pesticides. Adsorbents such as biochar, activated carbon, minerals, clays, and hybrid materials through metal organic frameworks (MOFs) (Abdelhameed et al. 2019) may be used especially for removal of pesticides from solutions and water bodies (Gill and Garg 2014). Biochar has been used in a variety of applications, including energy generation, water pollutants treatment, and other fields, due to its wide specific surface area, rich porous structure, plentiful surface functional groups, and high mineral content (Baharum et al 2020). Biochar is generated by pyrolyzing organic wastes under regulated oxygen and temperature conditions (about 300°C). It is a low-cost material that can be used for a variety of purposes, including soil amendment. Biochar is a porous substance that can hold a lot of water. It promotes aeration and creates favorable circumstances for microbial development, which reduces microorganism metabolic activity and improves pesticide breakdown (Beesley et al. 2010; Gaskin et al. 2008). Biochar alters soil characteristics, increasing microbial biomass and activity, enzyme activity, and the structure of microbial communities (Beesley et al. 2010; Tang et al. 2013; Van Zwieten et al. 2010; Wang et al. 2012; Zhu et al. 2017). Biochar can improve soil fertility, increase agricultural yields, reduce carbon emissions, and repair degraded land when used as a soil amendment (Cederlund et al. 2016; Laird 2008; Zhu et al. 2017). The stability of biochar in soil is critical for its utilization as a pesticide remedial material. Apart from pyrolysis settings and raw material type, H/C and O/C molar ratios have also been reported as key indications for evaluating biochar's potential for use in pollution bioremediation (EBC 2012). Furthermore, biochar can remove organic contaminants from water, such as dyes and pharmaceutical chemicals (Wu et al. 2020). Biochar is also considered a unique adsorbent capacity due to its high specific area and high carbonaceous nature. Therefore, mixing small amounts of biochar with soil could result in high adsorption and consequently decrease the bioavailability of contaminants to microbial communities, plants, earthworms, and other organisms in the soil. Furthermore, biochar can remove organic contaminants from water, such as dyes and pharmaceutical chemicals (Baharum et al. 2020; Wu et al. 2020). Biochar is produced from pyrolysis of organic wastes under controlled conditions of oxygen and temperature (<300°C). Biochar is relatively inexpensive with high water retention capacity (Tan et al. 2015; Zhu et al. 2017). It improves aeration conditions and provides suitable conditions for microbial growth, which in turn alleviates the metabolic activity of microorganisms and enhances the degradation of pesticides (Beesley et al. 2010; Gaskin et al.

2008). Biochar also induces changes in soil properties that increase microbial biomass activity, enzyme activity, and microbial community structure (Beesley et al. 2010; Van Zwieten et al. 2010; Tang et al. 2015; Wang et al. 2012; Zhu et al. 2017). Biochar as soil amendment agent can improve the soil fertility, increase crop yields, reduce carbon emissions, and restore degraded land (Cederlund et al. 2016; Laird 2008; Zhu et al. 2017). Because of its ability to limit the mobility of pollutants such as heavy metals (Pb^{2+}, Cr^{3+}, Cd^{2+}, Ni^{2+}, As^{5+}, and Cu^{2+}) and dissolved organic molecules (such as insecticides, polynuclear aromatic hydrocarbons, and dyes such as methylene blue), biochar has been utilized for environmental remediation (Beesley et al. 2011). Biochar's physicochemical qualities are determined by the initial organic material and the pyrolysis circumstances in which it is produced. Biomass feedstock is cooked at relatively low temperatures (700°C) in the absence or limited supply of oxygen in pyrolysis, a thermochemical conversion process, and rapidly changed into biochar, bio-oil, and syngas as bioenergy by-products (Song and Guo 2012). Producing biochars through pyrolysis requires a variety of feedstock sources, highest treatment temperatures (HTT), heating rates, processing time, holding time at HTT, ancillary inputs (nitrogen, oxygen, steam, etc.), partial pressures (1–10 bar gauge), reaction vessels, pretreatments (e.g., drying, chemical activation), post-treatments (e.g., crushing, de-ashing, phosphoric acid activation) (Kookana 2010; Lehmann and Joseph 2015). Variation of these parameters can improve production conditions and result in desired end by-products. The physical and chemical qualities of the original feedstock, as well as HTT, are found to be the most important parameters in influencing the quantity and quality of the resulting biochars. In order to assess the contribution of these two key parameters, Zhao et al. (2013) examined at a wide range of production temperatures (from 200 to 650°C) and biomass types (12 organic materials from six categories of wastes consisting of wood, crop, food, aquatic plant, municipal residues, and animal manures). To quantify the evaluations, heterogeneities based on feedstock and temperature were determined. The results revealed that total organic carbon, fixed carbon, and mineral elements of the end product are predominantly influenced by feedstock characteristics, whereas surface area, pH, and biochar recalcitrance are primarily influenced by HTT. The biomasses used in the production of biochar and its various environmental applications are illustrated in Figure 10.1 (Gautam et al. 2021).

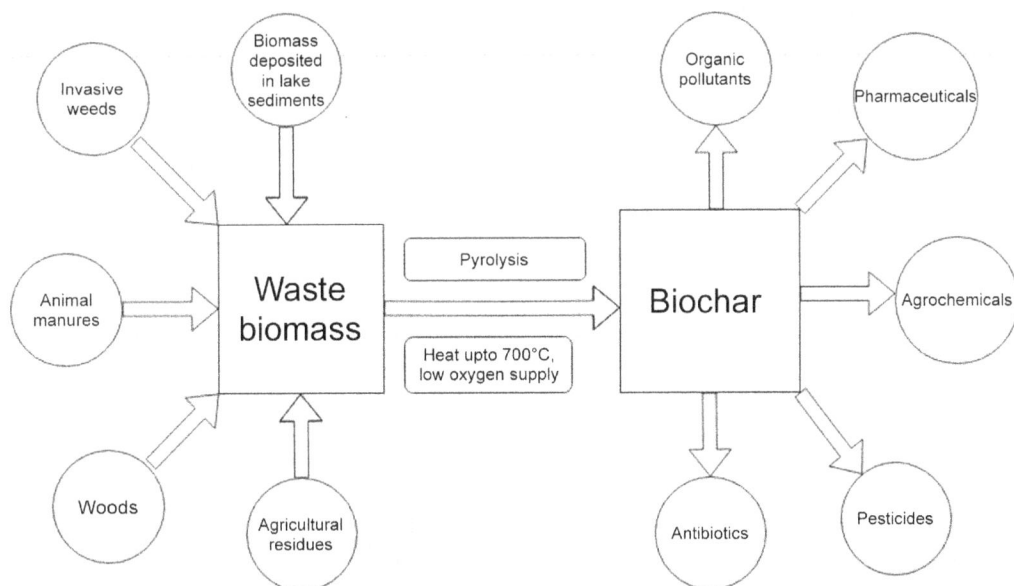

FIGURE 10.1 Biomasses used in production of biochar and its environmental applications.

Source: Gautam et al. (2021).

10.2 COMPOSITION OF BIOCHARS

Biochar's chemical composition varies depending on feedstock type and pyrolysis circumstances (e.g., residence time, temperature, heating rate, and reactor type); hence, no two biochars are alike, and defining the exact chemical composition of biochar is challenging (Lehmann and Joseph 2021). Biochars are mostly made up of carbon. Depending on the feedstock type, the organic portion of biochar contains a lot of carbon, whereas the inorganic portion mostly contains minerals including Ca, Mg, K, and inorganic carbonates (carbonate ion). The pyrolysis temperature used to make biochar does not produce much graphite, and aromatic rings in biochar are not stacked and aligned in the same way as they are in graphite. The size of these ring structures is determined by the temperature at which biochar is made. During biochar formation, more irregular carbon groupings are generated, which contain O and H (Lehmann and Joseph 2021).

10.2.1 SOLUBLE AND MINERALIZABLE CARBON

Biochars have water-soluble and mineralizable compounds, which can be food for microbes and can stimulate seeds and plants. This is referred to as labile C. When biomass depolymerizes (breaks down) in an oxygen-depleted environment, the majority of these chemicals are formed from the resultant volatile compounds. These can be divided into categories that include low–molecular-weight neutral compounds such as alcohols, aldehydes, ketones, sugars, low-molecular-weight acids; biopolymers (polysaccharides, proteins, and amino sugars), building blocks (polyphenolics/polyaromatic acids); large macromolecules similar to humic acids; solvent extractable organic molecules including carboxylic and benzoic acids, lactones, alkanes, urea, complex sugars, butenolide, glycerol, polyphenols, quinones, benzene and polycyclic aromatic hydrocarbons. The water-soluble compounds in the high-temperature biochars are mainly low-molecular-weight (LMW) acids.

10.2.2 POLYCYCLIC AROMATIC HYDROCARBON

There are maximum tolerable limits to total and bioavailable PAHs (polycyclic aromatic hydrocarbons). Naphthalene is the most abundant PAHs (approximately 90% of total PAHs) in biochar. They are produced by slow pyrolysis in the lab at various temperatures or in the field. Total PAH concentration and bioavailable PAH concentration are usually different. Strong solvents (e.g., toluene) are used to extract PAHs. Bioavailable PAHs are higher in some lower-temperature biochars.

10.2.3 ASH (INORGANIC COMPOUNDS)

Metals and nonmetals, the inorganic compounds in biochar, constitute ash. Some have crystalline structure and others are amorphous and have a dimension of <1 μm. Amorphous and crystalline compounds can include nitrates, chlorides, oxides, sulfates, sulfides, carbonates, and phosphates. Crystalline minerals can include rock phosphate TCP-$(Ca_3(PO_4)_2)$, salt $(NaCl)$, sylvite (KCl), struvite $(NH_4MgPO_4 \cdot 6H_2O)$, calcite $(CaCO_3)$, dolomite $(CaMg(CO_3)_2$, anatase (TiO_2), SiO_2, clays $(Al_2O_3.SiO_2.H_2O)$, $FeS/Fe_2O_3/Fe_3O_4$. They can be soluble (e.g., sodium chloride) or insoluble (e.g., calcium sulfate).

10.2.4 HEAVY METALS

Heavy metals are present in some biochars but are largely absent in plant-based biochars. Heavy metals are retained in biochars generated from heavy-metal-containing feedstock, such as sewage sludge. The amount of heavy metals that can be applied to land in most countries is regulated.

10.2.5 TOTAL AND AVAILABLE NUTRIENTS

Mineral concentration in low ash woody biochars is modest, ranging from 5 mg/kg (P in oak) to 5 g/kg total elements. Available amounts range from 24 mg/kg (Mg in oak and pine) to 1.6g/kg (K in hazelnut) (4–37%). Biochars generated from ash-rich biomass have ~10–100 times more total nutrients, from 1 to 70 g/kg. Except for Na in corn and K in paper waste, the accessible proportion in high-ash biochars is substantially greater, ranging from 15 to 98% (~1 g to 50 g/kg).

10.3   PHYSICAL PROPERTIES OF BIOCHARS

The physical properties of biochars influences its (1) suitability as natural habitat for soil bacteria and mycorrhizal fungi by allowing surface area and protection from predators, (2) interaction with water, minerals, and nutrients in the soil, and (3) its environmental mobility. The basic physical properties are bulk density, particle density, particle size, porosity and surface area, etc. These are all interrelated numerically, in measurement and in action.

10.3.1 DENSITY

The density of the biochars depends on the nature of the starting material and the pyrolysis process. The increase in solid density (or real density) of biochar is due to the loss of volatile and condensable chemicals from the disorganized phase of the biochars and the accompanying relative rise in the organized phase formed by graphite-like crystallites. The solid density of biochar increases with increasing process temperature and longer heating residence times. The maximum density of carbon in biochars has been reported to lie between 2.0 and 2.1 g/cm^3. Bulk density is also an important physical feature of biochars. The bulk densities of biochars generated from various types of woods treated in different types of traditional kilns ranged from 0.43 to 0.30 g/cm^3, according to Pastor-Villegas et al. (2006). Often, an increase in solid density is accompanied by a decrease in apparent densities as porosity develops during pyrolysis (Guo et al. 1998).

10.3.2 MACROPOROSITY

Despite the fact that micropore surface areas in biochars are much bigger than macropore surface areas, macropore volumes can be significantly larger than micropore volumes. As predicted from the regular size and arrangement of plant cells in most biomass from which biochars are derived, the macropore size distribution is composed of discrete groups of pores sizes instead of a continuum (Wildman et al. 1991).

10.3.3 NANOPOROSITY

Micropores (<2 nm in diameter) make up the majority of surface area of biochars and are responsible for the high adsorptive capacities. High temperatures cause micropores to widen in some cases, resulting in pore amplification (Zhang et al. 2004). As a result, the fraction of volume found in the micropore range decreases while the total pore volume increases. Biochars produced at atmospheric pressure with moderate heating rates primarily contain micropores, whereas biochars produced at high heating rates primarily include macropores because of melting. Biochar materials contain mesopores as well. Many liquid–solid adsorption processes require the presence of these holes. Pistachio nut shells, for example, include a combination of micro- and mesopores, with micropores predominating, indicating that these activated carbons can be employed for both gas and liquid adsorption (Lua et al. 2004).

10.3.4 Surface Area

Increasing the pyrolysis temperature increases the surface area and porosity of biochar. This depends on the breakdown of organic matter and the production of micropores. Furthermore, higher pyrolysis temperatures may result in increased surface area due to the degradation of aliphatic alkyls and ester groups, as well as the exposure of the aromatic lignin core. Biochar produced at a temperature higher than 400°C is more effective for organic and inorganic contaminant sorption due to its high surface area and considerable micropore development (Uchimiya et al. 2012). At low pyrolysis temperatures (100–300°C), organic and inorganic contaminants segment into noncarbonized biochar fractions.

10.3.5 Particle Size Distribution

The nature of the original material (feedstock), the manufacturing technique, and the temperature all influence the particle sizes of biochar produced by pyrolysis of organic material. The particle sizes of organic matter feedstock are larger than the resulting biochar due to shrinkage and attrition during pyrolysis. Depending on the mechanical intensity of the pyrolysis technology employed, a degree of attrition of the biomass particles occurs during processing. The loss of volatile stuff is accompanied by an increase in linear shrinkage of the particles that are being pyrolyzed. Freitas et al. (1997) found that raising the pyrolysis pressure (from atmospheric to 5, 10, and 20 bars) causes bigger biochar particles to form. Pelletizing or granulating can be used to enhance the particle size even more. Particle size can be reduced by grinding biochar, freeze-thaw cycles, ingesting and excretion by worms. Sieving can ensure a more uniform particle size. Particle size influences the pore volume between particles, the bulk density, the water holding capacity, hydraulic conductivity, the effectiveness in compaction remediation, the mobility and loss of biochar in wind and water, the ease of transport and ingestion by soil flora and fauna, and the speed of oxidation and aging

10.3.6 Mechanical Strength

Biochar's mechanical strength is proportional to its solid density. Mechanical strength is a criterion for determining the quality of activated carbon in biochar in terms of its capacity to endure wear and strain over time. As a result, pyrolyzed biomass has a stronger mechanical strength than the biomass feedstock from which it was formed due to the increased molecular order. Agricultural wastes, such as nut shells and fruit stones, are suitable as activated carbons due to their high mechanical strength and hardness, in turn due to high lignin and low ash content (Aygun et al. 2003).

10.3.7 Hydrophobicity

Tars (aliphatic chemicals) condensing on the charcoal surface during pyrolysis induce hydrophobicity. Hydrophobicity has an effect on the water uptake by biochar and therefore on the water holding capacity of biochar and microbial interactions. Biochars made at low temperatures are highly hydrophobic; however, longer pyrolysis times or washing biochar can lower hydrophobicity. Hydrophobicity may diminish as biochar reacts with soil.

10.4 CHEMICAL PROPERTIES OF BIOCHARS

10.4.1 Fertilizer Value

Although most manure and straw biochars have a high nitrogen concentration, when put to the soil, only a small amount of it is readily available in the form of nitrates or ammonium for plant absorption. Many additional elements are more concentrated in biochar than in biomass, except nitrogen.

The availability of phosphorus (P) and potassium (K) is high (36 and 54% of the total P and K), whereas the available N is low (<5 % of the total N).

10.4.2 pH AND LIMING VALUE

Biochars are mostly alkaline, having pH greater than 7, and their usage frequently raises the pH of acidic soils (called a liming effect) and influences cation mobility (Van Zwieten et al. 2010). At 350°C, most woody biochars have a pH of around 6, which rises to around 8 at 450°C and continues to rise more slowly over 450°C. The pH values of biochars are positively correlated with the formation of carbonates and the contents of inorganic alkali. Higher pH with increasing temperature has been associated with the increases in ash content and oxygen functional groups that occur during pyrolysis. The liming value of biochars is a measure of their ability to reduce soil acidity in terms of calcium carbonate ($CaCO_3$) equivalents, compared to the effectiveness of adding the same amount of pure $CaCO_3$ to the same soil (lime).

10.4.3 ELECTRICAL CONDUCTIVITY

When biochar is mixed with water, some of its salts dissolve, allowing the water to conduct electricity. Electrical conductivity is a measure of total dissolved salts (TDS), and it indicates whether nutrients are available or whether there is an excess of ash/salt. Making a slurry of ground biochar in distilled water and testing with an EC or TDS meter can be used to determine electrical conductivity, just like pH.

10.4.4 CATION EXCHANGE CAPACITY (CEC)

The ability of biochars to retain exchangeable cations like calcium (Ca^{2+}) and potassium (K^+) is measured by their cation exchange capacity (CEC). The oxygen functional groups on the surface of biochar account for the majority of its CEC. Low-temperature biochars have a higher CEC than high-temperature biochars. However, high-temperature biochars can adsorb more nutrients and organic matter (OM). Biochars typically have a lower CEC than fertile soils, but they can have a greater CEC than sandy and low-OM soils. Biochar's CEC can rise as it ages in soil (especially those with a low pH).

10.4.5 ANION EXCHANGE CAPACITY (AEC)

The ability of biochars to retain anions such as PO_4^{3-} (phosphate) and NO_3^- (nitrate) is critical for retaining nutrients and preventing leaching and associated water pollution. On exposure to soil, this property declines. AEC can be increased by (1) increasing temperature during pyrolysis, (2) precipitation with Fe, Ca, and Al for retaining anions such as phosphate, and (3) retention by trapping in pores for anions that do not precipitate (e.g., nitrate).

10.5 IMPACT OF BIOMASS ON THE SORPTIVE PROPERTIES OF BIOCHARS

The type and quality of precursor materials is the first major aspect that directly influences biochar recovery and its characteristics. The state of pyrolysis outputs is determined by the chemical composition of biomass, as well as its structure, size, shape, and feed density. Biomass composition has the greatest impact on biochar characteristics at low pyrolysis temperatures (500°C). The proportional abundance of cellulose, hemicelluloses, and lignin in biomass materials determines the characteristics of produced biochar. Thermal decomposition of hemicellulose occurs at low temperatures between 200 and 260°C. The temperature ranges from 240 to 350°C for cellulose and from 280 to 500°C for lignin (Lehmann and Joseph 2015). Thus the chemical, physical, and structural changes in biochars during processing are determined by the relative amounts of these components in the raw biomass. The mineral content in the original biomass can also influence the characteristics of biochar, affecting its ash level. De-ashed biochars have a greater affinity for phenanthrene, owing to an increase in hydrophobic sorption sites (Sun et al 2013). To avoid biochar structural loss caused by high ash content, a pretreatment approach to remove the biomass minerals before pyrolysis by washing it with acid may be beneficial (Rodrguez-Mirasol et al 1993). However, some findings suggest that biochars with a higher mineral ash content are more effective at absorbing some materials. Woody biomasses (e.g., wood, barks, and shells) typically include a high percentage of lignin, resulting in biochars with a high aromatic organic composition, low ash content, a large surface area and with greater porosity. Herbaceous feedstocks with high silica content, such as grass, husks, hulls, and straws, as well as animal manures, frequently yield porous biochars with high ash content. Animal manures are not very porous as a starting feedstock; however, pyrolysis has been demonstrated to improve their porosity (Joseph et al. 2007). The total organic carbon and surface area of plant-derived biochars were higher, whereas the surface carbon content, which is important for sorption of organic molecules, was higher in manure biochar (Qiu et al. 2014). The production, nutritional content, and CEC of seaweed-derived biochars are all high, while the organic carbon content and surface area are low. The size of the biomass particles can also affect the stability of the biochar generated depending on its particle size distribution. The capacity of biochar decreases as particle size increases. The particle size distribution is expected to alter during the pyrolysis process from a higher proportion of larger particles in the original feedstock to a higher proportion of finer particles in the resultant biochar. Particles larger than 4.75 mm were found in sawdust and wood chips in levels of around 2 and 15%, respectively. The particle size in the resultant biochar after pyrolysis at 700–750°C was reduced to almost zero in sawdust biochar and 5% in wood chips biochar. The sorption capacity of herbicides (diuron and artrazine) in biochars made from poultry, litter, and paper mill waste was compared. According to Clay and Malo (2012), pesticide removal efficiency was shown to be higher with hydrothermal biochars than with thermal biochars. Also, maize stover biochar has a higher affinity for atrazine (a weak cationic pesticide), whereas switchgrass biochar immobilizes 2,4-D (an anionic pesticide) significantly better. Thus the type of chemical affects the function of the feedstock.

10.6 HIGHEST TREATMENT TEMPERATURE (HTT) AND BIOCHAR SORPTIVE PROPERTIES

During pyrolysis, the highest treatment temperature (HTT) has been shown to be the most effective factor in determining the sorption capacity of the resulting biochars. Higher temperature enhances the conversion of noncarbonized (amorphous) material of biomass to carbonized domain. A nonlinear (competitive) and substantial surface adsorption is observed in the carbonized part of the biochar (Cornelissen et al. 2005). The enhanced sorption capacity of carbonaceous sorbents is mostly due to surface adsorption (Beesley et al. 2011). The sorption capacity of carbonized materials at high temperatures due to increased surface area is mainly a result of increased microporosity. The extent of structural changes in the starting biomass, such as microstructural rearrangement, attrition, and the formation of fractures, are important for surface area expansion and is mostly determined by processing temperatures (Lehmann and Joseph 2015)

10.7 BIOCHAR FOR PESTICIDE DECONTAMINATION

Pesticides are the chemicals used to keep pests at bay. Herbicides, insecticides, and other pesticides are all included in the term "pesticide" (which may comprise termiticides, insect growth regulators, etc.). Molluscicide, nematicide, piscicide, rodenticide, avicide, bactericide, animal repellent, insect repellent, fungicide, antimicrobial, and lampricide are some of the most common pesticides. Herbicides are the most frequent of them, accounting for over 80% of all pesticide use. Table 10.1 and Table 10.2 encompass the classification of pesticides based on sources of origin and intended use, respectively (Akashe et al. 2018).

TABLE 10.1
Classification of Pesticides Based on Sources of Origin (Akashe et al. 2018)

Biopesticides: These have an effect on the target pests as well as other closely related organisms. They disintegrate quickly, have low toxicity, and are only needed in small amounts.

Microbial pesticides: Microorganisms such as bacteria or fungi are the active element in these pesticides. Examples include *Bacillus thuringiensis* and *Bacillus sphaerius*, which create toxins that damage mosquito larvae and black fly larvae.

Plant incorporated protectants-: Plant-incorporated protectants are pesticidal chemicals produced by plants, as well as the genetic material required for the plant to create the substance, integrated together and introduced into the crop as a single unit: e.g., Bt toxin in corn.

TABLE 10.2
Classification of Pesticides Based on Intended Use (Akashe et al. 2018)

S. No.	Type	Use	Example
1	Rodenticides	Kill rats and other rodent pests	Anticoagulants, arsenic, strychnine
2	Acaricides	Kill ticks, mites, spiders, etc.	Organophosphorous compounds, carbamates, pyrethroids
3	Nematicides	Kill parasitic worms like roundworms and threadworms	Vydate, ethylene bromide
4	Insecticides	Kill or repel insects or related species	Aldrin, Chlordane, Chlordecone, Endosulfan, Endrin
5	Herbicide	Kill or hinder the growth of undesired plants such as weeds	Paraquat, Dinoseb, Diquat, Diclofop
6	Fungicide	Biocidal chemicals or biological creatures that are used to eliminate parasitic fungi, mold, and their spores	Propiconazole, metalaxyl, difenoconazole, hexaconazole
7	Molluscicides	Pesticides that kill molluscs	Metaldehyde, methiocarb, acetylcholinesterase inhibitors
8	Miscellaneous	–	Lead arsenate, copper sulfate

Sorption of a chemical to a sorbent is defined as a physicochemical transfer process by which the material is partitioned between the solid and aqueous phases (Wauchope et al. 2002). The sorbent properties (surface area, pores size, cation exchange capacity (CEC), and surface functional groups), the chemicals properties (molecular structure and size, acidity or basicity, hydrophobicity, solubility, polarity, and polarizability), and environmental conditions (temperature, moisture, light, and wind) can directly and indirectly influence the sorption process. Sorption is usually represented by a partition constant, K_d, which is a ratio of chemical concentration sorbed to a solid to the chemical remaining in solution after equilibration. K_d is determined at equilibrium condition in which the distribution between the dissolved and sorbed chemical becomes stable, usually over hours, days, or even weeks. Sorptive characteristics of a specific sorbent is its affinity for a specific material and is generally evaluated using the sorption isotherms (Nelson et al. 2000). A sorption isotherm model depicts the quantity of a material sorbed by a specific sorbent as a function of the equilibrium concentration of the material in solution at a constant temperature (Shirvani et al. 2006; Yavari et al. 2015). Batch equilibrium technique is frequently employed to determine sorption isotherms during which a series of solutions containing different concentrations of the sorbate are mixed with known amounts of sorbent and shaken at a specific temperature. After equilibrium, the concentration of the sorbate in each solution is measured. The amount of sorbed material can be calculated by subtracting the material concentration in the solution at equilibrium from the initial concentration. A relationship is then made between the amount of the sorbed material per unit weight of sorbent and the equilibrium concentration. The most commonly used models for pesticides sorption are the Freundlich and Langmuir sorption models.

The Freundlich sorption model is expressed as:

$$X/M = K_f C_e^{1/n}$$

where X/M is the ratio of pesticides to the sorbent mass, C_e is the pesticide concentration in the solution at equilibrium, and K_f and $1/n$ are Freundlich constants.

The Langmuir sorption equation is expressed as:

X/M $= A_{max}KC_e/(1 + KC_e)$
where X/M = ratio of pesticides to the sorbent mass,
A_{max} = maximum sorption of pesticide to the sorbent,
C_e = pesticide concentration in the solution at equilibrium, and
K = Langmuir constant.

The sorption and environmental persistence of the pesticides as an important group of agricultural chemicals are influenced by their physical and chemical properties. Although the properties of biochars can be modified by conditioning including post-treating the biochar with minerals, nutrients, and/or microorganisms and altering process conditions such as highest treatment temperature (HTT), etc., commercially viable biochars are almost always generated in high-C-containing biochar precursors through physical or chemical activation. Physical activation is achieved when the initial pyrolysis reactions, which occur in an inert atmosphere at moderate temperatures (400–800°C), are supplemented by a second stage in which the resulting biochars are partially gasified at a higher temperature (usually >900°C) with oxidizing gases such as steam, CO_2, air, or a mixture of these gases. This results in finished items with well developed interior pores that are readily accessible. The chemical and physical composition of biochars are diverse and heterogeneous, making it a suitable platform for pesticide removal (as given in Table 10.3) (Yavari et al. 2015).

10.8 FACTORS INFLUENCING SORPTIVE PROPERTIES OF BIOCHAR FOR REMOVAL OF PESTICIDES

The heating rate and flow rate of supplementary inputs are the most critical criteria to define the features of produced biochar in addition to the initial feedstock and pyrolysis temperature. Heating rate and pressure affect the volatile mass transfer during pyrolysis, and so these two factors control the formation of biochar pores and cracks, the thickness of cells walls, shape, and surface area of biochar particles and consequently their sorptive properties (Lua et al. 2004). As the heating rate and the temperature of the biomass is increased, degradation of the biomass components increases, thus affecting its structural properties. Cetin et al. (2004) established that a low heating rate (20°C/s) resulted in a well developed biochar microporous structure, but a high heating rate (500°C/s) resulted in plastic deformation of the biomass, resulting in a macroporous structure of the biochars formed. As a result of melting and sintering, the increased pyrolysis pressure creates bubbles, increases the size of the biochar particles, and decreases the volume of the biochar voids. However, the degree of heating rates and pressure affects in diverse biomasses has been reported to be variable: e.g., cane bagasse > spotted gum (hardwood) > radiata. Based on HTT, heating rate, and holding time at HTT, the pyrolysis method can be classified as quick, intermediate, or slow. In pesticide sorption process, slow pyrolysis is widely employed to create biochars with high sorption capacity (Wang et al. 2012). At HTT, the holding (residence) duration is another production variable

TABLE 10.3
Different Types of Biomass and Their Characteristics (Yavari et al. 2015)

Biomass	Biomass particle size	HTT (°C)	Heating rate (°C/min)	Holding time (h)	Ancillary inputs	Pyrolysis apparatus	Soil and water application percentage	Pesticide	Pesticide initial conc. (mg/L)
Red gum wood	<5 mm	450	7.08	2	Limited oxygen	Muffle furnace	5 % (w/w), sandy loam soil	Pyrimethanil	1–21
Wheat straw	Natural size	300	–	6	Limited oxygen	Muffle furnace	0.1% (w/w), sandy loam soil	MCPA	0.5–22
Poultry litter	NA	400	–	1	Limited oxygen	Muffle furnace	NA, aqueous solution	Fluridone	0.4–18
Paper mill sludge	NA	550	5–10	NA	NA	Continuous pilot slow-pyrolysis unit	0.5 %, highly permeable red ferrosol	Atrazine	1–10
Pig manure	<2 mm	350	–	2	Limited oxygen	Muffle furnace	1.25 g/L, aqueous solution	Carbaryl	1–40
De-ashed pig manure	<2 mm	350	–	2	Limited oxygen	Muffle furnace	1.25 g/L, aqueous solution	Carbaryl	1–40
Sugarcane bagasse	<2 mm	350	1	NA	NA	NA	10 g/L, aqueous solution	Metribuzin	0.1–25
Sewage sludge (4.9% moisture)	<2 mm	500	25 (K/min)	5	Nitrogen gas	NA	1 g/L, aqueous solution	Phenanthrene	0.0425–1.7
Rice straw	NA	400	NA	NA	Limited oxygen	NA	0.1% (w/w), Sandy loam soil	Pyrazosulfuron-ethyl	0.2–1

that influences biochar sorption potential. It permits the carbonization process to be accomplished and the highest possible surface area and porosity to be obtained. Thus biochars with porous structure and larger surface area can be achieved with high temperature and less residence time.

10.9   ADVANTAGES OF BIOCHAR AS ADSORBENT

Biochar is sustainable as it does not require chemical reagents and can be produced from waste materials (Bridgwater 2003). Making biochar is more cost-effective than activated carbon production. Besides, it allows a high energy recovery, generating a wide spectrum of bioenergy by-products with numerous applications. So it can be considered a self-sufficient process in terms of energy usage (Fernández and Menéndez 2011). Biochar on toxic material can be used in a similar manner with activated carbon for the in situ remediation of contaminated soils. The potential of biochar in pesticides sorption was reported for the first time in the 1960s when some Hawaiian soils containing burnt sugarcane residues were found to have high affinity for applied pesticides. Later in 1972, some researchers reported the increased sorption of atrazine with biochar application in soil and indicated that the sorption isotherm displayed a shift from linear to nonlinear shape, which shows that adsorption, instead of absorption (partitioning), is the dominant mechanism of sorption in the presence of these carbonized organic matters. Many studies have demonstrated the high potential of various biochars in the sorption of organic pesticides from water and soil environments (Wang et al. 2012).

10.10   AGING OF BIOCHARS AND ITS EFFECTS ON SORPTIVE PROPERTIES

Biochars undergo aging processes involving chemical, physical, and biological changes after application to the soil. Oxidation and hydrolysis are the most effective reactions in the aging process, which occur during first several months of incubation. Oxidation results in an increase in the oxygenated functional groups like carboxylate and phenolate on the external surface of biochar, which may cause specific interaction between biochar and organic molecules (Cheng et al. 2006). Oxidizing agents like ozone and high temperatures are useful in enhancing the oxidation process (Cheng et al. 2006). This process increases biochars' CEC and its surface hydrophilicity, which influence biochar sorption behavior and also make it susceptible to weathering. The biochar area and porosity decrease with biochar aging due to blocking of the pores and saturation of biochar active sites specifically in high organic matters. Physical processes like climatic changes may cause structure breaking, which may increase biochar area (Lehmann and Joseph 2015).

10.11   LIMITATIONS OF BIOCHAR

Despite its high stability, biochar has the potential to degrade over time. The decomposition of biochar allows the release of absorbed pesticides into the environment (Lehmann et al. 2005). Aging may also result in photochemical, microbial, and fungal decay of the biochar. Underdosing of herbicides over an extended period of time may be the cause of weed resistance, which occurs as a result of diminished herbicidal activity. Assessment of pesticide bioefficacy in soils with biochars has been done by Wang et al. (2012). According to Nag et al. (2011), adding wheat straw biochar to the soil can render herbicides like atrazine and trifluralin ineffective and lead to increased ryegrass (*Lolium rigidum*) seed germination and aboveground biomass. The propensity of biochar to retain pesticides can also have unintended consequences. Extending the duration that pesticides remain in the soil slows the rate of pesticide microbial breakdown. This can result in a localized concentration of pesticide toxicity in the soils over time (Nag et al. 2011). The compounds absorbed may be deteriorated as a result of biochar breakdown. The release of entrapped pesticide from saturated biochar is reliant on biochar characteristics, recalcitrance, organic material qualities, and environmental variables (Wu et al. 2015). Pyrolysis might be a source of pollution because it is an incomplete combustion process. Organic molecules such as PAHs, dioxins, and polychlorinated biphenyls (PCBs) as well as inorganic substances such as heavy metals have been found in the biochars produced in several investigations (Kloss et al. 2012). These chemicals progressively leak into the environment, posing harm to living things.

10.12   CONCLUSIONS

Pesticides are highly toxic substances that are classified as priority pollutants. Pesticide pollution has now become a global issue. Incineration (incinerators and open burning), chemical treatments (O_3/UV, hydrolysis, Fenton oxidation, and KPEG), physical treatments (inorganic absorbents, organic absorbents, and activated carbon), and biological treatments are currently employed for pesticide-containing wastewater (composting, bioaugmentation and phytoremediation). Strict laws governing the use of agrochemicals have required the development of eco-friendly products; for example, biochar very effectively adsorbs and degrades these pollutants. The characteristics of these adsorbents are directly influenced by production variables such as feedstock type and various other parameters as previously discussed. By optimizing the production variables, it is possible to manufacture designed biochars with excellent pesticide removal ability. Low-cost, long-lasting sorbent ability and potential to outperform activated carbon in terms of treatment efficiency make biochar a preferred choice. Environmental conditions, pesticide type, contact duration between biochar and substrate, biochar age, and stability, in addition to biochar qualities, can have a substantial impact on the sorption process and contribute to its complexity.

REFERENCES

Abdelhameed, Reda M., Mohamed Taha, Hassan Abdel-Gawad, Fathia Mahdy, and Bahira Hegazi. "Zeolitic imidazolate frameworks: Experimental and molecular simulation studies for efficient capture of pesticides from wastewater." Journal of Environmental Chemical Engineering 7, no. 6 (2019): 103499.

Akashe, Megha M., Uday V. Pawade, and Ashwin V. Nikam. "Classification of pesticides: A review." International Journal of Research in Ayurveda and Pharmacy 9, no. 4 (2018): 144–150.

Al Hattab, Mariam T., and Abdel E. Ghaly. "Disposal and treatment methods for pesticide containing wastewaters: critical review and comparative analysis." Journal of environmental protection 3, no. 5 (2012): 431–453.

Aygun, Ayşegül, Serpil Yenisoy-Karakaş, and I. Duman. "Production of granular activated carbon from fruit stones and nutshells and evaluation of their physical, chemical and adsorption properties." Microporous and Mesoporous Materials 66, no. 2–3 (2003): 189–195.

Baharum, Nor Atikah, Hanisah Mohmad Nasir, Mohd Yusoff Ishak, Noorain Mohd Isa, Mohd Ali Hassan, and Ahmad Zaharin Aris. "Highly efficient removal of diazinon pesticide from aqueous solutions by using coconut shell-modified biochar." Arabian Journal of Chemistry 13, no. 7 (2020): 6106–6121.

Beesley, Luke, Eduardo Moreno-Jiménez, and Jose L. Gomez-Eyles. "Effects of biochar and greenwaste compost amendments on mobility, bioavailability and toxicity of inorganic and organic contaminants in a multi-element polluted soil." Environmental Pollution 158, no. 6 (2010): 2282–2287.

Beesley, Luke, Eduardo Moreno-Jiménez, Jose L. Gomez-Eyles, Eva Harris, Brett Robinson, and Tom Sizmur. "A review of biochars' potential role in the remediation, revegetation and restoration of contaminated soils." Environmental pollution 159, no. 12 (2011): 3269–3282.

Bridgwater, Anthony V. "Renewable fuels and chemicals by thermal processing of biomass." Chemical Engineering Journal 91, no. 2–3 (2003): 87–102.

Carter, A. D., and A. I. J. Heather. "Pesticides in groundwater." In: Best GA, Ruthven AD (eds). *Pesticides - Developments, Impacts, and Controls*. Special Publication No. 174. London, UK: The Royals Society of Chemistry (1995): 113–123.

Cederlund, Harald, Elisabet Börjesson, Daniel Lundberg, and John Stenström. "Adsorption of pesticides with different chemical properties to a wood biochar treated with heat and iron." Water, Air, & Soil Pollution 227, no. 6 (2016): 203.

Cetin, Emre, Behdad Moghtaderi, Rajener Gupta, and T. F. Wall. "Influence of pyrolysis conditions on the structure and gasification reactivity of biomass chars." Fuel 83, no. 16 (2004): 2139–2150.

Cheng, Chih-Hsin, Johannes Lehmann, Janice E. Thies, Sarah D. Burton, and Mark H. Engelhard. "Oxidation of black carbon by biotic and abiotic processes." Organic geochemistry 37, no. 11 (2006): 1477–1488.

Clay, Sharon A., and Douglas D. Malo. "The influence of biochar production on herbicide sorption characteristics." In: Hasaneen MN (ed). *Herbicides-Properties, Synthesis and Control of Weeds*. Rijeka, Croatia: In Tech (2012): 59–74.

Cornelissen, Gerard, Örjan Gustafsson, Thomas D. Bucheli, Michiel TO Jonker, Albert A. Koelmans, and Paul CM van Noort. "Extensive sorption of organic compounds to black carbon, coal, and kerogen in sediments and soils: mechanisms and consequences for distribution, bioaccumulation, and biodegradation." Environmental Science & Technology 39, no. 18 (2005): 6881–6895.

EBC, European Biochar Certificate. 2012. Guidelines for a sustainable production of biochar. European Biochar Foundation (EBC), Arbaz, Switzerland. www.europeanbiochar.org/en/download. Version 8.3E of 1st September 2019, 10.13140/RG.2.1.4658.7043.

Fernández, Y., and J. A. Menéndez. "Influence of feed characteristics on the microwave-assisted pyrolysis used to produce syngas from biomass wastes." Journal of Analytical and Applied Pyrolysis 91, no. 2 (2011): 316–322.

Freitas, Jair C.C., Alfredo G. Cunha, and Francisco G. Emmerich. "Physical and chemical properties of a Brazilian peat char as a function of HTT." Fuel 76, no. 3 (1997): 229–232.

Gaskin, J. W, C. Steiner, K. Harris, K. C. Das, and B. Bibens. "Effect of low-temperature pyrolysis conditions on biochar for agricultural use." Transactions of the ASABE 51, no. 6 (2008): 2061–2069.

Gautam, Ravindra Kumar, Mandavi Goswami, Rakesh K. Mishra, Preeti Chaturvedi, Mukesh Kumar Awashthi, Ram Sharan Singh, Balendu Shekhar Giri, and Ashok Pandey. "Biochar for remediation of agrochemicals and synthetic organic dyes from environmental samples: A review." Chemosphere (2021): 129917.

Gill, Harsimran Kaur, and Harsh Garg. "Pesticide: Environmental impacts and management strategies." Pesticides-Toxic Aspects 8 (2014): 187.

Guo, Jia, and Aik Chong Lua. "Characterization of chars pyrolyzed from oil palm stones for the preparation of activated carbons." Journal of Analytical and applied Pyrolysis 46, no. 2 (1998): 113–125.

Joseph, S. D., A. Downie, P. Munroe, A. Crosky, and J. Lehmann. *Biochar for carbon sequestration, reduction of greenhouse gas emissions and enhancement of soil fertility; a review of the materials science*. In Proceedings of the Australian combustion symposium, pp. 130–133, 2007.

Lehmann, Johannes, and Stephen Joseph. *Biochar for Environmental Management: Science, Technology and Implementation*. London and New York: Routledge, Taylor and Francis Group, 2015.

Khorram, Mahdi Safaei, Qian Zhang, Dunli Lin, Yuan Zheng, Hua Fang, and Yunlong Yu. "Biochar: A review of its impact on pesticide behavior in soil environments and its potential applications." Journal of Environmental Sciences 44 (2016): 269–279.

Kloss, Stefanie, Franz Zehetner, Alex Dellantonio, Raad Hamid, Franz Ottner, Volker Liedtke, Manfred Schwanninger, Martin H. Gerzabek, and Gerhard Soja. "Characterization of slow pyrolysis biochars: effects of feedstocks and pyrolysis temperature on biochar properties." Journal of Environmental Quality 41, no. 4 (2012): 990–1000.

Kookana, Rai S. "The role of biochar in modifying the environmental fate, bioavailability, and efficacy of pesticides in soils: A review." Soil Research 48, no. 7 (2010): 627–637.

Laird, David A. "The charcoal vision: A win—win—win scenario for simultaneously producing bioenergy, permanently sequestering carbon, while improving soil and water quality." Agronomy Journal 100, no. 1 (2008): 178–181.

Lehmann, Johannes, and Stephen Joseph. *Biochar for Environmental Management: An Introduction*. London and New York: Routledge, Taylor and Francis Group, 2015.

Lehmann Johannes, and Stephen Joseph, eds. *Biochar for Environmental Management: Science, Technology and Implementation*. London and New York: Routledge, Taylor and Francis Group, 2021.

Lehmann, Johannes, Zhongdong Lan, Charles Hyland, Shinjiro Sato, Dawit Solomon, and Quirine M. Ketterings. "Long-term dynamics of phosphorus forms and retention in manure-amended soils." Environmental Science & Technology 39, no. 17 (2005): 6672–6680.

Lua, Aik Chong, Ting Yang, and Jia Guo. "Effects of pyrolysis conditions on the properties of activated carbons prepared from pistachio-nut shells." Journal of Analytical and Applied Pyrolysis 72, no. 2 (2004): 279–287.

Nag, Subir K., Rai Kookana, Lester Smith, Evelyn Krull, Lynne M. Macdonald, and Gurjeet Gill. "Poor efficacy of herbicides in biochar-amended soils as affected by their chemistry and mode of action." Chemosphere 84, no. 11 (2011): 1572–1577.

Nelson, S. D., W. J. Farmer, J. Letey, and C. F. Williams. Stability and mobility of napropamide complexed with dissolved organic matter in soil columns. Vol. 29, no. 6. American Society of Agronomy, Crop Science Society of America, and Soil Science Society of America, 2000.

Pastor-Villegas, J., J. F. Pastor-Valle, JM Meneses Rodríguez, and M. García García. "Study of commercial wood charcoals for the preparation of carbon adsorbents." Journal of Analytical and Applied Pyrolysis 76, no. 1–2 (2006): 103–108.

Qiu, Mengyi, Ke Sun, Jie Jin, Bo Gao, Yu Yan, Lanfang Han, Fengchang Wu, and Baoshan Xing. "Properties of the plant-and manure-derived biochars and their sorption of dibutyl phthalate and phenanthrene." Scientific Reports 4, no. 1 (2014): 1–10.

Rodríguez-Mirasol, J., T. Cordero, and J. J. Rodriguez. "Preparation and characterization of activated carbons from eucalyptus kraft lignin." Carbon 31, no. 1 (1993): 87–95.

Shirvani, Mehran, Mahmoud Kalbasi, Hosein Shariatmadari, Farshid Nourbakhsh, and Bijan Najafi. "Sorption—desorption of cadmium in aqueous palygorskite, sepiolite, and calcite suspensions: isotherm hysteresis." Chemosphere 65, no. 11 (2006): 2178–2184.

Song, Weiping, and Mingxin Guo. "Quality variations of poultry litter biochar generated at different pyrolysis temperatures." Journal of Analytical and Applied Pyrolysis 94 (2012): 138–145.

Sun, Ke, Mingjie Kang, Zheyun Zhang, Jie Jin, Ziying Wang, Zezhen Pan, Dongyu Xu, Fengchang Wu, and Baoshan Xing. "Impact of deashing treatment on biochar structural properties and potential sorption mechanisms of phenanthrene." Environmental Science & Technology 47, no. 20 (2013): 11473–11481.

Tang, Jingchun, Honghong Lv, Yanyan Gong, and Yao Huang. "Preparation and characterization of a novel graphene/biochar composite for aqueous phenanthrene and mercury removal." Bioresource Technology 196 (2015): 355–363.

Tang, Jingchun, Wenying Zhu, Rai Kookana, and Arata Katayama. "Characteristics of biochar and its application in remediation of contaminated soil." Journal of Bioscience and Bioengineering 116, no. 6 (2013): 653–659.

Tan, Xiaofei, Yunguo Liu, Guangming Zeng, Xin Wang, Xinjiang Hu, Yanling Gu, and Zhongzhu Yang. "Application of biochar for the removal of pollutants from aqueous solutions." Chemosphere 125 (2015): 70–85.

Uchimiya, Minori, Lynda H. Wartelle, and Veera M. Boddu. "Sorption of triazine and organophosphorus pesticides on soil and biochar." Journal of Agricultural and Food Chemistry 60, no. 12 (2012): 2989–2997.

Van Zwieten, Lukas, Stephen Kimber, S. Morris, Adriana Downie, E. Berger, Josh Rust, and Clemens Scheer. "Influence of biochars on flux of N2O and CO2 from Ferrosol." Soil Research 48, no. 7 (2010): 555–568.

Wang, Jinyang, Xiaojian Pan, Yinglie Liu, Xiaolin Zhang, and Zhengqin Xiong. "Effects of biochar amendment in two soils on greenhouse gas emissions and crop production." Plant and Soil 360, no. 1 (2012): 287–298.

Wauchope, R. Don, Simon Yeh, Jan B. H. J. Linders, Regina Kloskowski, Keiji Tanaka, Baruch Rubin, Arata Katayama et al. "Pesticide soil sorption parameters: theory, measurement, uses, limitations and reliability." Pest Management Science 58, no. 5 (2002): 419–445.

Wildman, Joe, and Frank Derbyshire. "Origins and functions of macroporosity in activated carbons from coal and wood precursors." Fuel 70, no. 5 (1991): 655–661.

Wu, Jia, Jianwei Yang, Pu Feng, Guohuan Huang, Chuanhui Xu, and Baofeng Lin. "High-efficiency removal of dyes from wastewater by fully recycling litchi peel biochar." Chemosphere 246 (2020): 125734.

Wu, Mengxiong, Qibo Feng, Xue Sun, Hailong Wang, Gerty Gielen, and Weixiang Wu. "Rice (Oryza sativa L) plantation affects the stability of biochar in paddy soil." Scientific Reports 5, no. 1 (2015): 1–10.

Yavari, Saba, Amirhossein Malakahmad, and Nasiman B. Sapari. "Biochar efficiency in pesticides sorption as a function of production variables—A review." Environmental Science and Pollution Research 22, no. 18 (2015): 13824–13841.

Zanella, Renato, Martha B. Adaime, Sandra C. Peixoto, Caroline do A. Friggi, Osmar D. Prestes, Sérgio LO Machado, Enio Marchesan, Luis A. Avila, and Ednei G. Primel. "Herbicides persistence in rice paddy water in southern Brazil." Herbicides-Mechanisms and Mode of Action (2011): 369–382.

Zhang, Tengyan, Walter P. Walawender, L. T. Fan, Maohong Fan, Daren Daugaard, and R. C. Brown. "Preparation of activated carbon from forest and agricultural residues through CO2 activation." Chemical Engineering Journal 105, no. 1–2 (2004): 53–59.

Zhao, Ling, Xinde Cao, Ondřej Mašek, and Andrew Zimmerman. "Heterogeneity of biochar properties as a function of feedstock sources and production temperatures." Journal of Hazardous Materials 256 (2013): 1–9.

Zhu, Xiaomin, Baoliang Chen, Lizhong Zhu, and Baoshan Xing. "Effects and mechanisms of biochar-microbe interactions in soil improvement and pollution remediation: A review." Environmental Pollution 227 (2017): 98–115.

11 Application of Biochar in the Removal of Organic and Inorganic Contaminants from Wastewater
Mechanism, Operating Conditions, and Modifications

Pranjal P. Das, Mukesh Sharma, Simons Dhara, and Mihir K. Purkait

CONTENTS

11.1 INTRODUCTION

One of the most important elements of life is water. Water quality improvement and preservation are becoming increasingly important in the present world. Many potentially harmful compounds have been released into water bodies as a result of population increase, urbanization, industrial development, and changes in people's lives. Contaminated water poses a number of risks to

DOI: 10.1201/9781003203438-11

biological species, humans, and the environment. For the removal of wastewater contaminants, a variety of conventional technologies are used around the world. However, application of such techniques at the industrial level with the advantages of cost-effectiveness, outstanding removal rate at lower concentrations, zero production of secondary pollutants, and quick treatment remains a challenging problem. Moreover, the design and operating procedures can still be regarded as a problem. Alternative approaches are also subject to technical and economic constraints when it comes to full-scale implementation. In this context, it is critical to develop long-term treatment solutions for heavy metal and dye removal before being released into the environment, as well as to meet the recommended water quality standards. The adsorption technique has shown promising results among all existing technologies for the degradation of dissolved inorganic and organic substances from various effluents (Jain et al. 2016). Adsorption can be termed a surface phenomenon that investigates the efficiency of certain solids to preserve specified substrates on its surface from either the aqueous or the gaseous phase. The adsorption process has been recognized as a promising technique for wastewater treatment, owing to its operational flexibility, design versatility, economic feasibility, and easy handling. As the effectiveness of adsorption technology is mainly dependent on the use of appropriate adsorbents, the need for nonhazardous, inexpensive, and recyclable adsorbent is a key concern for this approach. Nevertheless, parameters such as types and regeneration of adsorbents, selected adsorbates, process conditions, and its final disposal considerably influence the technical and economic aspects of the adsorption technique. Many adsorbent materials including activated carbons, nanoparticles, biosorbents, and biochars have been explored for the removal of a broad range of pollutants. Biochar is a solid fine-grained, carbon-rich porous substance, consisting of aromatic surfaces prepared via thermal breakdown of biomass in a low-oxygen environment (S and Paramasivan 2019). Biochar has emerged as a novel potential adsorbent when compared to commercially available activated carbon. Also, biochar has been proven to be a very efficient and cost-effective material for the sorption of a variety of contaminants. Porous structure distribution, high surface area, and improved surface chemical characteristics are some of the important parameters that contribute to biochar's specific sorption properties during the removal of pollutants from various effluents. The effective pollutant sorption characteristics of biochar are also based on factors such as ion exchange capacity, alkalinity, hydrophobicity and elemental compositions. Biochar may consist of noncarbonized fractions as well as a substantial number of functional groups due to carbonization under mild experimental conditions (Clemente et al. 2017). As a result, biochar can provide characteristics and features comparable to biosorbents or highly aromatic activated carbons, depending on the pyrolysis conditions utilized.

Different adsorbate species with varied characteristics, such as hazardous metals, emerging contaminants, dyes and pigments, have been explored in recent years. The surface characteristics and special features of biochar, namely aromatic carbon structure, higher surface area, high cation exchange capacity, oxygen-rich functional groups, and high mineral content, have been attributed to its effective use for various effluent treatment. Moreover, biochar has been extensively utilized as a novel adsorbent in the removal of both inorganic and organic contaminants from the environment. Figure 11.1 represents the characteristics, modification methods, and environmental remediations of designer biochar.

The persistent free radicals present in biochar were able to activate H_2O_2 or $S_2O_8^{2-}$ to produce active oxygen, thereby efficiently reducing inorganic and organic contaminants (Ruan et al. 2019). Adsorption takes place as a result of imbalanced forces in an adsorbent-solution-adsorbate system, which consist of complex degradation procedures with various molecular reactions occurring between adsorbents and adsorbates, thereby involving both chemical and physical interactions. Therefore, understanding the adsorption mechanisms, as well as identifying the interactions between biochar and adsorbates along with its implications, is crucial for the production of biochars as promising adsorbents and optimization of the adsorption process. As such, the aim of this chapter is to provide a deep insight on different modification techniques involved during the development

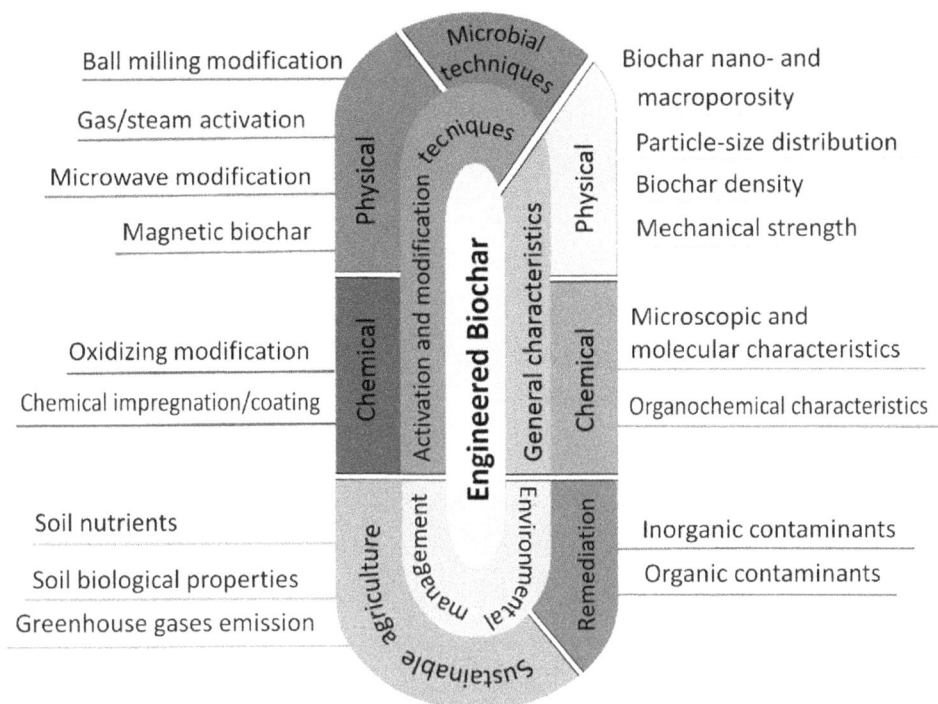

FIGURE 11.1 Characteristics, modification techniques, and environmental applications of engineered biochar.
Source: Obtained with permission from Kazemi Shariat Panahi et al. (2020) © Elsevier.

of biochars, apart from expanding the existing knowledge based on various operational factors and mechanisms that determine the adsorption of both inorganic and organic contaminants on their surface. A thorough understanding of such parameters, together with economic feasibility and regeneration efficiency, is very important in order to enhance the adsorption technique and production of biochars. Various challenges and recommendations associated with biochar are also presented in this chapter for future research and development.

11.2 BIOCHAR MODIFICATION

Physicochemical characteristics of biochar feedstocks can be altered using different modification methods. Table 11.1 depicts the various physicochemical properties of the most widely used biomass feedstocks. Biochar modification can significantly affect its specific surface area (SSA) and surface functional groups and can make alterations to the morphology and the distribution of pore sizes. The variation in treatment methods and modifiers can have an impact on biochar's adsorption effectiveness and mechanism toward contaminants. Thus it is very important to select suitable techniques of biochar activation/modification for the adsorption of emerging contaminants (ECs) based on physicochemical properties.

11.2.1 ACID AND ALKALINE MODIFICATION

Acid- and alkali-mediated biochar modifications are achieved through modifying agents of inorganic acids (sulfuric acid, hydrochloric acid, phosphoric acid, nitric acid), alkaline hydroxides

TABLE 11.1
Physicochemical properties of various biomass feedstocks (modified with permission from Madadi and Bester (2021) © Elsevier)

Properties	Wood	Sugarcane	Agricultural residues	Wheat straw	Rice straw	Livestock manure	Aquatic waste	Residual sludge	Nutshell/fruit peel
Physical properties									
Density (kg/m³)	1186	141.2	230	1233	200	658	189	69	387
Moisture content (%)	20	10.4	–	16	6	–	–	–	–
Ash content (%)	0.4–1	16.4	6.26	4	4.3	39	18	41.2	3.40
Volatile matter (%)	82	74	76.13	59	79	49.5	68	52.8	76.85
Fixed carbon (%)	17	13	17	21	10.7	7.61	13	3.70	19.72
Chemical properties									
Carbon (%)	51.6	43.2	44.4	48.5	48.3	26.6	38.6	29.5	46
Hydrogen (%)	6.3	6.7	5.5	5.5	5.32	3.75	9.65	4.62	6.90
Oxygen (%)	41.5	33.2	39.5	3.9	34.6	23.30	26.5	20.43	39.85
Nitrogen (%)	0.1	0.3	0.80	0.3	1.23	3.41	2.04	4.19	0.66
Fiber analysis									
Lignin (%)	25–30	23–32	14.23	15–20	18	9.43	13.3	–	18.8
Cellulose (%)	35–50	19–24	44.16	33–40	32.12	17.83	40.71	–	1.44
Hemicellulose (%)	20–30	32–48	–	20–25	24	–	–	–	–

(sodium hydroxide, potassium hydroxide), and mineral salts, respectively (zinc chloride, potassium carbonate). A comparative analysis between pristine biochar and acid- or alkaline-modified biochar with regard to adsorption characteristics is shown in Figure 11.2.

Acid modification of biochar is achieved by loading (1:10 loading ratio) the biochar into acid solutions at various temperatures ranging from ambient temperature to 120°C, where the time of reaction can normally vary from hours to days. Wan et al. (2016) performed the carbonization

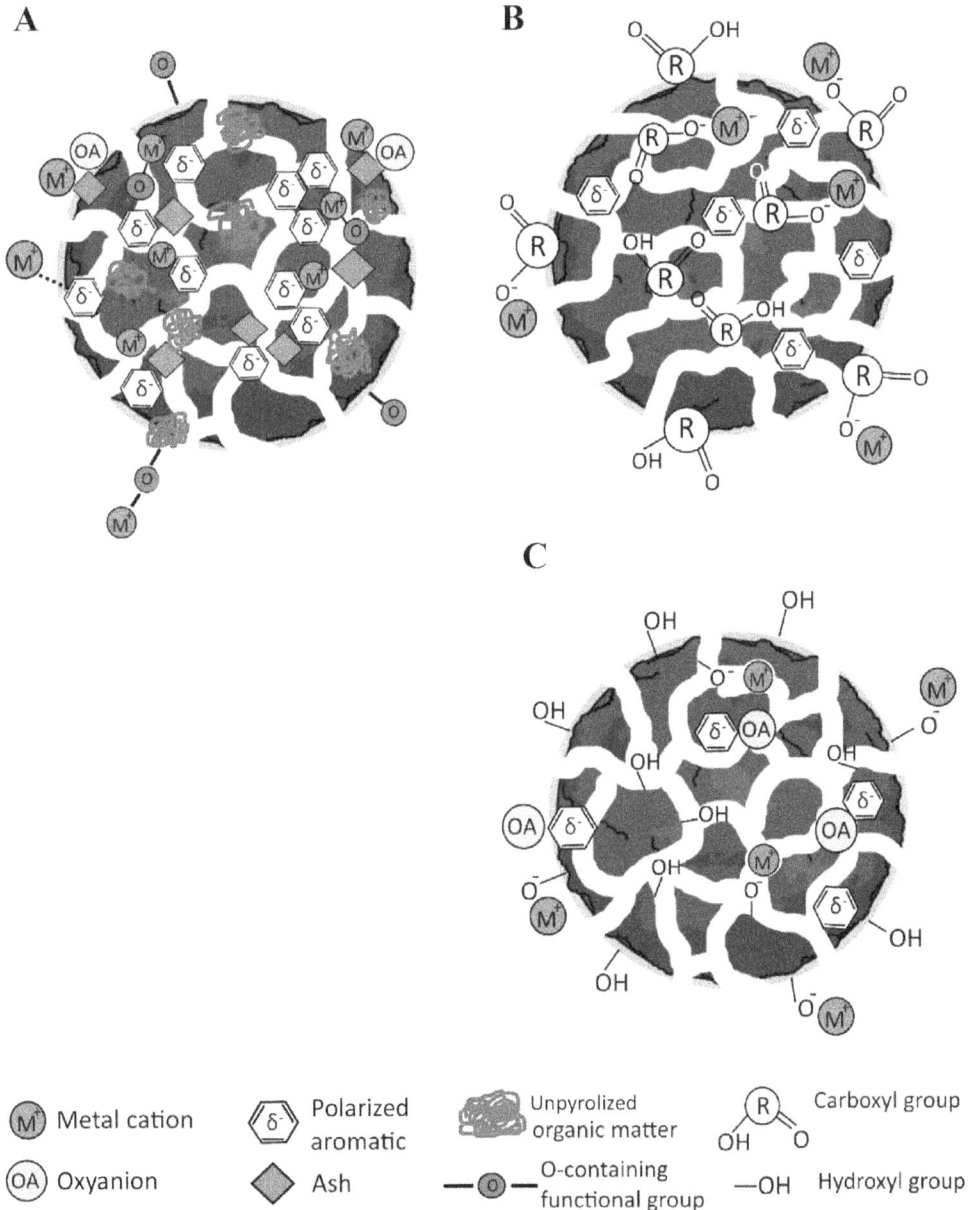

FIGURE 11.2 Comparison between modified and unmodified biochars based on sorption characteristics: (A) pristine biochar; (B) acid-modified biochar; and (C) alkali-modified biochar.

Source: Modified with permission from Kazemi Shariat Panahi et al. (2020) © Elsevier.

of eucalyptus-sawdust-based biomass in 85% phosphoric acid, at constant pyrolysis temperature with constant nitrogen supply. Total porosity and specific surface area of biochar were raised by 26.9 and 23.6 times, respectively, after activation using H_3PO_4. The surface area of kenaf-fiber-biomass-produced biochar increased from 289.497 to 346.57 m^2/g when treated with hydrochloric acid. Activated biochar's structure is extremely porous and shows honeycomb-like structures of different measurements. The biochar from the dry-up of *Opuntia ficus indica* cactus biomass was processed with the chemical stimulating mediator of HNO_3 at room temperature for 24 h, followed by reflux at 80°C for 3 h with simultaneous stirring. FT-IR spectra clarifies that the activation of biochar using nitric acid introduces carboxylic components on the biochar surface. In case of the H_2SO_4-mediated biochar modification, a H_2SO_4 concentration range of 5–30% has a productive aftermath of increased surface area, but higher acid concentration has a negative effect of decreased surface area. Cotton waste treatment with sulfuric acid brings drastic transformations to biochar's solid state. The sulfuric acid treatment of biochar tends to develop a high number of acid functional groups on its surface. The high proportion of oxygen leads to an increment in H/C and O/C atomic ratios, which sequentially points to the reduction in the hydrophobic nature of biochar. The activating agent of sodium hydroxide is used to treat biochar produced from poplar biomass using pyrolysis (Vithanage et al. 2015). Determined iodine number indicates a raise in the pyrolytic temperature, and with the alteration procedure, the surface area of biochar has increased significantly. Biochar produced from whitewood was activated using potassium hydroxide. A maximum of 1500 m^2/g surface area was observed, which is almost equivalent to the surface area of commercially available activated carbon. Activated biochar extracted from the shell powder of horse chestnut was soaked in the ratio of 1:1 into the aqueous potassium carbonate solution. The soaked biomass was carbonated at 700°C with the help of a stream of nitrogen gas. After activation, biochar surface area varied between 500 and 1500 cm^2/g. The depolymerization reaction takes place when treated with sodium hydroxide. Biochar modified with alkaline solution shows a substantial increase in surface area with high N/C, high H/C, and low O/C ratios. However, a reduced O/C ratio indicates the hydrophilic nature of acid-activated biochar, which is inversely related to the aromaticity of biochar. The higher ratio of N/C increased the number of functional groups containing nitrogen and head towards the development of elemental characteristics on the activated biochar surface. Apart from other activation agents, zinc chloride is also utilized for the activation of sesame-straw-derived biochar (Ma et al. 2014). Table 11.2 presents the influence of acid and alkaline modification on the properties of designer biochar.

TABLE 11.2
Effects of Acids and Bases on the Properties of Modified Biochar (modified with permission from Kazemi Shariat Panahi et al. (2020) © Elsevier

Chemical reagents	Enhanced properties
Acid modification	Availability of surface carboxyl groups
	Mineral composition tailoring
	Oxygen-containing functional groups
	Induced sorption capacity and heavy metals uptake
	Environmental persistency (half-life)
Alkaline modification	Surface graphite C and/or aromatic functional groups
	Improved surface area
	Increasing surface electrostatic attraction, π–π interaction, surface precipitation, and/or surface complexation

11.2.2 Modification Using Metal Oxide and Salts

Modification using metal salt and metal oxide can increase the sorption capacity of biochar surface as well as produce biochars with different sorption efficiencies for contaminants as per the variation of metal properties. Metal-oxide- and metal-salt-modified biochars have superior anion exchange capacity, electrostatic attraction ability, and precipitation, which improves its adsorption efficiency for emerging contaminants. Research has found that phosphorus-activated biochar (pBC-S-nZVI) removes more florfenicol when allocated with sulfurized nanoscale zero-valent iron (nZVI). The pBC-S-nZVI has a 4.3 times higher florfenicol removal rate than unmodified biochar. According to some studies, levofloxacin adsorption on Fe/Mn-biochar is extremely pH sensitive, and even after 5 cycles, Fe/Mn-modified biochar can sustain a definite adsorption capacity of levofloxacin. Ag/Fe nanoparticle (Ag/Fe/MB) reinforced biochar is capable of removing cephalexin from aqueous solutions. It is reported that 86% of cephalexin was removed under 90 min via reduction and adsorption. In comparison to unmodified biochar, Cu-assisted biochar showed a twofold doxycycline hydrochloride adsorption capacity (93.22%) in aqueous media. If we consider efficiency, cost, and its utilization in actual water, Cu-modified biochar is a possible candidate for enhancement of doxycycline hydrochloride exclusion from aqueous solution. Other research has confirmed that Fe-Cu-coated duple hydroxide biochar nanocomposite can be effectively manufactured by the hydrothermal process. Degradation of 97.6% cefazolin sodium (CFZ) from aqueous solution was achieved by biocatalytic performance (Cheng et al. 2021). Normally, activation of biochar using metal salt and oxide increases the chemical adsorption capacity by adding more functional groups on the surface to produce more chemical bonds. Along with adsorption capacity, biochar functionalized with metal salt and oxide has significantly enhanced magnetic and catalytic properties. For some specific contaminants, for instance anionic dyes development of metal salt/oxide, biochar nanocomposites are very crucial as these compounds are not properly absorbed due to the negative charge on the unmodified biochar surface. Metallic salts and oxides have been used for metal-biochar composite production due to their ability to enhance the catalytic characteristics of the substance for further persulfate modification methods. Iron, magnesium, manganese, and aluminum are some of the most common metals that have been used to synthesize metal salt/oxide-biochar. Research has shown that iron Fe(III) coating on rice-husk-produced biochar has significantly improved adsorption capacity to remove As(III) and As(V). Also, MnO_2-modified swine-derived biochar improved its heavy metals Cd(II) and Pb(II) removal capacity compared to pristine biochar (Amusat et al. 2021).

11.2.3 Modification with Nanomaterial

Blending the benefits of nanomaterials with the occurrence of various functional groups on biochar surface after pyrolysis, such as amino acids, carboxyl (-COOH), and hydroxyl (-OH), groups can improve the characteristics of biochar nanocomposites. Functional groups are crucial components of biochar application, mainly for water purification to remove contaminants. Besides, the vast specific surface area introduced by biochar and nanomaterial precursors formulate an extra efficient nanocomposite. Studies have found that the coalescence of biochar and nanomaterials enhances the physiochemical characteristics of its biochar nanocomposites. The discoveries have triggered a quantum leap to inhibit existing constraints to utilize unmodified biochar, therefore confronting the biochar applications for wastewater and water decontamination. Nevertheless, the stability of biochar nanocomposites is a matter of concern as after usage some nanomaterials leach out into purified water if not restrained correctly. To subdue those limitations, the use of magnetic biochar nanocomposites has been investigated. With this approach, magnetic nanocomposites can be separated with ease using just an ordinary magnet, and further nanocomposites can be recycled after use. The magnetic biochar nanocomposite synthesized through biomass pretreatment with ferric chloride composites was found to have remarkable ferromagnetism capabilities, allowing the composite to absorb arsenic pollutants (Zhang et al. 2013). Carbon nanotubes were also said to be integrated into biochar. CNT-biochar composites having a surface area of 359 m²/g and enhanced pollutant sorption efficiency were made possible by the combination of biochar's large surface area with CNTs. A comparable outcome was obtained once biomass waste of wheat straw was treated alongside graphene using a moderate pyrolysis process; the graphene coating improved the biochar surface properties. In comparison to the unmodified biochar's limited sorption characteristics, exceptional sorption of mercury and phenanthrene pollutants was noted when more functional groups were incorporated. Researchers have now been drawn toward the utilization of conjugated biochar-based magnetic nanocomposites due to the previously mentioned findings. Nanotechnology is now investigated considering it as a potential solution for wastewater treatment due to the significant advantages offered by nanoparticles. Wide surface area and exceptional capacity of interaction with a wide range of materials are two of the notable qualities of nanorange materials. Nanomaterials are considered excellent sorbents for organic and inorganic contaminants removal from polluted water because of these unique features and the simplicity of separation after use. Modifying nanomaterials using a variety of chemicals, metals and their oxides, and different synthesized and natural substances can boost adsorption capacity and improve their pollutant removal efficiency. Thus introducing nanomaterials like carbon nanotubes, graphene, ZnS nanocrystals, and chitosan into the surface morphology of sorbent should result in a cost-effective composite with improved adsorption capacities for a variety of pollutants. The benefits of biochar are combined with the characteristics of functionalized nanomaterials in this modification (X. fei Tan et al. 2016).

11.2.4 Ball Milling and Steam Activation

Ball milling is a technique for crushing samples, including biochar, down to nanoscale size. Biochar's pore structure, SSA, and surface functional groups can all be improved by ball milling. The physical sorption ability of biochar to emerging contaminants (ECs) can be enhanced by increasing SSA. Furthermore, using a grinding medium in the milling technique can provide biochar with various functional groups to enhance its sorption efficiency to ECs. During ball milling, the mass ratio of stabilized yttria-zirconia orbs to biochar and the grade of wet milling solvents have a direct impact on the physicochemical parameters of biochar. The results of Boehm titration indicates that ball-milled biochar using agate balls enhances the oxygen-incorporated functional groups (lactone, hydroxyl, and carboxyl) on the biochar surface. Singh et al. (2019) observed that the structure and optical properties can be modified by using different grinding media (dimethylformamide, isopropanol, ethanol, and deionized water). Just the form of the crushing medium can allow the feedstock to create a wide variety of effects. Another study used ball milling to develop ultrafine magnetic Fe_3O_4 biochar to absorb ECs. The hybrid adsorbent developed from 2 h of milling demonstrated an enhanced carbamazepine (CBZ) removal rate and simple magnetic segregation. A second study has found that biochar nanocomposite developed using the green method could remove up to 95% of carbamazepine from water, indicating that it may be used to remove trace contaminants. In a different study, biochar crushed with a ball mill was used to extract dual sulfonamide antibiotics (SAs) from wastewater: sulfapyridine (SPY) and sulfamethoxazole (SMX) (Cheng et al. 2021). Even though the ball-milled method is considered a green activation approach, only a few studies on ball-milled biochar and its sorption of ECs have been conducted. Because of its green and inexpensive approach, ball-milled biochar can grow as a significant research field in the future.

The steam activation technique is among the most used physical ways of biochar transformation. It is used to increase SSA and porosity while also assisting in the removal of surface contaminants such as organic debris, ash, and incomplete combustion products. This procedure includes submitting pyrolyzed biochar to partial gasification using steam as an activator, resulting in the fragmentation of some of the materials and the formation of crystalline carbon. Fragmenting the vapor that generates oxygen, which further interacts with drifting carbon on the material surface forming CO and H_2. It is the fundamental process of steam activation (Eq. 11.1). H_2 omitted during the course interacts with carbon atoms to form a feebly bonded complex of carbon and hydrogen (Eq. 11.2). Surface oxidation continuously generates hydrogen and carbon dioxide gas (Pandey, Daverey, and Arunachalam 2020).

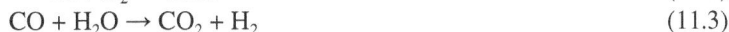

$$C + H_2O \rightarrow CO + H_2 \tag{11.1}$$
$$2C + H_2 \rightarrow 2CH \tag{11.2}$$
$$CO + H_2O \rightarrow CO_2 + H_2 \tag{11.3}$$

The surface area expands because the pores contained in the carbon structure open up as this procedure is repeated continuously. Wang et al. (2020) demonstrated modification of biochar using steam at 500°C for 45 min, with encouraging outcomes for extracting tetracycline (TC) and Cu^{2+}. Water-vapor-modified canola-straw-derived biochar had the maximum Pb^{2+} (195 mg/g) sorption, compared to pristine canola and wheat-straw-derived biochar, which had 109 and 108 mg/g, respectively. Rajapaksha et al. (2015) verified the enhanced adsorption capacity of steam-activated *Sicyos angulatus* biochar, with a 53% growth in adsorption efficiency compared to pristine biochar produced under similar working conditions. Yet Shim et al. speculated that biochar activation using steam would not be the best method for efficient copper adsorption and reduction of toxicity. Toxicity has been detected during this process, which might be because of an increment of aromaticity produced during steam modification.

11.3 ADSORPTION MECHANISM OF BIOCHAR

Adsorbents of a variety of sources, compositions, processes, attributes, and features exist as a result of extensive developments in adsorption study and materials development. Different pyrolysis temperatures (200–1000°C), activation, and modification processes, like reactor temperature, design, runtime, heating rate, flow rate, and solvents, can be used to generate biochar from different carbonaceous resources. As a result, biochar may be used to remove organic and inorganic contaminants through a variety of processes, such as chemical and physical interactions. Table 11.3 shows the effect of different pyrolysis temperatures on biochar characteristics. Physical sorption is a general procedure exhibiting feeble attractive force between the biochar surface and the adsorbate; multilayer sorption is feasible in this scenario. The physical processes, namely surface complexation, hydrogen bond, electrostatic and hydrophobic interaction, are primarily caused by the polarity of the adsorbent surface and the adsorbate molecule. Chemical adsorption, on the other hand, is more selective and includes stronger forces because of the formation of chemical bonds through electron distribution among the adsorbate and the biochar; monolayer adsorption can be seen in this scenario.

The adsorption procedure is not restricted to a solo technique; it may occur concurrently and can be mutually beneficial. Electrostatic attraction, ion exchange, surface complexation, and coprecipitation are recognized as major sorption processes for ionic species. Cui et al. (2016) reported ion exchange, surface complexation, and coprecipitation techniques for NH^{4+} and Cd^{2+} adsorption on biochar from wetland plants. For organic molecule adsorption, both hydrophobic interaction and hydrogen bonding interaction have proven significant. Different sorption techniques of sparfloxacin and ciprofloxacin on Fe_3O_4/graphene-activated biochar were observed by Zhou et al. (2019), including hydrogen bonds, π–π interactions, electrostatic and hydrophobic interactions. Furthermore, biochar properties such as oxygen-to-carbon-contents ratio (O/C ratio) and specific surface area (SSA) have been demonstrated to play key roles in the sorption of organic and inorganic molecules. generally, the SSA is significant for organic as well as for inorganic molecule adsorption, although greater O/C ratios improve inorganic adsorption as they are generally associated with a greater amount of oxygen-incorporated functional groups and superior capability to trade cations. Relatively high hydrophobicity and aromaticity (lower O/C ratio) are associated with increased sorption efficiency of organic substances, excluding ionic dye sorption and whenever decarbonized substance fractions are desirable.

TABLE 11.3
Effects of pyrolysis temperature on biochar properties (obtained with permission from Kamali et al. (2021) © Elsevier)

Types of biochar	Pyrolysis conditions	Change in properties
Whitewood biochar	High-temperature (1000°C) produced wood biochar	Absence of acidic functional groups
Sugarcane bagasse biochar	Pyrolysis temperature from 300 to 600°C	Decrease in average pore diameter of produced biochars with increasing pyrolysis temperature
Cow-manure-loaded biochar	Low-temperature biochar (250°C)	Presence of volatile organic compounds, which inhibit microbial activity
Miscanthus BC	Pyrolysis temperature above 360°C	Higher thermal and biological resistance to degradation
Coppiced woodlands biochar	High-temperature (800°C) crop residue and manure biochar	High ash content
Corn stover, red oak biochar	Biochar prepared under moderate temperature (500°C)	High cation exchange capacity but low anion exchange capacity, point of zero salt effect and zero net charge values
	Biochar prepared under high temperature (700°C)	Low cation exchange capacity and high anion exchange capacity, point of zero salt effect and zero net charge values
Pitch pine biochar	Increasing pyrolysis temperature from 300 to 500°C	Decrease in biochar yield from 60.7 to 14.4%
Hardwood-based wood pellets	Pyrolysis temperature from 400 to 600°C	Increased temperature leading to the conversion of labile carbon forms to aromatic carbon structures in biochar
Boiled radix isatidis residue biochar	Pyrolysis temperature from 300 to 700°C	Increasing temperature resulting in a reduction of volatile matter content, strengthens the carbon enrichment, and increasing the aromatic organic compounds

11.3.1 HYDROPHOBIC INTERACTION

Mutual repulsion between water molecules occurs when molecules contain nonpolar groups (like CeH). This is known as the hydrophobic interface. Moreover, such a sorption process is used for biochar sorption of ECs like tylosin (TYL), 17-estradiol (E2), estrogen, and certain SAs. Earlier researchers have studied the impact of microwave-mediated chemical activation approaches on biochar characteristics and how they may be used to take away 17-estradiol (E2) from water. The alkali-modified biochar showed a significantly better efficiency for E2 adsorption compared to other activated biochar, and that was attributable to its superior hydrophobicity and increased SSA. Several investigations on the sorption of SAs by alternatively activated biochar have been executed. To absorb SAs, researchers employed modified biochar made from cotton stalks. The SAs' adsorption behavior was shown to be influenced by van der Waals forces and hydrophobicity. Nano-Co_9S_8 and -CoO layered with sulfur- and nitrogen-incapacitated biochar (CoO/Co_9S_8@N-S-BC) were characterized and utilized to trigger peroxymonosulfate (PMS) to eliminate SMX. CoO/Co_9S_8@N-S-BC efficiently activated PMS to break down sulfamethoxazole, and 0.08 mM SMX was entirely decimated in under 10 min (Wang and Wang 2020). Dual SAs (SPY and SMX) were removed from wastewater using ball-milled biochar. In other experiments, nitrogen-doped biochar was utilized to adorn nickel foam electrodes and was applied in the electro-Fenton procedure to extract SMX. Further, a photochemical degeneration technique was utilized to handle it, where TiO_2 mixed with zinc was coated on the reeds straw biochar and acid was used to pretreat it (Zn-TiO_2/pBC). SMX photodegradation was used for further photocatalytic motion study of Zn-TiO_2/pBC. The antibiotic TYL is an abstruse blend of intimately associated components shaped using a streptomyces strain. It's being used to treat veterinarian diseases and boost poultry health. It has been found that novel goethite biochar composites have a higher TYL extraction ability from water. SEM image assessment of goethite nanoparticles on the surface of biochar revealed that they were effectively distributed. For TYL, the adsorbent provided quick and effective sorption. EDA interaction, hydrophobicity, hydrogen bonding, static electricity, and cation exchange were all responsible for the adsorption process (Guo et al. 2016). Current biochar activation techniques mostly employ clay minerals and metal oxides in TYL adsorption studies. The major reason is that the two ways of surface activation make it simpler to improve surface sorption. The pore size of biochar may be increased by clay mineral processing, and even the hydrophobicity can be improved.

11.3.2　Electrostatic Interaction

The solution pH can substantially influence the magnitude of ionization, the surface charge of biochar, and the morphology of the sorbate; that is why electrostatic interaction is so important for the adsorption of the ionizable molecules. The core of ionic bond formation is electrostatic interaction, which involves electrostatic repulsion and attraction. The surface of a solution with a lower pH compared to the adsorbent's point of zero charges (PZC) would be ionized with positive charges. Contrarily, for a higher pH solution, the adsorbent's surface would be charged negatively. Additionally, the pH of the solution modulates the adsorbate's ionization degree. As a result, electrostatic interaction between the adsorbent and the surface might occur under such circumstances. During past investigations, SAs, NPX, TC, and BPA sorption on biochar is dominated by electrostatic interactions. Under a controlled acidic environment, the sorption of SAs over modified biochar exhibits a good adsorption capacity. During the adsorption procedure, an electrostatic interface may arise. T. Chen et al. (2018) discovered that increasing the solution pH from 4.5 to 8.5 improved the sorption performance of activated biochar for TC, owing to the escalated electrostatic force connecting both biochar and TC. Some researchers have used a straightforward approach to make a Z-type photocatalyst using biochar@$CoFe_2O_4$/Ag_3PO_4. It was observed that 80.23% mineralization rate and 91.12% degradation efficiency were achieved when BPA was chosen as the target contaminant. Because PMS is quite hard to break down, iron biochar composites including permeable carbon with enriched functional groups were developed to stimulate PMS and simplify BPA elimination. The findings showed that biomass performance can be enhanced by the assistance of tenacious free radicals of biochar and electron transmission of nanofiber-mesoporous carbon assembly. Some research employed wild plum as source material to produce pristine biochar, which was subsequently submerged in KOH solution for various ECs. After that, active biochar (WpOH) was produced using the microwave at 700 W for 12 min. It was used to remove the ionizable medication naproxen (NPX). XRD, FTIR, and some other characterization techniques were used to investigate the basic adsorption processes, which revealed that the affinity of negative NPX ions toward the positive WpOH functional groups was the key point of electrostatic interaction (Paunovic et al. 2019).

11.3.3　Hydrogen Bonding Interaction and Functional Groups

The chemical characteristics of organic molecules are reckoned by functional groups, which are atoms or a bunch of atoms. Different functional groups may interact with one another, and few of them can even generate hydrogen bonds. They do have high binding energy and are difficult to separate due to the formation of hydrogen bonds. Environmental hormones including E2, EE2, BPA, and nonylphenol are the most susceptible to this adsorption process in several investigations. The primary processes that impact the adsorption of E2 by biochar are hydrophobicity and hydrogen bonding. A few studies have created magnetically improved biochar to adsorb E2, and the specific surface area of the produced biochar was approximately 14.5 times more than the pristine biochar, which improved the sorption efficiency of E2. The E2 adsorption capability of the magnetic biochar was found to be remarkable. The major adsorption interaction between magnetic biochar and E2 shifted from hydrogen bonding to π–π interaction as the pyrolysis temperature was increased. The magnetic biochar renewed by odorization, could well be exposed to E2 adsorption after a minimum of 5 times without losing much of its adsorption ability. Many studies have already examined the impact of biochar pre- and post-treatment. The chronology of KOH stimulation and its stages had a significant influence on the material. Pore volume, pore size, and SSA were all considerably enhanced after KOH treatment as compared to the original biochar (Cheng et al. 2021). Post-treatment of biochar increases its sorption efficiency to E2, as compared to direct-treatment, original, and prior-treatment methods. To eliminate E2, a higher pyrolysis temperature might be beneficial. External environmental variables (ionic strength, temperature, and pH) may have a significant impact on biochar's E2 adsorption ability. In an investigation, biochar generated from litchi was utilized as an adsorbent to restrict the transport of estrone in dirt and water. Estrone dissemination can be successfully administered by activated biochar (Fe-Mn-BC and Ca-BC), mainly by Fe-Mn-BC. Such discoveries were critical for regulating estrogen chemical movement in the soil and water ecosystem. Magnetic biochar (g-Fe2O3@BC) with a high SSA was produced using an easy and efficient hydrothermal technique to remove BPA. Without adjusting the pH, $g\text{-}Fe_2O_3@BC$ was shown to be highly efficient in degrading BPA, and it could be eliminated in 20 min. Heo et al. (2019) used a simplistic hydrothermal technique to create a novel biochar-category-assisted magnetic $CuZnFe_2O_4$ composite (CZF-biochar) to eliminate complex contaminants (SMX and BPA) from aqueous medium. Charge-mediated hydrogen bonding, hydrophobicity, and π–π EDA interactions were discovered to be the primary methods. Magnetic adsorbents drew a lot of interest as a way to make it easier to separate altered biochar from water because of their features, including cost-effectiveness, strong sorption capacity, and ease of use. Many researchers utilized magnetic modification to create quite basic spruce-sheet-shaped modified biochar and discovered that it was highly effective in the removal of BPA. A pseudo-second-order kinetic model could adequately represent the sorption mechanism, which was a spontaneous exothermic reaction. A novel zero-valent-iron-assisted biochar (BC-*n*ZVI) was produced using a sustainable strategy for continuous elimination of BPA and toxic metals. In the meantime, standalone BC-*n*ZVI and a combination of persulfate and BC-*n*ZVI (BC-*n*ZVI/PS) were used to eliminate BPA and Cu^{2+} concurrently. The findings revealed that the major free radical responsible for BPA degeneration was SO_4^{2-} ion. Also, two kinds of biochar-mediated MnO^{2-} composites with a 3-D framework were produced using hydrothermal procedures with timescale control, and their ultrasonic-aided Fenton-like catalytic activity (heterogeneous) was assessed. The study found that the sea-urchin-based improved biochar was likely to demonstrate good catalytic activity in the sonication-mediated heterogeneous Fenton-type process, enabling it to extensively treat different types of organic contaminants (Jung et al. 2019). In the coming days, further study in terms of sorption characteristics and processes will be required for understanding these phenolic compounds.

11.3.4 SURFACE COMPLEXATION AND COPRECIPITATION

The development of complexes on the surface of biochar is generally attributed to the sorption of ionic species like metal ions. Complexes are defined as polyatomic molecules having either one or more core atoms; encircled by ligands. Additionally, the creation of bonds on the biochar surface with the functional groups, surface compounds may lead to the omission of hydrogen and additional ions. Complexes are generated via coordinated chemical processes. Lewis acid-base complexes are those in which the transition metal ion functions as an electron acceptor (Lewis acid) and makes the coordinating covalent bonds with an electron-pair donor (Lewis base). Chelation is a sort of complexation that is described as an equilibrium process that involves the creation of a complex involving organic molecules and several functional groups (multibonded) with a centralized atom. Such a kind of interaction occurs when a metal ion reacts with a complexing agent, thereby resulting in a ring type formation. Surface complexation, which includes chelation along with coordination, is the key process for the elimination of ionic species, mostly metal ions, from biochar. Since sorption processes may include the utilization of hydrogen ions, the pH of the solution has a significant impact on complexation. Cui, Fang, et al. (2016) examined the adsorption of Cd^{2+} onto biochar, generated from *Canna indica*. It was reported that the main sorption mechanism of metal ions is surface complexation-assisted ionized O-enriched functional groups. Furthermore, the study revealed that, after subsequent adsorption of Cd^{2+} onto the surface of biochar, the pH of the solution decreases, indicating that surface complexation might be associated with the discharge of H^+ ions or any additional cations. Ifthikar et al. (2017) studied the method of Pb^{2+} ions sorption onto sewage-sludge-derived magnetic biochars. Because of the substantial changes in the FTIR analysis of pre- and postadsorption processes, the authors hypothesized that surface complexation was among the key mechanism of adsorption. Xu et al. (2013) discovered a similar trend while studying the sorption process of Cd^{2+}, Cu^{2+}, and Zn^{2+} onto modified biochar, derived from dairy dung. Apart from FTIR spectra, a few studies have also operated X-ray photoelectron spectroscopy (XPS) to evaluate the chemical components of adsorbent surfaces during pre- and postsorption processes, for instance, to observe the occurrence of the surface complexation mechanism during adsorption of Pb^{2+} onto anaerobically digested sludge-mediated biochar.

Precipitation occurs when a macrocomponent forms insoluble compounds or if a chemical approaches its solubility threshold. Such process is generally triggered by changes in the solution pH or the existence of chemicals onto the surface of biochar, like mineral, polymeric, and enzymatic compounds, that may induce condensation in the solution or on the surface. Though precipitation aids in the elimination of a few sorbates, in sorption studies, the threshold of solubility must not be surpassed since the adsorbate is not eliminated by sorption. As a result, the adsorption mechanism would be aided solely by condensation on the charcoal surface or coprecipitation. Cui, Fang, et al. (2016) have studied the sorption mechanism of Cd^{2+} onto biochar, derived from *Canna indica* and observed white granular crystals on its surface after adsorption, which had been confirmed as $CdCO_3$ by X-ray analysis and electron microscopy. Furthermore, once the preliminary content of Cd^{2+} was enhanced, the concentration of CO_3^{2-} emitted from the biochar was considerably reduced, implying that the Cd^{2+} ion concentration might be connected to the deposition of coprecipitate onto the biochar surface. Ho et al. (2017) reached the same conclusion when they studied the sorption process of Pb^{2+} ions onto anaerobic sludge-derived biochar. Researchers proposed surface coprecipitation as a result of the biochar surface's high phosphorus level, which combines with Pb^{2+} ions and produces Pb-phosphate precipitate. XRD spectra and X-ray elemental dot mapping corroborated these findings. Furthermore, Deng et al. (2021) reported that biochar altered with $MgCl_2$ and $AlCl_3$ can increase the phosphate sorption capacity as both Al^{3+} and Mg^{2+} ions play a crucial part during the coprecipitation process of phosphate adsorption. Also, R. Deng et al. (2019) reported similar findings when they studied the adsorption of Cu^{2+}, Cd^{2+}, and Pb^{2+} onto biochar derived from wood waste. The study suggested that the biochar modified using phosphorus and chlorine led to more precipitate development on the surface of biochar.

11.4 OPERATING PARAMETERS INFLUENCING THE ADSORPTION PROCESS

11.4.1 SOLUTION pH AND ADSORPTION TEMPERATURE

The surface activity of adsorbents, solution pH, and competitiveness of hydrogen ions with the intended pollutants are most certainly the critical parameters of adsorption mechanism. The solution chemistry is associated with chemical evolution and the ionization degree of the molecules in solution, implying that the adsorbate's features and qualities can substantially differ, depending on the solution pH. The surface charge and the functional groups present on the adsorbent surface are also affected by the pH, one that modifies its activities in the adsorption mechanism. When biochar is protonated, the active sites become positively charged, and when it is deprotonated, they become negatively charged. Furthermore, in acidic circumstances, large quantities of H^+ ions in the solution may enhance rivalry with the cationic groups like metal ions. Moreover, the influence of pH on the adsorption of Pb^{2+} onto the surface of anaerobic digestion sludge was investigated by Ho et al. (2017). The authors reported that the sorption efficiency of Pb^{2+} gets amplified within a solution pH of 1–6, since at pH 6, the zeta potential of biochar was found to be −20.4 mV and that was vital for the sorption of positive ions. Furthermore, significant presence of hydrogen ions in solution may compete for biochar's active sites, since the sorption efficiency dropped even more as the solution became acidic. Choi and Kan (2019) examined the influence of solution pH on the sorption of SMX and BPA onto biochar, extracted from alfalfa with a pHPZC of 8.1. The researchers claim that greater sorption capabilities were found at pH values within 3–5 because of positively charged biochar surface, while the SMX (pKa of 1.6 and 5.7) and BPA (pKa of 9.6) were without charge, indicating that electrostatic forces were insignificant. Heo et al. (2019) reached the same conclusion when they investigated the influence of pH on the sorption of SMX and BPA onto a magnetic $CuZnFe_2O_4$-biochar composite. The study revealed that the sorption efficiency of SMX and BPA was almost constant throughout the pH range of 3–9. Nevertheless, when the pH increases, the adsorption efficiency was steadily reduced as both the molecules and adsorption surface became negative.

The solution temperature can have two effects on the solid–liquid system: (1) it can change the adsorption rate, and (2) it can change the adsorption efficiency. Temperature increases the adsorbate mobility species and causes an increment of intraparticle dissemination rate to the pores on the surface of the adsorbent. This effect is attributable to a reduction in solution viscosity, which causes greater sorbate mobility and a reduction in mass transfer confrontation to the adsorbent surface. Furthermore, the solution temperature can affect sorption performance owing to the strength of interaction seen among the active sites of biochar and the adsorbate molecules, and that relies on whether the sorption procedure is endothermic or exothermic. Zhu et al. (2018) investigated the influence of solution temperature on the adsorption of methylene blue onto the surface of low-temperature biochar made of cow manure. Findings state that, at elevated temperature and for any particular sorption period, sorption efficiency increases. As soon as the solution temperature was enhanced from 25 to 35°C, the influence of temperature was reduced. The improved adsorption efficiency was due to the endothermic character of the adsorbent–adsorbate interface, while the adsorption rate was linked to the temperature-dependent change in water viscosity. Movasaghi, Yan, and Niu (2019) reported similar results when they studied the influence of solution temperature on the sorption of ciprofloxacin onto the surface of prepared oat hulls. Based on the analysis, it was observed that higher temperature enhances the permeation rate and decreases water viscosity, thereby facilitating adsorbate migration in the direction of the active areas on biochar and its insertion into pores. Ren et al. (2021) also investigated the impact of temperature on the sorption of nitrogen over activated carbon. According to the study, elevated temperature marginally reduced the sorption capability of ammonia nitrogen while constantly maintaining its sorption rate.

11.4.2 Particle Size and Dosage of Biochar

Adsorption and ion exchange mechanisms are influenced by particle size. Reducing particle size can enhance the adsorbent's surface of interaction, uncover a wider active surface area, shorten the diffusion path, and boost the sorption rate and efficiency. Moreover, due to the difficulties of retrieving or isolating the sorbent from the aqueous phase after the sorption step, the use of powdered biochar is not feasible. Crushed biochar can also exhibit low mechanical stability, poor hydraulic conductivity, and significant resistance to flow in fixed-bed columns. Furthermore, the determination of particle size is based on the adsorbent and adsorbate characteristics. Han et al. (2016) studied the influence of particle size on the adsorption of Cr^{6+} onto the surface of peanut-husk-extracted magnetically active biochar. Adsorption rates and performance of smaller size biochars were greater than those having large size. Nevertheless, only the influence of particle size was studied for the sorption kinetics, which later confirmed that sorption equilibrium was not entirely attained. Kang et al. (2018) studied the sorption of phenanthrene on granulated and powder biochar and stated that the biochar particle size can have a significant impact on the adsorption kinetics of hydrophobic organic contaminants. Similarly, Matsui et al. (2015) investigated the impact of particle size on 2-methylisoborneol and geosmin sorption with 9 distinct activated carbons. A significant increment in adsorption performance was observed when the particle size was reduced. Furthermore, due to the hydrophobicity of the sorbate and sorbent, certain carbons had only a little or moderate impact on the adsorption performance. Hence, for the sorption of hydrophobic microcontaminants onto less hydrophilic sorbents, smaller particle size in a few-micrometer range is suggested. However, reduced particle size did not show much effect on adsorption capacity when hydrophilic adsorbates or hydrophilic carbons were used.

The impact of primary sorbate concentration causes the sorption efficiency to thrive as the sorbate percentage in the system increases until the biochar reaches its saturation level. Consequently, the removal efficiency decreases with an increase in the initial sorbate concentration. On the other hand, as the biochar concentration is increased, the adsorption efficiency decreases, while at the same time removal capacity is enhanced. The first scenario may be described by the smaller number of active sites on the surface of biochar, while the second scenario can be attributed to an increment in the active sites. Fan et al. (2017) investigated the impacts of preliminary dosage and concentration of biochar (obtained from sewage sludge) on the adsorption of methylene blue onto its surface. The study reported that increasing the biochar dose from 2 to 10 g/L enhanced the methylene blue reduction effectively from 54 to 100%. Nevertheless, when the adsorbent dose increases, the adsorption efficiency decreases. Ho et al. (2017) noticed a similar trend and proposed that the stronger the biochar doses are, the better the accessibility of active sites will be. Thus, as the initial adsorbate concentration increases, so did the adsorption efficiency until the number of active sites became restrictive, i.e., a plateau was achieved. Contrarily, the rate of exclusion was reduced. Furthermore, Li et al. (2020) examined the influence of phosphate adsorption onto the surface of sewage-sludge-derived biochar. According to the authors, increasing the adsorbent doses leads to the formation of a driving force to surmount the mass transfer barrier at the solid–liquid interface, thereby increasing the sorption rate.

11.4.3 CONTACT TIME AND STIRRING TECHNIQUES

When a substrate comes into contact with an adsorbent, its concentration in the solution reduces over time until it reaches a steady state, i.e., the sorption equilibrium. Hence contact time is an essential parameter in the study of specific techniques and its utilizations as it impacts the number and sizes of reactors used in adsorption studies. The interaction time is highly influenced by both the biochar's and the adsorbent's features and qualities, as well as by the operational circumstances. Fan et al. (2016) evaluated the influence of contact time on the adsorption of methylene blue onto the surface of municipal-solid-waste-derived biochar, biochar derived from tea waste, as well as biochar from sewage sludge. The contact time necessary to establish adsorption equilibrium for the first case was 24 h, while the second biochar achieved adsorption equilibrium in 8 h. These findings suggest that the contact duration is affected by the biochar's features and attributes, such as porosity and surface charge, that could influence the rate-controlling stage. Zhou et al. (2019) investigated the impact of treatment time on sparfloxacin and ciprofloxacin sorption over Fe_3O_4/ graphene-activated biochar. Based on their study, the durations necessary to achieve sparfloxacin and ciprofloxacin adsorption equilibrium were 24 and 36 h, respectively. The properties and chemical heterogeneity of the adsorbates in solution, like ionic radius, complex formation, and molecular size, were attributed to these findings. Different stirring tactics, namely stirring apparatus, flasks variety, and stirring speed, can considerably affect the contact time. This is due to the fact that the outer mass transfer resistance can be easily controlled with the help of appropriate agitation, which in turn reduces the boundary layer or film diffusion thickness. As intraparticle diffusion and external mass transfer are often considered the rate-controlling stages in the sorption mechanism, appropriate stirring can improve the sorption rate while decreasing the time necessary to establish the adsorption equilibrium, as outlined by Kuśmierek and Swiatkowski (2015).

11.5 BIOCHAR IN THE REMOVAL OF WASTEWATER CONTAMINANTS

The production of biochar as a valuable commodity generated from biomass valorization has sparked a surge of interest over the last few years. Challenges associated with the generation and release of a substantial quantity of industrial wastewater along with its worldwide implications have instigated research efforts to explore effective and economical techniques to handle such issues. As a result, biochar has been employed as a viable option for dealing with various wastewater pollutants, including both traditional and emerging pollutants. Biochar and modified biochar have been the subject of extensive study, which shows that they are able to remove both the organic and inorganic pollutants present in wastewater. The adsorption behavior of modified biochar differs from that of biosorbent due to its superior properties such as porous surface structure, elemental composition, higher surface area, and anion–cation exchange capacity. One of the most significant applications of biochar is the mitigation or removal of environmental contaminants from polluted sites (Shengsen Wang et al. 2015). Biochar prepared from various waste source materials is commonly used to remove organic and inorganic pollutants from polluted wastewater, consisting of heavy metals, pharmaceutical compounds, pesticides, and metalloids. Table 11.4 depicts the application of designer biochar for the removal of both organic and inorganic contaminants.

11.5.1 Biochar Application in the Removal of Organic Contaminants

Significant efforts have been made to improve the biochar efficiency for the removal of different varieties of organic compounds: phenolics, dyes, antibiotics, halogenated hydrocarbons, and pesticides. In general, biochar adsorption of organic compounds is based on the pollutant types as well as surface characteristics of biochar. Consequently, the adsorption mechanisms for the removal of organic pollutants can be categorized as hydrophobic interactions, π–π interaction, complexes adsorption, cation exchange, pore filling, electrostatic attraction, and partition uncarbonized fraction. Designer biochar has been widely utilized for the removal of organic pollutants. One such instance is the improvement of levofloxacin adsorption on the surface of cerium-trichloride-mediated biochar. Treatment with cerium leads to an increment in the oxygen-containing functional groups on biochar surface. In addition, CeO_2 consist of a mesoporous surface structure, thereby enhancing the adsorption efficiency of biochar. Furthermore, Alghazwi et al. (2019) reported that biochar modified via the ball mill method showed higher adsorption efficiency for galaxolide as compared to the pristine biochar. Apart from improving the pore size and surface area, the surface polarity of biochar was also enhanced owing to the significant presence of oxygen-containing functional groups. Availability of such functional groups considerably assist in the adsorption of organic pollutants via complexation and H_2 binding between biochar and the organic compounds. The designer biochars are also effective for the mitigation of biological pollutants from wastewater. For instance, modifying wood biochar via H_2SO_4 oxidation with subsequent (twofold) improvement in biochar surface area eventually leads to an increase in the adsorption of *E. coli* from stormwater. Recently, the characteristics of pristine biochar prepared from pseudo-stem biomass of pyrolyzed banana at 600°C was substantially enhanced by the application of Fe_3O_4 coating. The modified biochar exhibited higher surface area with superparamagnetic features, which was effectively utilized for the degradation of furazolidone (antibiotic) from pharmaceutical wastewater. More precisely, chemisorption techniques were used to adsorb up to 37.86 mg furazolidone per gram of biochar. It should be noted that the reduction of such pharmaceutical compounds does not necessarily make the treated effluent harmless. For example, the formation of toxic metabolites, namely 3-amino-2-oxazolidone and ß-hydroxyethylhydrazine, may emerge from the biological degradation of furazolidone. This highlights the significance of utilizing biochar in further degrading them via the adsorption process. Also, modification of biochar led to a substantial improvement in its pesticide removal efficiency. In an effective field study, almond shell biochar (steam activated) after being gently pyrolyzed under nitrogen gas for 1 h at 650°C was successfully utilized for the degradation of

TABLE 11.4

Application of Modified Biochar for the Degradation of Organic and Inorganic Pollutants (modified with permission from Barquilha and Braga (2021) © Elsevier)

Wastewater contaminants	Adsorbate	Feedstock	Adsorption capacity (mg/g)	Surface area (m²/g)	O/C ratio	Optimal pH	Contact time (h)	Adsorption mechanism
Organic pollutants	17β-estradiol	Sugarcane bagasse	34.1–50.2	166.8–339	0.10–0.27	4	1	Hydrogen bond, π–π interaction
	Atrazine	Corn straw	58.3–72.2	7.8–15.1	0.10–0.41	9	24	Pore filling, π–π interaction
	Bisphenol A	Alfalfa	62.7	3.5–405	0.08–0.16	3	12.5	Hydrophobic, π–π interaction
	Cipro-floxacin	Sewage sludge and bamboo	62.5	47.5	1.63	4	12	Hydrogen bond, π–π interaction
	Methylene blue	Lychee seeds	124.5	154	0.40	>0.6	2	Electrostatic, π–π interaction
	Sulfamethoxazole	Alfalfa	90.4	3.5–405	0.08–0.16	3–5.7	12.5	Hydrophobic, π–π interaction
	Ibuprofen	Coffee	86.9	427.5	0.023	3	12	Pore filling, electrostatic interaction
Inorganic pollutants	Copper (Cu^{2+})	Corn straws	19.6–160.3	2.28–61	0.06–0.53	6	24	Ion exchange, metal-π interaction
	Lead (Pb^{2+})	Bamboo	177.3	4.6	0.58	>5	2.5	Electrostatic, complexation
	Nickel (Ni^{2+})	Rice straw	27.3–54.6	4–161.2	0.11–0.28	5.5	12	Ion exchange, coprecipitation,
	Arsenic (As^{3+})	Corn straws	2.9–8.3	60.9–208.6	0.20–0.59	4	1.5	Electrostatic, complexation
	Cadmium (Cd^{2+})	Canna indica	63.3–188.8	3.5–10.4	0.14–0.36	>5	3	Coprecipitation metal–π interaction
	Phosphate (PO_4^{3-})	Bagasse	128.2–263.2	66.9–106.7	0.19–1.57	>10	3–18	Coprecipitation, electrostatic interaction

dibromochloropropane from well fields. In addition, during the decolorization of water samples, it was reported that methylene blue (22.7 mg/g) was efficiently trapped in the honeycomb mesoporous surface structure of gently pyrolyzed (at 1000°C) acid-treated kenaf (*Hibiscus cannabinus*) biochar. This designer biochar exhibited a higher oxygen and fixed carbon content, as well as higher surface area, when compared to pristine biochars (Mahmoud et al. 2012). Biochar can also be utilized to remove solvents from water. For instance, biochar extracted from soybean stover has been widely used for the removal of solvent, i.e., trichloroethylene from water. The sorption efficiency of such biochar was substantially influenced by the pyrolysis temperature. As a matter of fact, biochar prepared at the highest pyrolysis temperature of 700°C exhibited a maximum trichloroethylene sorption efficiency of 32.02 mg/g. Biochar-assisted biofilters, biochar-mediated membrane filtration and permeable reactive barriers are all examples of hybrid water purification techniques that can be developed with the help of designer biochars. Furthermore, by utilizing *Corynebacterium variabile* HRJ4 and wheat-straw-biochar-assisted nano Fe-S composite, a novel biochem hybrid system, was developed in 2018. Compared to pristine biochars, this green synthesized biochar having higher stability was able to successfully remove trichloroethylene from water, while exhibiting a smaller hydrodynamic diameter. A TiO_2-assisted soft wood biochar composite after being pyrolyzed at 700°C could be effectively used to develop an organic–inorganic hybrid system. The photocatalytic decomposition capacity of phenol by the designer biochar was substantially enhanced, which resulted in 33.6 and 64.1% degradation under visible and UV light, respectively, when compared to pristine biochars and TiO_2 alone (Kazemi Shariat Panahi et al. 2020). Thus both standalone designer biochars and hybridization techniques could significantly enhance the degradation of pollutants from water and wastewater.

11.5.2 Biochar Application in the Removal of Inorganic Contaminants

Due to its microporous structure and higher surface area, utilization of biochar has shown tremendous potential during the mitigation of pollutants, notably heavy metals, from various wastewaters. However, it was found that pristine biochars were unable to remove higher pollutant concentrations from aqueous solution below the permissible limits. Moreover, due to its smaller diameter and lower density, the use of ball-milled biochar tends to complicate the different stages of the biochar separation process. As such, several studies have focused on the development of designer biochars with enhanced surface characteristics in order to overcome such constraints. Designer biochars have been extensively developed in recent years with properties such as improved surface-active sites, enhanced functional groups, higher surface area, and ease of separation to degrade a variety of wastewater pollutants. For increasing the removal rate of heavy metals from aqueous solution, various types of physicochemically designed biochars have been prepared. For instance, biochar coated with ultrafine particles of MnO_x exhibited a greater adsorption capacity for Cd(II), Cu(II), and Pb(II). Additional oxygen-containing functional groups, viz. carboxyl functional groups, phenolic hydroxyl, and hydroxyl groups, were found on the surface of MnO_x-loaded biochar, thereby providing more active sites for heavy metal adsorption. Ding et al. (2016) observed that biochar modified with NaOH exhibited a higher metal adsorption capacity of 2.6–5.8 times, as compared to pristine biochars. Alkali treatment enhances the surface area of biochar, which in turn increases porosity by improving the interactions between NaOH-C matrices. Furthermore, alkali modification leads to an increase in the cation exchange capacity, which could be linked to an increase in oxygen-containing functional groups. In addition, biochar can also minimize ammonia, phosphate, and nitrate contents from natural water bodies, thereby avoiding eutrophication, or the excessive development of photosynthetic microbes with regard to higher inorganic nutrient levels in the aquatic ecosystem. The use of designer biochar assisted with LDHs can considerably reduce the risks of eutrophication in waste streams containing phosphates. This can be achieved when the biomass undergoes pyrolysis in the presence of metal hydroxides, viz. Zn/Al, Mg/Al, and Ni/Fe. Nevertheless, biochar may also emit some quantity of phosphates/nitrates into the aquatic system. In general, the biochar adsorption of inorganic pollutants is significantly influenced by the solution pH, types of pollutants present, and the surface characteristics of the biochar. Yang et al. (2019) reported that cornstalk-Zn/Al-LDH exhibits a very high phosphate adsorption capacity of 152.1 mg/g along with quick Elovich kinetics of 5925 mg/g/h; however, the adsorption process entirely depends on the solution pH. Adsorption of phosphates on biochar having a negatively charged surface is often inhibited by high solution pH. When compared to pristine biochars, designer biochar can sometimes provide higher inorganic pollutants removal efficiency. For instance, pristine biochars could only reduce 10% of phosphate; however, MgO-assisted biochar resulted in a phosphate removal efficiency of 66.7%. The strong affinity of MgO for phosphate in an aquatic ecosystem could be the reason for increased adsorption rate. Mg-assisted biochar modification also exhibits higher sorption efficiency of >100 mg/g via surface deposition and precipitation of phosphates on the Mg. As the adsorbed phosphates remain bioavailable, slow-release phosphate fertilizers can be easily formulated. Furthermore, ball mill modification of biochar has also attracted considerable attention as a water treatment material. The development of pyrolyzed micro-/nanodesigned nitrogenous cow bone biochar via the ball milling method resulted in an effective removal of Pb (558.9 mg/g), CU(II) (287.6 mg/g), and Cd(II) (165.7 mg/g) from water. A similar study reported that ball-milling-assisted bamboo biochar led to an enhanced ammonium removal from water (22.9 mg/g) by more than three times compared to pristine biochars (7 mg/g). Furthermore, in recent years the development of magnetic biochar composites has attained significant focus due to its higher removal efficiencies as compared to non-magnetic biochar. It was observed that among all other metals, magnetic biochars provides higher sorption efficiency for Cu(II), Pb(II), Cd(II), and Zn(II) (Kazemi Shariat Panahi et al. 2020). In a study conducted by Z. Tan et al. (2017), incorporation of magnetic components (Fe^{3+}/Fe^{2+}) to rice straw before being pyrolyzed led to the development of hematite (γ-Fe_2O_3), while at the same time persevering the original functional groups of biochar, thereby enhancing the Cd(II) adsorption efficiency. Formation of hematite during pyrolysis also contributes to the adsorption of heavy metals. The study concluded that biochars produced under CO_2 pyrolysis condition shows greater adsorption efficiency for heavy metals, owing to the affinity of CO_2 to interact with both oxygenated and hydrogenated groups, which resulted in the formation of biochar with a higher specific surface area. The biochar adsorption mechanism for the removal of heavy metals and organic and inorganic contaminants is shown in Figure 11.3.

11.6 ECONOMIC ASPECTS AND REGENERATION PERFORMANCE OF BIOCHAR

The economic evaluation of biochar technologies for efficiency, adaptability, and environmental sustainability has been addressed in many studies. The substantial economic efficacy of biochar manufacturing processes varies based on the raw materials or feedstocks utilized, the conversion method used, and whether C-sequestration subsidies or carbon credits representing the intrinsic worth of greenhouse gas reduction have been included. In comparison to chemical fertilizer, biochar has an abiding ability to benefit agriculture. Biochar generated from various feedstocks was often investigated in the laboratories or tested in short field testing, leading to fewer or restricted products on its sustainable development. The utilization of new feedstock, rather than purpose-made biomass, enhances the biochar economy. The entire expenses of biochar manufacturing and its uses in wastewater remediation were investigated by Ahmed et al. (2016). One of the primary barriers in the marketing of biochar is the high manufacturing cost, which might hinder its quick implementation for real-world applications. Biochar-based processes have a fixed number of life cycle costs and economic analyses accessibility. For example, the estimated annual pyrolysis expense of 1 MWh unit is about $381,000, indicating that pyrolysis is the most extensive step of biochar production. The real cost of acquiring and shipping red oak (*Quercus rubra*) biochar for large-scale use is around $290/ton. Shackley et al. (2011) reported a price range of $222 to $584/ton for biochar manufactured in the United Kingdom, based on the raw materials and the complexity of the pyrolysis process. Maroušek

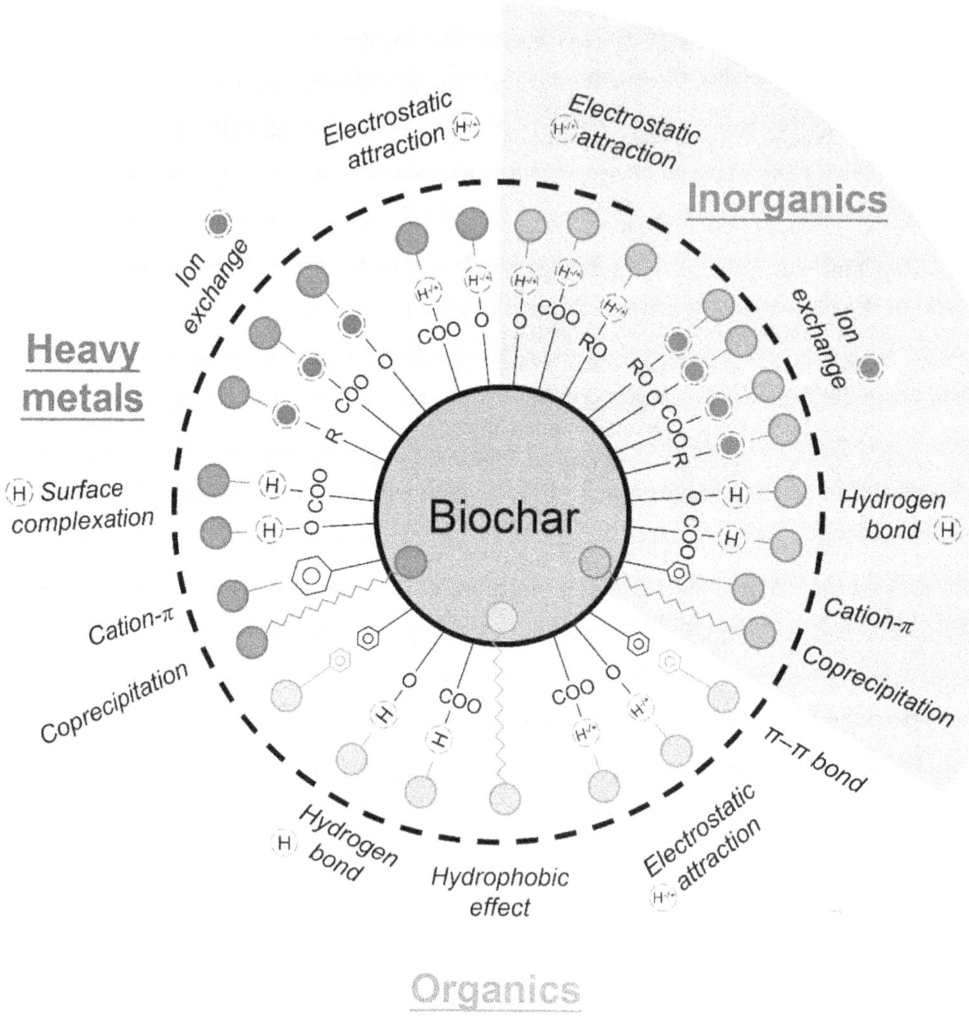

FIGURE 11.3 Adsorption mechanisms of biochar for the removal of heavy metals and inorganic and organic pollutants.

Source: Obtained with permission from Pan et al. (2021) © Elsevier.

(2014) estimated that unmodified biochar costs $500 per ton, whereas Rosales et al. (2017) estimated that biochar manufacturing costs are about $0.076/kg, or around 3–6% of the cost of other conventional carbon-based sorbents. Therefore, recent advancements in biochar production methods have significantly decreased biochar manufacturing costs, making them potential alternatives for restoration of contaminated wastewater. Efforts are being made to establish more cost-effective biochar manufacturing techniques. For example, Zhou et al. (2018) devised a cheap in situ approach known as the burning and soil covering (B-SC) technique, which farmers may use to treat agricultural wastes. Biochar manufacturing cost is also affected by energy prices, as noted by Pratt and Moran (2010), who pointed out that the electricity cost in Europe is substantially lower than in North America. Low-cost electricity in certain regions (like the Middle East) can make biochar manufacturing and its use more feasible. In this context, various innovative techniques have lately been created and implemented to lower manufacturing expenses. Solar energy is one such example. Giwa et al. (2019) used an intense solar thermal energy unit to pyrolyze date palm debris and concluded that this approach is a financially sustainable technique compared to other traditional techniques established for biochar preparation, with a recovery period of four years and 132 days. They also predicted a 14.8% expected rate of return and a 22.9% return on capital. According to recent research, unmodified biochar is not competent enough to cope with environmental pollutants. Biochar infused with metallic compounds has also shown promising results in terms of environmental pollutant removal. Recently, a variety of metallic compounds have been employed for this purpose. Q. Chen et al. (2018) demonstrated an excellent capacity (89.25%) for phosphorus removal using Mg-infused biochar at a comparatively cheap manufacturing cost of US$1.2/ton. The functionalized biochar had a minimal cost, which fulfilled the biochars' sustainability concerns. Many researchers have tried to estimate the overall expenditures of biochar for the remediation of contaminated wastewater. Poultry waste amine-assisted biochar was utilized to adsorb dimethyl sulfide from the aqueous solution, with an adsorption efficiency of 1.14 mg/g and a manufacturing cost of $1.60/kg including shipping, reagents, and electricity. In short, if compared to other innovative and traditional technologies, biochar may be regarded as the cheapest and most cost-effective method for dealing with contaminated wastewater. Kamali et al. (2021) reported about the sustainability issues regarding the applicability of engineered nanoparticles for the remediation of polluted wastewater leading to the conclusion that, in the current scenario, the development costs of $0.05–0.10/g for nanozero-valent iron particles and $0.03–1.21/g for TiO_2 nanoparticles can be predicted to be high as compared to the studied biochar.

Studies show that the sorption capacity of activated biochar is nearly equivalent to activated carbon, while activated carbon costs are almost double that of modified biochar. In terms of adsorption capacity, biochar surpasses low-cost adsorbents available in the market. In the current scenario, a major portion of raw materials of biochar preparation is sewage sludge, forestry and agricultural solid wastes, and different solid organic wastes. Preparation of biochar from solid scrap reduces the production cost, thereby realizing the resource utilization of solid waste. Additionally, biogas and bio-oils are derivatives formed during the preparation of biochar, which can be utilized for energy recovery, thereby reducing the production costs up to a certain level. Along with the continual development of pyrolysis processes and equipment, recycling of biogas and bio-oil can also be done. Many studies have conducted regeneration experiments to find out the regeneration performance of biochar. Methods like microwave elution, pyrolysis, and elution with eluent are extensively used for regeneration of biochars. Primarily elution with an eluent regeneration method is used for modified biochar with adsorbed antibiotics. Afzal et al. (2018), after studying the hydrogel beads of chitosan biochar and its regeneration efficiency, reported that keeping a higher level in the first three cycles resulted in the desorption and adsorption of chitosan biochar for 6 cycles. However, adsorption capacity starts to decline gradually from the fourth cycle. Similarly, biochar activated with $ZnCl_2$ can sustain a fixed adsorption performance in the first three cycles. Some studies have established that in case of acid-alkali-activated biochar, the regeneration performance somewhat decreases following the first five adsorption periods. Also, montmorillonite-assisted wheat straw biochar can also retain stability up to five cycles. Similar results were achieved by Dai et al. (2020), in which the magnetic activation of biochar managed to retain constant adsorption during the first five cycles, and maximum adsorption performance was reported during the first three initial cycles. In short, the majority of activated biochar exhibited an adequate regeneration performance and maintained a constant adsorption efficiency or adsorption capacity across three to five cycles. Thus excellent sustainability makes biochar a good prospect for various realistic applications.

11.7 CONCLUSION AND FUTURE PERSPECTIVES

For both water and wastewater treatment, biochar is a potential multifunctional adsorbent. The scope of research for various inorganic and organic contaminant adsorption onto modified biochars has steadily expanded in recent years. This chapter systematically examined some of the most widely used biochar modification techniques, along with its different adsorption effects, behaviors, and mechanisms for a number of extremely challenging inorganic and organic contaminants in wastewater. Also, the parameters that influence the degradation of contaminants from water and wastewater by modified biochar were thoroughly examined. In general, biochar that has been modified exhibits a better adsorption efficiency on wastewater contaminants. Surface characteristics such as functional groups, surface charge, surface area, pore volume, and pore distribution are expected to change considerably during the modification of biochar by different physical and chemical processes. As per statistical study, metal salt and metal oxide modifications are regarded as the superior modification techniques. The significance of the chemical activation method as a promising strategy to improve the adsorption capacity of biochar for both inorganic and organic pollutants has been highlighted in numerous research works. However, to avoid environmental contamination arising due to physical/chemical activation processes, extra attention must be given for mitigating the modification effects on biochar stability. To best utilize the beneficial properties of biochar while reducing the environmental constraints associated with its use, the physicochemical characteristics of biochar such as mechanical strength, surface chemistry, particle size distribution, organochemical properties, biochar density, and solid phases must be thoroughly investigated. Various chemical (i.e., functional nanoparticles coating, chemical impregnation, and acid/alkali modification) and/or physical (i.e., ball milling, magnetization, and steam activation) modification techniques could be utilized to upgrade the pristine biochars into designer biochar, depending on the surface properties. The useful traits of biochar, including surface area, separation rate, active sites, pore volumes, superficial functional groups availability and diversity, are greatly enhanced by these modification strategies for specific environmental application, namely inorganic and organic pollutant removal, while the undesirable traits of biochar are either minimized or kept under control.

However, a few challenges must still be addressed before biochars can be used more widely, such as treatment of complex effluents, reproducibility and repetition, reuse and regeneration of biochar. Furthermore, the biochar adsorption process can be highly complicated, with a variety of operating variables affecting adsorption performance and mechanisms, including chemical interactions and physical forces. Despite an increase in the number of biochar studies, a variety of research opportunities cover the existing knowledge gaps, enhance the adsorption efficiency, and prepare competitive biochars having desirable surface features and characteristics. Further studies are required on optimization strategies for achieving maximum biochar generation yield, economic investigation based on biochar production, dynamic analysis, and multicomponent study, as well as modeling and regeneration of biochar. Technoeconomic investigation of designer biochar must be carried out to better comprehend its long-term viability. As such, the potential ecological and environmental threats of designer biochars must be carefully assessed before being utilized for large-scale applications. Also, very little available detail is available about the large-scale manufacturing of designer biochars. As a result, more research is needed to substantiate the industrial applications of modified biochars. Moreover, to minimize the unwanted biomagnification and bioaccumulation of nanomaterials, the fate of designer biochars, particularly those consisting of nanoparticles, must also be evaluated. To limit the possibility of cross-contamination due to the addition of chemicals in nanoparticle preparation during both water and wastewater treatment, future studies should concentrate on the synthesis of nanoparticles via green/biological routes for the development of biochar-based nanocomposites. Furthermore, to prevent toxicity and environmental risks, specific focus should be given to the separation of dissolved biochar from treated water.

ACKNOWLEDGMENT

This work is partially supported by a grant (DST/TM/WTI/WIC/2 K17/84(G)) from the DST (Department of Science and Technology), New Delhi. Any opinions, findings, and conclusions expressed in this paper are those of the authors and do not necessarily reflect the views of DST, New Delhi.

REFERENCES

Afzal, Muhammad Zaheer, Xue Fei Sun, Jun Liu, Chao Song, Shu Guang Wang, and Asif Javed. 2018. "Enhancement of Ciprofloxacin Sorption on Chitosan/Biochar Hydrogel Beads." *Science of the Total Environment* 639. Elsevier B.V.: 560–569. doi:10.1016/j.scitotenv.2018.05.129

Ahmed, Mohammad Boshir, John L. Zhou, Huu Hao Ngo, and Wenshan Guo. 2016. "Insight into Biochar Properties and Its Cost Analysis." *Biomass and Bioenergy* 84. Elsevier Ltd: 76–86. doi:10.1016/j.biombioe.2015.11.002

Alghazwi, Mousa, Scott Smid, Ian Musgrave, and Wei Zhang. 2019. "In Vitro Studies of the Neuroprotective Activities of Astaxanthin and Fucoxanthin against Amyloid Beta (Aβ 1–42) Toxicity and Aggregation." *Neurochemistry International* 124 (December 2018). Elsevier: 215–224. doi:10.1016/j.neuint.2019.01.010

Amusat, Sefiu Olaitan, Temesgen Girma Kebede, Simiso Dube, and Mathew Muzi Nindi. 2021. "Ball-Milling Synthesis of Biochar and Biochar—Based Nanocomposites and Prospects for Removal of Emerging Contaminants: A Review." *Journal of Water Process Engineering* 41 (March). Elsevier Ltd: 101993. doi:10.1016/j.jwpe.2021.101993

Barquilha, Carlos E.R., and Maria C.B. Braga. 2021. "Adsorption of Organic and Inorganic Pollutants onto Biochars: Challenges, Operating Conditions, and Mechanisms." *Bioresource Technology Reports* 15 (May). Elsevier Ltd: 100728. doi:10.1016/j.biteb.2021.100728

Cheng, Ning, Bing Wang, Pan Wu, Xinqing Lee, Ying Xing, Miao Chen, and Bin Gao. 2021. "Adsorption of Emerging Contaminants from Water and Wastewater by Modified Biochar: A Review." *Environmental Pollution* 273. Elsevier Ltd: 116448. doi:10.1016/j.envpol.2021.116448

Chen, Qincheng, Jiaolong Qin, Zhiwen Cheng, Lu Huang, Peng Sun, Lu Chen, and Guoqing Shen. 2018. "Synthesis of a Stable Magnesium-Impregnated Biochar and Its Reduction of Phosphorus Leaching from Soil." *Chemosphere* 199. Elsevier Ltd: 402–408. doi:10.1016/j.chemosphere.2018.02.058

Chen, Tingwei, Ling Luo, Shihuai Deng, Guozhong Shi, Shirong Zhang, Yanzong Zhang, Ouping Deng, Lilin Wang, Jing Zhang, and Luoyu Wei. 2018. "Sorption of Tetracycline on H3PO4 Modified Biochar Derived from Rice Straw and Swine Manure." *Bioresource Technology* 267 (May). Elsevier: 431–437. doi:10.1016/j.biortech.2018.07.074

Choi, Yong Keun, and Eunsung Kan. 2019. "Effects of Pyrolysis Temperature on the Physicochemical Properties of Alfalfa-Derived Biochar for the Adsorption of Bisphenol A and Sulfamethoxazole in Water." *Chemosphere* 218. Elsevier Ltd: 741–748. doi:10.1016/j.chemosphere.2018.11.151

Clemente, Joyce S., Suzanne Beauchemin, Ted MacKinnon, Joseph Martin, Cliff T. Johnston, and Brad Joern. 2017. "Initial Biochar Properties Related to the Removal of As, Se, Pb, Cd, Cu, Ni, and Zn from an Acidic Suspension." *Chemosphere* 170. Elsevier Ltd: 216–224. doi:10.1016/j.chemosphere.2016.11.154

Cui, Xiaoqiang, Hulin Hao, Changkuan Zhang, Zhenli He, and Xiaoe Yang. 2016. "Capacity and Mechanisms of Ammonium and Cadmium Sorption on Different Wetland-Plant Derived Biochars." *Science of the Total Environment* 539. Elsevier B.V.: 566–575. doi:10.1016/j.scitotenv.2015.09.022

Cui, Xiaoqiang, Siyu Fang, Yiqiang Yao, Tingqiang Li, Qijun Ni, Xiaoe Yang, and Zhenli He. 2016. "Potential Mechanisms of Cadmium Removal from Aqueous Solution by Canna Indica Derived Biochar." *Science of the Total Environment* 562. Elsevier B.V.: 517–525. doi:10.1016/j.scitotenv.2016.03.248

Dai, Jiawei, Xiangfu Meng, Yuhu Zhang, and Yunjie Huang. 2020. "Effects of Modification and Magnetization of Rice Straw Derived Biochar on Adsorption of Tetracycline from Water." *Bioresource Technology* 311 (April). Elsevier: 123455. doi:10.1016/j.biortech.2020.123455

Deng, Rui, Danlian Huang, Jia Wan, Wenjing Xue, Lei Lei, Xiaofeng Wen, Xigui Liu, et al. 2019. "Chloro-Phosphate Impregnated Biochar Prepared by Coprecipitation for the Lead, Cadmium and Copper Synergic Scavenging from Aqueous Solution." *Bioresource Technology* 293 (July). doi:10.1016/j.biortech.2019.122102

Deng, Yu, Min Li, Zhan Zhang, Qiao Liu, Kele Jiang, Jingjie Tian, Ying Zhang, and Fuquan Ni. 2021. "Comparative Study on Characteristics and Mechanism of Phosphate Adsorption on Mg/Al Modified Biochar." *Journal of Environmental Chemical Engineering* 9 (2). Elsevier Ltd: 105079. doi:10.1016/j.jece.2021.105079

Ding, Zhuhong, Xin Hu, Yongshan Wan, Shengsen Wang, and Bin Gao. 2016. "Removal of Lead, Copper, Cadmium, Zinc, and Nickel from Aqueous Solutions by Alkali-Modified Biochar: Batch and Column Tests." *Journal of Industrial and Engineering Chemistry* 33. The Korean Society of Industrial and Engineering Chemistry: 239–245. doi:10.1016/j.jiec.2015.10.007

Fan, Shisuo, Jie Tang, Yi Wang, Hui Li, Hao Zhang, Jun Tang, Zhen Wang, and Xuede Li. 2016. "Biochar Prepared from Copyrolysis of Municipal Sewage Sludge and Tea Waste for the Adsorption of Methylene Blue from Aqueous Solutions: Kinetics, Isotherm, Thermodynamic and Mechanism." *Journal of Molecular Liquids* 220. Elsevier B.V.: 432–441. doi:10.1016/j.molliq.2016.04.107

Fan, Shisuo, Yi Wang, Zhen Wang, Jie Tang, Jun Tang, and Xuede Li. 2017. "Removal of Methylene Blue from Aqueous Solution by Sewage Sludge-Derived Biochar: Adsorption Kinetics, Equilibrium, Thermodynamics and Mechanism." *Journal of Environmental Chemical Engineering* 5 (1). Elsevier B.V.: 601–611. doi:10.1016/j.jece.2016.12.019

Giwa, Adewale, Ahmed Yusuf, Oluwole Ajumobi, and Prosper Dzidzienyo. 2019. "Pyrolysis of Date Palm Waste to Biochar Using Concentrated Solar Thermal Energy: Economic and Sustainability Implications." *Waste Management* 93. Elsevier Ltd: 14–22. doi:10.1016/j.wasman.2019.05.022

Guo, Xuetao, Hao Dong, Chen Yang, Qian Zhang, Changjun Liao, Fugeng Zha, and Liangmin Gao. 2016. "Application of Goethite Modified Biochar for Tylosin Removal from Aqueous Solution." *Colloids and Surfaces A: Physicochemical and Engineering Aspects* 502. Elsevier B.V.: 81–88. doi:10.1016/j.colsurfa.2016.05.015

Han, Yitong, Xi Cao, Xin Ouyang, Saran P. Sohi, and Jiawei Chen. 2016. "Adsorption Kinetics of Magnetic Biochar Derived from Peanut Hull on Removal of Cr (VI) from Aqueous Solution: Effects of Production Conditions and Particle Size." *Chemosphere* 145. Elsevier Ltd: 336–341. doi:10.1016/j.chemosphere.2015.11.050

Heo, Jiyong, Yeomin Yoon, Gooyong Lee, Yejin Kim, Jonghun Han, and Chang Min Park. 2019. "Enhanced Adsorption of Bisphenol A and Sulfamethoxazole by a Novel Magnetic CuZnFe2O4—Biochar Composite." *Bioresource Technology* 281 (December 2018). Elsevier: 179–187. doi:10.1016/j.biortech.2019.02.091

Ho, Shih Hsin, Yi di Chen, Zhong kai Yang, Dillirani Nagarajan, Jo Shu Chang, and Nan qi Ren. 2017. "High-Efficiency Removal of Lead from Wastewater by Biochar Derived from Anaerobic Digestion Sludge." *Bioresource Technology* 246 (June). Elsevier: 142–149. doi:10.1016/j.biortech.2017.08.025

Ifthikar, Jerosha, Ting Wang, Aimal Khan, Ali Jawad, Tingting Sun, Xiang Jiao, Zhuqi Chen, et al. 2017. "Highly Efficient Lead Distribution by Magnetic Sewage Sludge Biochar: Sorption Mechanisms and Bench Applications." *Bioresource Technology* 238. Elsevier Ltd: 399–406. doi:10.1016/j.biortech.2017.03.133

Jain, M., V. K. Garg, K. Kadirvelu, and M. Sillanpää. 2016. "Adsorption of Heavy Metals from Multi-Metal Aqueous Solution by Sunflower Plant Biomass-Based Carbons." *International Journal of Environmental Science and Technology* 13 (2): 493–500. doi:10.1007/s13762-015-0855-5

Jung, Kyung Won, Seon Yong Lee, Young Jae Lee, and Jae Woo Choi. 2019. "Ultrasound-Assisted Heterogeneous Fenton-like Process for Bisphenol a Removal at Neutral PH Using Hierarchically Structured Manganese Dioxide/Biochar Nanocomposites as Catalysts." *Ultrasonics Sonochemistry* 57 (April). Elsevier: 22–28. doi:10.1016/j.ultsonch.2019.04.039

Kamali, Mohammadreza, Lise Appels, Eilhann E. Kwon, Tejraj M. Aminabhavi, and Raf Dewil. 2021. "Biochar in Water and Wastewater Treatment—A Sustainability Assessment." *Chemical Engineering Journal* 420 (P1). Elsevier B.V.: 129946. doi:10.1016/j.cej.2021.129946

Kang, Seju, Jihyeun Jung, Jong Kwon Choe, Yong Sik Ok, and Yongju Choi. 2018. "Effect of Biochar Particle Size on Hydrophobic Organic Compound Sorption Kinetics: Applicability of Using Representative Size." *Science of the Total Environment* 619–620. Elsevier B.V.: 410–418. doi:10.1016/j.scitotenv.2017.11.129

Kazemi Shariat Panahi, Hamed, Mona Dehhaghi, Yong Sik Ok, Abdul Sattar Nizami, Benyamin Khoshnevisan, Solange I. Mussatto, Mortaza Aghbashlo, Meisam Tabatabaei, and Su Shiung Lam. 2020. "A Comprehensive Review of Engineered Biochar: Production, Characteristics, and Environmental Applications." *Journal of Cleaner Production* 270. Elsevier Ltd: 122462. doi:10.1016/j.jclepro.2020.122462

Kuśmierek, Krzysztof, and Andrzej ̄wiątkowski. 2015. "The Influence of Different Agitation Techniques on the Adsorption Kinetics of 4-Chlorophenol on Granular Activated Carbon." *Reaction Kinetics, Mechanisms and Catalysis* 116 (1): 261–271. doi:10.1007/s11144-015-0889-1

Li, Jing, Bing Li, Haiming Huang, Ning Zhao, Mingge Zhang, and Lu Cao. 2020. "Investigation into Lanthanum-Coated Biochar Obtained from Urban Dewatered Sewage Sludge for Enhanced Phosphate Adsorption." *Science of the Total Environment* 714. Elsevier B.V.: 136839. doi:10.1016/j.scitotenv.2020.136839

Madadi, Rozita, and Kai Bester. 2021. "Fungi and Biochar Applications in Bioremediation of Organic Micropollutants from Aquatic Media." *Marine Pollution Bulletin* 166 (November 2020). Elsevier Ltd: 112247. doi:10.1016/j.marpolbul.2021.112247

Mahmoud, Dalia Khalid, Mohamad Amran Mohd Salleh, Wan Azlina Wan Abdul Karim, Azni Idris, and Zurina Zainal Abidin. 2012. "Batch Adsorption of Basic Dye Using Acid Treated Kenaf Fibre Char: Equilibrium, Kinetic and Thermodynamic Studies." *Chemical Engineering Journal* 181–182. Elsevier B.V.: 449–457. doi:10.1016/j.cej.2011.11.116

Maroušek, Josef. 2014. "Significant Breakthrough in Biochar Cost Reduction." *Clean Technologies and Environmental Policy* 16 (8): 1821–1825. doi:10.1007/s10098-014-0730-y

Matsui, Yoshihiko, Soichi Nakao, Asuka Sakamoto, Takuma Taniguchi, Long Pan, Taku Matsushita, and Nobutaka Shirasaki. 2015. "Adsorption Capacities of Activated Carbons for Geosmin and 2-Methylisoborneol Vary with Activated Carbon Particle Size: Effects of Adsorbent and Adsorbate Characteristics." *Water Research* 85. Elsevier Ltd: 95–102. doi:10.1016/j.watres.2015.08.017

Ma, Ying, Wu Jun Liu, Nan Zhang, Yu Sheng Li, Hong Jiang, and Guo Ping Sheng. 2014. "Polyethylenimine Modified Biochar Adsorbent for Hexavalent Chromium Removal from the Aqueous Solution." *Bioresource Technology* 169. Elsevier Ltd: 403–408. doi:10.1016/j.biortech.2014.07.014

Movasaghi, Zahra, Bei Yan, and Catherine Niu. 2019. "Adsorption of Ciprofloxacin from Water by Pretreated Oat Hulls: Equilibrium, Kinetic, and Thermodynamic Studies." *Industrial Crops and Products* 127 (October 2018). Elsevier: 237–250. doi:10.1016/j.indcrop.2018.10.051

Pandey, Deepshikha, Achlesh Daverey, and Kusum Arunachalam. 2020. "Biochar: Production, Properties and Emerging Role as a Support for Enzyme Immobilization." *Journal of Cleaner Production* 255. Elsevier Ltd: 120267. doi:10.1016/j.jclepro.2020.120267

Pan, Xuqin, Zhepei Gu, Weiming Chen, and Qibin Li. 2021. "Preparation of Biochar and Biochar Composites and Their Application in a Fenton-like Process for Wastewater Decontamination: A Review." *Science of the Total Environment* 754. Elsevier B.V.: 142104. doi:10.1016/j.scitotenv.2020.142104

Paunovic, Olivera, Sabolc Pap, Snezana Maletic, Mark A. Taggart, Nikola Boskovic, and Maja Turk Sekulic. 2019. "Ionisable Emerging Pharmaceutical Adsorption onto Microwave Functionalised Biochar Derived from Novel Lignocellulosic Waste Biomass." *Journal of Colloid and Interface Science* 547. Elsevier Inc.: 350–360. doi:10.1016/j.jcis.2019.04.011

Pratt, Kimberley, and Dominic Moran. 2010. "Evaluating the Cost-Effectiveness of Global Biochar Mitigation Potential." *Biomass and Bioenergy* 34 (8). Elsevier Ltd: 1149–1158. doi:10.1016/j.biombioe.2010.03.004

Rajapaksha, Anushka Upamali, Meththika Vithanage, Mahtab Ahmad, Dong Cheol Seo, Ju Sik Cho, Sung Eun Lee, Sang Soo Lee, and Yong Sik Ok. 2015. "Enhanced Sulfamethazine Removal by Steam-Activated Invasive Plant-Derived Biochar." *Journal of Hazardous Materials* 290. Elsevier B.V.: 43–50. doi:10.1016/j.jhazmat.2015.02.046

Rangabhashiyam, S, and Balasubramanian Paramasivan. 2019. "The Potential of Lignocellulosic Biomass Precursors for Biochar Production: Performance, Mechanism and Wastewater Application—A Review." *Industrial Crops and Products* 128 (May 2018). Elsevier: 405–423. doi:10.1016/j.indcrop.2018.11.041

Ren, Zhijun, Biao Jia, Guangming Zhang, Xiaolin Fu, Zhanxin Wang, Pengfei Wang, and Longyi Lv. 2021. "Study on Adsorption of Ammonia Nitrogen by Iron-Loaded Activated Carbon from Low Temperature Wastewater." *Chemosphere* 262. Elsevier Ltd: 127895. doi:10.1016/j.chemosphere.2020.127895

Rosales, Emilio, Jessica Meijide, Marta Pazos, and María Angeles Sanromán. 2017. "Challenges and Recent Advances in Biochar as Low-Cost Biosorbent: From Batch Assays to Continuous-Flow Systems." *Bioresource Technology* 246. Elsevier Ltd: 176–192. doi:10.1016/j.biortech.2017.06.084

Ruan, Xiuxiu, Yuqing Sun, Weimeng Du, Yuyuan Tang, Qiang Liu, Zhanying Zhang, William Doherty, Ray L. Frost, Guangren Qian, and Daniel C.W. Tsang. 2019. "Formation, Characteristics, and Applications of Environmentally Persistent Free Radicals in Biochars: A Review." *Bioresource Technology* 281 (December 2018). Elsevier: 457–468. doi:10.1016/j.biortech.2019.02.105

Shackley, Simon, Jim Hammond, John Gaunt, and Rodrigo Ibarrola. 2011. "The Feasibility and Costs of Biochar Deployment in the UK." *Carbon Management* 2 (3): 335–356. doi:10.4155/cmt.11.22

Singh, Jasvir, Shivani Sharma, Sumedha Soni, Sandeep Sharma, and Ravi Chand Singh. 2019. "Influence of Different Milling Media on Structural, Morphological and Optical Properties of the ZnO Nanoparticles Synthesized by Ball Milling Process." *Materials Science in Semiconductor Processing* 98 (December 2018). Elsevier Ltd: 29–38. doi:10.1016/j.mssp.2019.03.026

Tan, Xiao fei, Yun guo Liu, Yan ling Gu, Yan Xu, Guang ming Zeng, Xin jiang Hu, Shao bo Liu, Xin Wang, Si mian Liu, and Jiang Li. 2016. "Biochar-Based Nano-Composites for the Decontamination of Wastewater: A Review." *Bioresource Technology* 212. Elsevier Ltd: 318–333. doi:10.1016/j.biortech.2016.04.093

Tan, Zhongxin, Yuanhang Wang, Alfreda Kasiuliené, Chuanqin Huang, and Ping Ai. 2017. "Cadmium Removal Potential by Rice Straw-Derived Magnetic Biochar." *Clean Technologies and Environmental Policy* 19 (3): 761–774. doi:10.1007/s10098-016-1264-2

Vithanage, Meththika, Anushka Upamali Rajapaksha, Ming Zhang, Sören Thiele-Bruhn, Sang Soo Lee, and Yong Sik Ok. 2015. "Acid-Activated Biochar Increased Sulfamethazine Retention in Soils." *Environmental Science and Pollution Research* 22 (3): 2175–2186. doi:10.1007/s11356-014-3434-2

Wang, Rong Zhong, Dan Lian Huang, Yun Guo Liu, Chen Zhang, Cui Lai, Xin Wang, Guang Ming Zeng, Qing Zhang, Xiao Min Gong, and Piao Xu. 2020. "Synergistic Removal of Copper and Tetracycline from Aqueous Solution by Steam-Activated Bamboo-Derived Biochar." *Journal of Hazardous Materials* 384 (March 2019). Elsevier: 121470. doi:10.1016/j.jhazmat.2019.121470

Wang, Shengsen, Bin Gao, Yuncong Li, Ahmed Mosa, Andrew R. Zimmerman, Lena Q. Ma, Willie G. Harris, and Kati W. Migliaccio. 2015. "Manganese Oxide-Modified Biochars: Preparation, Characterization, and Sorption of Arsenate and Lead." *Bioresource Technology* 181. Elsevier Ltd: 13–17. doi:10.1016/j.biortech.2015.01.044

Wang, Shizong, and Jianlong Wang. 2020. "Peroxymonosulfate Activation by Co9S8@ S and N Co-doped Biochar for Sulfamethoxazole Degradation." *Chemical Engineering Journal* 385 (December 2019). Elsevier: 123933. doi:10.1016/j.cej.2019.123933

Wan, Shungang, Zulin Hua, Lei Sun, Xue Bai, and Lu Liang. 2016. "Biosorption of Nitroimidazole Antibiotics onto Chemically Modified Porous Biochar Prepared by Experimental Design: Kinetics, Thermodynamics, and Equilibrium Analysis." *Process Safety and Environmental Protection* 104. Institution of Chemical Engineers: 422–435. doi:10.1016/j.psep.2016.10.001

Xu, Xiaoyun, Xinde Cao, Ling Zhao, Hailong Wang, Hongran Yu, and Bin Gao. 2013. "Removal of Cu, Zn, and Cd from Aqueous Solutions by the Dairy Manure-Derived Biochar." *Environmental Science and Pollution Research* 20 (1): 358–368. doi:10.1007/s11356-012-0873-5

Yang, Fan, Shuaishuai Zhang, Yuqing Sun, Daniel C.W. Tsang, Kui Cheng, and Yong Sik Ok. 2019. "Assembling Biochar with Various Layered Double Hydroxides for Enhancement of Phosphorus Recovery." *Journal of Hazardous Materials* 365 (October 2018). Elsevier: 665–673. doi:10.1016/j.jhazmat.2018.11.047

Zhang, Ming, Bin Gao, Sima Varnoosfaderani, Arthur Hebard, Ying Yao, and Mandu Inyang. 2013. "Preparation and Characterization of a Novel Magnetic Biochar for Arsenic Removal." *Bioresource Technology* 130: 457–462. doi:10.1016/j.biortech.2012.11.132

Zhou, Qifa, Benjamin A. Houge, Zhaohui Tong, Bin Gao, and Guodong Liu. 2018. "An In-Situ Technique for Producing Low-Cost Agricultural Biochar." *Pedosphere* 28 (4). Soil Science Society of China: 690–695. doi:10.1016/S1002-0160(17)60482-X

Zhou, Yue, Shurui Cao, Cunxian Xi, Xianliang Li, Lei Zhang, Guomin Wang, and Zhiqiong Chen. 2019. "A Novel Fe3O4/Graphene Oxide/Citrus Peel-Derived Bio-Char Based Nanocomposite with Enhanced Adsorption Affinity and Sensitivity of Ciprofloxacin and Sparfloxacin." *Bioresource Technology* 292 (July). Elsevier: 121951. doi:10.1016/j.biortech.2019.121951

Zhu, Yao, Baojun Yi, Qiaoxia Yuan, Yunlian Wu, Ming Wang, and Shuiping Yan. 2018. "Removal of Methylene Blue from Aqueous Solution by Cattle Manure-Derived Low Temperature Biochar." *RSC Advances* 8 (36). Royal Society of Chemistry: 19917–19929. doi:10.1039/c8ra03018a

12 Removal of Pharmaceutical Active Compounds (PhACs) from Wastewater by Designer Biochar

Pranjal P. Das, Mukesh Sharma, Abhik Bhattacharjee, Piyal Mondal, and Mihir K. Purkait

CONTENTS

DOI: 10.1201/9781003203438-12

12.1 INTRODUCTION

Rapid population growth has triggered industrialization and urbanization in the last few decades, resulting in water scarcity and a considerable spike in the concentration of emerging contaminants (ECs). Emerging contaminants are a class of extremely complex organic chemical pollutants found in water that have the potential to effect both human health and the environment. The most common sources of emerging contaminants are pharmaceutical active compounds (PhACs), endocrine-disrupting chemicals, disinfection by-products, personal care products, and persistent organic pollutants. These pollutants last for a long time and are persistent in the environment. Researchers have utilized various waste source as feedstock for the preparation of modified biochar and other value-added products to mitigate such contaminants. More than 30 types of emerging contaminants have been detected in treated effluent, untreated effluent, agricultural rainwater, urban rainfall, and freshwater in previous studies. Among these contaminants, pharmaceutical active compounds and personal care products were widely found in numerous water samples. In the environment, such contaminants are constantly circulated, migrated, and transformed. Although such contaminants are found in moderate concentration in water, they may have a significant influence on both human health and environment via the food chain after being accumulated by different organisms (Tran et al. 2019). As such, the question of how to successfully remove such contaminants from water has gained a lot of attention. Some of the most commonly utilized methods for the removal of emerging contaminants from water includes adsorption methods, microbial methods, chemical oxidation processes, electrochemical methods, and membrane processes. Among these techniques, adsorption is the most widely used method particularly for the removal of PhACs from the environment, due to its benefits such as high efficiency, cost-effectiveness, and wide processing range. Activated carbon, absorbent resin, silica gel, polyacrylamide, zeolite, and alumina are some of the most commonly utilized adsorbents. Adsorption mechanism and characteristics differ depending on the physicochemical properties of different adsorbents. Although activated carbon possess a very high specific surface area (SSA) and adsorption efficiency, it has a high production cost compared to other adsorbents like adsorption resin, alumina, and silica gel. Narrow adsorption surface and ineffectiveness in manufacturing polyacrylamide are some of the drawbacks of activated carbon. Thus it is crucial to find highly efficient, sustainable, and cost-effective adsorbents for the degradation of emerging contaminants. Also, it is very important to select the right adsorbent for removing specific pollutants from water. When the physicochemical characteristics of an adsorbent, such as functional groups, chemical structure, and adsorption parameters (pore size and specific surface area,) are known, it can be matched to the type of contaminants for acquiring the most selective and efficient adsorption process. The adsorbents must also be sustainable from a practical standpoint (Cheng et al. 2021).

As a novel adsorbent, biochar/modified biochar has been extensively utilized in the degradation of pharmaceutical compounds from the environment. Biochar-based adsorbents have attained popularity in recent years, owing to their low cost and wide range of modification possibilities. The persistent free radicals present in biochar were found to be capable of activating H_2O_2 or $S_2O_8^{2-}$ to produce active oxygen, thereby successfully removing the inorganic contaminants from water (Ruan et al. 2019). Biochar is a carbon-rich substance, prepared by heating at a low temperature (<700°C) in the presence of little or no air. Biochars can be prepared from a variety of organic feedstocks, viz. food processing residues, agricultural and livestock waste, municipal waste, and waste biomass, to name a few. The important thermochemical techniques used for the preparation of biochar include slow pyrolysis, fast pyrolysis, gasification, and flash carbonization. Due to deoxygenation and dehydration of biomass, although the biochars generated at high temperatures of 600–700°C exhibit a strong aromatic property with well structured carbon layers, they possess very few O and H functional groups, potentially resulting in reduced ion exchange capacity. On the contrary, the biochars generated at lower temperatures of 300–400°C consist of a wide range of organic structures, including cellulose and aliphatic-like structures as well as a significant number of C-H and C=O functional groups. Table 12.1 depicts the effect of different pyrolysis conditions on

TABLE 12.1

Operational Parameters and Overall Yield of Different Pyrolysis Conditions (obtained with permission from X. Li et al. (2020) © Elsevier)

Pyrolysis conditions	Residence time (s)	Temperature (K)	Heating rate (K/s)	Overall yield (%)		
				Biochar	Bio-oil	Biogas
Slow pyrolysis	450–550	550–950	0.1–10	35	30	35
Fast pyrolysis	0.5–1.0	850–1250	10–200	50–70	10–30	15–20
Flash pyrolysis	< 0.5	1050–1300	< 1000	60	40	5–15

the operating parameters and overall yield of produced outputs. The physical and chemical composition of biochar is heterogeneous and complex, making it an excellent platform for the degradation of contaminants via the adsorption mechanism (Rajapaksha et al. 2016). As such, studies on the adsorption of pharmaceutical active compounds such as antibiotics, sulfonamide, and tetracycline with biochar have increased in recent years.

The current chapter investigates the chemical and physical properties of biochar and compares different available strategies, viz. chemical, physical, and biological techniques, for producing designer biochars. The application of chemical and physical modification techniques for designing biochars and the improvements conferred (e.g., porosity, surface area, functional groups, and active site) are also examined. It also illustrates the utilization of different biochar-based adsorbents, viz. pristine biochar, modified biochar, and biochar composites for the adsorption and degradation of pharmaceutical active compounds. As such, the application of designer biochars for the removal of antibiotics and other PhACs is extensively discussed and summarized. This chapter also includes a number of biochar-related suggestions and recommendations for further research and development.

12.2 BIOCHAR: GENERAL CHARACTERISTICS

12.2.1 Physical Properties

Generally, the nature of biomass and the conditions such as pretreatment of biomass and handling directly control the physical properties of biochar. The biomass's original structure is altered during the pyrolysis processes to various degrees by undergoing attrition, structural rearrangements, and cracks formation. Due to the release of volatile organics during pyrolysis, feedstock mass is reduced, resulting in disproportionate shrinkage in biomass. Carbon and mineral skeletons form the basic porous structure of the original biomass at the end of the pyrolysis step. Biochar macroporosity is mostly contributed by the cellular plant components generally present in coals and wood-derived biochar. The main controlling parameters during pyrolysis for shaping the biochar are heating rate, biomass pretreatment, reactor type, reaction temperature, pressure, purge gas flow rate, and postmodifications. The chemical composition of the biomass directly controls the physical characteristics. Over 120°C, organic materials of the biomass such as hemicellulose (200–260°C), cellulose (240–350°C), and lignin (280–500°C) start decomposing (Downie, Crosky, and Munroe 2009).

12.2.1.1 Nano- and Macroporosity

Pore size distribution plays a vital role among the other physical properties, since it determines their potential industrial applications. The total pore volume of the biochar is mainly composed of macropores (>50 nm), mesopores (2–50 nm), and micropores (<2 nm). The microporous nature of the biochar contributes the most toward better surface area and adsorption efficiency for minute molecules, whereas, for liquid–solid adsorption, the mesopores play a significant role. During pyrolysis, a higher reaction temperature can enhance the microporous volume due to the availability of the required activation energy and reaction time, which results in improved structural organization of the biochar molecule. Moreover, the number of pores of biochar is enhanced due to the release of volatile matter (feedstocks) when pyrolysis occurs at higher temperatures. Also, porosity of the woody biochar increases due to slow decomposition of lignin through the slow pyrolysis process (Shaaban et al. 2014). When pyrolysis temperature exceeds 400°C, formation of micropores occurs due to the condensation of organic matter, which enhances the specific surface area hugely. Eventually, at higher temperatures of 800–1000°C, generally for slow heating rate the enhancement in surface area becomes insignificant. Moreover, due to shrinkage of feedstock above 850°C, the overall porosity of the biochar decreases. The micropore volume of the biochar is reduced due to the enlargement of the pores, but on the other hand, the total pore volume is enhanced. The heating rate of the pyrolysis process plays an important role in porosity enhancement. Due to the melting of biomass at the higher heating rate, macropores are formed, whereas micropore volume is enhanced in the biochar due to the low heating rate (T. Zhang et al. 2004).

12.2.1.2 Density and Mechanical Strength

The physical properties of biochar consists of bulk/apparent density and solid density. Bulk density is found to vary inversely with solid density. Generally, the solid density of biochar denotes the packing degree of the carbon structure and is found to be higher than the raw biomass. Such a result is denoted, due to the formation of graphitic crystallites, by the release of the volatile compounds during pyrolysis. With the increase in pyrolysis temperature, the solid density also enhances, which causes more shrinkage of raw biomass, along with more carbonization. Similarly, the solid density of the biochar is found to be less than the precursor wood due to biomass shrinkage and carbonization. Generally, for gas adsorption and dye removal, biochar preferably has a bulk density in the range of 0.50–0.75 g/cm³. An increase in density of the biochar occurs due to the formation of more micropores than macropores and mesopores (Kazemi Shariat Panahi et al. 2020).

The mechanical stability of the biochar is found to be directly related to density and inversely proportional to porosity. Lower stiffness and higher strength are generally found in carbonized wood biochar, compared to the raw wood. Both moduli and hardness are found to increase with increasing temperature, and finally a plateau is obtained at 700–2000°C. Further, any increase in the pyrolysis temperature exerts a negative impact on the produced biochar's properties. An initial decrement of both compressive and impact strengths was obtained by Kumar, Verma, and Gupta (1999), when biochar was obtained using eucalyptus and acacia as raw material at temperature elevation up to 600°C. Moreover, after increasing the temperature, an improvement in those properties was observed. A better mechanical strength was observed for biochar obtained from eucalyptus than acacia. The study confirmed that low heating rate (4°C/min) during pyrolysis enhances the compressive strength, when compared to higher heating rate (30°C/min). Such an occurrence can be denoted due to the rapid release of volatile matters and excess evaporation during a high heating rate causing the formation of more cracks. Furthermore, the moisture content and the nanocomposite characteristics of biomass such as high lignin content and high density can also considerably effect its mechanical strength as these parameters remain mostly unaltered during the pyrolysis process. Due to the higher molecular order, biochar has higher mechanical strength, accompanied with lower compressive strength and anisotropy, compared to the raw biomass. Thus, for environmental application purposes, mechanical strength plays an important role in withstanding both wear and tear stresses (Weber and Quicker 2018).

12.2.2 Chemical Properties

During any pyrolysis process, the biomass is chemically converted to biochar with a wide range of chemical properties. Type of feedstock and pyrolysis temperature mainly decide the biochar properties. Biochar yield is found to decrease with an increase in pyrolysis temperature, whereas the pH value of the biochar has been found to increase linearly. The pH increase can be attributed to the fact that at higher temperatures within the biochar, the decomposition of hydroxyl and other weak bonds takes place. Li et al. (2019) studied the relation between the cation exchange capacity and pyrolysis temperature of the biochar and concluded their inverse relationship. Since biosolids consist of very high amount of mineral content (e.g., K, P, Na, Mg, and Ca), biosolid-derived biochars shows higher cation exchange property. Such mineral content has the tendency of forming biochar surface-mediated O-containing functional groups during pyrolysis. Volatile content is reduced linearly with increasing pyrolysis temperature, unlike ash content. Ash formation occurs due to the decomposition of hydrogen, oxygen, and carbon, which are found in the remaining inorganic minerals of the biomass. Such decomposition signifies the cleavage/breaking of hydroxyl and other weak bonded groups during the temperature increase (Agrafioti et al. 2013).

12.2.2.1 Microchemical Characteristics

The superficial sorption properties of biochar can be affected by its microchemical characteristics. The composition and nature of biochar is the main controlling parameters for such characteristics, along with the surface functional groups of biochar and the entrapped oils. The surface functional groups and other properties become adhered due to the thermal decomposition of biomass. Due to the hemolytic cleavage of covalent bonds during pyrolysis of dried biomass, numerous free radicals are released within biomass from impurities that are inorganic in nature and structural oxygen (O). Free radicals are produced during the initial pyrolysis stages, obtained at low O_2 levels from the atmosphere. In the later stages, carboxyl and carbonyl group formation occurs, followed by its cleavage in to CO_2 and CO.

Solid phases and its distribution: The distribution of solid phases during pyrolysis due to the lignocellulosic biomasses thermal decomposition leads toward four regions: (1) dehydration, (2) pyrolysis, (3) graphene nucleation, and (4) carbonization. The transition generally occurs at atmospheric pressure and slow heating rate (2°C/min) with temperatures ranging at 250°C, 350°C, and 600°C. To be specific at <250°C, less mass loss occurs due to the dehydration of the cellulosic content in biomass. In the first region, when heated above 150°C, the carbon (C) concentration of the biomass was found to increase progressively. Utilizing the 13C-NMR technique, it was obtained that at 200°C, weaker signal intensity was obtained for biochar compared to raw biomass, which was denoted due to the hemicellulosic and cellulosic carbon content (Knicker et al. 2005). Moreover, stronger lignin carbon content was obtained by the characterization of biochar than the raw biomass. Structural transformation of O-alkyl to aryl-C was found in biochar through characterization at 250°C. Due to higher loss in oxygen and hydrogen compared to carbon, the hydrogen-to-carbon ratio decreases in the region of 250–350°C, confirming the formation of unsaturated carbon structures. In the second region, complete depolymerization of cellulose, followed by decrease in volatile content, and the formation of carbon matrix amorphous in nature are obtained. During pyrolysis at 300°C, biomass weight reduction was obtained about 81 wt%. The third region marks the transformation of amorphous carbon at two different temperature ranges: (1) at ~330°C, aromatic carbon formation occurs, and (2) at > 350°C, polyaromatic graphene forms. The fourth region, marked at > 600°C, mostly removes the noncarbon atoms remaining in the biomass through carbonization. Moreover, lateral growth of the graphene sheets occurs within the biochar and ultimately coalesce together, developing electrical continuity throughout the material (Kazemi Shariat Panahi et al. 2020; Sontakke and Purkait 2021).

Surface chemistry: Due to the variation in composition within the biochar, enriched surface chemistry is obtained, which exhibits numerous acidic and basic, along with hydrophobic and hydrophilic properties. Such immense chemical variation depends on the nature and composition of raw biomass along with pyrolysis conditions. Aromatic rings within the biomass consist of several heteroatoms such as phosphorus, sulfur, nitrogen, oxygen, and hydrogen, which form numerous surface functional groups over the graphene sheets. Surface chemical heterogeneity was found due to the electronegativity difference between the C atoms and other heteroatoms. Both electron acceptor (NO_2, (C=O) H, or (C=O) OH) and electron donor (O(C=O) R, OR, NH_2) functional groups are found on the biochar surface. Carbonyls and phenols within biochar are generally regarded as weak acidic groups, and carboxyl groups are categorized as strong Brønsted acids. The basic functional groups are composed of pyrones and chromenes (Rimstidt and Vaughan 2003). Within biochar pores, active sites contain both acidic and basic properties along with its outer surface, due to the diversity in mineral composition within biochar surface. Amphoteric sites could be formed due to the presence of mineral oxides on the surface of the biochar. Such a phenomenon causes modification in the charge and pH due to the presence of both positive and negative charges. Heavy metal incorporation is generally found in the functional groups due to the high mineral ash content of the biochar. Moreover, biochar prepared from manures and sewage sludge usually consists of various sulfur- and nitrogen-based functional groups. The nitrogen-based functional groups are generally found due to the protein content of the biomass, which are being modified during the pyrolysis process (Leng et al. 2019).

12.2.2.2 Organochemical Characteristics

The characteristic property of hydrogen-to-carbon ratio of lignocellulosic biomass drops from ~1.5 to <0.5 during a pyrolysis temperature of 400°C. Due to the concentration change in elemental nitrogen, oxygen, hydrogen, and carbon during thermal decomposition, the hydrogen–to-carbon ratio is thus found to decrease. The degree of aromaticity and maturation is generally denoted by the oxygen-to-carbon and hydrogen-to-carbon ratios, which typically decrease at higher pyrolysis temperatures. With an increase in graphitic structures, the hydrogen–to-carbon ratio decreases. Moreover, through chemical treatment, the hydrogen-to-carbon and oxygen-to-carbon ratios could be modified, apart from depending on pyrolysis conditions and biomass properties (Graetz and Skjemstad 2003). Reduction in organic contents of biochar was found with increased temperature during artificial aging by means of heat incubation treatment. In an oxygen-reduced environment, when such a process was carried out, the organic content was found to be enriched by the increasing temperature. The organic content in biochar was generally found to be >500 mg/g, but in some cases it can be less than <500 mg/g, for example grass- and manure-based biochars. Generally, carbon content from the alkyl, O-alkyl, and aromatic (aryl) groups increases during pyrolysis, which thus decreases the hydrogen-to-carbon ratio in the biochar. Further, at higher temperatures of >500°C, the hydrogen-to-carbon ratio again decreases, and mass loss occurs very much due to the formation of aryl carbon structures, which remove the original carbon-containing functional groups of plant biomass (O-alkyl C) (Kazemi Shariat Panahi et al. 2020). Figure 12.1 represents the effect of feedstocks and pyrolysis temperature on both the physical and the chemical properties of biochar.

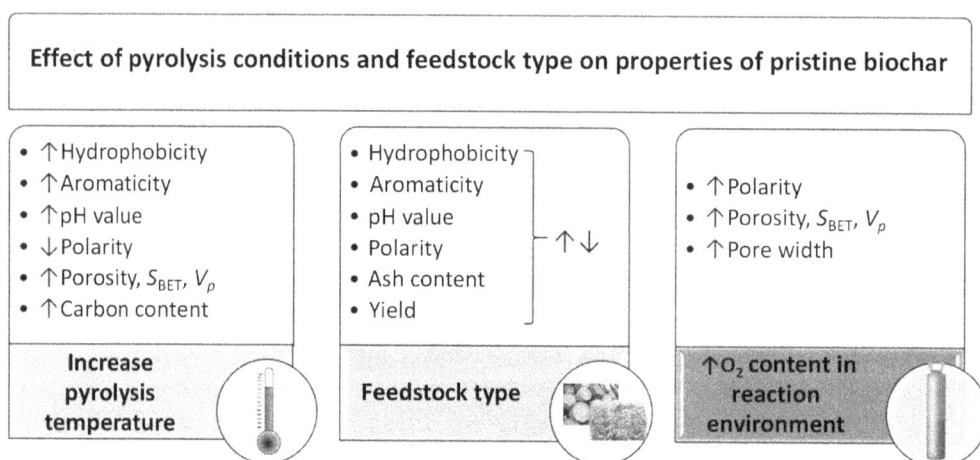

Effect of pyrolysis conditions and feedstock type on properties of pristine biochar

- ↑Hydrophobicity
- ↑Aromaticity
- ↑pH value
- ↓Polarity
- ↑Porosity, S_{BET}, V_p
- ↑Carbon content

Increase pyrolysis temperature

- Hydrophobicity
- Aromaticity
- pH value
- Polarity
- Ash content
- Yield

↑↓

Feedstock type

- ↑Polarity
- ↑Porosity, S_{BET}, V_p
- ↑Pore width

↑O_2 content in reaction environment

FIGURE 12.1 Simplified diagram showing the effects of pyrolysis conditions and feedstock type on physical and chemical properties of biochar.

Source: Obtained with permission from Krasucka et al. (2021) © Elsevier.

12.3　ENGINEERING BIOCHAR MODIFICATION TECHNIQUES

Biochar could be engineered as an adsorbent for a variety of environmental issues. Its physico-chemical characteristics, like surface area, pore geometry, and surface functional groups, can be modified to better fit its intended environmental application. This modification could be carried out using a variety of means, including physical, biological, and chemical. The detailed modification methods of biochar per the literature are described next. Figure 12.2 depicts a comparison between different modification techniques for the preparation of designer biochars with regard to adsorption properties. Some of the most commonly used modification/activation techniques for enhancing the properties of pristine biochar, viz. chemical techniques (functional group, chemical oxidation, chemical coating, and surfactant modification), physical techniques (gas purging, steam activation, mineral oxide impregnation, and magnetic modification), and biological techniques are extensively discussed and summarized in the following section.

12.3.1　Chemical Modification Techniques

In these techniques, biochar is activated by the existence of an inert gas atmosphere and various chemicals. Acid or alkaline solution could be employed to oxidize the biochar. As a result, several properties such as porosity, ion exchange capability, and diverse functional groups present in the biochar are improved. In addition, biochar can be blended with clay materials, carbon-based compounds, nanomaterials, and metal oxides to produce composite. However, biochar with large

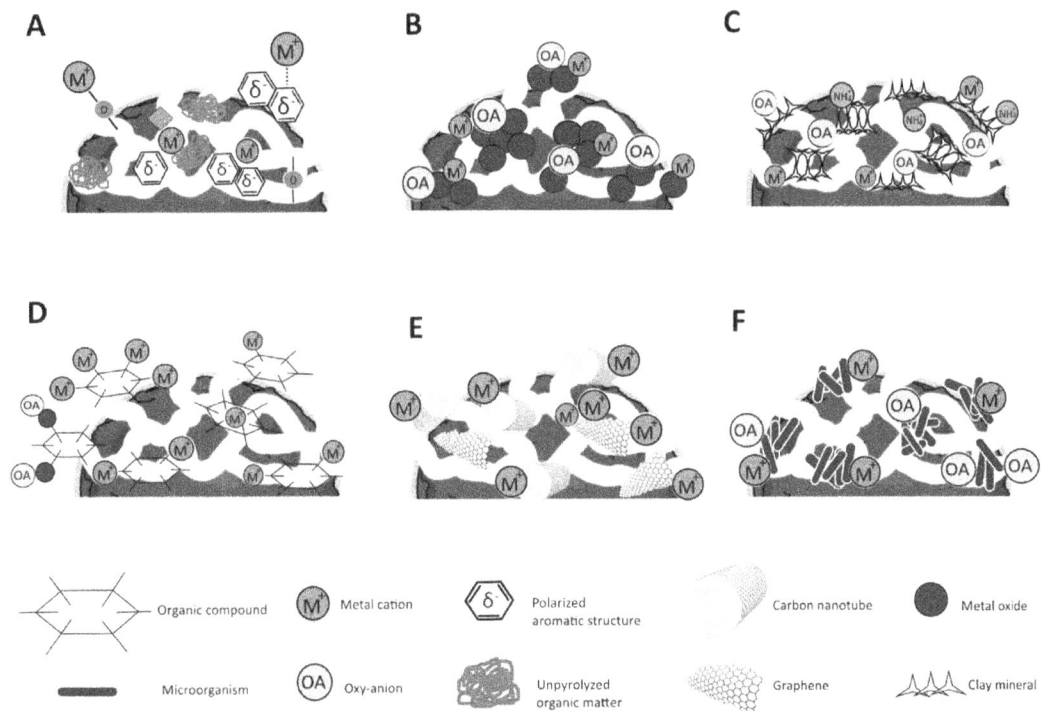

FIGURE 12.2　Comparison of different modification methods for engineering biochar with respect to adsorption characteristics: (A) pristine biochar; chemical impregnation/coating by (B) metal oxides, (C) clay minerals, (D) organic compounds such as amino groups, chitosan, or graphene oxide, and (E) carbon nanotubes, and (F) microbial modification.

Source: Obtained with permission from Kazemi Shariat Panahi et al. (2020) © Elsevier.

surface area is desirable for this kind of modifications as the biochar surface acts as a support for materials deposition. Several chemical modification techniques of biochar are given next.

12.3.1.1 Modification of Functional Groups

Biochar's hydrophilicity and surface functional groups can be altered for pollutant elimination from water and contaminated land. Usually, biochar synthesized at relatively low temperatures (250–400°C) has more carbine (C-H) and alkene (C=C) functional groups. At relatively low temperatures, chemical oxidation using H_3PO_4, $KMnO_4$, HNO_3, $KMnO_4$, H_2O_2, or a mixture of HNO_3/H_2SO_4 may incorporate acidic groups such as lactonic, carbonyl, phenolic, carboxylic groups on the biochar. In addition, HNO_3 can generate a greater amount of acidic groups on the surface as compared to $KMnO_4$, due to its stronger oxidizing capacity. Acids like H_3PO_4 or H_2SO_4 partially oxidize the carbonized surface due to their inherent oxidizing properties. As a result, carboxylic groups are enriched in the matrix. This surface carboxylation is commonly achieved by a one-stage oxidation process (Qian et al. 2015). Similarly, functional groups containing nitrogen (like pyrrolic, amide, pyridinic, and lactame) serve a major role in environmental uses due to their stronger affinities with Cu, Cd, and Zn metals. Typically, these groups are introduced by nitration reaction on biochar. In the nitration step, HNO_3 is dissociated and produces highly active nitronium ions (NO_2^+). These active ions combine with the aromatic substances on the biochar surface and form nitrated products (-NO_2). As the rate of nitration reaction is quite slow, concentrated H_2SO_4 is required to increase the development of active nitronium ions on the biochar surface. The nitration step occurs through electrophilic aromatic exchange, and subsequently nitrogen groups are introduced on the aromatic sides of biochar. Later, the N-functional groups can be reduced to amino groups to enhance the basic characteristics of the biochar as an adsorbent. Sodium dithionite ($N_2S_2O_4$) is used to convert the nitro groups into amino groups. The amine functional group was introduced by Zhou et al. (2013) on the char surface using chitosan to increase the adsorption capacity for heavy metals, such as Cu^{2+}, Cd^{2+}, and Pb^{2+}. The findings of the research revealed that covering biochar surfaces with chitosan might improve the contaminated soil. In addition, this chitosan-modified biochar composite is an eco-friendly, economical, and effective sorbent for heavy metal removal. The amine groups can have strong interactions with metal ions and thus improve the metal uptake capacity of modified biochar. Yang and Jiang (2014) found that the separation of Cu^{2+} ions from water can be effectively performed by amino-modified biochar with enhanced functional groups. It is believed that Cu^{2+} is conjugated with the surface amino groups. Similarly, hydrogen peroxide (H_2O_2) is employed for alteration of the peanut hull biochar surface and imparts the carboxylic or O-containing functional groups. Biochar with a high amount of carboxylic groups is reported to have an effective sorption capacity for Ni^{2+}, Pb^{2+}, Cd^{2+}, and Cu^{2+} heavy metal cations. Furthermore, the addition of KOH to biochar increased the number of oxygen groups (C=O, O-H, COOH, and C-O groups) on the surface, which improved tetracycline sorption capability. The O-containing groups in modified biochar aided the formation of hydrogen bonds with the adsorbate at neutral pH and thus increased the tetracycline sorption capacity (Liu et al. 2012).

12.3.1.2 Chemical Oxidation Modification

This technique could be used to chemically modify biomass before or after pyrolysis (postmodification). Biochar's sorption capacity for heavy metals and other pollutants is greatly improved by this technique in the presence of a suitable oxidant. These improvements in sorption capacity of biochar might be defined as the enhanced affinity of negatively charged carbonyl groups for binding with positively charged metal ions. In chemical-based modification techniques, numerous acids (e.g., H_2SO_4, H_3PO_4, HCl, H_2O_2, and HNO_3), alkali (e.g., NaOH and KOH), and oxidizing agents (such as $FeCl_3.6H_2O$ and $KMnO_4$ salts) have been already used to modify the biochar with desirable properties. In addition, careful selection of the oxidizing agent is required for the production of carboxyl-group-rich biochar for the long-term environmental remediation of heavy metals. In general, alkali-treated biochar has a greater surface area than pristine or acid-treated biochar. Acid-treated biochar possesses more carboxyl groups, which is suitable for Cu, Pb, Zn, sulfamethazine, Cd, and oxytetracycline adsorption (Uchimiya, Bannon, and Wartelle 2012). On the other hand, alkaline treatment of biochar produces more graphite surface and/or aromatic groups (e.g., hydroxyl groups). Moreover, alkaline-treated biochar is better for arsenic, cadmium, tannic acid, and chloramphenicol sorption by increasing its π–π interaction, surface electrostatic attraction, surface precipitation, and/or surface conjugation with adsorbate. Alkaline-treated biochars are also used as supercapacitors. For utilization of biochar as super-capacitors, the raw powder must be activated at high temperatures (~960°C) in the existence of KOH and an inert environment for several hours. After that, the alkaline-treated biochar is chemically washed and dried extensively to produce porous nanostructures with a uniform pore size distribution across a large surface area. However, the economic feasibility of such a method is poor, posing a considerable barrier to the production of biochar-based supercapacitors. Biochar development with oxygen plasma is a relatively modern technology for activating biochar at lower temperatures (<150°C) in a quick and cost-effective manner (Jiang et al. 2013). In oxygen plasma technology, a controlled flow rate of O_2 gas is supplied into the plasma chamber and is ionized to O^* and O^+. Plasma is then generated by an excitation radio frequency antenna under the influence of a strong electromagnetic field. The procedure is carried out under vacuum at a preferred temperature for successful ion implantation of the samples. In general, the biochar samples are kept on a plate beneath the antenna, and the textural properties (e.g., surface area and pore size) are modified. Gupta et al. (2015) improved the supercapacitor properties (171.4 F/g) by activating yellow pine biochar with the help of oxygen plasma for 5 min. These improvements are 185 and 72.3%, respectively, as compared to pristine and conventional-based-activated biochars. Biochar's metal sorption capacity was generally increased after acid pretreatment and chemical oxidation steps. For example, the quantity of carboxyl groups on the surface of biochar increased after it was activated with H_2O_2. These functional groups provided extra active sorption sites for Pb^{2+} and Hg^{2+} surface conjugation, which was comparable to or even better than commercial activated carbons. The Pb^{2+} sorption capacity for H_2O_2-modified biochar was found at 22.8 mg/g per the literature, which was 20-fold higher than plain biochar (0.88 mg/g). The presence of extra carboxyl groups in the modified biochar surface leads to higher Pb^{2+} sorption capacity.

12.3.1.3 Chemical Coating Modification

Biochar has lately been coated with various metal oxides to improve its characteristics and, as a result, increase its sorption capacity. For instance, the adsorption of anionic dyes is very inefficient due to electrostatic repulsion forces between dye molecules and the negatively charged biochar surface. Thus the surface properties of biochar may be altered by coating.

Coating with metal oxides: Wang et al. (2020) created bamboo charcoal coated with cobalt (Co) and iron (Fe) metals for the separation of heavy metals from water. First, the commercial bamboo charcoal was mixed in a 100 mL aqueous solution comprising cobalt nitrate hexahydrate or iron chloride salts with 9 molar concentrations of HNO_3 acid. The dipping process was followed by microwave heating at 640 W power with a frequency of 2.45 GHz for few minutes. In addition, magnesium hydroxide was coated on wheat straw charcoal using magnesium chloride and sodium hydroxide (molar ratio of OH/Mg ~1.5). Fe-coated biochar substantially consisted of more hydroxyl groups than unmodified biochar due to the development of iron oxide layers on the biochar surface (X. N. Zhang et al. 2014). Zhao et al. (2013) reported Fe_3O_4-coated biochar as a sorbent for removing PBDEs from water via solid-phase microextraction. First, a solution was prepared with the help of ferric chloride ($FeCl_3.6H_2O$) and other additives, such as urea and tetrabutylammonium bromide and then raw bamboo, were added to that solution. In a second stage, bamboo-charcoal/Fe_3O_4 composites were made by heating the solution's residual solid at 450°C for 3 h under N_2. The results revealed that Fe_3O_4-treated bamboo charcoal was more efficient than other adsorbents at removing PBDEs. Additionally, using $FeCl_3$ as an iron source, Fe(III)-coated biochar was produced. This Fe(III) coating on the biochar enhanced As^{3+} and As^{5+} uptake capacities per the literature.

Coating biochar with carbon nanotubes (CNTs): CNTs are very efficient sorbents for pollutant elimination owing to their nanostructure and enormous surface area. However, the expensive cost and inconvenient nature of engineering applications have limited their use. As a result, composites of bio-char and CNTs could be used as more effective and reusable adsorbents for sewage water treatment where biochar acts as a microporous or mesoporous carrier of CNTs. For example, a hybrid multi-walled CNT was coated on biochar by a dip-coating technique. The coating of biomass was done by varying the concentration of carboxyl-functionalized CNT solutions (weight percentages of 0.01 and 1) on biochar. Later, the mixture was pyrolyzed in a tubular furnace at 600°C for 1 h at a N_2 flow rate of 10°C/min (Inyang et al. 2014). J. Zhang et al. (2009) reported a chemical vapor deposition (CVD) technique, which was used to grow CNTs on bamboo biochar. In this process, the fresh bio-char produced under N_2 atmosphere, was pretreated with ferrocene solution. Later, a peristaltic pump was used to feed treated biochar into a CVD furnace at 820°C for numerous minutes while a mixture of H_2 and Ar flowed through the system. Huang et al. (2012) proposed a similar CVD approach to manufacture CNT-coated bamboo biochar utilizing ferric nitrate as a catalyst. In comparison to uncoated biochar, hybrid multiwalled CNT-coated biochar had bigger surface areas, higher porosity, and better thermal stability than uncoated biochar. For example, surface area for CNT loaded biochar was found around 120/m^2g (0.01 wt% of CNT) and 390/m^2g (1 wt % of CNT) as compared to pure bagasse biochar (9/m^2g). So it was a clear observation that the rise in CNT loading on biochar led to a considerable improvement in surface area and pore volume. Similar findings related to pore volume and surface area were witnessed for metal-coated biochar. For example, cobalt-coated bamboo bio-char had a greater surface area and a larger pore volume compared to uncoated biochar, resulting in a higher Cr(VI) uptake capacity. Similarly, increasing the amount of Fe coating on bamboo charcoal resulted in a substantial increase in textural properties. In addition, the role of surfactant (SDBS) in the case of the CNTs-biochar composite was also observed. The surfactant SDBS was used for excel-lent spreading and distribution of CNTs on the biochar surface during composite synthesis. Thus the loading capacity of CNTs-biochar composites for various pollutants, such as sulfapyridine (SPY) and Pb^{2+}, was significantly higher as compared to composites prepared without SDBS. In this case, Pb^{2+} and sulfapyridine showed no evident competition, demonstrating that these two contaminants are absorbed in a site-specific manner on the CNT-biochar surface (Inyang et al. 2015).

12.3.1.4   Surfactant Modifications

Surfactants are classed as cationic, anionic, and nonionic as per the type of hydrophilic group present in the structure. In general, surfactants are employed as an additive in the industrial making of washing detergents. Recently, surfactants have been employed for the modification of the surface properties of various materials, such as zeolite and bentonite. Cationic surfactant could easily have conjugated with negatively charged biochar via electrostatic attraction. Due to this, various cations such as Na^+, K^+, Mg^{2+} can exchange with the biochar surface, resulting in the formation of a surfactant biochar complex. Saleh (2006) reported that cetylpyridinium chloride (CPC) sorption on granular charcoal occurred at low concentrations via ion exchange. First, a partial monolayer of CPC is generated with a rise in surfactant concentration. Later, after a further rise in surfactant concentration, the CPC adsorption on the biochar surface is increased due to hydrophobic interaction between the charcoal and hydrophobic sites of CPC. H. Li et al. (2014) investigated the adsorption of TX-100, a nonionic surfactant, onto bio-char; the surfactant was loaded onto charcoal in the amount of 300 mg/g. On the other hand, micellar and monomolecular anionic surfactants are difficult to adsorb on the biochar surface due to electro-static repulsion. Cetyltrimethyl ammonium bromide (CTAB) was chosen as a cationic surfactant in this case. CTAB was added to the PCP solution along with biochar. So the biochar was modified through an ion exchange process with CTAB, and uptake of PCP occurred simultaneously on the biochar-CTAB conjugates. As a result, cationic surfactants can be used to boost biochar's ability to remove anionic contaminants. Biochar could also be modified by the nonionic surfactant. The process is called physi-sorption as evidenced by modest free energy changes in sorption (Rajapaksha et al. 2016). Moreover, the study discovered that after modification with CTAB, the PCP sorption ability of biochar was reduced due to a rise in surfactant concentrations in water. This could be due to CTAB sorption inter-fering with hydrophobic sorption sites. In addition, the mobility and solubility of PCP in the solution caused by CTAB may result in a decrease in PCP uptake on biochar. The adsorption of the amphiphilic phenothiazine medication thioridazine hydrochloride (THCl) on activated biochar was also decreased by CTAB. The competitiveness between cationic THCl and cationic surfactant decreases the THCl sorption on biochar, in addition to the sorption-site-hindrance mechanism. On the other hand, anionic surfactant decreased THCl adsorption, but TX-100, a nonionic surfactant, enhanced THCl sorption. As previously described, adsorption of nonionic surfactant on biochar is governed by a physisorption process. Thus, during the sorption process, THCl drug might be loaded into the preoccupied nonionic surfactant (i.e., co-adsorption) (Erdinç, Göktürk, and Tunçay 2010).

12.3.2 Physical Modification Techniques

The improvement in the porous structure and functional groups of biochar can be done by various physical techniques without the addition of any impurity. The most common physical techniques for improving biochar properties include gas or steam activation, ball milling, magnetic modification, and microwave. The modified biochar had an excellent sorption capacity for heavy metals, organic pollutants, and nutrient elements owing to increased surface area and the presence of micro-/meso-pores. Physical techniques are safe and produce cost-effective biochars. Those sorbents are free of contaminants as compared to chemically modified biochars, as clean and easily regulated agents are used in physical techniques.

12.3.2.1 Gas Purging

In this method, biochar can be modified by treating the sorbent with a high-temperature CO_2-ammonia mixture. Xiong et al. (2013) synthesized NH_3- and CO_2-improved biochar at various temperatures. Fresh biochar was thermally treated up to a temperature of 500–900°C in a reactor at N_2 flow. Later, NH_3 or CO_2 was introduced into the reactor for purging. The N-containing functional groups in the modified biochar can be increased by up to 3.91 wt.% due to the introduction of NH_3 (ammonification), whereas CO_2 treatment could aid pore creation and increase the microporosity in the biochar, allowing for greater gas adsorption capacity. It was found that CO_2-modified biochar had a substantially higher pore volume and larger surface area than untreated biochar. The increment in pore volume and surface area is due to the modified biochar's microporous structures, which may emerge as a result of CO_2 reacting with the carbon surface and forming CO (i.e., hot corrosion). The gas uptake capability of CO_2-modified biochar was substantially higher than plain biochar at ambient temperature. The modified biochar's CO_2 sorption capacity was shown to be linearly proportional to the micropore volume of the sorbent, and it was recognized as physical sorption (Yao et al. 2014).

12.3.2.2 Modification by Steam Activation

Conventional biochar might be upgraded to activated biochar with enriched carbonaceous structures and a large surface area using this technique. Steam activation of biochar occurred in the absence of oxygen at modest temperatures (400–800°C). After that, the treated biochar was partially gasified with steam in a second phase. This process removed undesirable volatile components and unreacted monomer from the char and promoted crystallinity in the biochar. The reactions occurring in steam activation are given next. In the first stage, oxygen from water molecules may exchange with carbon surface site C (f) to create a surface oxide C (O) and H_2 gas (Lussier, Zhang, and Miller 1998). The produced H_2 may react with the C surface to form surface hydrogen (H) complexes [C(H)] (Eqs. 1–3). Steam oxidizes C surface sites, producing CO_2 and H_2, which may be used for biochar activation while inhibiting C site gasification (Rajapaksha et al. 2016).

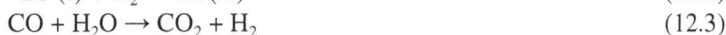

$$C\ (f) + H_2O \rightarrow C(O) + H_2 \tag{12.1}$$
$$2C\ (f) + H_2 \rightarrow 2C(H) \tag{12.2}$$
$$CO + H_2O \rightarrow CO_2 + H_2 \tag{12.3}$$

In comparison to nonactivated biochar, research has been done to check the possible application of steam-activated biochar for wastewater treatment and soil remediation. Lima, Boateng, and Klasson (2010) used waste biomass to make steam-activated biochar at 800°C for 45 min under 1 atm pressure. Water was injected at 3 mL/min with N_2 flow ($0.1 m^3/h$) into the heated vessel to accomplish the steam activation process. The treated samples were purified by 0.1 mol/L HCl solution and water (three times) to remove surface ash and other impurities. The washed samples were dried at 80°C overnight. Invasive plant biomass and tea waste biochar were steam activated for 45 min at temperatures of 300 and 700°C, respectively, using a flow rate of 5 mL/min. These biochars were used to remove veterinary medicines from aqueous solution without any postsynthesis modifications. Only a small literature is available where key parameters of the steam activation process are optimized to improve the physical characteristics of the biochar. It has been observed that steam-activated biochar has excellent sorption capacity for inorganic and organic pollutants. For example, Rajapaksha et al. (2015) synthesized biochar from invasive plant biomass and tea waste using steam activation for the removal of veterinary waste. The activated biochar made from invasive plants at 700°C had a 55% higher sorption capacity compared to conventional biochar made at the same temperature. The rise in surface area for the steam-treated biochar might be the reason for such an increment in sorption capacity for veterinary wastes.

12.3.2.3 Magnetic Modification

Biochar's multifunctional properties suggest that it could be an excellent adsorbent for pollutants removed from wastewater. However, it is tough to remove the powdered biochar from the aqueous solution, making it unlikable to users. As a result, research is focused on the development of magnetic biochar materials in order to remove adsorbents from the mixture and improve the separation process. As biochar surfaces are usually negatively charged, adsorption of anionic pollutants is negligible. Therefore, biochar has been modified by imparting magnetism to improve its anionic pollutant sorption. Chen, Chen, and Lv (2011) suggested a novel method that combines the manufacture of biochar and iron oxide in one step. In addition, the process also reduced the cost of imparting the magnetism in biochar. The magnetic biochar was synthesized in a single step by coprecipitation of Fe^{3+}/Fe^{2+} ions on orange peel powder, and then the pyrolysis of that powder mixture was done at numerous temperatures (250, 400, and 700°C). The surface area of prepared magnetic biochar was less compared to nonmagnetic biochar. Nonmagnetic biochar, on the other hand, has a smaller average pore diameter than magnetic biochar. This is likely due to the fact that magnetic biochar has a high concentration of iron oxide, which has a low surface area and a large number of transitional pores (2–50 nm). This magnetic biochar's hybrid sorption capability allows for simultaneous separation of phosphate and organic contaminants. The long-term stability of magnetism in the modified biochar established the potential of magnetic separation after the sorption process. It is a significant benefit for the water treatment process. In another study, $FeCl_3$-treated biomass was fed to pyrolysis at 600°C to create magnetic biochar. In the prepared magnetic biochar, the nanosized γ-Fe_2O_3 particles are implanted into the porous biochar structure. The saturation magnetization of modified biochar was 69.2 emu/g, and it was close to the magnetization of pure γ-Fe_2O_3 (76.0 emu/g) (M. Zhang et al. 2013). Furthermore, magnetic biochar is capable of adsorbing As^{3+} in solution. As the adsorbent possesses ferromagnetic characteristics, the spent adsorbent can be easily collected by applying an external magnetic field. To prevent magnetic nanoparticles from falling out of the biochar matrix, a suitable coupling agent may be required for effective encapsulation. Accordingly, the amino-terminated 3-triethoxysilylpropylamine is used as a coupler for the attachment of Fe_3O_4 particles to the biochar surface. Therefore, functionalization of such coupling agents is beneficial for imparting magnetic particles and compensates for the reduction in surface area caused by the magnetic particles. Functional groups contain certain elements (such as O, N, or S) that can easily conjugate with metal ions or other environmental contaminants (X. Zhou et al. 2018).

12.3.2.4 Mineral Oxides Impregnation

Biochar can also be modified or prepared by impregnation of minerals. Clay minerals are renowned for pollutant removal due to their surface charge, cation exchange capability, and structure composition. Clay minerals such as montmorillonite, gibbsite, and kaolinite are often utilized as cost-effective sorbents. Clay particles (montmorillonite and/or kaolin) were dispersed throughout the biochar matrix by Yao et al. (2014) to improve its functioning. First, the clay powders were mixed with various biomass raw materials (bagasse, bamboo, etc.) and then pyrolysis was done in an N_2 atmosphere for 1 h at a temperature of 600°C. Biochar acts as a good porous support for these biochar-clay composites for the distribution of fine clay particles. M. Zhang et al. (2012) utilized various biomass feedstocks to produce porous MgO-biochar composites comprising of 20 nm thick flakes of MgO. First, the biomass resources were mixed with $MgCl_2$, and then the mixture was heated for 1 h at 600°C under an N_2 atmosphere to prepare the composite. The use of N_2 is beneficial for the elimination of by-products such as HCl, allowing the production of MgO on the biochar surface. Mg/Ca-enhanced tomato tissues are employed in another scenario to make Mg-rich biochar, which comprises nanosized $Mg(OH)_2$ and MgO particles inside the matrix. Manganese oxide (MnO_x) composites and their various phases (b-MnO_2, d-MnO_2, etc.) are known for their heavy metal removal and phosphate binding capabilities. Therefore, Song et al. (2014) reported MnO_x-biochar by thermal treatment of biochar and $KMnO_4$ at 600°C. A well dispersed layered structure of MnO_x was observed in the composite when 10% $KMnO_4$ was present in the raw mixture. The oxidation states of Mn in those composites were found to be Mn^{3+} and Mn^{4+}. Magnesium-biochar composites produced from Mg-rich plants were reported as a good adsorbent for phosphate in aqueous solution. Also, the highest phosphate uptake capacities of the Mg-biochar composites synthesized from different biomass were higher (> 100 mg/g) than conventional sorbents (20 mg/g). The $MgO/Mg(OH)_2$ present on the Mg-biochar surface is formed a strong interaction with phosphate and thus increased the phosphate uptake capacity. However, the presence of $Fe(OH)_3$ in the Mg-biochar composite reduced its phosphate sorption capacity, possibly due to the blockage of MgO particles by $Fe(OH)_3$.

12.3.3 Modification by Biological Methods

Various metabolic ways existing in microorganisms allow them to absorb a variety of organic molecules (pollutants and wastes) and convert them into value-added products, such as medicinal compounds. They can penetrate through tiny pores and form a biofilm that prevents them from flushing out, owing to their microscopic size. Microbes can form biofilms on the biochar surface and biochemically improve the biochar surface by colonizing it. Thus biochar with the desired characteristics is formed due to this postmodification technique. More particularly, the simultaneous adsorption and breakdown of organic pollutants by biochar scaffold and inoculated microorganisms was reported. It's worth noting that some microbes are excellent bioadsorbents for a variety of heavy metals. So the colonization of microorganisms on biochar helped in the elimination of heavy metals and the degradation of naphthenic acids from the environment (Frankel et al. 2016). In addition, biochar can be used as a suitable scaffold for microbial biofilms as compared to conventional sand-active-biofilm. The sorption and biodegradation capabilities of these two biofilm scaffolds were evaluated on four pharmaceutically active compounds (i.e., carbamazepine, ranitidine, caffeine, and metoprolol). The high uptake of carbamazepine (> 98%) was observed for biochar-active biofilm compared to sand-active biofilm (7% removal) after 22 weeks of study. The removal percentages for ranitidine and caffeine were relatively similar in both scaffolds. The leftovers of the biomass anaerobic digestion (AD) process may be utilized for the synthesis of engineered biochar owing to their redox potential and suitable pH. The biochar formed from AD-treated biomass possesses the following physiochemical characteristics, such as larger surface area, negative surface charge, and good ion exchange capability (Dehhaghi, Mohammadipanah, and Guillemin 2018). This biologically modified biochar might be utilized as an ion exchanger for the detection of phosphates and heavy metals. Table 12.2 represents the various modification techniques along with the obtained biochar properties.

TABLE 12.2
Properties of Designer Biochars Prepared from Different Modification Methods (modified with permission from Rajapaksha et al. (2016) © Elsevier)

Modification techniques	Feedstock	Pyrolysis temperature (°C)	Ash (%)	pH	Element content (%)				Surface area (m²/g)	Pore volume (cm³/g)
					C	H	O	N		
Steam-activated biochar	Tea waste	300	6.4	8.6	71.5	4.8	18.2	5.5	1.5	0.004
MnOₓ loaded biochar	Pine wood	600	14.0	–	80.0	1.9	14.6	0.3	463.1	0.022
Magnetic biochar	Orange peel	250	42.4	–	35.1	3.6	–	1.1	41.2	0.052
Clay-biochar composites	Bagasse	600	–	–	75.3	2.2	18.9	0.7	407.0	–
Steam-activated biochar	Sicyos angulatus L. plant	700	70.7	11.7	50.6	1.7	44.9	2.5	7.1	0.038
MnOₓ-loaded biochar	Corn straw	700	13.1	–	73.0	0.3	10.9	0.7	3.2	0.006
Clay-biochar composites	Bamboo	600	–	–	83.3	2.3	12.4	0.2	408.1	–
Magnetic biochar	Orange peel	400	35.0	–	29.4	2.2	–	0.5	23.4	0.042
Steam-activated biochar	Tea waste	700	16.7	10.5	82.4	2.1	11.6	3.9	576.1	0.109
Methanol modified biochar	Rice husk	450	0.9	–	71.3	3.6	23.4	0.8	66.0	0.051
Zinc-nitrate-modified biochar	Pine cones	500	2.1	4.0	71.2	3.0	20.4	0.5	11.5	0.028
Clay-biochar composites	Hickory chips	600	–	–	80.9	2.2	15.1	0.3	376.1	–
Magnetic biochar	Orange peel	700	95.7	–	0.4	0.2	–	0.2	19.4	0.033
Ammonia-CO₂ modification	Cotton stalks	600	–	–	–	–	–	3.5	251.9	0.08

12.4 BIOCHAR-BASED COMPOSITE MATERIALS FOR THE REMOVAL OF PHARMACEUTICAL ACTIVE COMPOUNDS (PhACs)

Biochar is a carbon-based substance made from the pyrolysis of organic matter. It has a strong affinity for both inorganic and organic pollutants. In this section, applications of engineered or modified biochar for the removal of antibiotics and active compounds are discussed.

12.4.1 MODIFIED BIOCHAR APPLICATION FOR THE REMOVAL OF ANTIBIOTICS

One of the greatest achievements in medical science of the twentieth century was the introduction of antibiotics into therapeutic use. However, many antibiotics are currently designated as contaminants, posing a risk to flora and fauna, including humans. Antibiotics are mostly consumed by humans and animals, apart from being used in agriculture and plant nutrition. As such, these pollutants are accumulated in soil or surface waters. Sorption is one of the efficient methods for removal of these contaminants. Recently, engineered biochar has emerged as a low-cost sorbent material for antibiotic adsorption (Krasucka et al. 2021). The various sources of antibiotics along with their distributions are illustrated in Figure 12.3. Zhu et al. (2018) reported physically modified wood biochar (*Pinus radiata*) for the removal of tetracycline antibiotics. After pyrolysis, the biochar was modified by thermal treatment in a tube furnace with a mixture of air and nitrogen at a temperature of 600–800°C. Before being injected into the furnace, air with varying flow rates (0–90 mL/min) was mixed with N_2 in a mixer. The presence of the air in gas mixture was responsible for development of surface area and mesopore in biochar matrix. The highest mesopore surface area and pore volume were tabulated at around 316 m^2/g and 0.284 cm^3/g respectively, when biochar was processed at a temperature of 700°C. The air flux of 50–90 mL/min increased the mesopore surface area, and the mesopore size distribution was found to be in the range of 20–60 Å. The tetracycline sorption was enhanced by mesopore creation in the biochar matrix, and it was 5.5–9.2 times higher than biochar developed only in the N_2 atmosphere. In another study, a composite of montmorillonite-biochar (MT-BC) was investigated for the elimination of norfloxacin (NOR). The composite was prepared at 400°C in a muffle furnace under a N_2 blanket for 6 h. The effects of Cu^{2+} ions, pH, humic acids, and other variables on NOR adsorption were investigated. The porous structure in biochar was enriched by montmorillonite, with pore volume and surface area of 0.604 cm^3/g and 112.6 m^2/g, respectively. The composite MT-BC had a maximum uptake capacity of 25.53 mg/g, which was 2.41 times that of normal biochar. However, in the presence of Cu^{2+} ions and humic acids, the norfloxacin uptake was reduced due to the blockage of active sorption sites in adsorbents (Jinghuan Zhang et al. 2018). Xiang et al. (2019) reported a cost-effective, eco-friendly, highly efficient ferromanganese-improved biochar (Fe/Mn-BC) for the adsorption of levofloxacin (LEV). Iron (Fe) and manganese (Mn) salts were coprecipitated on vinasse wastes to create the composite. The resulting sample was then pyrolyzed at 800°C in the presence of N_2. The findings of the experiments indicated that LEV adsorption was pH dependent, with maximal adsorption occurring at pH 5. In addition, the LEV adsorption was increased with an increase in ionic strength in the solution due to a salting out effect, and nonelectrostatic forces occurred between the adsorbent and the adsorbate. Fe/Mn-BC was magnetic, with saturation magnetism of 20 emu/g, and could be recycled for 5 cycles with a specified LEV adsorption capacity. Ma et al. (2020) investigated the use of [Fe/Zn or H_3PO_4 or (Fe/Zn + H_3PO_4)] chemically engineered sludge biochar for the removal of several fluoroquinolone drugs from aqueous solution, including ofloxacin (OFL), ciprofloxacin (CIP), and norfloxacin (NOR). In the presence of Fe/Zn$^+$ H_3PO_4, the physiochemical properties such as surface area, pore volume, pore width, and oxygen-comprising functional groups in the composites were enhanced. The maximum adsorption capacity for Fe/Zn + H_3PO_4 modified biochar was 25.4, 39.3, and 83.7 mg/g for the antibiotics OFL, CIP, and NOR, respectively. It was 20-fold higher than the sorption capacity of pure Fe/Zn-modified and H_3PO_4-modified biochar. Hu et al. (2021) reported cobalt (Co) and gadolinium (Gd) metal-modified biochar for the separation of tetracycline (TC)

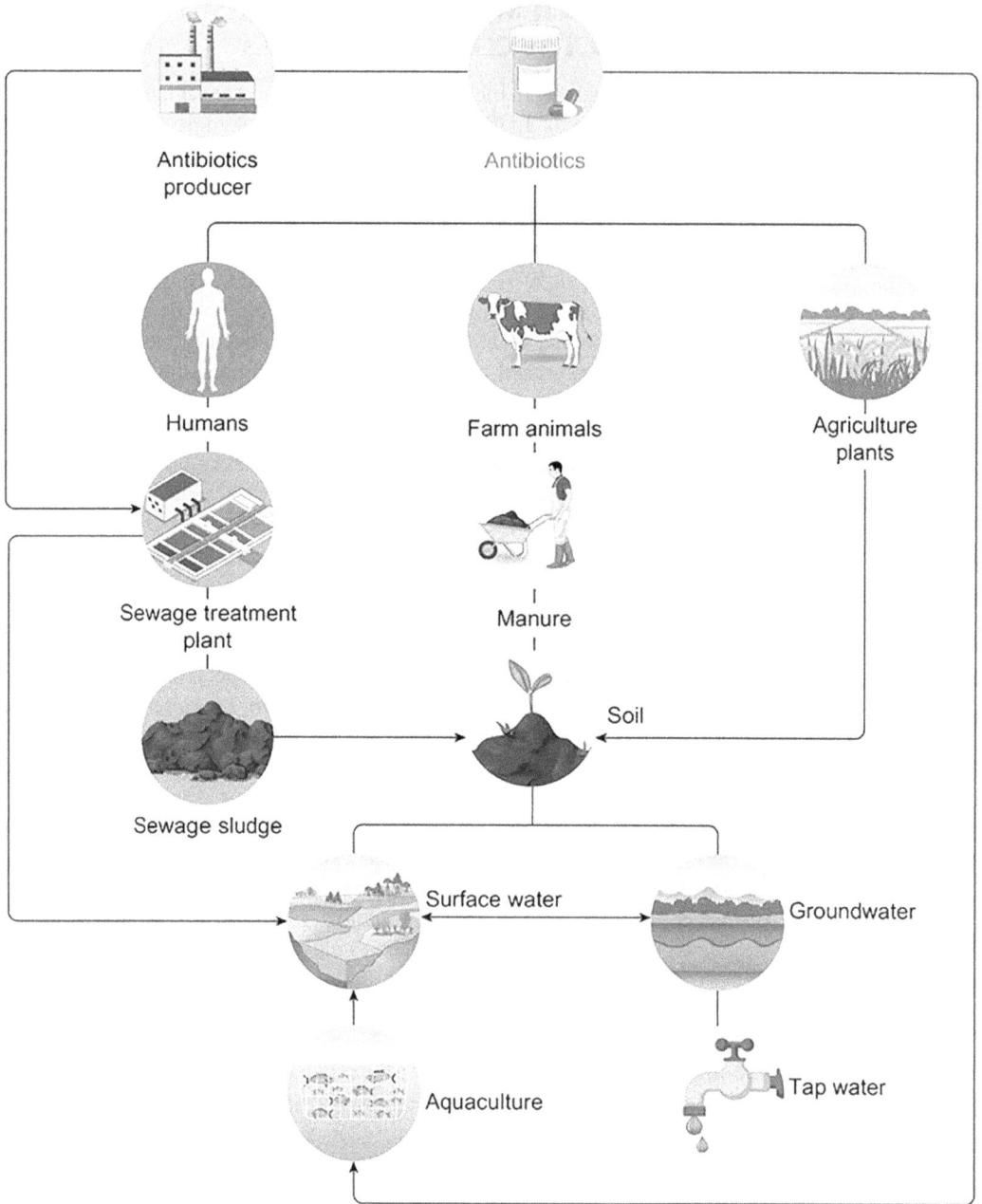

FIGURE 12.3 Sources and distribution of antibiotics in the environment.

Source: Obtained with permission from Krasucka et al. (2021) © Elsevier.

and ciprofloxacin (CIP) in single and binary blends. The modified biochar had a sponge-like structure with a specific surface area and pore volume of 370.37 m²/g and 0.1991 cm³/g respectively. In comparison to pure biochar, the addition of metals doubled the surface area and pore volume of the modified one. The composite's sorption capacity was 119.05 mg/g for TC and 44.44 mg-g for CIP, respectively, and was controlled via chemisorption. However, the adsorption capacity was slightly decreased in the case of the binary competitive system.

12.4.2   MODIFIED BIOCHAR APPLICATION FOR THE REMOVAL OF OTHER PHARMACEUTICAL ACTIVE COMPOUNDS

In this section, the utility of modified biochar for the sorption of a few pharmaceutical drugs (other than antibiotics) are discussed. These are mainly anti-inflammatory, antifungal, antidepression drugs. The use of these drugs is increasing with the size of the human population and its distribution with age. As a result, the number of these active compounds in the water grows dramatically, necessitating specific attention for their removal.

Chakraborty et al. (2018) studied steam-activated modified biochar for the removal of the anti-inflammatory drug ibuprofen from the water. The biochar (wood apple) was physically activated under superheated steam at pressures of 1.5 kg/cm at 800°C for 1.5 h at a steam flow rate of 1–2 kg/h. The active sites in the biochar were boosted by physical treatment, and this was confirmed by morphological studies. The maximum removal of ibuprofen was found at around 90 and 95% for biochar and steam-activated biochar from the water at temperatures of 15 and 20°C, respectively. The existence of the ibuprofen in the adsorbents was confirmed by elemental analysis by observing the enhancement in carbon and oxygen percentage. The elemental analysis showed the presence of ibuprofen in the adsorbents by monitoring the increment in carbon and oxygen percentages in the ibuprofen-loaded biochars. The same group investigated adsorption capabilities of physically and chemically activated *Cocos nucifera* shell biochar for the removal of ibuprofen (IBP). The physical activation process was processed under superheated steam at a flow rate of 1.5–2 kg/cm/h and at 550°C for 45 min. On the other hand, the carbonized biochar was treated with ortho-phosphoric acid (H_3PO_4) at a weight ratio of 1.5 to make chemically modified biochar. The maximal uptake of IBP was 9.69 mg/g for physically active biochar and 12.16 mg/g for chemically activated biochar, respectively (Chakraborty et al. 2019). In addition, both the adsorbents were cost-effective and could be recycled multiple times. In another study, magnetic/chitosan-activated biochar was prepared for the removal of naproxen (NPX), diclofenac (DCF), and ibuprofen (IBP) from the aqueous solution. In order to make the adsorbent, the magnetic fluid was first made by coprecipitating Fe^{3+} and Fe^{2+} salt solutions in a 3:2 molar ratios. Later, agricultural residue biochar was chemically treated with magnetic fluid and chitosan to produce cross-linked magnetic/chitosan-activated biochar (CMCAB). The removal rates of NPX, DCF, and IBP were 95.2, 96.4, and 98.8%, respectively, at pH 6.0. The rapid uptake of these anti-inflammatory medications could be due to the diminishing nature of H^+ ions in the solution at pH 6.0. The sorbent CMCAB could be recycled at least 8 times without losing its effectiveness. Caffeine (CFN), a psychoactive drug, was adsorbed by steam-activated tea waste biochar (TWBC-SA) by Keerthanan et al. (2020). The biochar was activated under 5 mL/min of stream flow rate at 700°C for 45 min. The surface characteristics of the biochar, such as porosity and aromatic contents, were improved by steam activation, as confirmed by FT-IR and XPS analysis. The maximum CFN adsorption capacity was observed at 15.4 mg/g at pH 3.5. Finally, it was found that chemisorption via electrostatic interactions and nucleophilic attraction controlled the loading of CFN on TWBC-SA. Table 12.3 depicts the adsorption capacity of designer biochars for different pharmaceutical active compounds (PhACs).

12.5   TECHNOECONOMIC PROSPECTS OF BIOCHAR

To study the performance and application-based analysis of biochars, the economic assessment for preparation of biochar plays an important role. Such economic perspective depends hugely on the type of raw material used, pyrolysis technique, and condition utilized, or the inclusion of any carbon sequestration techniques involved or not. It has been studied that practicing smart agriculture techniques benefits crop yields and gross production, which results in the economic prosperity of the farmers. In comparison to inorganic fertilizer, biochar has the advantage of long-term capability toward agricultural improvement. Studies should be focused on utilization of novel feedstock rather than only using high-carbon-content biomass. Such a process would help in creating an

TABLE 12.3
Adsorption of Pharmaceutical Active Compounds (PhACs) to Modified Biochars (modified with permission from Cheng et al. (2021) © Elsevier

Modification techniques	Feedstocks	Pyrolysis temperature (°C)	pH	PhACs	Adsorption capacity (mg/g)
Magnetic biochar	Astragalus residue	700	6.0	Ciprofloxacin	68.9
MnOx-loaded biochar	Cornstalks	350	4.0	Oxytetracycline	39.92
Anaerobic digestion	Bagasse	600	2.6	Sulfamethoxazole	54.38
Mineral oxide impregnation	Camphor leaf	650	4.0	Ciprofloxacin	449.40
Magnetic biochar	Coffee grounds	250	2.0	Tetracycline	96
MnOx-loaded biochar	Date palm waste	600	5.0	Chlortetracycline	89.05
Steam activated biochar	Tea waste	700	4.0	Sulfamethazine	33.81
Mineral oxide impregnation	Coconut and walnut shells	500	6.0	Tetracycline	94.2
Clay-biochar composites	Straw	400	7.0	Norfloxacin	25.53
Methanol-modified biochar	Rice husk	500	2.0	Tetracycline	95
Mineral oxide impregnation	Wood processing residues	700	6.0	Tylosin	118.83
Steam-activated biochar	*Sicyos angulatus L.* plant	700	3.0	Sulfamethazine	37.73
Chemical coating modification	Grapefruit	450	3.0	Ciprofloxacin	36.72
Clay-biochar composites	Wasted sludge	750	9.0	Tetracycline	183.01

economically cheap outcome. It was reported that when biochar is utilized for soil quality improvement, the cost scenarios that come into play are (1) the total value/cost procured/incurred for utilizing biochar as an energy source, and (2) the costs incurred from the biochar's expected break-even value. Due to the nutrient contents of the utilized biochar, it can also be utilized in organic farming. Studies also analyzed the minimum selling price of such biochars and cost-reduction strategies to make it more economical for farmers to use them in agriculture. Studies also showed that biochar, woodchips briquettes (both torrefied and nontorrefied) are cost-effective and could be utilized for making value-added products (Khan et al. 2021). Cocomposted biochar utilization was studied by Pandit et al. (2020), and it was found that when such a product was applied in 60 tons per hectare, 243% enhancement in maize production was obtained. Moreover, the economic benefits of biochar were also studied by Shaheen et al. (2019) with respect to the treatment of contaminated water. It was obtained that biochar was more economical, about \$91–329/ton, when compared to zeolite ~\$6000/ton and activated carbon ~\$1500/ton. Environment-related applications and its advantages for biochar have not yet been well investigated by researchers, and thus its commercial point of view is still to be looked at and studied. Furthermore, there is a lack of awareness among farmers regarding the applicability of biochars and their advantages over productivity, along with soil fertility restoration.

12.6　CONCLUSION AND FUTURE PERSPECTIVES

Biochar can act as a promising compound for tackling the environmental remediation issues in today's society. Biochar is found to be an efficient CO_2-mitigating agent, in addition to providing a solution to sustainable environment management. Further, biochar could improve both the biological and nutritional properties of soils, in addition to being utilized for pollution remediation purposes. In order to utilize biochar effectively for minimizing environmental hazardous substances, its physicochemical properties play an important role and must be studied carefully. The pore size distribution and average pore size, along with surface chemistry involving functional groups and its pH value, play a vital role during a sorption study of the removal of contaminants. The major properties that make biochar beneficial for environmental remediation purposes depend mainly on the raw feedstock composition and pyrolysis conditions. Various physical (gas/steam activation, magnetization, etc.) and chemical treatment technologies (oxidization, functional nanoparticles-coating, etc.) are utilized to modify the pristine biochar to obtain upgraded physicochemical properties for proper utilization. Such strategies help to modify the pore volumes, functional properties, surface area, and availability of active sites that are useful for effective separation applications.

Moreover, technoeconomic analysis of the modified biochar should be performed to analyze their sustainability aspect. Thus prior to large-scale production, the toxicology and ecological risk analysis must be performed. Apart from resolving the negative impacts of the environment, if implemented inappropriately, the surface-modified biochars can also behave negatively toward the environment. Hence future research must concentrate on the negative impacts of biochars from a specific feedstock type, their treatment conditions, and importantly the modification techniques. Future studies should also focus on the shelf life of the engineered biochar and its recycling and collecting techniques. Importantly, when nanoparticles are embedded within biochars during modification, special care must be taken for the fate of such materials, which could lead to unavoidable bioaccumulation and biomagnification processes. The activation technique that includes physical and chemical methods for modifying the stability of biochars must be minimized since it results in environmental contamination. Engineered modification involving magnetic property inclusion is thought to be a better solution for the separation and recovery of modified biochar after application. Therefore, investigation and research should be focused on the applications of engineered biochar in real-world fieldwork, and hence its large-scale production and commercial viability thus can be evaluated.

ACKNOWLEDGMENT

This work is partially supported by a grant (DST/TM/WTI/WIC/2 K17/84(G)) from the DST (Department of Science and Technology), New Delhi. Any opinions, findings, and conclusions expressed in this paper are those of the authors and do not necessarily reflect the views of DST, New Delhi.

REFERENCES

Agrafioti, Evita, George Bouras, Dimitrios Kalderis, and Evan Diamadopoulos. 2013. "Biochar Production by Sewage Sludge Pyrolysis." *Journal of Analytical and Applied Pyrolysis* 101. Elsevier B.V.: 72–78. doi:10.1016/j.jaap.2013.02.010

Chakraborty, Prasenjit, Soumya Banerjee, Sumit Kumar, Sutonu Sadhukhan, and Gopinath Halder. 2018. "Elucidation of Ibuprofen Uptake Capability of Raw and Steam Activated Biochar of Aegle Marmelos Shell: Isotherm, Kinetics, Thermodynamics and Cost Estimation." *Process Safety and Environmental Protection* 118: 10–23. doi:https://doi.org/10.1016/j.psep.2018.06.015

Chakraborty, Prasenjit, Sumona Show, Wasi Ur Rahman, and Gopinath Halder. 2019. "Linearity and Non-Linearity Analysis of Isotherms and Kinetics for Ibuprofen Remotion Using Superheated Steam and Acid Modified Biochar." *Process Safety and Environmental Protection* 126: 193–204. doi:https://doi.org/10.1016/j.psep.2019.04.011

Chen, Baoliang, Zaiming Chen, and Shaofang Lv. 2011. "A Novel Magnetic Biochar Efficiently Sorbs Organic Pollutants and Phosphate." *Bioresource Technology* 102 (2). Elsevier Ltd: 716–723. doi:10.1016/j.biortech.2010.08.067

Cheng, Ning, Bing Wang, Pan Wu, Xinqing Lee, Ying Xing, Miao Chen, and Bin Gao. 2021. "Adsorption of Emerging Contaminants from Water and Wastewater by Modified Biochar: A Review." *Environmental Pollution* 273. Elsevier Ltd: 116448. doi:10.1016/j.envpol.2021.116448

Dehhaghi, Mona, Fatemeh Mohammadipanah, and Gilles J. Guillemin. 2018. "Myxobacterial Natural Products: An under-Valued Source of Products for Drug Discovery for Neurological Disorders." *NeuroToxicology* 66. Elsevier B.V.: 195–203. doi:10.1016/j.neuro.2018.02.017

Downie, Adriana, Alan Crosky, and Paul Munroe. 2009. "Physical Properties of Biochar." In *Biochar for Environmental Management*, 1st Editio, 20. Taylor and Francis group.

Erdinç, Neşe, Sinem Göktürk, and Melda Tunçay. 2010. "A Study on the Adsorption Characteristics of an Amphiphilic Phenothiazine Drug on Activated Charcoal in the Presence of Surfactants." *Colloids and Surfaces B: Biointerfaces* 75 (1): 194–203. doi:10.1016/j.colsurfb.2009.08.031

Frankel, Mathew L., Tazul I. Bhuiyan, Andrei Veksha, Marc A. Demeter, David B. Layzell, Robert J. Helleur, Josephine M. Hill, and Raymond J. Turner. 2016. "Removal and Biodegradation of Naphthenic Acids by Biochar and Attached Environmental Biofilms in the Presence of Co-Contaminating Metals." *Bioresource Technology* 216. The Author(s): 352–361. doi:10.1016/j.biortech.2016.05.084

Graetz, R. D., and J. O. Skjemstad. 2003. "The Charcoal Sink of Biomass Burning on the Australian Continent." *CSIRO Atmospheric Research* 64: 1–69.

Gupta, Rakesh Kumar, Mukul Dubey, Parashu Kharel, Zhengrong Gu, and Qi Hua Fan. 2015. "Biochar Activated by Oxygen Plasma for Supercapacitors." *Journal of Power Sources* 274. Elsevier B.V: 1300–1305. doi:10.1016/j.jpowsour.2014.10.169

Huang, Zheng Hong, Fangzhen Zhang, Ming Xi Wang, Ruitao Lv, and Feiyu Kang. 2012. "Growth of Carbon Nanotubes on Low-Cost Bamboo Charcoal for Pb(II) Removal from Aqueous Solution." *Chemical Engineering Journal* 184. Elsevier B.V.: 193–197. doi:10.1016/j.cej.2012.01.029

Hu, Bin, Yuhong Tang, Xinting Wang, Lieshan Wu, Jiajing Nong, Xiaona Yang, and Jianqiang Guo. 2021. "Cobalt-Gadolinium Modified Biochar as an Adsorbent for Antibiotics in Single and Binary Systems." *Microchemical Journal* 166 (April). Elsevier B.V.: 106235. doi:10.1016/j.microc.2021.106235

Inyang, Mandu, Bin Gao, Andrew Zimmerman, Ming Zhang, and Hao Chen. 2014. "Synthesis, Characterization, and Dye Sorption Ability of Carbon Nanotube-Biochar Nanocomposites." *Chemical Engineering Journal* 236. Elsevier B.V.: 39–46. doi:10.1016/j.cej.2013.09.074

Inyang, Mandu, Bin Gao, Andrew Zimmerman, Yanmei Zhou, and Xinde Cao. 2015. "Sorption and Cosorption of Lead and Sulfapyridine on Carbon Nanotube-Modified Biochars." *Environmental Science and Pollution Research* 22 (3): 1868–1876. doi:10.1007/s11356–014–2740-z

Jiang, Junhua, Lei Zhang, Xinying Wang, Nancy Holm, Kishore Rajagopalan, Fanglin Chen, and Shuguo Ma. 2013. "Highly Ordered Macroporous Woody Biochar with Ultra-High Carbon Content as Supercapacitor Electrodes." *Electrochimica Acta* 113. Elsevier Ltd: 481–489. doi:10.1016/j.electacta.2013.09.121

Kazemi Shariat Panahi, Hamed, Mona Dehhaghi, Yong Sik Ok, Abdul Sattar Nizami, Benyamin Khoshnevisan, Solange I. Mussatto, Mortaza Aghbashlo, Meisam Tabatabaei, and Su Shiung Lam. 2020. "A Comprehensive Review of Engineered Biochar: Production, Characteristics, and Environmental Applications." *Journal of Cleaner Production* 270. Elsevier Ltd: 122462. doi:10.1016/j.jclepro.2020.122462

Keerthanan, S., Amit Bhatnagar, Kushani Mahatantila, Chamila Jayasinghe, Yong Sik Ok, and Meththika Vithanage. 2020. "Engineered Tea-Waste Biochar for the Removal of Caffeine, a Model Compound in Pharmaceuticals and Personal Care Products (PPCPs), from Aqueous Media." *Environmental Technology & Innovation* 19: 100847. doi:10.1016/j.eti.2020.100847

Khan, Nawaz, Pankaj Chowdhary, Edgard Gnansounou, and Preeti Chaturvedi. 2021. "Biochar and Environmental Sustainability: Emerging Trends and Techno-Economic Perspectives." *Bioresource Technology* 332: 125102. doi:10.1016/j.biortech.2021.125102

Knicker, Heike, Kai Uwe Totsche, Gonzalo Almendros, and Francisco J. González-Vila. 2005. "Condensation Degree of Burnt Peat and Plant Residues and the Reliability of Solid-State VACP MAS 13C NMR Spectra Obtained from Pyrogenic Humic Material." *Organic Geochemistry* 36 (10): 1359–1377. doi:10.1016/j.orggeochem.2005.06.006

Krasucka, Patrycja, Bo Pan, Yong Sik Ok, Dinesh Mohan, Binoy Sarkar, and Patryk Oleszczuk. 2021. "Engineered Biochar—A Sustainable Solution for the Removal of Antibiotics from Water." *Chemical Engineering Journal* 405 (August 2020). Elsevier: 126926. doi:10.1016/j.cej.2020.126926

Kumar, M., B. B. Verma, and R. C. Gupta. 1999. "Mechanical Properties of Acacia and Eucalyptus Wood Chars." *Energy Sources* 21 (8): 675–685. doi:10.1080/00908319950014425

Leng, Lijian, Siyu Xu, Renfeng Liu, Ting Yu, Ximeng Zhuo, Songqi Leng, Qin Xiong, and Huajun Huang. 2019. "Nitrogen Containing Functional Groups of Biochar: An Overview." *Bioresource Technology* 298 (August 2019). Elsevier: 122286. doi:10.1016/j.biortech.2019.122286

Li, Helian, Ronghui Qu, Chao Li, Weilin Guo, Xuemei Han, Fang He, Yibing Ma, and Baoshan Xing. 2014. "Selective Removal of Polycyclic Aromatic Hydrocarbons (PAHs) from Soil Washing Effluents Using Biochars Produced at Different Pyrolytic Temperatures." *Bioresource Technology* 163. Elsevier Ltd: 193–198. doi:10.1016/j.biortech.2014.04.042

Lima, Isabel M, Akwasi A Boateng, and Kjell T Klasson. 2010. "Physicochemical and Adsorptive Properties of Fast-Pyrolysis Bio-Chars and Their Steam Activated Counterparts." *Journal of Chemical Technology & Biotechnology* 85 (11). John Wiley & Sons, Ltd: 1515–1521. doi:10.1002/jctb.2461

Li, Simeng, Scott Harris, Aavudai Anandhi, and Gang Chen. 2019. "Predicting Biochar Properties and Functions Based on Feedstock and Pyrolysis Temperature: A Review and Data Syntheses." *Journal of Cleaner Production* 215. Elsevier Ltd: 890–902. doi:10.1016/j.jclepro.2019.01.106

Liu, Pei, Wu Jun Liu, Hong Jiang, Jie Jie Chen, Wen Wei Li, and Han Qing Yu. 2012. "Modification of Bio-Char Derived from Fast Pyrolysis of Biomass and Its Application in Removal of Tetracycline from Aqueous Solution." *Bioresource Technology* 121. Elsevier Ltd: 235–240. doi:10.1016/j.biortech.2012.06.085

Li, Xiangping, Chuanbin Wang, Jianguang Zhang, Juping Liu, Bin Liu, and Guanyi Chen. 2020. "Preparation and Application of Magnetic Biochar in Water Treatment: A Critical Review." *Science of The Total Environment* 711: 134847. doi:10.1016/j.scitotenv.2019.134847

Lussier, M. G., Z. Zhang, and Dennis J. Miller. 1998. "Characterizing Rate Inhibition in Steam/Hydrogen Gasification via Analysis of Adsorbed Hydrogen." *Carbon* 36 (9): 1361–1369. doi:10.1016/S0008–6223(98)00123–7

Ma, Yongfei, Ping Li, Lie Yang, Li Wu, Liuyang He, Feng Gao, Xuebin Qi, and Zulin Zhang. 2020. "Iron/Zinc and Phosphoric Acid Modified Sludge Biochar as an Efficient Adsorbent for Fluoroquinolones Antibiotics Removal." *Ecotoxicology and Environmental Safety* 196 (April). doi:10.1016/j.ecoenv.2020.110550

Pandit, Naba Raj, Hans Peter Schmidt, Jan Mulder, Sarah E. Hale, Olivier Husson, and Gerard Cornelissen. 2020. "Nutrient Effect of Various Composting Methods with and without Biochar on Soil Fertility and Maize Growth." *Archives of Agronomy and Soil Science* 66 (2). Taylor & Francis: 250–265. doi:10.108 0/03650340.2019.1610168

Qian, Kezhen, Ajay Kumar, Hailin Zhang, Danielle Bellmer, and Raymond Huhnke. 2015. "Recent Advances in Utilization of Biochar." *Renewable and Sustainable Energy Reviews* 42. Elsevier: 1055–1064. doi:10.1016/j.rser.2014.10.074

Rajapaksha, Anushka Upamali, Meththika Vithanage, Mahtab Ahmad, Dong Cheol Seo, Ju Sik Cho, Sung Eun Lee, Sang Soo Lee, and Yong Sik Ok. 2015. "Enhanced Sulfamethazine Removal by Steam-Activated Invasive Plant-Derived Biochar." *Journal of Hazardous Materials* 290. Elsevier B.V.: 43–50. doi:10.1016/j.jhazmat.2015.02.046

Rajapaksha, Anushka Upamali, Season S. Chen, Daniel C.W. Tsang, Ming Zhang, Meththika Vithanage, Sanchita Mandal, Bin Gao, Nanthi S. Bolan, and Yong Sik Ok. 2016. "Engineered/Designer Biochar for Contaminant Removal/Immobilization from Soil and Water: Potential and Implication of Biochar Modification." *Chemosphere* 148. Elsevier Ltd: 276–291. doi:10.1016/j.chemosphere.2016.01.043

Rimstidt, Donald D., and David J. Vaughan. 2003. "Pyrite Oxidation: A State-of-the-Art Assessment of the Reaction Mechanism." *Geochimica et Cosmochimica Acta* 67 (5): 873–880. doi:10.1016/S0016–7037(02)01165-1

Ruan, Xiuxiu, Yuqing Sun, Weimeng Du, Yuyuan Tang, Qiang Liu, Zhanying Zhang, William Doherty, Ray L. Frost, Guangren Qian, and Daniel C.W. Tsang. 2019. "Formation, Characteristics, and Applications of Environmentally Persistent Free Radicals in Biochars: A Review." *Bioresource Technology* 281 (December 2018). Elsevier: 457–468. doi:10.1016/j.biortech.2019.02.105

Saleh, Mahmoud M. 2006. "On the Removal of Cationic Surfactants from Dilute Streams by Granular Charcoal." *Water Research* 40 (5): 1052–1060. doi:10.1016/j.watres.2005.12.032

Shaaban, A., Sian Meng Se, M. F. Dimin, Jariah M. Juoi, Mohd Haizal Mohd Husin, and Nona Merry M. Mitan. 2014. "Influence of Heating Temperature and Holding Time on Biochars Derived from Rubber Wood Sawdust via Slow Pyrolysis." *Journal of Analytical and Applied Pyrolysis* 107. Elsevier B.V.: 31–39. doi:10.1016/j.jaap.2014.01.021

Shaheen, Sabry M., Nabeel Khan Niazi, Noha E. E Hassan, Irshad Bibi, Hailong Wang, Daniel C W Tsang, Yong Sik Ok, Nanthi Bolan, and Jörg Rinklebe. 2019. "Wood-Based Biochar for the Removal of Potentially Toxic Elements in Water and Wastewater: A Critical Review." *International Materials Reviews* 64 (4). Taylor & Francis: 216–247. doi:10.1080/09506608.2018.1473096

Song, Zhengguo, Fei Lian, Zhihong Yu, Lingyan Zhu, Baoshan Xing, and Weiwen Qiu. 2014. "Synthesis and Characterization of a Novel MnOx-Loaded Biochar and Its Adsorption Properties for Cu2+ in Aqueous Solution." *Chemical Engineering Journal* 242. Elsevier B.V.: 36–42. doi:10.1016/j.cej.2013.12.061

Sontakke, Ankush D., and Mihir K. Purkait. 2021. "A Brief Review on Graphene Oxide Nanoscrolls: Structure, Synthesis, Characterization and Scope of Applications." *Chemical Engineering Journal* 420 (P1). Elsevier B.V.: 129914. doi:10.1016/j.cej.2021.129914

Tran, Ngoc Han, Martin Reinhard, Eakalak Khan, Huiting Chen, Viet Tung Nguyen, Yiwen Li, Shin Giek Goh, Q. B. Nguyen, Nazanin Saeidi, and Karina Yew Hoong Gin. 2019. "Emerging Contaminants in Wastewater, Stormwater Runoff, and Surface Water: Application as Chemical Markers for Diffuse Sources." *Science of the Total Environment* 676. Elsevier B.V.: 252–267. doi:10.1016/j.scitotenv.2019.04.160

Uchimiya, Minori, Desmond I. Bannon, and Lynda H. Wartelle. 2012. "Retention of Heavy Metals by Carboxyl Functional Groups of Biochars in Small Arms Range Soil." *Journal of Agricultural and Food Chemistry* 60 (7): 1798–809. doi:10.1021/jf2047898

Wang, Rong Zhong, Dan Lian Huang, Yun Guo Liu, Chen Zhang, Cui Lai, Xin Wang, Guang Ming Zeng, Qing Zhang, Xiao Min Gong, and Piao Xu. 2020. "Synergistic Removal of Copper and Tetracycline from Aqueous Solution by Steam-Activated Bamboo-Derived Biochar." *Journal of Hazardous Materials* 384 (March). Elsevier: 121470. doi:10.1016/j.jhazmat.2019.121470

Weber, Kathrin, and Peter Quicker. 2018. "Properties of Biochar." *Fuel* 217 (December 2017). Elsevier: 240–261. doi:10.1016/j.fuel.2017.12.054

Xiang, Yujia, Zhangyi Xu, Yaoyu Zhou, Yuyi Wei, Xingyu Long, Yangzhou He, Dan Zhi, Jian Yang, and Lin Luo. 2019. "A Sustainable Ferromanganese Biochar Adsorbent for Effective Levofloxacin Removal from Aqueous Medium." *Chemosphere* 237. Elsevier Ltd: 124464. doi:10.1016/j.chemosphere.2019. 124464

Xiong, Zhang, Zhang Shihong, Yang Haiping, Shi Tao, Chen Yingquan, and Chen Hanping. 2013. "Influence of NH3/CO2 Modification on the Characteristic of Biochar and the CO2 Capture." *Bioenergy Research* 6 (4): 1147–1153. doi:10.1007/s12155-013-9304-9

Yang, Guang Xi, and Hong Jiang. 2014. "Amino Modification of Biochar for Enhanced Adsorption of Copper Ions from Synthetic Wastewater." *Water Research* 48 (1). Elsevier Ltd: 396–405. doi:10.1016/j. watres.2013.09.050

Yao, Ying, Bin Gao, June Fang, Ming Zhang, Hao Chen, Yanmei Zhou, Anne Elise Creamer, Yining Sun, and Liuyan Yang. 2014. "Characterization and Environmental Applications of Clay-Biochar Composites." *Chemical Engineering Journal* 242. Elsevier B.V.: 136–143. doi:10.1016/j.cej.2013.12.062

Zhang, Jinghuan, Mingyi Lu, Jun Wan, Yuhuan Sun, Huixia Lan, and Xiaoyan Deng. 2018. "Effects of PH, Dissolved Humic Acid and Cu2+ on the Adsorption of Norfloxacin on Montmorillonite-Biochar Composite Derived from Wheat Straw." *Biochemical Engineering Journal* 130. Elsevier B.V.: 104–112. doi:10.1016/j.bej.2017.11.018

Zhang, Jiangnan, Zheng Hong Huang, Ruitao Lv, Quan Hong Yang, and Feiyu Kang. 2009. "Effect of Growing CNTs onto Bamboo Charcoals on Adsorption of Copper Ions in Aqueous Solution." *Langmuir* 25 (1): 269–274. doi:10.1021/la802365w

Zhang, Ming, Bin Gao, Sima Varnoosfaderani, Arthur Hebard, Ying Yao, and Mandu Inyang. 2013. "Preparation and Characterization of a Novel Magnetic Biochar for Arsenic Removal." *Bioresource Technology* 130: 457–462. doi:10.1016/j.biortech.2012.11.132

Zhang, Ming, Bin Gao, Ying Yao, Yingwen Xue, and Mandu Inyang. 2012. "Synthesis of Porous MgO-Biochar Nanocomposites for Removal of Phosphate and Nitrate from Aqueous Solutions." *Chemical Engineering Journal* 210. Elsevier B.V.: 26–32. doi:10.1016/j.cej.2012.08.052

Zhang, Tengyan, Walter P. Walawender, L. T. Fan, Maohong Fan, Daren Daugaard, and R. C. Brown. 2004. "Preparation of Activated Carbon from Forest and Agricultural Residues through CO2 Activation." *Chemical Engineering Journal* 105 (1–2): 53–59. doi:10.1016/j.cej.2004.06.011

Zhang, X. N., G. Y. Mao, Y. B. Jiao, Y. Shang, and R. P. Han. 2014. "Adsorption of Anionic Dye on Magnesium Hydroxide-Coated Pyrolytic Bio-Char and Reuse by Microwave Irradiation." *International Journal of Environmental Science and Technology* 11 (5): 1439–1448. doi:10.1007/s13762-013-0338-5

Zhao, Ru Song, Yan Long Liu, Xiang Feng Chen, Jin Peng Yuan, Ai Ying Bai, and Jia Bin Zhou. 2013. "Preconcentration and Determination of Polybrominated Diphenyl Ethers in Environmental Water Samples by Solid-Phase Microextraction with Fe3O4-Coated Bamboo Charcoal Fibers Prior to Gas Chromatography-Mass Spectrometry." *Analytica Chimica Acta* 769. Elsevier B.V.: 65–71. doi:10.1016/j. aca.2013.01.027

Zhou, Xiaohui, Jianjun Zhou, Yaochi Liu, Jing Guo, Jialin Ren, and Fang Zhou. 2018. "Preparation of Iminodiacetic Acid-Modified Magnetic Biochar by Carbonization, Magnetization and Functional

Modification for Cd(II) Removal in Water." *Fuel* 233 (March). Elsevier: 469–479. doi:10.1016/j.fuel.2018.06.075

Zhou, Yanmei, Bin Gao, Andrew R. Zimmerman, June Fang, Yining Sun, and Xinde Cao. 2013. "Sorption of Heavy Metals on Chitosan-Modified Biochars and Its Biological Effects." *Chemical Engineering Journal* 231. Elsevier B.V.: 512–518. doi:10.1016/j.cej.2013.07.036

Zhu, Xiaoxiao, Chunyan Li, Jianfa Li, Bin Xie, Jinhong Lü, and Yimin Li. 2018. "Thermal Treatment of Biochar in the Air/Nitrogen Atmosphere for Developed Mesoporosity and Enhanced Adsorption to Tetracycline." *Bioresource Technology* 263 (March). Elsevier: 475–482. doi:10.1016/j.biortech.2018.05.041

13 Recent Advances in Dye Removal Technologies by Designer Biochar

Prangan Duarah, Pranjal P. Das,
Mukesh Sharma, and Mihir K. Purkait

CONTENTS

13.1 INTRODUCTION

The growth of industries without special regard for wastewater treatment has produced large-scale water and soil contaminations. The textile industries are regarded as a major pollutant for reservoirs of ground and surface water as they use over 8000 chemicals and a significant quantity of water. Several studies revealed that for the manufacture of around 8000 kg of textiles, the sector consumes approximately 1.6 million gallons of water a day (Behera et al. 2021; Purkait et al. 2005). During wet processing, the textiles sector uses significant volumes of water and thus creates a considerable amount of wastewater including a vast number of diluted dyestuffs and other products, such as salts, heavy metals, parasites, fertilizers, and emulsifiers. According to some studies, about 20,000 tonnes of toxic textile dyes are introduced to the environment every year due to an inefficient dyeing procedure in which not all of the colors are adequately absorbed into the dying material (Patel 2018).

The dye compounds contain lipophilic architectures with azobenzene- or anthraquinone-based functional groups. This causes several diseases that may transmit to the human body by inhalation, ingestion, or even skin contact due to the nonbiodegradable and toxic nature of these compounds.

DOI: 10.1201/9781003203438-13

There have been reports showing increase in lungs and nasal carcinoma. Others have identified azo-dyes-specific health risks such as carcinoma of the larynx and lung parenchyma. There are also reports disclosing that insoluble dyes and surfactants have carcinogenic and mutagenic consequences. According to some studies, inhaling azo amine dyes as dust or by skin contact has been linked to cardiovascular and respiratory illnesses, urinary tract and lung cancer, detrimental consequences on red blood cells, and allergic skin responses (Tounsadi et al. 2020). Therefore, removal of these contaminants from the aquatic environment is a major environmental concern. Current water treatment techniques include chemical precipitation, ion exchange, and adsorption to remove pollutants from water. In addition, other techniques that are effective for the removal of impurities are sophisticated water treatment technologies, including advanced oxidation, electrocoagulation, nanofiltration, and reverse osmosis. Most of these procedures are highly expensive and often cause secondary chemical pollutants. Therefore, more effective, eco-friendly, low-cost processes have to be employed to remove the new class of developing pollutants with easily available material. Adsorption remains the most suitable technique due to its ease of design, low cost, and efficiency in removing pollutants; nevertheless, the selection of a good adsorbent is a critical aspect of the adsorption process (Haldar, Duarah, and Purkait 2020; Bedia et al. 2018; Duarah, Haldar, and Purkait 2020).

Biochar is widely utilized as a new adsorbent for the elimination of organic and inorganic contaminants from the environment. Biochar is a solid material with a high level of carbon content, generated in the presence of a limited amount of oxygen by thermochemical transmutation of biomass at temperatures below 700°C. Biochar has a structure with a large number of pores, a large specific surface area, and a substantial concentration of surface functional groups as well as inorganic elements, which all contribute to the effective adsorption of many water contaminants (Haeldermans et al. 2020). Biochar has therefore been acclaimed in the water treatment sector as a potential alternative adsorbent. However, the drawbacks of certain pristine biochars, such as poor aromaticity and limited sorption capacity for pollutants, have led the path for biochar to be improved by creating design biochar or biochar-based composites, which involve employing various advanced materials as supporting material to attach contaminants owing to its improved surface area, availability of the functional groups, and large active sites.

Over the last few years, much emphasis has been paid to the application of design biochar for dye removal from wastewater. In view of that, a brief review of the chemistry of several kinds of dyes prevalent in polluted water is featured in this chapter. Several aspects of the biochar synthesis and activation process for the elimination of these contaminants are explicitly discussed. An inclusive understanding of the mechanism of different interactions of biochar with different types of dyes found in wastewater is presented. Finally, different views and problems associated with process scale-up are fully described with an eye toward future advancements.

13.2 BIOCHAR SYNTHESIS AND ACTIVATION TECHNIQUES

As previously stated, different biochar-based compounds have been investigated for use in a variety of sectors such as drug delivery, wastewater treatment, catalysis, agriculture, and so on. As a result, many techniques for producing biochar from biomass are being investigated. This section highlights several ways of producing biochar.

13.2.1 PYROLYSIS

The most prevalent technique of biochar production is conventional pyrolysis (CPS). Biochar is manufactured in an inert environment by externally heating biomass. All the elements impacting the quality of biochar are the kind of biomass, the pyrolysis, and the heating rate (Zhang et al. 2020). On the basis of the residence duration, heating rate, and mode of heating, pyrolysis technology may be classified as slow or rapid and as carbonation or microwave assisted.

During slow pyrolysis, the temperature range is 400–600°C, and the residence duration is high. This technique produces 30–60% biochar and has a surface area of around 400 m^2/g. Fast pyrolysis is often performed at 450–600°C, with a higher heating rate (200°C/min) and a shorter residence period than slow pyrolysis. Because this technique occurs in a short stretch of time, the impacts of heat and mass transfer, dynamics, and other variables have a substantial influence on product yield and process efficiency. The poor biochar yield (10–20%) is a result of the fast pyrolysis operating parameters. The short residence time may contribute to the poor calorific value and high oxygen content of biochar produced by fast pyrolysis. There are several advantages of using flash carbonization over traditional carbonization, including a high biochar yield (28–32%) and a rapid reaction time (30 min). The raw materials are initially loaded into a packed bed reactor in the flash carbonization process. With the help of air, it is then compressed to 1–2 bar and then heated by a flame at its bottom. The flame is propelled higher by air flowing downstream, which lasts for less than 30 min (Li et al. 2020).

Microwave processing is a relatively new pyrolysis technique. The difference is mainly attributable to the differing heating systems between the CPS and the microwave-assisted pyrolyze (MWPs). After all, MWP is based on dielectric heating, which transfers electromagnetic energy through various processes into thermal energy. Heat is created by electrofield components such as depolarization, ionic migration, and polarization with Maxwell–Wagner, on the one hand and by magnetic field components such as loss of eddy current, loss of resonance, and loss of hysteresis on the other hand. Due to heating mechanism differences, MWP offers numerous benefits over CPS. First, with MWP, heat production is generally quicker and more selective. Second, microwave heating can decrease pretreatments such as desiccation, as the polar character of water can efficiently be removed during processing. Finally, immediate heating ensures quick and simple start-up and shutdown. The disadvantages are the lack of flexibility of input materials and unpredictability in the economic scale-up as compared to CPS (Haeldermans et al. 2020).

The temperature and time of pyrolysis can influence the characteristics of biochar. Increased pyrolysis temperature lowers pH, fixed carbon, and the amount of the core functional groups in biochar, whereas increased temperature reduces overall yield and the numbers of functional groups, notably acidic ones, in biochar. The pyrolysis residence time, on the other hand, has no discernible influence on the pH or production of the biochar (Sun et al. 2017). Other research found that increasing the pyrolysis temperature from 350 to 900°C resulted in a substantial increase in the pH of biochar by 35.41%, confirming the release of hydrogen ions due to the thermal cracking of biomass (Zhang et al. 2017). The biochar produced at higher temperatures, according to Sun et al. (2012), had a greater C/H ratio, larger aromatic ring clusters, and a larger surface area than biochar synthesized at lower temperatures (Sun et al. 2012).

13.2.2 HYDROTHERMAL CARBONIZATION

Hydrothermal carbonization (HTC) is an exothermic technique that decreases the level of oxygen and hydrogen in the feedstock (as measured by the molecular oxygen-to-carbon and hydrogen-to-carbon ratios). Biomass and water solution is heated for a couple of hours at a saturation pressure maintaining a temperature of 180–220°C. The solid phase HTC-biochar may be removed from the water before the completion of the procedure. In the solid phase (HTC-biochar), the bulk of carbon input is located at 75–80%; around 15–20% is liquified, and about 5% is transformed to gas (mostly CO_2) (Oliveira, Blöhse, and Ramke 2013). Instead of using chemical reagents, water is used as the carbonization medium in HTC, which is a low-cost and environmentally benign alternative to standard high-temperature pyrolysis methods. The oxygen-containing functional groups and microporous or mesoporous structure of biochar generated by the HTC of biomass make it ideal for adsorption of contaminants from wastewater (Wu et al. 2020).

13.2.3 ACTIVATION OF BIOCHAR

Physical and/or chemical modifications are commonly employed to enhance the surface properties of biochar following pyrolysis. Figure 13.1 illustrates routes for the synthesis and activation process of biochar. Chemical modification techniques have always been the most widely used form of modification, including acidic/alkaline modification of biochar or modification with metal salts. Physicochemical characteristics of biochar that have been changed using these approaches have been reported to be substantially improved. It is possible to enhance the porous structure and oxygenic functional groups in biochar without adding any contaminants. For biochar characteristics enhancement, the most common physical techniques include ball milling, magnetic alterations, gas or steam activation, and microwave. Due to the presence of high specific surface area with an excellent number of mesopores and micropores, the physically modified biochar shows greater adsorption efficiency for organic contaminants, nutrients, and heavy metal ions. The characteristics of biochar, such as active area of the surface for inorganic and organic ions adsorption, may be manipulated by physical ball milling. Through an optimized planetary ball mill operation, the surface area of corn-stover-based biochar may be increased by more than 3.2 times, to reach 194 m^2/g (Peterson et al. 2012). Furthermore, nanosized biochar may be generated via ball milling, with performance equivalent to activated carbon and carbon nanotube (CNT) in terms of organic

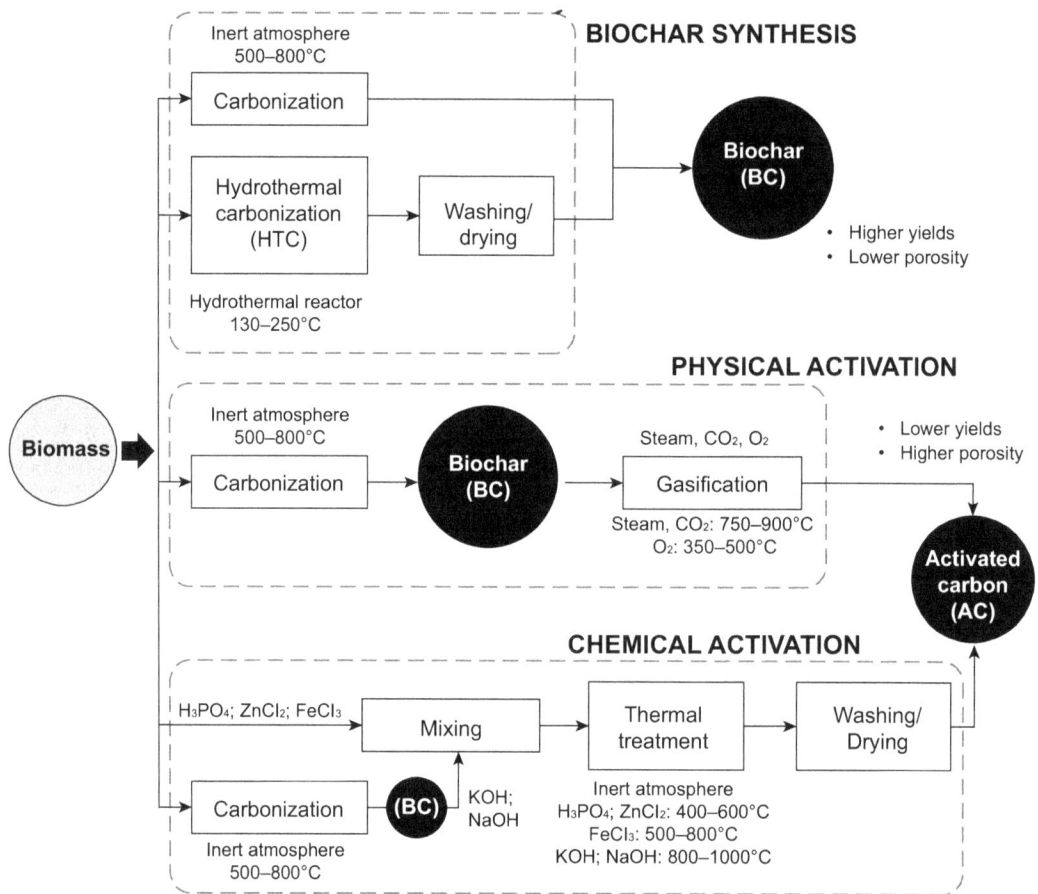

FIGURE 13.1   Biochar synthesis and activation route from biomass.

Source: Reproduced with permission from Bedia et al. (2018).

and inorganic pollutants elimination (Shan et al. 2016, Sontakke and Purkait 2021). When activated by gas/steam, such as air, water vapor, CO_2, etc., the advantageous qualities of biochar can be practically doubled. This process produces biochar that retains nutrients better, which might be absorbed by plants. This enhancement might be due to the production of activated carbon, a process (700–1100°C) that induces porosity formation in biochar, increases surface area, and significantly increases surface reactivity (Wang, Gao, and Fang 2017). Microwave activation of biochar is another promising physical method for improving its properties. Electromagnetic base radiation, such as microwave, might be administered effectively and controlled easily. At low reaction temperatures of about 200–300°C, microwave pyrolysis may produce biochar yields of more than 60% by weight. Because of the dipole orientation that happens when polar molecules are exposed to microwave radiation, the electrons around atoms or nuclei are displaced trillions of times each second. Molecular rotation produces heat as molecules rub against one another (Yek et al. 2020).

To activate biomass/biochar, chemicals and inert gas are used in chemical activation techniques. Acids or bases might be employed to oxidize biochar, enhancing its micropores, surface area, cation exchange capacity, and functional groups. This form of modification often employs biochar with a large surface area as a scaffold to assist the deposition of components. As a result of such a technique, new biochar surface functional groups are produced. Alkali substances such as NaOH and KOH can activate raw biochar. The biochar is heated under inert gas flow at temperatures ranging from 500 to 1000°C after being impregnated with KOH or NaOH. The porosity of activated biochar may be tuned by adjusting the amount of alkaline and the impregnation period. In general, a greater alkaline concentration can result in increased porosity, as well as increased surface area and pore volume (Liu, Jiang, and Yu 2015). During biomass pyrolysis, in situ catalytic pore creation is another approach for modifying the pore structure of biochar. Activation via chemical route can be done in a single step by thermally decomposing the raw material and then using chemical reagents to activate it. Phosphoric acid (H_3PO_4), zinc chloride ($ZnCl_2$), and potassium hydroxide/carbonate (KOH/K_2CO_3) are the most often utilized compounds. The high dehydration capability of $ZnCl_2$ following heat treatment, which may substantially enhance the development of open pores, was discovered to be the major cause of $ZnCl_2$ activation (Uçar et al. 2009). The acid medication process is popular for producing biochar for the catalysis process. These groups ($-SO_3H$) make up a large portion of solid acidic compounds. Utilizing concentrated H_2SO_4 or its derivatives (such as oleum and chlorosulfonic acid), biochar is sulfonated on its surface to produce biochar-based solid acids (Nakajima and Hara 2012). Further, many researchers have reported that acid treatment causes additional carboxyl groups on biochar, which allows for greater adsorption of Cu, Pb, and Zn as well as Cd, sulfamethazine, and oxytetracyclines (Kazemi Shariat Panahi et al. 2020).

13.3 BIOCHAR AS A PROMISING ADSORBENT

To eliminate inorganic and organic contaminants from the environment, biochar has been widely employed as a new adsorbent. Researchers discovered that biochar may degrade organic and inorganic contaminants by generating active oxygen from the free radicals that remain in the biochar. Pristine biochar, on the other hand, was unable to properly eliminate high pollutant concentrations in an aqueous solution. Furthermore, the use of ground biochar (i.e., biochar created by the ball milling process) complicates biochar separation considering the tiny size of the particles and density. This has led to the development of engineered biochar with enhanced surface characteristics. A variety of pollutants may be removed using designed biochar as a result of its exceptional surface area, enhanced functional groups, and expanded surface-active sites. Yoon et al. (2017) demonstrated that biochar prepared by paper mill sludge pyrolysis under CO_2 atmosphere exhibit excellent absorption capability for Cd(II) and As(V) (Yoon et al. 2017). Contaminant adsorption activity of biochar is mostly determined by its number of surface-active functional groups available, availability of surface area, and cation exchange capacity. For biochar to interact with contaminants, mineral components and functional groups are key structures. The major physical characteristics that impact biochar's capacity to absorb heavy metals are porosity and specific surface area. The adsorption process is greatly influenced by pore size. Biochar, for example, cannot capture large adsorbates due to its tiny pore size, regardless of its charge or polarity. Using hydrothermal carbonization, Regmi et al. (2012) synthesized switchgrass-derived biochar and studied a cold activation method utilizing KOH at room temperature to improve the porosity structure and sorption characteristics of the produced biochar. Modified biochar was compared to unmodified biochar in terms of its ability to adsorb Cd^{2+} and Cu^{2+} in water. According to the researchers, the specific surface area of modified biochar was 2.4 times more than that of pristine biochar (Regmi et al. 2012). Using hickory wood and $KMnO_4$ solution, Wang et al. (2015) synthesized modified biochar by pyrolyzing the mixture at 600°C. The produced biochar was then utilized to eliminate heavy metal ions. As a result of the increased quantity of -OH and -COOH, modified biochar was shown to have a greater capacity for removing Pb^{2+}, Cu^{2+}, and Cd^{2+} than nonmodified biochar (Wang et al. 2015). Along with eliminating inorganic contaminants, biochar has the ability to remove harmful organic chemicals, such as dyes and antibiotics, and pesticides. Generally speaking, the adsorption of organic molecules on biochar is dependent on the kind of contaminants and the surface characteristics; pore filling, electrostatic attraction, hydrophobic interactions, electrostatic cation exchange, and partitioned uncarbonized fraction of biochar are some of the processes for organic pollutants removal that may be categorized (Kazemi Shariat Panahi et al. 2020).

13.4　CLASSIFICATION OF DYE

The use of dyes is widespread in everyday life, including painting, dyeing paper, skin, and clothes. Unsaturated and aromatic chemical compounds make up the majority of the dyes. Chromophores are unsaturated chemical groups that give them their color. Table 13.1 summarizes the name of various types of dyes used in textile industries. The classification of dyes can be made according to their source materials, chemical composition, or their ironic charge.

13.4.1　BASED ON SOURCE MATERIALS

The source of the dye is a very common way to classify them. Chemically, most dyes are complex organic compounds. Prior to azulene synthesis, dyes were created from pigments. Therefore, the

TABLE 13.1
Utilization of Different Types of Dyes in Textile Industries

Type of dye	Industrial products	Types of chemicals	pH range	Fixation (%)	Loss (%)	References
Acid	Nylon, paper, wool, silk, inks, and leather	Anthraquinone, nitro, triphenylmethane, azine, xanthene, and nitroso	2–7	80–95	20–5	Patel (2018)
Basic	Paper, polyacrylonitrile inks, modified nylon, and polyester	Diazacarbocyanine, xanthene, diazahemicyanine, cyanine, hemicyanine, azo, diphenylmethane, oxazine, triarylmethane,	–	95–100	5–0	Patel (2018)
Azo	Cotton, polyester cellulose acetate, and rayon	Azo, o-aminoazotoluene, Sudan azo	2–10	80–95	20–5	Selvaraj et al. (2021); Patel (2018)
Direct	Cotton, viscose, silk, rayon, and leather	Azo, stilbene, oxazine, and phthalocyanine,	7	70–95	30–5	Aspland (1991); Patel (2018)
Disperse	Polyester, acrylic, plastics, polyamide, and acetate	Azo, nitro, styryl, anthraquinone, and benzodifuranone	4.5–5.5	90–100	10–0	(Patel 2018; Aspland 1992)
Reactive	Cotton, silk, wool, and nylon	Azo, anthraquinone, phthalocyanine, formazan, oxazine, and basic	-	50–90	50–10	(Patel 2018)
Vat	Cotton, linen, viscose	Anthraquinone and indigoids	12–14	80–95	20–5	(Patel 2018)

dye can be classed as either natural or synthetic, depending on the source materials. Colorants have been utilized in clothing, food, ceramics, leather, and buildings for thousands of years. For painting, pigments acquired from colorful rocks or minerals were utilized, while dyes from animals and plants were used for dyeing. Natural dyes are some of the most prevalent dyes derived from natural sources. Anion is the colored portion of a molecule in most natural dyes. Positively charged natural dyes do exist, although they are extremely rare (Mani and Bharagava 2018).

13.4.2 BASED ON CHEMICAL STRUCTURE

Chemical-structure-based classification and more specifically their chromophore groupings are used to anticipate the chemical interactions between the dye and reducing agents, oxidizing agents, or any number of other chemicals. Azo dyes, indigo dyes, nitrosated dyes, anthraquinone dyes, phthalocyanine dyes, nitrated dye, polymethine dyes diphenylmethane dyes, and triphenylmethane dyes are the types of dye that are classified based on their chemical structure.

It is estimated that 70% of synthetic dyes are made from azo dyes, which is the largest family of synthetic dyes in existence. All three dyes are distinguished by the occurrence of one or more azo groups, notably monoazo, diazo, and triazo dyes. Azo dye is generated by treating fibers with a combination of diazoic and coupling components. The final color of the textile is dictated by the diazoic and coupling components employed in the process, making this method of dyeing textiles unique. After azo dyes, anthraquinone dyes are the most significant class. A quinone nucleus can be linked to the chromophore group, according to their general formulae, which are derived from anthracene. As the name suggests, indigo dyes are derived from indigo. Because of these hypochromic effects caused by the selenium, Sulphur, and oxygen homologues of indigo blue, the hues span from orange to turquoise. Similarly, the fluorescence of xanthene dyes is strong, with fluorescein being the most well-known molecule. The phthalocyanine dyes are a group of dyes made by reacting dicyanobenzene with Cu, Ni, Co, or Pt in the presence of a metal. The phthalocyanine nucleus provides the molecule with excellent lightfastness. Copper phthalocyanine is the most frequently used dye in this class due to its chemical stability. There are just a few dyes in the nitrated or nitrosated family, and they are rather old dyes. Because of their low price and simple chemical structure defined by the presence of an electron donor group ($-NO_2$) in orthogonal position, they are still in use today. The least important dye among these dyes are diphenylmethane and triphenylmethane dyes. Diphenylmethane is obtained from auramine, whereas the most important synthetic dyes such as fuchsin and malachite green come from triphenylmethane (Berradi et al. 2019).

13.4.3 BASED ON ION

Based on ions, dyes can be categorized as ionic and nonionic. The ionic dyes can be further categorized as anionic and cationic dyes. Similarly, nonionic dyes can be classified as vat and dispersed dyes. Ionic dyes are frequently utilized in acrylic, wool, nylon, and silk dyeing to create cationic colors. Cationic dyes have a variety of chemical structures based on aromatic substituents. As hazardous colorants, these dyes can induce allergic dermatitis, skin rashes, and cancer. Based on positive ions like hydrochloride or zinc chloride, these colors are known as basic dyes. To produce anionic dyes, a negative ion is required. Anionic dyes comprise a wide range of compounds from a wide range of dye classes, each with a unique structure, but all have water-soluble ionic substituents. From a chemical viewpoint, a high fraction of anionic dyes is reactive (Salleh et al. 2011).

13.5 APPLICATION OF BIOCHAR TO REMOVE DYES

Over the last several years, various significant applications of design/modified biochar have been studied for different kinds of dye removal. Table 13.2 illustrates various reported modified biochars for dye removal.

13.5.1 APPLICATION OF AZO DYE REMOVAL

Methylene blue (MB) and orange II (OR-II) have complex aromatic molecular structures in the textile printing and tinting industry, which can create issues in the renal, brain, and liver systems as well as in central neurological systems. High-efficiency technology is thus needed for the treatment of MB- and OR-II-polluted wastewater. Mu et al. (2021) used NaOH-modified tea biochar (NaOH/TRP) to eliminate MB and OR-II. The produced biochar was effective in removing MB and OR-II from water. In order to achieve the optimum elimination capacity of MB and OR-II, the starting pH was set at 10 and 2, respectively, with a contact duration of 60 min, a NaOH/TRP mass ratio of 10% w/w, and a pyrolysis temperature of 700°C. A 250 mg/L starting concentration yielded optimum elimination capacities of 91.68 and 105.27 mg/g for OR-II and MB dyes, respectively (Mu and Ma 2021). Similarly, by coating okra mucilage on biochar surfaces (BBC) and using it to extract MB from water, Nath et al. (2021) created a promising adsorbent. The efficiency of modified biochar was compared to that of pristine biochar. The comparison research indicated that Langmuir isotherm and pseudo-second-order kinetic models could represent MB adsorption on unmodified and modified biochar. The modified biochar had a Langmuir lead sorption capacity of 78.13 mg/g, which was around 284% more than pristine biochar.

Liu, Li, and Singh (2021) modified biochar with different oxidation number manganese compounds ($KMnO_4$, $MnSO_4$, and MnO_2). Biochar loaded with MnO_2 (MnO_2/WLB) displayed the optimum adsorption efficiency. Compared to unmodified biochar, the highest adsorption rate of 248.96 mg/g and the removal efficiency of 99.73% were shown by the modified biochar (Liu, Li, and Singh 2021). Further, it was reported that with increasing pH, the removal efficiency and adsorption capacity improved. MB had poor clearance effectiveness of 77% at pH 2. The absorbability of the MB dye achieved 100% when the pH hit 11, and the highest elimination rate was 248.96 mg/g. Similarly, the influence of initial concentration on the adsorption of MB is illustrated in Figure 13.2A. It can be observed that removal efficiency decreased with an increase in the initial concentration. However, a reverse trend is observed with the adsorption capacity. A study by Cuong Nguyen et al. (2021) examined the adsorption of methyl orange (MO) dye via three biochars made from agro-waste and invading plants. Wattle bark-based biochar (WBBC) and mimosa-based biochar (MBBC) had the highest specific surface area of 393.15 m²/g and 285.53 m²/g, respectively, whereas CHBC had the lowest specific surface area (2.62m²/g). When the pH is between 2 and 7, the adsorption effectiveness of MO is constant (95–96%), but it degrades at pH 7–12. The adsorption efficiency of MO between 0 and 30 min was >82%, and by 120 min, the adsorption equilibrium had been reached. The MBBC had a maximum adsorption capacity of 12.3 mg/g (Cuong Nguyen et al. 2021). After freeze-drying lotus root biochar, Hou et al. (2021) produced N-doped biochar with surface areas up to 694 m²/g and partial mesopores for MO removal from the aquatic environment. Maximum elimination capability of up to 449 mg/g has been recorded. The adsorption process is reported to be a physisorption process that allows the adsorption of MO in the pores and on the surface of biochar (Hou et al. 2021).

Silk, wool, and nylon fibers can be dyed with eriochrome black T (EBT), an ionic azo dye. The presence of sulfonate groups in the EBT structure renders it extremely harmful to the ecosystem because of its chemical composition. Toxicologically, EBT can cause cancer and mutagenesis. It also has long-term, catastrophic impacts on the ecosystem. Additionally, exposure to EBT dye is harmful and can lead to serious health issues. CuFe-layered double hydroxide composites (B-CuFe) made from biochar were developed by Zubair et al. (2021) and used as adsorbents to remove EBT dye from the water phase. The highest EBT adsorption capacity on B-CuFe composite at pH 2.5 was 565.32 mg/g (Zubair et al. 2021). Shaikh et al. (2021) synthesized a biochar-based silver nanocomposite (Ag-nBC) using *Shorea robusta* leaf extract through mild thermal pyrolysis (300°C). It was reported that the dye removal tests were spontaneously exothermic, represented by Freundlich isotherm and a pseudo-second-order reaction kinetic model, with the Ag-nBC showing >90% elimination for Congo red (CR) and rhodamine B (RhB). However, RhB's adsorption process relied only on

TABLE 13.2
Removal of Various Types of Dyes Using Design/Modified Biochar

Type of dyes	Specific names	Adsorbent	Q* (mg/g)	pH	Time (min)	References
Azo dye	Methylene blue	NaOH/TRP	105.27	10	60	Mu and Ma (2021)
	Orange II		91.68	2		
	Methylene orange	MBBC	12.3	2–7	30	Cuong Nguyen et al. (2021)
		LBC	449	–	–	(Hou et al. 2021)
	Eriochrome Black T	B-CuFe	565.32	2	90	Zubair et al. (2021)
	Congo red	$CoFe_2O_4$-BC	899.44	3	–	(Wu et al. 2021)
		Ag-nBC	23.55	–	50	Shaikh, Islam, and Chakraborty (2021)
	Rhodamine B		23.39	–	60	
	Indosol black	Wood waste biochar	185	2	180	Kelm et al. (2019)
	Reactive brilliant red	FWD/PMS	–	–	10	Huang et al. (2020)
	Tartrazine		85.47	2	30	Mahmoud et al. (2020)
	Sunset yellow		74.07			
Acid dye	Acid red 97	ZnO-BC	–	7	37	Leichtweis, Silvestri, and Carissimi (2020)
	Acid orange 7	M-BC	38.48	5	–	Jung et al. (2016)
		MS-450	110.27	2	160	Santhosh et al. (2020)
	Orange acid dye 52	BCCLS	500	7	–	(El Farissi et al. 2018)
		BCCLSh	358.48		–	
	Acid chrome blue K	CTAB-MC	33.82	3		Wang, Wang, and Gao (2020)
Basic dye		CSB	10	–	–	
	Basic red 09	GnSB	46.29	–	–	Praveen et al. (2021)
		RHB	44	–	–	
	Basic blue 41	SGBC	1216	8	–	Sewu, Woo, and Lee (2021)
	Basic blue 12	WWBC	80.41	–	180	do Nascimento et al. (2021)
	Basic red 46	CMRSB220	53.19	10	–	Yang et al. (2021)
Direct dye	Direct red 31	WWSD/BC	–	3	180	Behl et al. (2019)
Anthraquinone dye	Remazol brilliant blue R	SSB	80.6	10	90	Raj et al. (2021)
Triphenylmethane	Malachite green	S@CS-CT	53.35	8	120	Vigneshwaran et al. (2021)
Diazo	Congo red	nZVMn/PBC	117.647	5.8	120	Iqbal et al. (2021)

surface complexation via a particular electrostatic attraction and H-bonding. As a result of its exceptional pH stability and reusability, this heterogeneous Ag-nBC system has significant promise for dye effluent remediation (Shaikh, Islam, and Chakraborty 2021). Similarly, Wu et al. (2021) prepared a polyethyleneimine-modified magnetic porous biochar from waste bamboo powder ($CoFe_2O_4$-BC) for selective elimination of CR. The adsorption capacity of CR was remarkable in the biochar, with an optimum adsorption capacity of 435.9 mg/g. Likewise, in the Elovich model, chemisorption was the primary factor in controlling the adsorption process in this study. The CR adsorption on $CoFe_2O_4$-BC adhered to the multi-site adsorption process, according to adsorption isotherms that suited the Freundlich model well. This adsorption rate of the material exhibited 899.44 mg/g at pH 3, and its adsorption efficiency increased considerably as the ionic strength increased, except for CO_3^{2-} (Wu et al. 2021). Huang et al. (2020) synthesized food waste digested biochar and performed its activation by using peroxymonosulfate (PMS). Within 10 min, wastewater treated with 0.5% biochar and 0.1% PMS was shown to have had >99% of reactive brilliant red removed. The excellent removal efficiency of the prepared biochar is owing to the presence of catalytic sites that oxidize PMS to create reactive oxygen species ($^1O_2$, O_2^{-}, OH, and SO_4^{-}), which were then used to remove the pollutants (Huang et al. 2020). Nanobiochar (NCB) from corncob was synthesized by Mahmoud et al. (2020) and functionalized with triethylenetetramine (TA) to generate (NCB-TA). To improve surface characteristics and create flexible nonabsorbent, the modified biochar was treated with sulfuric acid to generate positively charged amine nanobiochar (NCB-TA-PC). It was observed that tap water had a high elimination percentage of tartrazine and sunset yellow of 93.73 and 95.42%, seawater had a high removal efficiency of 94.29 and 94.52%, and industrial wastewater had a high elimination percentage of 93.31 and 92.06%, respectively (Mahmoud et al. 2020).

13.5.2 Application of Acid Dye Removal

For a possible catalyst for the breakdown of acid red 97 from industrial effluent, the novel compound biochar was generated from pecan nutshell with ZnO (ZnO-BC). Composites of various ZnO ratios were stabilized by mechanical mixing, followed by pyrolysis of biochars in a nitrogen reduction environment at 650°C. In only 67 min (30 min biosorption +37 min photocatalysis), the composite with a higher quantity of ZnO reinforced in the biochar (N20Z) eliminates 100% of the targeted dye. This outstanding performance might be attributed to a lower electron-hole recombination rate, lower bandgap energy, and a larger photocatalyst surface area (Leichtweis, Silvestri, and Carissimi 2020). The effect of varied N20Z composite concentrations on acid red 97 elimination effectiveness was investigated, as illustrated in Figure 13.2B. In 60 min of reaction, the removal efficiency improved from 58 to 100%, accompanied by a rise in catalyst concentration. However, the breakdown effectiveness of acid red 97 decreased to 83% when the catalyst dosage was raised to 2.0 g/L. Jung et al. (2016) utilized an electromagnetization technique to generate marine macroalgae-based magnet biochar/Fe_3O_4 nanocomposites (M-BC) with a remarkable elimination capacity of. Due to the improved porosity and functionality of M-BC in comparison to conventional pyrolyzed (nonmagnetic) biochar, it achieved high removal capability.

Furthermore, the magnetic characteristic of M-BC enables easy separation during the postadsorption step using a bar magnet (Jung et al. 2016). The performance of different types of magnetic biochar nanocomposites generated via a coprecipitation technique has been studied by Santhosh et al. (2020). Two biochar materials, a sewage sludge biochar and woodchips biochar, were created at two different temperatures of 450 and 700°C. Fe_3O_4 was added to the biochars after they were produced. The maximum monolayer adsorption capacities of MS-450 (magnetic sewage sludge biochar prepared at 450°C) and MWC-700 (magnetic woodchip biochar prepared at 700°C) were detected as 110.27 mg/g and 80.96 mg/g for acid orange 7 dye and Cr (VI) (Santhosh et al. 2020). Orange acid 52 was removed from a solution by adsorption on biochar seeds (BCCLS) and shells of *Cistus laniferous* (BCCLSh) by El Farissi et al. (2018). For both the prepared adsorbents, the influence of pH on the adsorbed quantity was quite significant. As the pH increases from 1.2 to 4, the quantity of BCCLS and BCCLSh adsorbed rapidly increases from 176.4 to 265.05 mg/g and from 201.31 to 290 mg/g for the two materials, respectively. Since the quantity of proton H+ increases when hydrogen bonds are formed on adsorbent surfaces, the number of active sites increases as well. The adsorbed amount remains nearly constant (minimal fluctuation) when the pH rises from 4 to 8, indicating that the biochar of the two materials is saturated. At 8 pH, the quantity adsorbed by the BCCLS and BCCLSh became quickly depleted, with the BCCLS amount adsorbed, decreasing from 258.07 to 105.23 mg/g and the BCCLSh's amount adsorbed decreasing from 283 to 53.23 mg/g. This significant change might be attributed to an increase in HO- ions, which impede adsorption

(A)

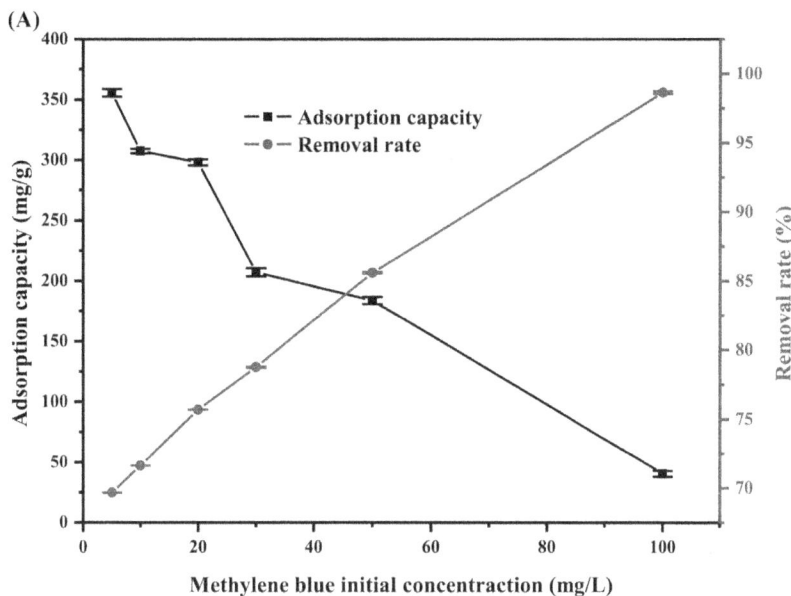

(B)

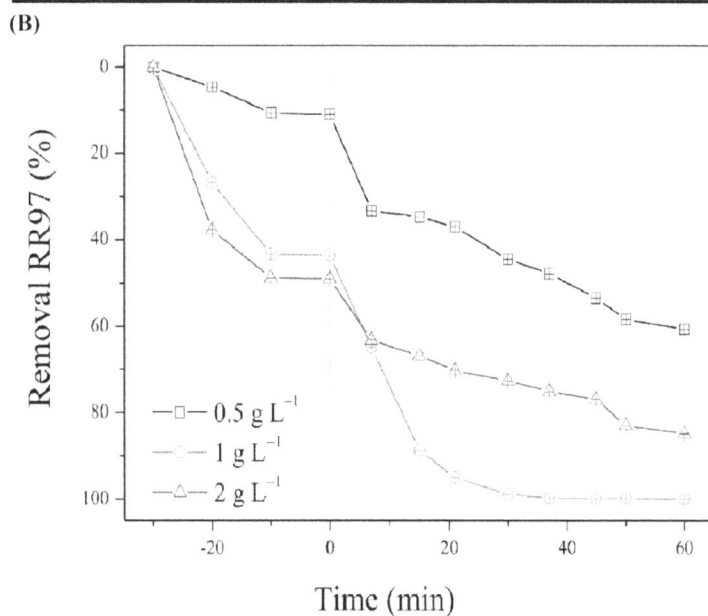

FIGURE 13.2 (A) Influence of initial concentration on adsorption of MB. (B) Impact of varying composite amounts on the percentage of removal from acid red 97 with N20Z composite.

Source: (A) Reproduced with permission from Liu, Li, and Singh (2021). (B) Reproduced with permission from Leichtweis, Silvestri, and Carissimi (2020).

by occupying active sites in the adsorbate (El Farissi et al. 2018). Biochar made from pine nut shells was pyrolyzed at 700°C in a horse fluoride furnace and then modified using cetyl trimethyl ammonium bromide to create magnetic biochar (CTAB-MC) by Wang et al. (2020). The results of the investigation showed that adsorption capacity was impacted by solution pH, adsorption duration, temperature, starting concentration, and ionic strength, among other parameters. A greater adsorption of acid chrome blue K by CTAB-MC was observed, which was up to 40% higher than that of MC (Wang, Wang, and Gao 2020).

13.5.3 APPLICATION OF BASIC DYE REMOVAL

The economic study of agro-waste biochars for basic red 09 sorption from the aqueous environment is being investigated by Praveen et al. (2021). At equilibrium batch, adsorptions of coconut shell (CSB), rice husk (RHB), and groundnut shell (GnSB) biochar exhibited maximum adsorption capacities of 10, 46.3, and 44 mg/g, respectively. Pyrolysis trials of 5 different temperatures at 300, 350, 400, 450, and 500°C produced the biochars. When the pyrolysis temperature rose from 300 to 500°C, the biochar production was reduced from 60 to 40%. The discharge of more volatile materials was blamed for the decrease in biochar production. Additionally, it was observed that increasing the pyrolysis duration at a steady temperature affects the surface and interior structure of biochar (Praveen et al. 2021). Sewu et al. (2021) conducted a series of experiments to investigate the potential for synergistically increasing the carbon fraction of low lignin-based biomass, *Saccharina japonica*, by copyrolysis with goethite (α-FeOOH) in percentages of 5, 10, and 20% to create biochar. The prepared carbonaceous adsorbent exhibited excellent removal capacity for basic blue 41 (Sewu, Woo, and Lee 2021). Nascimento et al. (2021) removed basic blue 12 and reactive black 5 from the wastewater by means of biochar synthesized from the gasification of wood residues as adsorbent (WWBC). Langmuir isotherm is the best fit for balancing data at 34.17 mg/g (reactive black 5) and 83.85 mg/g maximal adsorption capacities (basic blue 12). Regeneration experiments demonstrated that, after three adsorption/desorption cycles, the regenerated biochar still adsorbs both dyes. However, the removal efficiency fell to close to 50%, demonstrating the firmly adsorbed RB5 and BB12 into the biochar. Moreover, the process of regeneration with NaOH had the better outcomes compared to HCl, which was obvious for RB5 (do Nascimento et al. 2021). To eliminate basic red 46 dye from the aquatic environment, Yang et al. (2021) performed a hydrothermal carbonization approach to generate *Chrysanthemum morifolium* Ramat straw biochar (CMRSB) at three different temperatures: 180°C (CMRSB180), 200°C (CMRSB200), and 220°C (CMRSB220). All three kinds of CMRSB were found to have mesoporous and microporous morphologies, with the CMRSB generated at 200°C having the greatest specific surface area. The adsorption response of BR46 by CMRSB was demonstrated using the pseudo-second-order model. CMRSB200 and CMRSB220, on the other hand, adhered to the Elovich model to some extent. The elimination procedure followed Langmuir and Temkin adsorption isotherms, according to the kinetics experiment. The three prepared biochars had maximum adsorption capacities of 32.26, 49.50, and 53.19 mg/g, respectively (Yang et al. 2021).

13.5.4 REMOVAL OF OTHER TYPES OF DYE

Apart from the dyes just mentioned, the removal of a variety of additional dyes has been documented in the literature. For instance, Iqbal et al. (2021) investigated the creation of a composite of mesoporous nanozero-valent manganese (nZVMn) and *Phoenix dactylifera* leaves biochar (PBC) for the elimination of CR from an aqueous environment. CR was absorbed by the nZVMn/PBC at a rate of 117.647 mg/g vs. 25.316 mg/g. The significant CR removal induced by nZVMn/PBC might be attributed to its large surface area and hence the abundance of maximal adsorption sites. The dyes are said to be good electron acceptors, and the nZVMn in nZVMn/PBC loses electrons that the dye accepts, resulting in dye degradation (Iqbal et al. 2021). Vigneshwarana et al. (2021) was efficiently generated by sulfur-treated adsorbent chitosan-tapioca peel biochar (S-CS@TB) composites. As a starting point for the biochar matrix, the tapioca peel was used. With S-CS@TB composite, the organic dyes MG and Rh B were eliminated, and adsorption efficiencies of 53.35 and 40.86 mg/g were achieved correspondingly (Vigneshwaran et al. 2021).

13.6   CHALLENGES AND FUTURE PERSPECTIVES

Despite the potential benefits of biochar for environmental applications, several concerns have been expressed about environmental effects associated with biochar production, as well as the behavior of the synthesized biochar when released into the environment. The pyrolysis process can result in the emission of pollutants into the environment (CO_2, NO_x, and particulate matter). In addition, economic concerns are a top priority for commercializing laboratory-scale technology, especially when it comes to raw material prices and biochar manufacturing processes. Biochar-based technologies have a limited number of life cycle costs and economic analyses accessible. Furthermore, we observe that biochar is employed under highly particular doses, temperatures, pH, contact times, and other circumstances in most of the research that has been cited, all of which influence how the process takes place. As a result, additional steps must be taken to eliminate the biochar from the wastewater before it can be reused for other purposes, potentially increasing the treatment cost.

It's also worth noting that, while design biochar might mitigate some of the negative effects of pure biochar, if used incorrectly, it could exacerbate them. Further study is needed on how the contaminants are absorbed on modified biochar. Many previous studies have explained this through various methods of interaction. However, the main mechanisms and the contributions they make to this process are still unknown. Improving the adsorption capability and ambient uses of modified biochar is very important in clarifying such concerns.

13.7   CONCLUSIONS

Biochar has made great progress as a possible adsorbent for dye removal in environmental remediation in recent years. In general, modified biochar has a greater adsorption impact on dyes. For processing, numerous types of raw materials (agricultural waste, plant residue, algal biomass, and sludge) for biochar synthesis, pyrolysis, and hydrothermal carbonization have been the most favored methods. Chemisorption is present in the majority of biochar-based adsorption methods. Kinetic experiments were conducted solely on these systems. It was observed that most of the reported experimental data fits well with pseudo-first- or second-order kinetic models. Various reports have exposed that the dye removal efficiency of biochar is influenced by a variety of factors, including pH, initial concentration of the dye, loading of supporting materials, operating temperature, and contact time.

The preceding literature indicates that the majority of the research in batch mode is carried out in the laboratory under controlled circumstances; thus the potential of biochar in actual environmental systems at a wider industry scale has to be explored urgently. The pH of the solution has an important influence on the dye removal process because it influences biochar active sites. Biochars that function at neutral pH in the treatment of diverse environmental contaminants are thus in high demand. Efforts should be undertaken to synthesize and functionalize biochar-based nanoparticles that can function at neutral pH in a shorter equilibrium time. Despite the fact that the usage of biochar as adsorbents is growing, this study revealed a number of research gaps and ambiguities. More relevant investigations in the future study are required to fill these knowledge gaps.

ACKNOWLEDGMENT

This work is partially supported by a grant (DST/TM/WTI/WIC/2 K17/84(G)) from the DST (Department of Science and Technology), New Delhi. Any opinions, findings, and conclusions expressed in this paper are those of the authors and do not necessarily reflect the views of DST, New Delhi.

REFERENCES

Aspland, J.R. 1991. "Direct dyes and their application." *Textile Chemist and Colorist* 23 (11):41–45.
Aspland, J.R. 1992. "Disperse dyes and their application to polyester." *Textile Chemist and Colorist* 24:18–18.

Bedia, Jorge, Manuel Peñas-Garzón, Almudena Gómez-Avilés, Juan J. Rodriguez, and Carolina Belver. 2018. "A Review on the Synthesis and Characterization of Biomass-Derived Carbons for Adsorption of Emerging Contaminants from Water." *C* 4 (4):63.

Behera, Meerambika, Jayato Nayak, Shirsendu Banerjee, Sankha Chakrabortty, and Suraj K. Tripathy. 2021. "A Review on the Treatment of Textile Industry Waste Effluents Towards the Development of Efficient Mitigation Strategy: An Integrated System Design Approach." *Journal of Environmental Chemical Engineering* 9 (4):105277. https://doi.org/10.1016/j.jece.2021.105277

Behl, Kannikka, Surbhi Sinha, Mahima Sharma, Rachana Singh, Monika Joshi, Amit Bhatnagar, and Subhasha Nigam. 2019. "One-Time Cultivation of Chlorella Pyrenoidosa in Aqueous Dye Solution Supplemented with Biochar for Microalgal Growth, Dye Decolorization and Lipid Production." *Chemical Engineering Journal* 364:552–561. https://doi.org/10.1016/j.cej.2019.01.180

Berradi, Mohamed, Rachid Hsissou, Mohammed Khudhair, Mohammed Assouag, Omar Cherkaoui, Abderrahim El Bachiri, and Ahmed El Harfi. 2019. "Textile Finishing Dyes and Their Impact on Aquatic Environs." *Heliyon* 5 (11):e02711. https://doi.org/10.1016/j.heliyon.2019.e02711

Cuong Nguyen, X., T. Thanh Huyen Nguyen, T. Hong Chuong Nguyen, Quyet Van Le, T. Yen Binh Vo, T. Cuc Phuong Tran, D. Duong La, Gopalakrishnan Kumar, V. Khanh Nguyen, S. Woong Chang, W. Jin Chung, and D. Duc Nguyen. 2021. "Sustainable Carbonaceous Biochar Adsorbents Derived from Agro-Wastes and Invasive Plants for Cation Dye Adsorption from Water." *Chemosphere* 282:131009. https://doi.org/10.1016/j.chemosphere.2021.131009

do Nascimento, Bruna Figueiredo, Caroline Maria Bezerra de Araujo, Alisson Castro do Nascimento, Gabriel Rodrigues Bezerra da Costa, Brener Felipe Melo Lima Gomes, Maryne Patrícia da Silva, Ronald Keverson da Silva Santos, and Maurício Alves da Motta Sobrinho. 2021. "Adsorption of Reactive Black 5 and Basic Blue 12 using biochar from gasification residues: Batch tests and fixed-bed breakthrough predictions for wastewater treatment." *Bioresource Technology Reports* 15:100767. https://doi.org/10.1016/j.biteb.2021.100767

Duarah, Prangan, Dibyajyoti Haldar, and Mihir Kumar Purkait. 2020. "Technological Advancement in the Synthesis and Applications of Lignin-Based Nanoparticles Derived from Agro-Industrial Waste Residues: A Review." *International Journal of Biological Macromolecules* 163:1828–1843. https://doi.org/10.1016/j.ijbiomac.2020.09.076

El Farissi, H., R. Lakhmiri, A. Albourine, and M. Safi. 2018. "The Adsorption of the Orange Acid Dye 52 in Aqueous Solutions by the Biochar of the Seeds and Shells of Cistus Ladaniferus." *International Journal of Scientific & Engineering* 9:563–571.

Haeldermans, T., L. Campion, T. Kuppens, K. Vanreppelen, A. Cuypers, and S. Schreurs. 2020. "A Comparative Techno-Economic Assessment of Biochar Production from Different Residue Streams Using Conventional and Microwave Pyrolysis." *Bioresource Technology* 318:124083. https://doi.org/10.1016/j.biortech.2020.124083

Haldar, Dibyajyoti, Prangan Duarah, and Mihir Kumar Purkait. 2020. "MOFs for the Treatment of Arsenic, Fluoride and Iron Contaminated Drinking Water: A Review." *Chemosphere* 251:126388. https://doi.org/10.1016/j.chemosphere.2020.126388

Hou, Yanrui, Ye Liang, Hongbo Hu, Yinping Tao, Jicheng Zhou, and Jinjun Cai. 2021. "Facile Preparation of Multi-Porous Biochar from Lotus Biomass for Methyl Orange Removal: Kinetics, Isotherms, and Regeneration Studies." *Bioresource Technology* 329:124877. https://doi.org/10.1016/j.biortech.2021.124877

Huang, Simian, Teng Wang, Kai Chen, Meng Mei, Jingxin Liu, and Jinping Li. 2020. "Engineered Biochar Derived from food Waste Digestate for Activation of Peroxymonosulfate to Remove organic Pollutants." *Waste Management* 107:211–218. https://doi.org/10.1016/j.wasman.2020.04.009

Iqbal, Jibran, Noor S. Shah, Murtaza Sayed, Nabeel Khan Niazi, Muhammad Imran, Javed Ali Khan, Zia Ul Haq Khan, Aseel Gamal Suliman Hussien, Kyriaki Polychronopoulou, and Fares Howari. 2021. "Nano-Zerovalent Manganese/Biochar Composite for the Adsorptive and Oxidative Removal of Congo-red Dye from Aqueous Solutions." *Journal of Hazardous Materials* 403:123854. https://doi.org/10.1016/j.jhazmat.2020.123854

Jung, Kyung-Won, Brian Hyun Choi, Tae-Un Jeong, and Kyu-Hong Ahn. 2016. "Facile Synthesis of Magnetic Biochar/Fe3O4 Nanocomposites using Electro-Magnetization Technique and its Application on the Removal of Acid Orange 7 from Aqueous Media." *Bioresource Technology* 220:672–676. https://doi.org/10.1016/j.biortech.2016.09.035

Kazemi Shariat Panahi, Hamed, Mona Dehhaghi, Yong Sik Ok, Abdul-Sattar Nizami, Benyamin Khoshnevisan, Solange I. Mussatto, Mortaza Aghbashlo, Meisam Tabatabaei, and Su Shiung Lam. 2020.

"A Comprehensive Review of Engineered biochar: Production, Characteristics, and Environmental Applications." *Journal of Cleaner Production* 270:122462. https://doi.org/10.1016/j.jclepro.2020.122462

Kelm, Miguel Antônio Pires, Mário José da Silva Júnior, Sávio Henrique de Barros Holanda, Caroline Maria Bezerra de Araujo, Romero Barbosa de Assis Filho, Emerson Jaguaribe Freitas, Diogo Rafael dos Santos, and Maurício Alves da Motta Sobrinho. 2019. "Removal of Azo Dye From water via Adsorption on Biochar Produced by the Gasification of Wood Wastes." *Environmental Science and Pollution Research* 26 (28):28558–28573. https://doi.org/10.1007/s11356-018-3833-x

Leichtweis, Jandira, Siara Silvestri, and Elvis Carissimi. 2020. "New Composite of Pecan Nutshells Biochar-ZnO for Sequential Removal of Acid red 97 by Adsorption and Photocatalysis." *Biomass and Bioenergy* 140:105648. https://doi.org/10.1016/j.biombioe.2020.105648

Liu, Wu-Jun, Hong Jiang, and Han-Qing Yu. 2015. "Development of Biochar-Based Functional Materials: Toward a Sustainable Platform Carbon Material." *Chemical Reviews* 115 (22):12251–12285. https://doi.org/10.1021/acs.chemrev.5b00195

Liu, Xu-Jing, Ming-Fei Li, and Sandip K. Singh. 2021. "Manganese-Modified Lignin Biochar as Adsorbent for Removal of Methylene Blue." *Journal of Materials Research and Technology* 12:1434–1445. https://doi.org/10.1016/j.jmrt.2021.03.076

Li, Yunchao, Bo Xing, Yan Ding, Xinhong Han, and Shurong Wang. 2020. "A Critical Review of the Production and Advanced Utilization of Biochar via Selective Pyrolysis of Lignocellulosic Biomass." *Bioresource Technology* 312:123614. https://doi.org/10.1016/j.biortech.2020.123614

Mahmoud, Mohamed E., Amir M. Abdelfattah, Rana M. Tharwat, and Gehan M. Nabil. 2020. "Adsorption of Negatively Charged Food Tartrazine and Sunset Yellow Dyes onto Positively Charged Triethylenetetramine Biochar: Optimization, Kinetics and Thermodynamic Study." *Journal of Molecular Liquids* 318:114297. https://doi.org/10.1016/j.molliq.2020.114297

Mani, Sujata, and Ram Naresh Bharagava. 2018. "Textile Industry Wastewater: Environmental and Health Hazards and Treatment Approaches." In *Recent Advances in Environmental Management*, 47–69. CRC Press.

Mu, Yongkang, and Hongzhu Ma. 2021. "NaOH-Modified Mesoporous Biochar Derived from tea Residue for Methylene Blue and Orange II Removal." *Chemical Engineering Research and Design* 167:129–140. https://doi.org/10.1016/j.cherd.2021.01.008

Nakajima, Kiyotaka, and Michikazu Hara. 2012. "Amorphous Carbon with SO3H Groups as a Solid Brønsted Acid Catalyst." *ACS Catalysis* 2 (7):1296–1304. https://doi.org/10.1021/cs300103k

Nath, Hariprasad, Ankumoni Saikia, Prasanta Jyoti Goutam, Binoy K. Saikia, and Nabajyoti Saikia. 2021. "Removal of Methylene Blue from Water Using Okra (Abelmoschus Esculentus L.) Mucilage Modified Biochar." *Bioresource Technology Reports* 14:100689. https://doi.org/10.1016/j.biteb.2021.100689

Oliveira, Ivo, Dennis Blöhse, and Hans-Günter Ramke. 2013. "Hydrothermal Carbonization of Agricultural Residues." *Bioresource Technology* 142:138–146. https://doi.org/10.1016/j.biortech.2013.04.125

Patel, Himanshu. 2018. "Charcoal as an Adsorbent for Textile Wastewater Treatment." *Separation Science and Technology* 53 (17):2797–2812. https://doi.org/10.1080/01496395.2018.1473880

Peterson, Steven C., Michael A. Jackson, Sanghoon Kim, and Debra E. Palmquist. 2012. "Increasing Biochar Surface Area: Optimization of Ball Milling Parameters." *Powder Technology* 228:115–120. https://doi.org/10.1016/j.powtec.2012.05.005

Praveen, Saravanan, Ravindiran Gokulan, Thillainayagam Bhagavathi Pushpa, and Josephraj Jegan. 2021. "Techno-Economic Feasibility of Biochar as Biosorbent for Basic Dye Sequestration." *Journal of the Indian Chemical Society* 98 (8):100107. https://doi.org/10.1016/j.jics.2021.100107

Purkait, M. K., D. S. Gusain, S. DasGupta, and S. De. 2005. "Adsorption Behavior of Chrysoidine Dye on Activated Charcoal and Its Regeneration Characteristics by Using Different Surfactants." *Separation Science and Technology* 39 (10):2419–2440. https://doi.org/10.1081/SS-120039347

Raj, Abhay, Ashutosh Yadav, Abhay Prakash Rawat, Anil Kumar Singh, Sunil Kumar, Ashutosh Kumar Pandey, Ranjna Sirohi, and Ashok Pandey. 2021. "Kinetic and thermodynamic investigations of sewage sludge biochar in removal of Remazol Brilliant Blue R dye from aqueous solution and evaluation of residual dyes cytotoxicity." *Environmental Technology & Innovation* 23:101556. https://doi.org/10.1016/j.eti.2021.101556

Regmi, Pusker, Jose Luis Garcia Moscoso, Sandeep Kumar, Xiaoyan Cao, Jingdong Mao, and Gary Schafran. 2012. "Removal of Copper and Cadmium from Aqueous Solution Using Switchgrass Biochar Produced via Hydrothermal Carbonization Process." *Journal of Environmental Management* 109:61–69. https://doi.org/10.1016/j.jenvman.2012.04.047

Salleh, Mohamad Amran Mohd, Dalia Khalid Mahmoud, Wan Azlina Wan Abdul Karim, and Azni Idris. 2011. "Cationic and Anionic Dye Adsorption by Agricultural Solid Wastes: A Comprehensive Review." *Desalination* 280 (1):1–13. https://doi.org/10.1016/j.desal.2011.07.019

Santhosh, Chella, Ehsan Daneshvar, Kumud Malika Tripathi, Pranas Baltrėnas, TaeYoung Kim, Edita Baltrėnaitė, and Amit Bhatnagar. 2020. "Synthesis and Characterization of Magnetic Biochar Adsorbents for the Removal of Cr(VI) and Acid Orange 7 Dye from Aqueous Solution." *Environmental Science and Pollution Research* 27 (26):32874–32887. https://doi.org/10.1007/s11356-020-09275-1

Selvaraj, V., T. Swarna Karthika, C. Mansiya, and M. Alagar. 2021. "An Over Review on Recently Developed Techniques, Mechanisms and Intermediate Involved in the Advanced Azo Dye Degradation for Industrial Applications." *Journal of Molecular Structure* 1224:129195. https://doi.org/10.1016/j.molstruc.2020.129195

Sewu, Divine Damertey, Seung Han Woo, and Dae Sung Lee. 2021. "Biochar from the Co-Pyrolysis of Saccharina Japonica and Goethite as an Adsorbent for Basic Blue 41 Removal from Aqueous Solution." *Science of The Total Environment* 797:149160. https://doi.org/10.1016/j.scitotenv.2021.149160

Shaikh, Wasim Akram, Rafique Ul Islam, and Sukalyan Chakraborty. 2021. "Stable Silver Nanoparticle Doped Mesoporous Biochar-Based Nanocomposite for Efficient Removal of Toxic Dyes." *Journal of Environmental Chemical Engineering* 9 (1):104982. https://doi.org/10.1016/j.jece.2020.104982

Shan, Danna, Shubo Deng, Tianning Zhao, Bin Wang, Yujue Wang, Jun Huang, Gang Yu, Judy Winglee, and Mark R. Wiesner. 2016. "Preparation of Ultrafine Magnetic Biochar and Activated Carbon for Pharmaceutical Adsorption and Subsequent Degradation by Ball Milling." *Journal of Hazardous Materials* 305:156–163. https://doi.org/10.1016/j.jhazmat.2015.11.047

Sontakke, Ankush D., and Mihir K. Purkait. 2021. "A Brief review on Graphene Oxide Nanoscrolls: Structure, Synthesis, Characterization and scope of Applications." *Chemical Engineering Journal* 420:129914. https://doi.org/10.1016/j.cej.2021.129914

Sun, Hao, William C. Hockaday, Caroline A. Masiello, and Kyriacos Zygourakis. 2012. "Multiple Controls on the Chemical and Physical Structure of Biochars." *Industrial & Engineering Chemistry Research* 51 (9):3587–3597. https://doi.org/10.1021/ie201309r

Sun, Junna, Fuhong He, Yinghua Pan, and Zhenhua Zhang. 2017. "Effects of Pyrolysis Temperature and Residence Time on Physicochemical Properties of Different Biochar Types." *Acta Agriculturae Scandinavica, Section B—Soil & Plant Science* 67 (1):12–22. https://doi.org/10.1080/09064710.2016.1214745

Tounsadi, Hanane, Yousra Metarfi, M. Taleb, Karima El Rhazi, and Zakia Rais. 2020. "Impact of Chemical Substances Used in Textile Industry on the Employee's Health: Epidemiological Study." *Ecotoxicology and Environmental Safety* 197:110594. https://doi.org/10.1016/j.ecoenv.2020.110594

Uçar, Suat, Murat Erdem, Turgay Tay, and Selhan Karagöz. 2009. "Preparation and Characterization of Activated Carbon Produced from Pomegranate Seeds by ZnCl2 Activation." *Applied Surface Science* 255 (21):8890–8896. https://doi.org/10.1016/j.apsusc.2009.06.080

Vigneshwaran, Sivakumar, Palliyalil Sirajudheen, Manuvelraja Nikitha, Krishnapillai Ramkumar, and Sankaran Meenakshi. 2021. "Facile Synthesis of Sulfur-Doped Chitosan/Biochar Derived from Tapioca Peel for the Removal of Organic Dyes: Isotherm, Kinetics and Mechanisms." *Journal of Molecular Liquids* 326:115303. https://doi.org/10.1016/j.molliq.2021.115303

Wang, Bing, Bin Gao, and June Fang. 2017. "Recent Advances in Engineered Biochar Productions and Applications." *Critical Reviews in Environmental Science and Technology* 47 (22):2158–2207. https://doi.org/10.1080/10643389.2017.1418580

Wang, Hongyu, Bin Gao, Shenseng Wang, June Fang, Yingwen Xue, and Kai Yang. 2015. "Removal of Pb(II), Cu(II), and Cd(II) from Aqueous Solutions by Biochar Derived from KMnO4 Treated Hickory Wood." *Bioresource Technology* 197:356–362. https://doi.org/10.1016/j.biortech.2015.08.132

Wang, Huan, Shan Wang, and Yihong Gao. 2020. "Cetyl Trimethyl Ammonium Bromide Modified Magnetic Biochar from Pine Nut Shells for Efficient Removal of Acid Chrome Blue K." *Bioresource Technology* 312:123564. https://doi.org/10.1016/j.biortech.2020.123564

Wu, Jia, Jianwei Yang, Guohuan Huang, Chuanhui Xu, and Baofeng Lin. 2020. "Hydrothermal Carbonization Synthesis of Cassava Slag Biochar with Excellent Adsorption Performance for Rhodamine B." *Journal of Cleaner Production* 251:119717. https://doi.org/10.1016/j.jclepro.2019.119717

Wu, Zhengde, Ximo Wang, Jing Yao, Siyan Zhan, Hui Li, Jian Zhang, and Zumin Qiu. 2021. "Synthesis of Polyethyleneimine Modified CoFe2O4-Loaded Porous biochar for Selective Adsorption Properties Towards Dyes and exploration of Interaction Mechanisms." *Separation and Purification Technology*:119474. https://doi.org/10.1016/j.seppur.2021.119474

Yang, Xinyuan, Wenfang Zhu, Yali Song, Haifeng Zhuang, and Haojie Tang. 2021. "Removal of Cationic Dye BR46 by Biochar Prepared from Chrysanthemum Morifolium Ramat Straw: A Study on Adsorption Equilibrium, Kinetics and Isotherm." *Journal of Molecular Liquids*:116617. https://doi.org/10.1016/j.molliq.2021.116617

Yek, Peter Nai Yuh, Wanxi Peng, Chee Chung Wong, Rock Keey Liew, Yee Ling Ho, Wan Adibah Wan Mahari, Elfina Azwar, Tong Qi Yuan, Meisam Tabatabaei, Mortaza Aghbashlo, Christian Sonne, and Su Shiung Lam. 2020. "Engineered Biochar via Microwave CO2 and Steam Pyrolysis to Treat Carcinogenic Congo Red Dye." *Journal of Hazardous Materials* 395:122636. https://doi.org/10.1016/j.jhazmat.2020.122636

Yoon, Kwangsuk, Dong-Wan Cho, Daniel C. W. Tsang, Nanthi Bolan, Jörg Rinklebe, and Hocheol Song. 2017. "Fabrication of Engineered Biochar from Paper Mill Sludge and Its Application into Removal of Arsenic and Cadmium in Acidic Water." *Bioresource Technology* 246:69–75. https://doi.org/10.1016/j.biortech.2017.07.020

Zhang, Hanzhi, Chengrong Chen, Evan M. Gray, and Sue E. Boyd. 2017. "Effect of Feedstock and Pyrolysis Temperature on Properties of Biochar Governing end Use Efficacy." *Biomass and Bioenergy* 105:136–146. https://doi.org/10.1016/j.biombioe.2017.06.024

Zhang, Xiaoxiao, Peizhen Zhang, Xiangru Yuan, Yanfei Li, and Lujia Han. 2020. "Effect of Pyrolysis Temperature and Correlation Analysis on the Yield and Physicochemical Properties of Crop Residue Biochar." *Bioresource Technology* 296:122318. https://doi.org/10.1016/j.biortech.2019.122318

Zubair, Mukarram, Hamidi Abdul Aziz, Ihsanullah Ihsanullah, Mohd Azmier Ahmad, and Mamdouh A. Al-Harthi. 2021. "Biochar Supported CuFe Layered Double Hydroxide Composite as a Sustainable Adsorbent for Efficient Removal of Anionic Azo Dye From Water." *Environmental Technology & Innovation* 23:101614. https://doi.org/10.1016/j.eti.2021.101614

14 Restoration of Contaminated Agricultural Soils Using Modified Biochar

Mukesh Sharma, Kumar Abhishek, Pranjal P. Das, Arun Chakraborty, Jayanarayanan Kuttippurath, and Mihir K. Purkait

CONTENTS

14.1 INTRODUCTION

Biochar is prepared from biomass through the thermochemical process of pyrolysis, wherein the organic residues are heated in the oxygen-free environment under ambient pressure for a specified period to carbonize them into charcoal, with the bio-oils and syngas as by-products of pyrolysis (Zama et al. 2018; Guo, Xiao, and Li 2019; Guo, Song, and Tian 2020; R.-H. Zhang et al. 2017). The most common feedstock materials for the preparation of biochar are forest wood, crop residues, food wastes, and sewage sludge/biosolids. The characteristics and functional capability of biochar products, when added to soil, depend largely on the nature of the feedstock material. The composition and structure of biochar products are a function of organic and ash compositions of feedstock and pyrolysis temperature. Therefore, these parameters are also used to control the quality characteristics of the final product. It has been reported that manipulations in the carbonization conditions through changes in the peak or endpoint temperature, residence time, and heating rate result in notable differences in the properties of biochar materials (Guo, Song, and Tian 2020). The temperature of pyrolysis used to produce biochar is typically between 300 and 600°C. In a similar study by Huang et al. (2021), the impact of the type of feedstock, pyrolysis temperature, and compaction on water retention of biochar-remediated soil was discussed. An increase in the temperature enhances the carbonization process, allowing for a more thorough pyrolytic transformation of biomass in less time (Guo, Song, and Tian 2020; Y. Zhang et al. 2020). Pyrolysis must be accomplished completely to convert all feedstock OC to carbonized, pyrogenic OC (i.e., altered, amorphous carbon structure). The time required for the full pyrolysis is governed by the rate of heating and the pyrolysis temperature, which can range from seconds to days. The rate of heating can range from 1 to >200°C s^{-1}, depending on the pyrolysis temperature, feed properties, and feed mass flow (Guo, Song, and Tian 2020). Additionally, biochar products obtained from incomplete pyrolysis retain significant amounts of uncarbonized carbon (Keiluweit et al. 2010).

Biochar is porous, comprised of a rough morphological surface and anatomical or other irregular features resembling honeycombs under the microscope. Figure 14.1 briefly illustrates the micrographs (native and exfoliated) for the biochar prepared from different agricultural wastes.

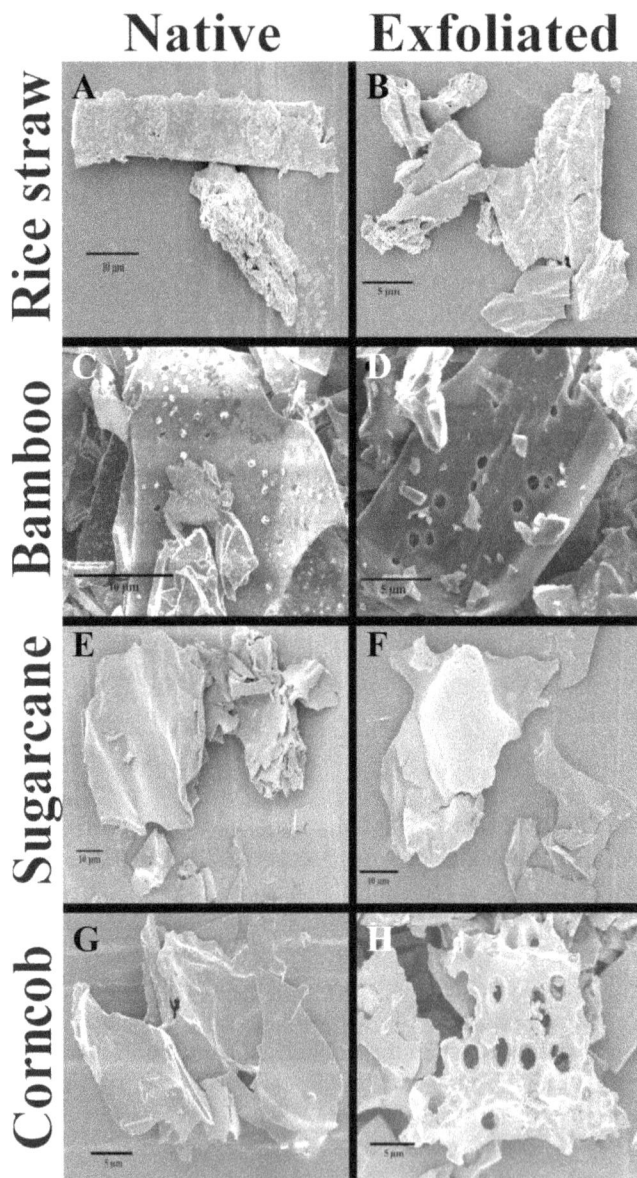

FIGURE 14.1 Micrographs of the native (BC) and exfoliated (Ex-BC) biochar samples prepared from various agricultural waste rice straw (A, B), bamboo (C, D), sugarcane waste (E, F), and corn cob (G, H).

The micrographs of the biochars prepared from several sources including rice straw, bamboo, sugarcane, and corncob have been demonstrated. Further, Figure 14.1 (a), (c), (e) and (g) shows the formation of the native biochar from the mentioned sources, whereas, the Figure 14.1 (b), (d), (f) and (h) demonstrates the exfoliated biochars from the same sources. The intrinsic micropores are having variable dimensions usually ranging in diameter from 0.8 to 235 mm (Hardie et al. 2014). Biochar is predominantly made of amorphous, aromatic carbon and has an abundance of surface functional groups including oxygen. As the pyrolysis temperature is increased, biochars aromaticity enhances, but the surface functionality declines (Zhao et al. 2016; Yuan et al. 2013; R. Xu et al. 2011). This is mostly due to the gradual loss of carboxyl, aliphatic C-H, carbonyl, olefinic C=C, and hydroxyl groups as pyrolysis temperature increases (Chen et al. 2014; Tan et al. 2015). Biochar is a result of both fast pyrolysis (aiming for bio-oil) and gasification (aiming for syngas). The carbonization conditions can be optimized for different thermochemical procedures, even with the same feedstock, to obtain different products varying in OC and ash content.

14.2   SOIL PHYSICOCHEMICAL PROPERTIES AND BIOCHAR PROPERTIES

14.2.1   PROPERTIES OF SOIL

The chemical and structural variations in biochar are responsible for the increase in its range of behavior in the soil. Modifications can be evident through the physicochemical characteristics such as the aggregate stability, bulk density, and water availability, which play a significant role in improving the development and yield of the plant (Atkinson, Fitzgerald, and Hipps 2010; Abdelhafez, Li, and Abbas 2014). The utilization of biochar in the soil also helps build its alkalization, CEC as well as its EC values (electrical conductivity) and dietary repute (Herath et al. 2015). Even though biochar alone hardly ever includes excessive concentrations of available nutrients (with few exemptions, for example potassium). The significance of biochar in plant nutrition is ascribed to its capacity to hold plant supplements in soil solution (for example, fertilizer additions) and consequently minimize volatilization and leaching issues, bringing about increased nutrient accessibility and absorption and a better harvest yield. However, for soils that are contaminated by HM, the possible positive effect of biochar is through its ability to lessen the bioavailability, versatility, and distribution of HM (Zhao et al. 2016; K. Zhang et al. 2018; Cao et al. 2018; Zhao et al. 2016). The upgrades in the soil's physical characteristics rely upon the physical nature of the soil (Fryda and Visser 2015). The contrasts in the pore sizes and porosity levels of biochar are significant characteristics in determining the rate of absorption of HMs. The pyrolysis temperature of biochar regulates its surface and structural properties (for instance, size of the pore, pore volume, and the surface area that by and large rises with the pyrolysis temperature), and these impact the distinctive HM adsorption potential of the biochar (Méndez et al. 2013; Yuan et al. 2013; R. Xu et al. 2011; Zama et al. 2018; Chen et al. 2014). In a study with sewage-sludge-derived biochar, Chen et al. (2014) observed that for adsorption of Cd by ion exchange and surface precipitation, a pyrolysis temperature of 900°C was ideal; thus connecting the pyrolysis temperature with biochar's surface properties and their subsequent adsorption behavior of HM. Similarly, the characteristics of rapeseed biochars at varying pyrolysis temperature was discussed (Angın and Şensöz 2014) (Table 14.1). The properties of the rapeseed biochars synthesized at several pyrolysis temperatures, with the proximate and ultimate analyses, have been illustrated in Table 14.1. The microstructural porosity of biochar, together with its particulate macrostructural developments of total soil stability, directly enhances the porosity of the soil (Hardie et al. 2014). The quality of clay soil is improved by the increased total stability, water holding capacity (WHC), and pore size dispersion that are induced by biochar (P. Xu et al. 2016; K. Zhang et al. 2018; Zama et al. 2018), whereas sandy soils display advanced physical characteristics, for example total porosity and bulk density (O'Connor, Peng, Zhang, et al. 2018) and plant water availability (Atkinson, Fitzgerald, and Hipps 2010). Since the impact capacity is often found to be more than the additive, the relationship between these variables is also quite complex.

TABLE 14.1

Characteristics of rapeseed biochars prepared at varying pyrolysis temperatures (Angın and Şensöz 2014)

	Pyrolysis temperature (°C)				
	400	**450**	**500**	**550**	**700**
Proximate analysis (dry, wt %)					
Volatile m atter	25.14 ± 0.63	16.18 ± 0.57	13.62 ± 0.38	10.42 ± 0.22	7.76 ± 0.21
Ash	19.28 ± 0.48	21.40 ± 0.51	23.35 ± 0.55	24.94 ± 0.64	26.08 ± 0.72
Fixed carbon*	55.58 ± 0.83	62.42 ± 0.91	63.03 ± 0.88	64.64 ± 0.97	66.16 ± 0.95
Ultimate analysis (wt %)					
C	57.95	59.77	61.98	67.29	70.41
H	3.43	2.36	1.92	1.75	1.24
N	5.43	5.12	4.32	4.75	3.24
O*	33.16	32.75	31.78	26.21	25.11
H/C	0.71	0.47	0.37	0.31	0.21
O/C	0.43	0.41	0.38	0.29	0.27
Higher heating value (MJ kg^{-1})	28.15	28.89	29.22	30.03	30.47
pH	8.33	8.44	8.66	10.70	10.84

*By difference.

14.2.2 PROPERTIES OF BIOCHAR

Properties of biochar are highly dependent on the pyrolysis conditions and feedstock used. The pH in typical biochar ranges from 5.4 to 12.4 in 1:5 solid/water extract, and the mineral ash level ranges from 1.1 to 82.0%. Usually, the pH of biochar depends on the balance between water-soluble base cations and organic acids. These base cations come from the mineral ash in the feedstock, while the organic acids come from biomass pyrolysis. Complete pyrolysis produces biochar with few organic acids and a pH of >7.5 due to the presence of basic ash minerals. However, the lower endpoint temperature during pyrolysis lowers the ash content of biochars and pH values. The role of biochar pH is significant in regulating the mobility of cations and anions in soil. In addition to regulating the mobility of ions in the soil, biochar also supplies available nutrients to altered soils. Biochar contains slow-release nutrients like N (nitrogen), P (phosphorus), K (potassium), and other plant nutrients that encourage plant growth. The observed N, P, and K concentrations in biochars range from 0.5–48.0, 0.1–198.0, and 1.4–91.5 g kg^{-1}, which are closely associated with ash content in biochar. However, the majority of the feedstock N is lost to pyrolysis vapors, especially at a temperature higher than 500°C for pyrolysis (Guo, Song, and Tian 2020; Lin et al. 2019). Biochar organic carbon (OC) is classified as labile, intermediate stable, and stable (Y. Zhang et al. 2020). However, most OC in biochars with a H/OC molar ratio of 0.7 or less is expected to survive in natural soils for more than 100 years (Joseph et al. 2018). The International Biochar Initiative (IBI) recommends the determination of different properties like particle size distribution, germination inhibition assay, total and accessible plant nutrients, and the presence of inorganic and organic contaminants as quality parameters of biochar. To be classified as biochar, the OC concentration must surpass 10%, and the H/OC molar ratio must be 0.7. Therefore, in order to be certified, biochar must contain 50% TC at 0.4 O/C and 0.7 H/OC molar ratios with Zn, Pb, Cu, Cr, Ni, Cd, and Hg being maintained at concentrations less than 400, 150, 100, 1.5, 90, 50, and 1.5 mg kg^{-1}, respectively. EBC further provides guidelines regarding PAHs, PCBs, and PCDD/Fs, which should be limited to 12, 0.2, and 0.00002 mg kg^{-1}. It is suggested that when the TC content and H/OC molar ratio criteria are met, the biochars selected for soil remediation should be weighted more in pH, lime equivalency, cation exchange capacity, and nutrient (particularly P) content.

14.3 VARIATION IN SOIL PROPERTIES

The variations in biochar chemistry and structure are attributed to its behavior in the soil. The plant growth and yield improvements after the application of biochar can be related to changes in several soil physical properties, such as bulk density, aggregate stability, and accessible water content (Atkinson, Fitzgerald, and Hipps 2010). These improvements in soil physical features are contingent on the soil's physical characteristics (Fryda and Visser 2015). However, physical properties of biochar, like pore size and porosity, also influence important chemical features of biochar, i.e., adsorption and desorption. It also has a direct impact on soil porosity, as well as particle and macrostructural aggregate stability (Hardie et al. 2014). Higher aggregate stability, higher water holding capacity (WHC), and improved pore size distribution add up to more than the sum of their individual impacts (Zama et al. 2018; K. Zhang et al. 2018; P. Xu et al. 2016). Biochar boosts the alkalinity of the soil, CEC, and electrical conductivity (EC) levels and also modulates nutritional status (Herath et al. 2015). On the other hand, biochar is rarely found to have large concentrations of plant-available nutrients (with few exceptions, e.g., K). The purported usefulness of biochar is linked to its capacity to store nutrients available in soil solution (such as after fertilizer applications) and, as a result, lowers leaching and volatilization losses, increases nutrient availability and uptake, resulting in improved crop yields (Y. Xu et al. 2018; Y. Zhang et al. 2020; R. Xu et al. 2011; Lin et al. 2019; Pan et al. 2021; M. Zhang et al. 2019; P. Xu et al. 2016; R.-H. Zhang et al. 2017, 2017).

14.4 SOIL REMEDIATION USING BIOCHAR

Biochars improve the physical, chemical, and biological characteristics of soil, thus benefitting overall soil health (Guo, Song, and Tian 2020). Besides, plant uptake of heavy metals (including Pb, Al, Cr, Mn, Fe, Co, Ni, Cu, Zn, Cd, and Tl) in contaminated soils has been demonstrated to be reduced by biochar (X. Yang et al. 2016; Mandal et al. 2021) through immobilization (Yu et al. 2019; Liu et al. 2018). Additionally, soil particles and biochar may also interact with heavy metals and alter their mobility and chemistry. The biochar functional groups, in combination with charged particles in the soil, have a strong binding affinity for heavy metals, and thus their uptake in plants is reduced. However, the process of nonspecific sorption is reversible, and so the sorbed heavy metals in the soil solution can also be released back into the soil (Campillo-Cora et al. 2020). Additionally, some experiments involving the elements Cu, Pb, Ni, and Zn demonstrate that greater retention can be observed in the presence of any of these elements compared to their omnipresence. The research presented to date, however, does not prove that biochar can completely remove a soil's heavy metals. This might be related to a suboptimal stoichiometric ratio between soil heavy metal load and biochar dosage (Abdelhafez, Li, and Abbas 2014). Therefore, more studies on dose appropriation of biochar resulting in variable surface area and density of heavy metal bands are required (Seneviratne et al. 2017).

Biochar has also been utilized in soil-phytoremediation in order to assist in the heavy metal absorption process, along with the living plants that act as sinks for heavy metals. Earlier results showed that *alfalfa* phytoremediation of Cd-polluted soil resulted in a significant drop of 90 g Cd ha^{-1} in soil, and this occurred owing to root exudates and reduced root Cd uptake (M. Zhang et al. 2019; Y. Zhang et al. 2020). This highlights the significance of synergistic effects of biochar addition in soil remediation. The effectiveness of biochar in facilitating the breakdown of certain organic pollutants in soils has also been proven. Biochar immediately interacts with organic pollutants and soil microbes when incorporated into contaminated soil after vigorous mixing. On biochar surfaces and pores, the organic pollutants are stabilized and thereafter might be degraded by bacteria driven by biochar modification. The porous, functional, and aromatic-condensed surface of the biochar can adsorb diverse organic molecules via various methods. Biochar interacts with organic molecules by London dispersion forces, dipole–dipole forces, and van der Waals forces. Although van der

Waals forces are fairly trifling but rise with the surface area of the particles, biochar particles of lower size can be dominated by van der Waals forces (J. Yang et al. 2015; S. Yang and Evans 2007; Xiong et al. 2013; Tan et al. 2015). Another method of electrostatic attraction, other than van der Waals forces, is called H-bonding (Tan et al. 2015; Yu et al. 2019; Chen et al. 2014). Polar organic molecules can interact with the biochar through the O-, N-, and H-containing functional groups, and the H-bonds are formed as a result. Functional groups in biochar might become dissociated or protonated depending on pH; therefore biochar is formed with a net negative charge and a net positive charge (Fidel, Laird, and Spokas 2018). Biochar's surface charges allow it to attract polar organic molecules and ionized organic contaminants, all having a counter-electric field. Ionized organic substances chemically react with biochar functional groups and are then maintained in the inner sphere (R. Xu et al. 2011). H-bonding and van der Waals interactions are often much weaker than this sort of attractive force and are traditionally known as electrostatic attraction. Biochar may form complexes with polar and ionized organic molecules through metal ions like Fe, Al, Ca, and Mg and deposits on the biochar's surface or soil. Few studies suggest that "chemisorption" is an important mechanism in organic–biochar interactions, based on the sorption kinetics of numerous organic compounds on biochar (Tan et al. 2015; Chen et al. 2014). However, the kinetics of the processes like electrostatic attraction, specific interaction, and surface precipitation depend on solution pH and the ionic strength of the medium.

14.5   REMEDIAL MECHANISM IN HEAVY METAL CONTAMINATED SOIL

Biochar offers a wide range of sorption capacity and mechanisms; therefore, the stabilizing capacities of biochar for soil contaminants also vary. For instance, hydrophobic, nonpolar organic pollutants, for instance petroleum hydrocarbons and PAHs, are adsorbed by pore partition and filling, while the organic contaminants that are polar and ionized, like most antibiotics and pesticides, are adsorbed through hydrogen bonding, electrostatic interaction, surface precipitation, and specific interaction. Biochar is composed of heterogeneous reactive moieties such as acidic, basic, and hydrophobic groups, enabling a wide range of enhanced reactivity to soil-based compounds and heavy metals (Atkinson, Fitzgerald, and Hipps 2010). Because biochar does not completely eliminate but rather stabilizes the heavy metals present in the soil, turning the toxic elements into a less soluble and bioaccessible conformation, biochar amendment can be considered an effective method of managing heavy metal toxicity in soil than popular methods of adding treatment solutions such as phosphates and carbonates to contaminated soils. However, the general method is relatively similar to the chemical stabilization process in which the treatment agents react with pollutants to precipitate them. In biochar-facilitated soil remediation, the biochar product is thoroughly mixed with the contaminated soil at heavy-metal-polluted locations. In this way, biochar takes up and binds to heavy metals, reducing their bonding with water molecules, which leads to a decrease in the possibility that plant roots absorb or accumulate heavy metals (Ahmad et al. 2014). Cationic heavy metals are stabilized principally through sorption and chemical precipitation in biochar (Tan et al. 2015; Chen et al. 2014). Biochar electrical charges result from dissociation or protonation of surface functional groups, which further depends on solution pH. As the pH of the solution increased between 3.0 and 8.0, the zeta potential of biochars became more negative, indicating a higher negative charge density on the biochar surface (R. Xu et al. 2011). Now such biochars can adsorb heavy metal cations via electrostatic attraction through its negatively charged surface functional groups. Furthermore, through π–π^* donor–acceptor interactions, the aromatic biochar surface, which is rich in electrons, could attract electron-deficient metal cations. The heavy metal ions present in the soil solution can also be adsorbed to biochar through ion exchange by cation substitution (e.g., Mg^{2+}, Ca^{2+}, K^+, H^+, and Na^+) that were previously connected with the functional groups on the surface (Fidel, Laird, and Spokas 2018).

14.5.1 Factors Governing the Efficacy of Remediation

While biochar is extremely efficient at treating contaminated soil, this efficiency is highly dependent on the type of biochar, the size of the biochar particles, the dosage of the amendment, the type of pollutant, and the soil type. A host of variables, including the source of biochar, amendment rate, metal species, soil type, and placement in soil, further affect the efficacy of biochar application for immobilizing heavy metals (Ahmad et al. 2014; O'Connor, Peng, Li, et al. 2018). Biochars can be observed with wide changes in their pH, mineral ash concentration, composition, surface functionality, and CEC when produced from different biomass feedstocks using varied pyrolysis settings. To achieve greater outcomes of heavy metal stabilization, the biochar amendment rate should be optimized between 5 and 10% of soil weight. Although coexisting cations in the soil and clay content affect the biochar's efficiency in stabilizing toxic metals, dose optimization still has huge potential in achieving biochar-mediated soil remediation goals. It has been observed that biochar made from animal sources is more efficient at quarantining soils' heavy metals than biochar made from wood. As a rule, soil adjustment with biochar at a rate of >2.0 wt% balances out the cationic heavy metals (for example Pb^{2+}, Zn^{2+}, Ni^{2+}, Cu^{2+}, and Cd^{2+}) and diminishes the bioaccessible piece and the bioaccumulation of such harmful components in soil. The viability is more noteworthy for biochar products having higher CEC, ash content, and pH levels that are more obvious in an acidic, low-OM, and coarse-textured soil and that lessen in the long run. Biochars that are derived from manure are commonly more effective than the biochars derived from wood in settling the heavy metals of soil. Furthermore, biochar amendment is effective in stabilizing cationic hazardous elements while mobilizing anionic toxic elements (J. Yang et al. 2015). In a literature review, Ahmad et al. (2014) indicated that the immobilization of inorganic and organic contaminants in soil and water is significantly enhanced with biochars having more O-containing functional groups. Manure-derived, low-pyrolysis biochar products (i.e., <500°C) might therefore be more effective in reducing heavy-metal pollution of soil. Due to their higher pyrolysis temperatures, steamed biochars tend to have a higher specific surface area, porosity, and aromaticity levels and thus may be preferred for sorption-based organic contaminant stabilization. Since the biochar does not remove all heavy metals from soil, its application is limited to mildly contaminated soils. However, the application rate and thorough incorporation to soil must be considered in practice, with regular monitoring of the remedial effect. A similar study reported that mechanical mixing to incorporate biochar into soil improved the distribution of biochar and thereby increased PCB retention, as compared to hand and shovel mixing (Denyes et al. 2014). Even so, the biochar amendment had a positive effect on soil health and a stimulating effect on the microbial population, as demonstrated by the growth of sunflower (*Helianthus annuus*) in pot soils supplemented with charcoal (Sneath, Hutchings, and de Leij 2013).

14.5.2 Immobilization of Heavy Metals

Biochar immobilizes and decreases the phytoavailability (Kim et al. 2016), and HMs absorption by the plant (Y. Xu et al. 2018; P. Xu et al. 2016; R. Xu et al. 2011), including Al, Cr, Pb, Fe, Co, Ni, Mn, Cd, Cu, Tl, and Zn (Fellet et al. 2011). Various techniques have been put forward to comprehend the soil remediation potential of biochar. Such mechanisms include the exchange of ions, surface sorption (Egene et al. 2018), electrostatic interactions, chemical complexion, and precipitation (Abdelhafez, Li, and Abbas 2014). For instance, biochar derived from peat moss decreases the movement and bioavailability of Cd, Pb, and Cu by metal electrons coordination to form π-electron C=C bonds (Wei et al. 2019). Occurrences of the functional groups of biochar, together with HMs' binding affinity, decrease the bioavailability of biochar. Additionally, biochar has been utilized in the phytoremediation of soil in order to commend living plants going about as absorption sinks for HMs. It was observed that the alfalfa phytoremediation of polluted soil decreases the soil Cd at a rate of about 90 g Cd ha^{-1} as a result of the root exudations that complexes with Cd, thereby decreasing the absorption of Cd by the roots (M. Zhang et al. 2019). However, these experiments fail to exhibit the utilization of biochar to completely remediate the HM problem of soil. Such issues might be due, at least to some degree, to an inappropriate stoichiometric proportion between the HM load of soil and the biochar dose that is accomplished (Abdelhafez, Li, and Abbas 2014; Guo, Song, and Tian 2020). This is by and large evident, in any case, that higher rates of utilization give more HM bonding sites and a noteworthy surface area (Seneviratne et al. 2017). Likewise, there may also be stability and complex interaction with soil particles and biochar and their effect on the movement and chemistry of HM ions. It has been also suggested that nonspecific sorptions are reversible and that the HMs can be delivered to the soil solution (Campillo-Cora et al. 2020). Tests with Pb, Ni, Zn, and Cu exhibit that the aforementioned elements possess higher maintenance when they are independent (singly present) as compared to when present together in a mix, emphasizing the significance of cooperative effects in the mobility of HM. Nonetheless, it can be reasoned that the application of biochar alone does not regulate the bioavailability and mobility of HM in soil. Research has additionally shown that biochar is fit for holding and, in any event, for advancing the disintegration of different organic impurities present in soils. Modification of biochar might be a useful way to deal with and improve soils contaminated by pesticides, herbicides, PCBs, PAHs, anti-infection agents, different POPs, and oil hydrocarbons (Ahmad et al. 2014; Ahmad et al. 2014).

14.5.3 MECHANISMS

Biochar swiftly associates with the organic pollutants and the microorganisms of the soil as soon as it is introduced into the soil. The organic pollutants are balanced at the surface of the biochar and at the pores. It further can decay via microbes that get invigorated by biochar modification. The availability of a sufficient amount of functional groups, pores, and the aromatic C-condensed biochar surface is facilitated to soak several organic compounds via different methods. As the biochar adsorbs the organic pollutants available in the contaminated soil, their compositions tend to decrease in the soil, and bioaccessibility to the various soil organisms, along with the roots of the plants, is minimized. Furthermore, the biochar alteration increases the complete soil health by ameliorating the chemical, biological, and physical characteristics of the soil (Guo, Song, and Tian 2020). Biochar has the ability to fix pollutants through several chemical and physical sorption mechanisms (Figure 14.2). The mechanism involved by biochar for stabilizing the contaminants of the soil is demonstrated in Figure 14.2. As demonstrated in the figure, the biochar associates with the organic molecules through van der Waals forces, the unique electrostatic attraction allying nonpolar and the polar molecules. Even though the van der Waals forces are weak, they increase with the surface area of the molecules and could be prominent with the reduction in the size of the particles of the biochar (S. Yang and Evans 2007). The major constituents of the biochar include N, O, H, C, and several other mineral elements. It has the ability to mix with the polar organic compounds via the formation of the H-bonds through the functional groups consisting of O-, N-, and H-. The charges available at the surface of the biochar facilitate the biochar to attract electrically the polar organic molecules and the ionized organic contaminants having opposite charges. The involved electrostatic attraction is conventionally considered as powerful as H-bonding and the van der Waals linkages. The escalated electrostatic pull could cause inner-sphere adsorption wherein the ionized organic compounds chemically react along with the available functional groups at the surface of the biochar (R. Xu et al. 2011).

It can be concluded that various biochar products exhibit different adsorption ranges for adsorbing the available organic contaminants in the soil for the remediation process. The nonpolar, water-rejecting organic contaminants like PAHs and other petroleum hydrocarbons are soaked up by biochar via pore filling, followed by partition, and the hydrophobic effect. Similarly, the ionized polar organic contaminants, such as pesticides and antibiotics, are adsorbed through specific interaction, hydrogen bonding, electrostatic interaction, and surface precipitation.

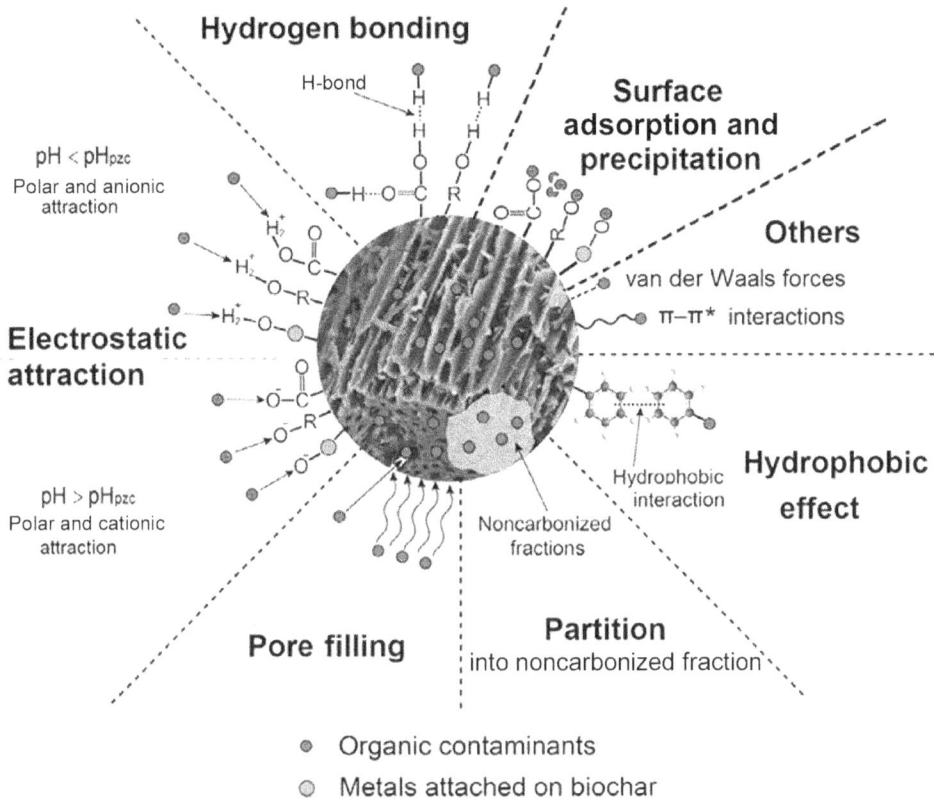

FIGURE 14.2 Involved mechanisms by which biochar stabilizes the contaminants present in soil.
Source: Guo, Song, and Tian (2020).

14.5.4 EFFICACY VARIATIONS

The productivity of modification of biochar in order to alleviate organic toxins and enable their removal from the soil is case specific, differing with the source of biochar, size of the particle, alteration dosage, chemical nature of the contaminant, and type of soil. Biochars that are obtained from rice hull, peanut straw char, canola straw, and soybean straw char through a similar pyrolysis technique of 350°C, for example, clearly displayed unique sorption capabilities for the removal of methyl violet contained in the water (R. Xu et al. 2011). Utilizing the incubation experiments for 30 days, Kołtowski et al. (2016) observed that, at an alteration rate of 5 wt%, wheat-straw-, coconut-shell-, and willow-derived biochars decreased the bioaccessible (silicon poles extractable) and easily dissolved (POM strips extractable) parts of soil PAHs gathered from three separate industrial locations by 28 to 87% compared with the unaltered controls (Kołtowski et al. 2017). Activation of steam enormously improved the SSA and pore volume of biochar and consequently the capacity of sorption with respect to PAHs (Kołtowski et al. 2017). The PAHs content of biochar (up to 65 mg kg^{-1}) (Keiluweit et al. 2010; Quilliam et al. 2013), which is mostly insignificant and poses a low-ecological danger (R.-H. Zhang et al. 2017; Weidemann et al. 2018). Altering the clay (OC 2.4%; pH 5.8) and sandy loam (OC 0.54%; pH 5.7) soils with sugarcane-residue-derived biochar (SSA 58.9 m^2 g^{-1}; pH 8.6; CEC 113.7 cmol$_c$ kg^{-1}) at about 5 wt % upgraded the 17α-ethinylestradiol (EE2, a type of estrogen chemical) sorption rate on the two soils, followed by a decrease in their mineralization of microbes (Wei et al. 2019). In fact, biochars that are derived from plant and wood residue from a higher range of pyrolysis temperatures along with steam-activated biochars generally have greater porosity, aromaticity levels, and greater SSA and are thus preferred over biochars derived from manure for sorption-related adjustment of organic pollutants. In the long term, change in biochar might work with the mineralization as well as the possible elimination of organic toxins from the soil through further development of microbial processes of soil. The processes through which the change in biochar enhances the health and microbial activities of the soil can be observed elsewhere (Guo, Xiao, and Li 2019; Guo, Song, and Tian 2020).

14.5.5 BIOCHAR NANOCOMPOSITES

Incorporating/stacking the nanomaterial(s) with biochar in order to produce nanocomposites of biochar is an alternative inventive method of accomplishing greater biochar viability for the elimination of HM (Tan et al. 2015; Mandal et al. 2021). The nanocomposites of biochar display enhanced physiochemical properties comparative with normal biochar, for example surface areas, functional groups, pore properties (Chen et al. 2014), and more noteworthy constancy (Pan et al. 2021; Pan et al. 2021). "Smart" nanocomposites of biochar are manufactured by choosing a proper feedstock and the nanomaterial(s). Contingent upon the stacking/doping strategy for the nanomaterial contained in biochar, the method of producing biochar nanocomposites is either as a pretreatment or as a post-treatment (Pan et al. 2021). Various investigations have been accomplished using biochar nanocomposites for HM remediation (counting Hg, Pb, As, and Cd) tainted soils. A new investigation regarding Fe-Mn-altered biochar, in soil with As contamination, displayed remarkable results in terms of a decrease in As contamination, along with changes in the redox potential and soil pH (X. Yang et al. 2016; R.-H. Zhang et al. 2017). Despite much research on the utilization of biochar nanocomposites for remediation of water contaminated by HM, similar research has also been carried out on HM-debased soil remediation.

14.6 BIOCHAR MODIFICATIONS

The common investigated technological variations in the biochar include chemical changes, physical changes, mineral oxides doping, and variations in magnetic properties (Ahmad et al. 2014; Rajapaksha et al. 2016). Even though the physical variations process is simple and financially feasible, the conventional approach for the physical activation of the biochar is the "steam" activation process done at the initial pyrolysis. This method involves pyrolysis to be performed in two steps, involving the introduction of charcoal to restricted steam gasification inside the pyrolysis chamber at stage two. The obtained biochar exhibits a bigger surface and better carbonic structures (Rajapaksha et al. 2016). The process involving the ejection of gas was found to enhance the surface area of the biochar along with the volume of the pore compared to the normal biochar (Xiong et al. 2013). The chemical amendment method is basically a method involving thermochemical activation involving the chemical reagents (Sakhiya, Anand, and Kaushal 2020). Previous investigations demonstrate that involved chemical amendments provide a better scope of obtaining efficient and effective biochar possessing with greater surface area and dedicated sorption sites, surface complexation, enhanced effective specific groups at the surface, and intensified electrostatic attraction (Ahmad et al. 2014; Rajapaksha et al. 2016). Further, several used techniques like the inclusion of chemical oxidation and acid/base treatment, exercising organic solvent, functional group, and surfactant altering, and the coating of the biochar have been tested (Ahmad et al. 2014; Rajapaksha et al. 2016). Another interesting methodology to attain enhanced potent biochar is to nanomaterial mixing of the biochar. The nanocomposites of biochars obtained in this way demonstrate superior physicochemical characteristics as compared to the conventionally used biochar, considering the functional groups, enhanced stability surface sites, and the pores (Pan et al. 2021).

The technique to develop the nanocomposites is designated as pretreatment/post-treatment on the basis of the loading/doping process of the nanoparticles in the biochar. Few studies conducted on the biochar nanocomposite remediation of the adulterated soils exposed the considerable variations in the redox potential, pH, and reduction in adulteration (X. Yang et al. 2016; R.-H. Zhang et al. 2017, 2017; Lin et al. 2019; Zhao et al. 2016). However, a huge number of investigations have been done considering the biochar-nanocomposites for treatment of the polluted water, but comparatively fewer investigations have been done for treatment of adulterated soils.

14.7 CONCLUSIONS

Biochars are truly magical material when it comes to their ability to remediate ill soils. They not only boost soil health but also reduce the bioavailability of heavy metals. The role of feedstock material, pyrolysis conditions, and pre-/postmodulation hold key information on the tailored mitigative product. Among various available methods for activation of biochar to make it more efficient in reducing a load of heavy metals in soil, the formation of biochar nanocomposites needs more research. Early results show that such products are promising in building soil health as well as soil remediation. The role of metal-oxide-impregnated and magnetically modified biochar also needs systematic attention and more research. Nevertheless, biochar can be referred to as "black gold" owing to its proven capabilities and hidden potentials.

REFERENCES

Abdelhafez, Ahmed A, Jianhua Li, and Mohamed H H Abbas. 2014. "Feasibility of Biochar Manufactured from Organic Wastes on the Stabilization of Heavy Metals in a Metal Smelter Contaminated Soil." *Chemosphere* 117: 66–71. doi:10.1016/j.chemosphere.2014.05.086

Ahmad, Mahtab, Anushka Upamali Rajapaksha, Jung Eun Lim, Ming Zhang, Nanthi Bolan, Dinesh Mohan, Meththika Vithanage, Sang Soo Lee, and Yong Sik Ok. 2014. "Biochar as a Sorbent for Contaminant Management in Soil and Water: A Review." *Chemosphere* 99: 19–33. doi:10.1016/j.chemosphere.2013.10.071

Angın, Dilek, and Sevgi Şensöz. 2014. "Effect of Pyrolysis Temperature on Chemical and Surface Properties of Biochar of Rapeseed (Brassica Napus L.)." *International Journal of Phytoremediation* 16 (7–8). Taylor & Francis: 684–693. doi:10.1080/15226514.2013.856842

Atkinson, Christopher J, Jean D Fitzgerald, and Neil A Hipps. 2010. "Potential Mechanisms for Achieving Agricultural Benefits from Biochar Application to Temperate Soils: A Review." *Plant and Soil* 337 (1): 1–18. doi:10.1007/s11104-010-0464-5

Campillo-Cora, Claudia, Manuel Conde-Cid, Manuel Arias-Estévez, David Fernández-Calviño, and Flora Alonso-Vega. 2020. "Specific Adsorption of Heavy Metals in Soils: Individual and Competitive Experiments." *Agronomy.* doi:10.3390/agronomy10081113

Cao, Yune, Yanming Gao, Yanbin Qi, and Jianshe Li. 2018. "Biochar-Enhanced Composts Reduce the Potential Leaching of Nutrients and Heavy Metals and Suppress Plant-Parasitic Nematodes in Excessively Fertilized Cucumber Soils." *Environmental Science and Pollution Research* 25 (8): 7589–7599. doi:10.1007/s11356-017-1061-4

Chen, Tan, Yaxin Zhang, Hongtao Wang, Wenjing Lu, Zeyu Zhou, Yuancheng Zhang, and Lulu Ren. 2014. "Influence of Pyrolysis Temperature on Characteristics and Heavy Metal Adsorptive Performance of Biochar Derived from Municipal Sewage Sludge." *Bioresource Technology* 164: 47–54. doi:10.1016/j.biortech.2014.04.048

Denyes, Mackenzie J, Michèle A Parisien, Allison Rutter, and Barbara A Zeeb. 2014. "Physical, Chemical and Biological Characterization of Six Biochars Produced for the Remediation of Contaminated Sites." *Journal of Visualized Experiments : JoVE* 93 (November). MyJove Corporation: e52183—e52183. doi:10.3791/52183

Egene, C E, R Van Poucke, Y S Ok, E Meers, and F M G Tack. 2018. "Impact of Organic Amendments (Biochar, Compost and Peat) on Cd and Zn Mobility and Solubility in Contaminated Soil of the Campine Region after Three Years." *Science of The Total Environment* 626: 195–202. doi:10.1016/j.scitotenv.2018.01.054

Fellet, G, L Marchiol, G Delle Vedove, and A Peressotti. 2011. "Application of Biochar on Mine Tailings: Effects and Perspectives for Land Reclamation." *Chemosphere* 83 (9): 1262–1267. doi:10.1016/j.chemosphere.2011.03.053

Fidel, Rivka B, David A Laird, and Kurt A Spokas. 2018. "Sorption of Ammonium and Nitrate to Biochars Is Electrostatic and PH-Dependent." *Scientific Reports* 8 (1): 17627. doi:10.1038/s41598–018–35534-w

Fryda, Lydia, and Rianne Visser. 2015. "Biochar for Soil Improvement: Evaluation of Biochar from Gasification and Slow Pyrolysis." *Agriculture* 5 (4): 1076–1115. doi:10.3390/agriculture5041076

Guo, Mingxin, Pengli Xiao, and Hong Li. 2019. "Valorization of Agricultural Byproducts Through Conversion to Biochar and Bio-Oil." *Byproducts from Agriculture and Fisheries*. Wiley Online Books. doi:10.1002/9781119383956.ch21

Guo, Mingxin, Weiping Song, and Jing Tian. 2020. "Biochar-Facilitated Soil Remediation: Mechanisms and Efficacy Variations." *Frontiers in Environmental Science* 8 (October). doi:10.3389/fenvs.2020.521512

Hardie, Marcus, Brent Clothier, Sally Bound, Garth Oliver, and Dugald Close. 2014. "Does Biochar Influence Soil Physical Properties and Soil Water Availability?" *Plant and Soil* 376 (1): 347–361. doi:10.1007/s11104–013–1980-x

Herath, I, P Kumarathilaka, A Navaratne, N Rajakaruna, and M Vithanage. 2015. "Immobilization and Phytotoxicity Reduction of Heavy Metals in Serpentine Soil Using Biochar." *Journal of Soils and Sediments* 15 (1): 126–138. doi:10.1007/s11368-014-0967-4

Huang, H, N G Reddy, X Huang, P Chen, P Wang, Y Zhang, Y Huang, P Lin, and A Garg. 2021. "Effects of Pyrolysis Temperature, Feedstock Type and Compaction on Water Retention of Biochar Amended Soil." *Scientific Reports* 11: 7419.

Joseph, Stephen, Claudia I Kammann, Jessica G Shepherd, Pellegrino Conte, Hans-Peter Schmidt, Nikolas Hagemann, Anne M Rich, et al. 2018. "Microstructural and Associated Chemical Changes during the Composting of a High Temperature Biochar: Mechanisms for Nitrate, Phosphate and Other Nutrient Retention and Release." *Science of The Total Environment* 618: 1210–1223. doi:10.1016/j.scitotenv.2017.09.200

Keiluweit, Marco, Peter S Nico, Mark G Johnson, and Markus Kleber. 2010. "Dynamic Molecular Structure of Plant Biomass-Derived Black Carbon (Biochar)." *Environmental Science & Technology* 44 (4). American Chemical Society: 1247–1253. doi:10.1021/es9031419

Kim, Hyuck-Soo, Kwon-Rae Kim, Jae E Yang, Yong Sik Ok, Gary Owens, Thomas Nehls, Gerd Wessolek, and Kye-Hoon Kim. 2016. "Effect of Biochar on Reclaimed Tidal Land Soil Properties and Maize (*Zea mays* L.) Response." *Chemosphere* 142: 153–159.

Kołtowski, Michał, Barbara Charmas, Jadwiga Skubiszewska-Zięba, and Patryk Oleszczuk. 2017. "Effect of Biochar Activation by Different Methods on Toxicity of Soil Contaminated by Industrial Activity." *Ecotoxicology and Environmental Safety* 136: 119–125. doi:10.1016/j.ecoenv.2016.10.033

Lin, Lina, Zhongyang Li, Xuewei Liu, Weiwen Qiu, and Zhengguo Song. 2019. "Effects of Fe-Mn Modified Biochar Composite Treatment on the Properties of As-Polluted Paddy Soil." *Environmental Pollution* 244: 600–607. doi:10.1016/j.envpol.2018.10.011

Liu, Jing, Yong Jun Liu, Yu Liu, Zhe Liu, and Ai Ning Zhang. 2018. "Quantitative Contributions of the Major Sources of Heavy Metals in Soils to Ecosystem and Human Health Risks: A Case Study of Yulin, China." *Ecotoxicology and Environmental Safety* 164: 261–269. doi:10.1016/j.ecoenv.2018.08.030

Mandal, Sandip, Shengyan Pu, Sangeeta Adhikari, Hui Ma, Do Heyoung Kim, Yingchen Bai, and Deyi Hou. 2021. "Progress and Future Prospects in Biochar Composites: Application and Reflection in the Soil Environment." *Critical Reviews in Environmental Science and Technology* 51 (3). Taylor & Francis: 219–271. doi:10.1080/10643389.2020.1713030

Méndez, A, A M Tarquis, A Saa-Requejo, F Guerrero, and G Gascó. 2013. "Influence of Pyrolysis Temperature on Composted Sewage Sludge Biochar Priming Effect in a Loamy Soil." *Chemosphere* 93 (4): 668–676. doi:10.1016/j.chemosphere.2013.06.004

O'Connor, David, Tianyue Peng, Guanghe Li, Shuxiao Wang, Lei Duan, Jan Mulder, Gerard Cornelissen, Zhenglin Cheng, Shengmao Yang, and Deyi Hou. 2018. "Sulfur-Modified Rice Husk Biochar: A Green Method for the Remediation of Mercury Contaminated Soil." *Science of The Total Environment* 621: 819–826. doi:10.1016/j.scitotenv.2017.11.213

O'Connor, David, Tianyue Peng, Junli Zhang, Daniel C W Tsang, Daniel S Alessi, Zhengtao Shen, Nanthi S Bolan, and Deyi Hou. 2018. "Biochar Application for the Remediation of Heavy Metal Polluted Land: A Review of in Situ Field Trials." *Science of The Total Environment* 619–620: 815–826. doi:10.1016/j.scitotenv.2017.11.132

Pan, Xuqin, Zhepei Gu, Weiming Chen, and Qibin Li. 2021. "Preparation of Biochar and Biochar Composites and Their Application in a Fenton-like Process for Wastewater Decontamination: A Review." *Science of The Total Environment* 754: 142104. doi:10.1016/j.scitotenv.2020.142104

Quilliam, Richard S, Sally Rangecroft, Bridget A Emmett, Thomas H Deluca, and Davey L Jones. 2013. "Is Biochar a Source or Sink for Polycyclic Aromatic Hydrocarbon (PAH) Compounds in Agricultural Soils?" *GCB Bioenergy* 5 (2). John Wiley & Sons, Ltd: 96–103. doi:10.1111/gcbb.12007

Rajapaksha, Anushka Upamali, Season S Chen, Daniel C W Tsang, Ming Zhang, Meththika Vithanage, Sanchita Mandal, Bin Gao, Nanthi S Bolan, and Yong Sik Ok. 2016. "Engineered/Designer Biochar for Contaminant Removal/Immobilization from Soil and Water: Potential and Implication of Biochar Modification." *Chemosphere* 148: 276–291. doi:10.1016/j.chemosphere.2016.01.043

Sakhiya, Anil Kumar, Abhijeet Anand, and Priyanka Kaushal. 2020. "Production, Activation, and Applications of Biochar in Recent Times." *Biochar* 2 (3): 253–285. doi:10.1007/s42773-020-00047-1

Seneviratne, Mihiri, Lakshika Weerasundara, Yong Sik Ok, Jörg Rinklebe, and Meththika Vithanage. 2017. "Phytotoxicity Attenuation in Vigna Radiata under Heavy Metal Stress at the Presence of Biochar and N Fixing Bacteria." *Journal of Environmental Management* 186: 293–300. doi:10.1016/j.jenvman.2016.07.024

Sneath, Helen E, Tony R Hutchings, and Frans A A M de Leij. 2013. "Assessment of Biochar and Iron Filing Amendments for the Remediation of a Metal, Arsenic and Phenanthrene Co-Contaminated Spoil." *Environmental Pollution* 178: 361–366. doi:10.1016/j.envpol.2013.03.009

Tan, Xiaofei, Yunguo Liu, Guangming Zeng, Xin Wang, Xinjiang Hu, Yanling Gu, and Zhongzhu Yang. 2015. "Application of Biochar for the Removal of Pollutants from Aqueous Solutions." *Chemosphere* 125: 70–85. doi:10.1016/j.chemosphere.2014.12.058

Weidemann, Eva, Wolfram Buss, Mar Edo, Ondřej Mašek, and Stina Jansson. 2018. "Influence of Pyrolysis Temperature and Production Unit on Formation of Selected PAHs, Oxy-PAHs, N-PACs, PCDDs, and PCDFs in Biochar—a Screening Study." *Environmental Science and Pollution Research* 25 (4): 3933–3940. doi:10.1007/s11356–017–0612-z

Wei, Zhuo, Jim J Wang, Amy B Hernandez, Andrea Warren, Jong-hwan Park, Yili Meng, Syam K Dodla, and Changyoon Jeong. 2019. "Effect of Biochar Amendment on Sorption-Desorption and Dissipation of 17α-ethinylestradiol in Sandy Loam and Clay Soils." *Science of The Total Environment* 686: 959–967. doi:10.1016/j.scitotenv.2019.06.050

Xiong, Zhang, Zhang Shihong, Yang Haiping, Shi Tao, Chen Yingquan, and Chen Hanping. 2013. "Influence of NH3/CO2 Modification on the Characteristic of Biochar and the CO2 Capture." *BioEnergy Research* 6 (4): 1147–1153. doi:10.1007/s12155-013-9304-9

Xu, Ping, Cai-Xia Sun, Xue-Zhu Ye, Wen-Dan Xiao, Qi Zhang, and Qiang Wang. 2016. "The Effect of Biochar and Crop Straws on Heavy Metal Bioavailability and Plant Accumulation in a Cd and Pb Polluted Soil." *Ecotoxicology and Environmental Safety* 132: 94–100. doi:10.1016/j.ecoenv.2016.05.031

Xu, Ren-kou, Shuang-cheng Xiao, Jin-hua Yuan, and An-zhen Zhao. 2011. "Adsorption of Methyl Violet from Aqueous Solutions by the Biochars Derived from Crop Residues." *Bioresource Technology* 102 (22): 10293–10298. doi:10.1016/j.biortech.2011.08.089

Xu, Yilu, Balaji Seshadri, Binoy Sarkar, Hailong Wang, Cornelia Rumpel, Donald Sparks, Mark Farrell, Tony Hall, Xiaodong Yang, and Nanthi Bolan. 2018. "Biochar Modulates Heavy Metal Toxicity and Improves Microbial Carbon Use Efficiency in Soil." *Science of The Total Environment* 621: 148–159. doi:10.1016/j.scitotenv.2017.11.214

Yang, Jingli, Zhuo Chen, Songquan Wu, Yi Cui, Lu Zhang, Hao Dong, Chuanping Yang, and Chenghao Li. 2015. "Overexpression of the Tamarix Hispida ThMT3 Gene Increases Copper Tolerance and Adventitious Root Induction in Salix Matsudana Koidz." *Plant Cell, Tissue and Organ Culture (PCTOC)* 121 (2): 469–479. doi:10.1007/s11240-015-0717-3

Yang, S, and J R G Evans. 2007. "Metering and Dispensing of Powder; the Quest for New Solid Freeforming Techniques." *Powder Technology* 178 (1): 56–72. doi:10.1016/j.powtec.2007.04.004

Yang, Xing, Jingjing Liu, Kim McGrouther, Huagang Huang, Kouping Lu, Xi Guo, Lizhi He, et al. 2016. "Effect of Biochar on the Extractability of Heavy Metals (Cd, Cu, Pb, and Zn) and Enzyme Activity in Soil." *Environmental Science and Pollution Research* 23 (2): 974–984. doi:10.1007/s11356-015-4233-0

Yuan, Haoran, Tao Lu, Dandan Zhao, Hongyu Huang, Kobayashi Noriyuki, and Yong Chen. 2013. "Influence of Temperature on Product Distribution and Biochar Properties by Municipal Sludge Pyrolysis." *Journal of Material Cycles and Waste Management* 15 (3): 357–361. doi:10.1007/s10163-013-0126-9

Yu, Haowei, Weixin Zou, Jianjun Chen, Hao Chen, Zebin Yu, Jun Huang, Haoru Tang, Xiangying Wei, and Bin Gao. 2019. "Biochar Amendment Improves Crop Production in Problem Soils: A Review." *Journal of Environmental Management* 232: 8–21. doi:10.1016/j.jenvman.2018.10.117

Zama, Eric F, Brian J Reid, Hans Peter H Arp, Guo-Xin Sun, Hai-Yan Yuan, and Yong-Guan Zhu. 2018. "Advances in Research on the Use of Biochar in Soil for Remediation: A Review." *Journal of Soils and Sediments* 18 (7): 2433–2450. doi:10.1007/s11368-018-2000-9

Zhang, Kaikai, Peng Sun, Marie Christine A S Faye, and Yanrong Zhang. 2018. "Characterization of Biochar Derived from Rice Husks and Its Potential in Chlorobenzene Degradation." *Carbon* 130: 730–740. doi:10.1016/j.carbon.2018.01.036

Zhang, Manyun, Jun Wang, Shahla Hosseini Bai, Yaling Zhang, Ying Teng, and Zhihong Xu. 2019. "Assisted Phytoremediation of a Co-Contaminated Soil with Biochar Amendment: Contaminant Removals and Bacterial Community Properties." *Geoderma* 348: 115–123. doi:10.1016/j.geoderma.2019.04.031

Zhang, Run-Hua, Zhi-Guo Li, Xu-Dong Liu, Bin-cai Wang, Guo-Lin Zhou, Xing-Xue Huang, Chu-Fa Lin, Ai-hua Wang, and Margot Brooks. 2017. "Immobilization and Bioavailability of Heavy Metals in Greenhouse Soils Amended with Rice Straw-Derived Biochar." *Ecological Engineering* 98: 183–188. doi:10.1016/j.ecoleng.2016.10.057

Zhang, Yaping, Zhenyan Chen, Weiwei Xu, Qilin Liao, Huiyan Zhang, Shefeng Hao, and Sihui Chen. 2020. "Pyrolysis of Various Phytoremediation Residues for Biochars: Chemical Forms and Environmental Risk of Cd in Biochar." *Bioresource Technology* 299: 122581. doi:10.1016/j.biortech.2019.122581

Zhao, Ling, Xinde Cao, Wei Zheng, John W Scott, Brajendra K Sharma, and Xiang Chen. 2016. "Copyrolysis of Biomass with Phosphate Fertilizers To Improve Biochar Carbon Retention, Slow Nutrient Release, and Stabilize Heavy Metals in Soil." *ACS Sustainable Chemistry & Engineering* 4 (3). American Chemical Society: 1630–1636. doi:10.1021/acssuschemeng.5b01570

15 Use of Biochar for the Remediation of Heavy-Metal-Contaminated Soil

Mukesh Sharma, Pranjal P. Das, Niladri Shekhar Samanta, Arun Chakraborty, and Mihir K. Purkait

CONTENTS

15.1 INTRODUCTION

Heavy metals (HMs) end up in the soil, whether they come from natural or anthropogenic sources. Fundamentally, the primary anthropogenic activities leading to the mixing of HMs in the soil are exploitation of mineral resources and smelting, paint and coating processing, metal electroplating, farmland sewage irrigating, electronic equipment manufacturing, and nonstandard application of pesticide and chemical fertilizer (Shu, Wang, and Zhong 2016; Dong et al. 2009; Beesley, Moreno-Jiménez, and Gomez-Eyles 2010). In the last half century, about 30,000 tons of chromium and around 8 million tons of lead have been discharged into the environment, of which the maximum portion is absorbed by the soil (Lee et al. 2010). Until now, cadmium, lead, and arsenic pollution, as well as the accompanying ecological health concerns, have been more severe in southeast China than in northwest China, and industrial regions have been worse than agricultural regions (Li et al. 2014; Dong et al. 2009). As per the latest national soil survey

DOI: 10.1201/9781003203438-15

report published in 2014, the rates of cadmium (7.0%), mercury (1.6%), arsenic (2.7%), copper (2.1%), lead (1.5%), chromium (1.1%), zinc (0.9%), and nickel (4.8%) are available in the soils of China. These HMs are involved in the food chain through the soils where agriculture is practiced and further get accumulated in the bodies of the organisms via nutrition, skin contact, lung inhalation, and several other pathways, causing substantial detrimental impacts on human health either directly or indirectly (Kambo and Dutta 2015; Cheng et al. 2020; Li et al. 2014). The most urgent warning is the emergence of cancer communities produced by HMs. Soil contamination and reduction in cultivable land due to HMs pollution illustrate the requirement of adopting the range of the effective in situ and ex situ remediation strategies in order to minimize the effects of HMs and to enhance the quality and security of cultivable land. Physical, chemical, and biological remediation strategies have been incorporated in the past few years (Li et al. 2014). However, every individual process has its own advantages and disadvantages leading to its applicability. This basically includes the involved procedure, economic feasibility, efficiency, and other relevant parameters (Kambo and Dutta 2015). By implementing the necessary changes to the HM-contaminated soil is currently considered the best in situ remediation method (Li et al. 2014). Liming materials, clay minerals, phosphate compounds, organic composts, coal fly ash, biochar, and metal oxides are among the most commonly utilized soil additives (Chou, Wey, and Chang 2010). Apart from several techniques for environmental remediation, membrane based and electrochemical process have been reported (Sinha and Purkait 2015; Purkait, Bhattacharya, and De 2005; Sinha and Purkait 2014; Nandi, Uppaluri, and Purkait 2009; Nandi et al. 2009; Emani, Uppaluri, and Purkait 2013; Bulasara et al. 2011; Sharma et al. 2021). In summary, HMs are immobilized primarily via the complexation, adsorption, precipitation, and reduction processes, leading to the distribution of HMs to solid phases from liquid phases of the soil and thereby limiting their transportation and bioavailability (Kambo and Dutta 2015).

Biochar is rich in carbon and exhibits the pores along with pH in the range of neutral to alkaline, greater specific surface area, negative charges in the surface and high proton exchange capacity (Li et al. 2014; W. J. Liu, Jiang, and Yu 2015). Seed germination, crop yields, microbial activity and population, and plant growth (Tong et al. 2011; Cheng et al. 2020), have been found to be available in higher quantity in the HM-contaminated soil treated with biochars, according to numerous studies. In the meantime, many similar studies and field trials (Li et al. 2014) have validated the effects of biochar on the immobilization/mobilization of various types of HMs. Furthermore, the biochar manufacturing process is recognized as an effective management strategy for disposing of a huge number of organic wastes, with significant economic and feasibility advantages. Surprisingly, there are still some possible hazards in the field application of biochar, which could stymie its growth.

Henceforth, the chapter illustrates the various aspects of biochar like the different types of preparation techniques, their performance and influencing parameters, the mechanisms in the involved interaction with various HMs, the effects on the growth of the plants, and lastly the involved potential threats in the field of several engineering applications.

15.2   PREPARATION OF BIOCHAR

Wood branches/chips, agricultural residue, sewage sludge, other organic waste, animal manure, and other woody biomass are the key materials for biochar production. Table 15.1 illustrates several thermochemical conversion technics such as pyrolysis, intermediate pyrolysis, hydrothermal carbonization, gasification, and torrefaction that are involved in the preparation of biochar. Bio-oil, biochar, and syngas are the main products of these technologies, and they are classified based on residence time, different heating rates, peak temperature, reaction atmospheres, and other variables (W. J. Liu, Jiang, and Yu 2015). Generally, different operating conditions are highly affected on biochar yield; various conventional thermochemical conversion methods and their estimated product yields are summarized in Table 15.1.

TABLE 15.1

Reaction Conditions and Product Distribution of Various Thermochemical Conversion Technologies (reprinted with permission Maguyon-Detras et al. (2020))

Mode	Process conditions		Product distribution (%)		
	Peak temperature	Vapor residence time	Char	Liquid	Gas
Slow	Moderate (~500°C)	Long (5–30 min)	35%	30% (70% water)	35%
Intermediate	Moderate (~500°C)	Moderate (10–20 s)	20–25%	50% (50% water)	25–30%
Fast	Moderate (~500°C)	Short (<2 s)	12%	75% (25% water)	13%
Gasification	High (>800°C)	Moderate (10–20 s)	10%	5% tar (55% water)	85%

As shown by fast pyrolysis, biomass breakdown proceeds thermochemically with low energy density at a sufficient pyrolysis temperature in the presence or absence of oxygen. This technique produces bio-oil having syngas, high energy density, a low energy density, and a tiny quantity of biochar due to the greater rate of heat energy (>200 K min^{-1}) and lowers residence period (<2 s), as shown in Table 15.1. On the other hand, slow pyrolysis is the most usual kind of pyrolysis that promotes the production of biochar instead of the formation of liquid and gaseous components. The pyrolization of biowaste is carried out in this process over a wide range of carbonization temperatures and heating rates ranging from 0.1 to 1°C s^{-1} for treatment times ranging from a few hours to days (W. J. Liu, Jiang, and Yu 2015). The fixed carbon amount in biochar is increased with peak temperature, with a temperature range of 400–500°C, which is a consequence of slow pyrolysis. The solid–liquid product yield is balanced by the operating environment of intermediate pyrolysis that lies between rapid and slow pyrolysis. In addition to that, pyrolysis treatment resulting in a brittle kind of biochar that does not comprise reactive tar and that can be utilized for soil amendments, solid fuel, and fertilizer. Gasification is the oxidation of dry biomass in direct contact with steam, carbon dioxide, oxygen, air, nitrogen, or a combination of these substances. The principal yield of gasification is an inflammable gas that is rich in CH_4, H_2, and, CO, whereas low-yield biochar includes a high concentration of hazardous chemicals such as polyaromatic hydrocarbons, alkaline earth metals, and alkali that are yielded after high-temperature reaction (Kambo and Dutta 2015). The carbonization of biochar is conducted by means of conventional hydrothermal treatment under a predetermined pressure and temperature in the range of 2–6 MPa and 180–300°C, respectively, where drying pretreatment is not required because in this treatment method wet or dry biomass is directly blended with water. Hydrothermal carbonization possesses have a wide range of applications unlike other methods, namely gasification, pyrolysis, etc. The high yield and high purity hydrochar show a greater level of surface functional groups and aromatization. Moreover, hydrochar comprises reduced alkali and alkaline earth and heavy metal content, as well as a greater amount of carbon and heating value. Torrefaction, also known as moderate pyrolysis, results in a 30% mass reduction of biomass. Torrefaction generates organic carbon compounds with high specific energy density; however, they cannot be called biochar since torrefaction is only the first step in the pyrolysis technique (Kambo and Dutta 2015). As a result, the physicochemical characteristics of the product from torrefaction are somewhere between biochar and biomass, containing some volatile organic components.

Pyrolysis is the traditional method for making biochar. In this chapter, the characteristics of pyrolyzed biochar are discussed first. In addition, the strong relationship between biomass raw materials and the performance features of biochar, residence period, and pyrolysis temperature is also explained. The performance characteristics are discussed in depth in the next chapter based on the aforementioned parameters (various synthesis conditions) that gives a theoretical framework for describing the biochar–soil HMs interaction mechanism (W. J. Liu, Jiang, and Yu 2015).

15.3 PERFORMANCE CHARACTERIZATION OF BIOCHAR

15.3.1 Metal Biochar Interactions in Soil: Mechanisms

The various raw materials or feed materials sources and various pyrolysis circumstances employed in biochar synthesis result in different biochar properties, which may impact biochar–metal interaction. Organic carbon content, specific surface area (SSA), cation exchange capacity (CEC), pH, active functional groups, microporous structure, and mineral content are only a few examples. These changes, in combination with variations in soil's characteristics, could affect biochar–metal bindings in soils, affecting metal motility and bioavailability following biochar addition to soil, as stated by Qi et al. (2017). As a result, it's difficult to anticipate and comprehend the mechanism of biochar–metal combinations in various clay. As illustrated in Figure 15.1, biochar influences metal motility and biological availability through two mechanisms: (1) metals and biochar direct interactions and (2) altering soil properties and therefore indirectly increasing metal attainability. As shown in Figure 15.1, the direct interaction between the metals and the biochar is done using electrostatic attraction, precipitation, complexation, and ion exchange mechanisms.

15.3.1.1 Direct Interactions

Different feedstocks such as manure, farm waste, animal bodies, or woody plant for biochar production have significant effects on heavy metal (HMs) immobilization in contaminated soils, as demonstrated by Meng et al. (2018). The proposed mechanisms, namely ion exchange, electrostatic attraction, complexation, and precipitation, are described next.

15.3.1.1.1 Electrostatic Attraction

Zeta potential values are often applied to define the adsorbent material's electrostatic potential. According to recent findings, positively charged ions are attracted to the highly electronegative charge density biochar surface due to electrostatic force, as stated by Ahmad et al. (2018). Chou et al. (2010)

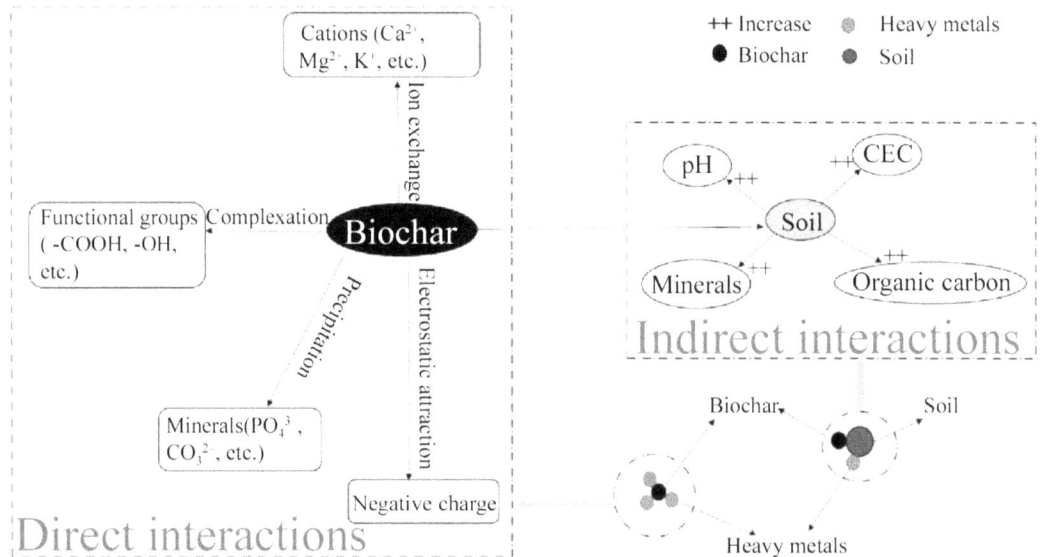

FIGURE 15.1 Influence of biochar on metal motility and biological availability via (1) direct interactions between metals and biochar and (2) altering soil properties and therefore indirectly increasing metal attainability.

Source: Reprinted with permission (He et al. 2019).

and Ahmad et al. (2018) also revealed that the negatively charged surface group presents in the surface, resulting in a high surface charge intensity that may increase at higher pH. Tong et al. (2011) and Ahmad et al. (2018) examined that the Cu (II) uptake capability was influenced by increasing pH over synthesized biochar. With increasing HMs concentrations, the electrostatic sorption also increases.

15.3.1.1.2   Ion Exchange Property

Biochars have an excellent CEC property and can release cations like Mg (II) and Ca (II) that interact with metal ions on biochar surfaces (Li et al. 2014). The recent study also proposed that the enhanced CEC of biochar may raise its HMs adsorption capacity. For instance, biochar derived from animal manure comprises a high amount of Ca (II) compared to plant-extract biochars; therefore, animal-based biochars have an efficient role in the removal of Cu (II) and Cd (II). In addition, according to Ho et al. (2017) and Ho, Zhu, and Chang (2017), oxygen-facilitated functional groups in biochar, particularly carboxyl groups (-COOH), absorb metal ions through ion exchange techniques.

15.3.1.1.3   Complexation

HMs can be immobilized by surface complexation of biochar's surface operational groups, especially in biochars with low mineral content. As an example, surface complexation of crop–residue-modified biochar predominantly takes up HMs (Tong et al. 2011). The findings also demonstrated that biochar's functional groups like -OH, -COOH, C N, and -C O- act as binding sites for HMs, allowing them to form composites that improve metal-specific sorption. Bashir et al. (2018) observed a significant change in band intensity in the range of 2349–2360 cm^{-1} and 2341 cm^{-1} in FTIR spectra when biochar was applied to cadmium- (II)-polluted soil, implying metal–biochar complexation.

The key mechanism in HM immobilization is due to the interaction of the HMs-biochar functional group, which accounts for 38.2–42.3% of total Pb (II), as reported by Lu et al. (2012). The capability of metal–biochar complexation improves when Mn (II), Fe (II), and carbonate concentrations rise, leading to the creation of a metal complex that is stable and insoluble (Ahmad et al. 2014). They also stated that inorganic ions in biochar, such as Cl, S, and Si, can form complexes with HMs like Cd (II), therefore reducing their mobility in soil.

15.3.1.1.4   Precipitation

Biochar's mineral components may precipitate with metal, resulting in insoluble precipitates. In comparison to bamboo-based biochar, Qin et al. (2018) proposed that adding 8.5% P content pig-based biochars to soil reduced Cd and Pb leaching. The sorption of inorganic P from P-enriched biochar resulted in the precipitation of $Pb_{10}(PO_4)_6(OH)_2$ and $Ca_2Pb_8(PO_4)_6(OH)_2$, as suggested by Liang et al. (2014). Furthermore, precipitation of chloride, Pb oxide, and sulfate may occur depending on the minerals present in particular biochars Meng et al. (2014

15.3.1.2   Indirect Interactions

As previously discussed, direct interactions that occur among metal and biochar, such as complexation, electrostatic attraction, precipitation, and ion exchange, have received a lot of attention. Despite this, changes in biochar have indirect effects on bioavailability and metal mobility, such as changes in soil characteristics and hence metal–soil complexation, which is undervalued. Biochar addition may cause changes in soil properties such as pH, sulfate, dissolved organic carbon, and CEC, as revealed by Abbas et al. (2021).

15.3.1.2.1   Biochar Amendment Impacts Soil pH and Thus Metal Mobility/Bioavailability

Dong et al. (2009) revealed that the pH of the soil is a major determinant of metal speciation and mobility. Biochar is typically alkaline; therefore, increasing application rates might raise soil pH, especially in acid soils. Table 15.2 illustrates a few of the physicochemical amendment parameters. As illustrated in the table, the oxidizable transformation and remaining portions of heavy metals are sped up by increasing their hydrolysis, increasing their adsorption by soil, and increasing their

TABLE 15.2
Important amendment parameters (reprinted with permission from Venegas et al. 2016)

Parameter		MOW2	BF	BS
pH		8.1	8.6	9.4
ANC (meq kg^{-1})		4180	420	725
Moisture (%)		6.0	4.5	6.0
FC (%)		175	116	100
LOI (%)		60	95	95
DOC (mg L^{-1})		475	10	220
CEC (cmol$_c$ kg^{-1})		65	25	55
Cd (mg kg^{-1})	TC	6.4	1.2	0.6
	WS	0.4	0.8	0.1
Cu (mg kg^{-1})	TC	150	33	20
	WS	7.3	0.4	0.2
Ni (mg kg^{-1})	TC	75	15	1.6
	WS	0.9	<l.q.	<l.q.
Pb (mg kg^{-1})	TC	175	4.2	1.7
	WS	1.9	1.6	1.0
Zn (mg kg^{-1})	TC	430	75	105
	WS	2.1	9.2	0.8
Effect of aging in selected amendment properties				
pH	T0	8.1	8.6	9.4
	T2	8.0	8.3	8.7
ANC (meq kg^{-1})	T0	4180	420	725
	T2	4915	515	790
CEC (cmol$_c$ kg^{-1})	T0	65	25	55
	T2	85	30	60
DOC (mg L^{-1})	T0	475	10	220
	T2	390	6	245

ANC (acid neutralization capacity), *FC* (field capacity), *LOI* (loss on ignition), *DOC* (dissolved organic carbon), *CEC* (cation exchange capacity), *TC* (total content), *WS* (water soluble)

adsorption by soil (Dong et al. 2009). It was also proposed by Jiang et al. (2012) that increased pH may also promote HMs complexation, resulting in a reduction of Pb (II)from soils. The increased pH in the soil as a consequence of the biochar amendment thus reduces the HMs' risk.

15.3.1.2.2   *Biochar Amendment Impacts Soil CEC and Thus Metal Mobility/Bioavailability*

Increasing the CEC value of biochar might raise the soil CEC. The study also confirmed that the highest number of cation exchange sites found in biochar reduces the leachability, solubility, and concentration of toxic metals in soil. As an example, Jiang et al. (2012) attributed that 30 days of biochar modification shows an increasing CEC value, enhancing the Pb (II) sorption capacity. They also determined that better sorption capacity for Cu (II) and Pb (II) in biochar-embedded soil increased the CEC of soil.

15.3.1.2.3   *Biochar Amendment Impacts Soil Mineral Composition and Metal Mobility/Bioavailability*

Uchimiya et al. (2011) stated that biochar comprises a substantial amount of minerals such as Ca, P, K, Na, Mg, which may be released into solids as a result of biochar modification. Therefore, discharged minerals from mineral phases to biochar surfaces induce metal adsorption from the soil. For example, the quantity of P in soils grows with higher biochar application rates, resulting in Pb (II) retention owing to the production of stable phosphate minerals (Cao et al. 2009). The laser ablation–inductively coupled plasma–mass spectrometric (LA-ICP-MS) analysis of biochar after mitigation confirmed the reduction of minerals like Mg, K, and P from biochar, while Pb (II) and Cu (II) were significantly increased, indicating the possibility of increasing HM adsorption in solids by releasing minerals from biochar (Bian et al. 2014). Biochar also contains Ca, Mn, and Si oxides, all of which can be partially diminished and can provide high-energy sorption sites for metal cations in soil (Bian et al. 2014).

15.3.1.2.4   Biochar Amendment Impacts Soil Organic Carbon Content and Metal Mobility/Bioavailability

The presence of the metal-oxygen functional group in the modified biochar could discharge the dissolved organic carbon, thus enhancing the organic carbon content (OC) of the soil; as a consequence, bioavailability and metal mobility may be reduced (Ibrahim et al. 2016). The addition of biochar increases soil OC, which can convert unstable Pb (II) into less mobile components, viz. organically bound fractions, reducing plant's adsorption capacity of HMs (Abdelhafez, Li, and Abbas 2014). Meanwhile, the added biochar raises the soil's OC, which might result in a variety of outcomes. For instance, despite the concentrations of As (total) and Cu (II) in pore water rising by greater than 30 times, the amount of Cd (II), and Zn (II) concentration dropped considerably in response to the increase in OC and soil pH following charcoal addition (Beesley, Moreno-Jiménez, and Gomez-Eyles 2010).

15.3.2   Modification Methods

The use of biochar as a technique of treating the polluted soil and water has been proposed. To meet remediation standards, traditional biochar's practical applications for pollutant immobilization and removal need to be improved. As a result, biochar modification has received a lot of interest recently. To improve biochar's remediation efficacy, chemical and physical methods including the activation, magnetization, digestion, and oxidation are commonly applied. Steam activation is a reliable method of activation. The surface of the biochar exhibits more complicated pore structure and a higher volume density postactivation process. Hass et al. (2012) looked at how pyrolysis temperature and steam activation affected biochar characteristics (Hass et al. 2012). Steam activation was discovered to raise the pH and SSA of the biochar. However, the steam activation of the biochar prepared at 350°C has a liming effect equal to the one prepared at 700°C. Another technique for enhancing the sorption capacity of the biochar is magnetization. Magnetic biochar exhibits raised ferromagnetic capacity, permitting the exhausted/spent biochar to be separated easily. Chen et al. (2011) investigated three magnetic biochars made by chemical coprecipitation of iron (II) (Fe2+) and iron (III) (Fe3+) on orange peel powder and discovered that the prepared biochar (magnetized) has a greater pore diameter and is better at removing organic pollutants and phosphate from the water (Chen, Chen, and Lv 2011). Additionally, the γ-iron (III) oxide (γ-Fe$_2$O$_3$) nanoparticles at the surface of the carbon served as sorption sites via electrostatic interactions. The as-prepared magnetized biochar can be isolated easily and eliminated using external magnets. Further, they may be employed in several arsenic (As) contaminant removal techniques (Wang et al. 2015).

To increase the adsorption capabilities of biochar, several oxidants including potassium permanganate (KMnO$_4$), nitric acid (HNO$_3$) and hydrogen peroxide (H$_2$O$_2$) are incorporated. Biochar's surface postoxidation has a huge amount of functional groups of acids. Further, reports suggest nitric acid to be more effective than the KMnO$_4$ potassium permanganate. Biochar's adsorption capability can also be improved through digestion. Biochar extracted from the biomass feedstocks, including beetroot (Yao et al. 2011), dairy manure, and bagasse (Inyang et al. 2010) and treated using the anaerobic digestion process, exhibit better adsorption capacity as compared to the biochar generated without digestion treatment. Furthermore, the adsorption capacity of these biochars can be compared to activated carbon, which is available commercially. The digested bagasse biochar exhibits raised pH levels, CEC, surface area, and the water loving property. It also has a raised negative charge at the surface (Inyang et al. 2010). The CEC, functional groups, SSA, pH, and several other physicochemical characteristics of the biochar are all affected differently by the various alteration processes. These changes are linked to the effectiveness of rehabilitation of the soil mixed with the toxic HMs. The major markers for evaluating biochar's heavy metal remediation capabilities are its porous structure, alkalinity, and the available oxygen functional groups at the surface.

15.4 EFFECT OF BIOCHAR ON HEAVY-METAL-CONTAMINATED SOIL

15.4.1 MOBILITY

HMs can be adsorbed by biochar, reducing environmental threats in soil. Several batch processes, such as column leaching and the toxicity characteristic leaching procedure (TCLP), are involved in examining heavy metal's leachability in contaminated soils. Various biomass raw materials generated distinguished biochar materials that influence heavy metal's movability in soil. On the other hand, biochar has the potential to mobilize HMs from soil granules into soil slurries. For instance, phosphorous (P) and As both exhibit competitive sorption characteristics on the biochar surface. Figure 15.2 demonstrates the effects of biochar on the mobility of HMs along with the bioavailability in the contaminated farmlands in China. The figure further illustrates the impact on the soil, yield, and the plants, along with the influencing factors such as the location, HMs, and the biochars. The incorporation of biochar causes the enhancing of phosphorous content in the soil, hence forcing more arsenic to be leached out as demonstrated by Hartley et al. (2009). By reducing heavy metals, biochar can also immobilize them. For instance, chromium is found in two forms in the environment: hexavalent (chromium (VI)) and trivalent (chromium (III)). Chromium (VI) is very hazardous and tightly movable, unlike nonhazardous chromium (III), which is actively linked to the soil components. Choppala et al. (2012) stated that reduction of chromium (III) can diminish the threat of chromium (VI). The biochar surface contained oxygen groups that initiate a redox reaction between soil and chromium (VI), hence reducing its movability substantially. Choppala et al. (2012) also examined the diminishing of hexavalent chromium (chromium (VI)) with different concentrations of 600 mg Cr (VI)/Kg and 454 mg Cr (VI)/Kg, from contaminated soil over synthesized carbon

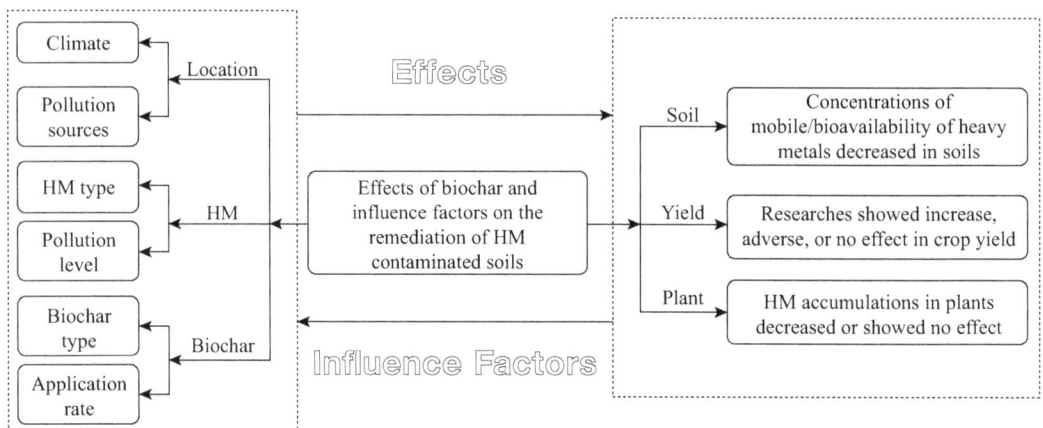

FIGURE 15.2 Effects of biochar on the mobility of HMs along with bioavailability in contaminated farmlands in China.

Source: Reprinted with permission (He et al. 2019).

materials (black carbon and chicken manure biochar). The investigation revealed that chromium (VI) was decreased considerably, with the consequent fixation of the chromium (III) in soils, using prepared carbon materials. The chromium (IV) reduction rate in the black loam/restoration sandy loam was up to 75%. In soil under aerobic circumstances, arsenic (III) possesses the highest motility and noxiousness rather than arsenic (V). Biochar has the potential effect to convert arsenic (V) to arsenic (III), hence raising the mobility of arsenic (Ahmad et al. 2018). It was also proposed that the use of biochar can result in enhanced arsenic movability. Hence further changes are required to compensate for the detrimental impact. In addition, Warren et al. (2003) proposed that, in soil, the iron-oxide-embedded biochar decreases the motility of arsenic through anion exchange. Similar studies also confirmed that the kinds of heavy metals and biocharcoal characteristics can immobilize or mobilize heavy-metals-contaminated soil.

The efficacy of biochar is influenced by the nature of the soil. A considerable reduction of zinc and nickel from sandy polluted soil was found using hardwood-based biochar, as reported by Shen et al. (2019). On the other hand, no substantial effect on the mobility of the lead has been observed after employing the similar biochar to Pb-polluted kaolin containing 35% clay (Shen et al. 2019). Shen et al. (2019) did the same thing using biochar prepared from wheat straw in clay-rich polluted soil. Results showed that the addition of biochar moderately reduced the number of heavy metals (Pb, Cd, and copper (Cu)), that the reduction was not substantial, and that the immobilization of these metals deteriorated with accelerated aging. Consequently, the efficacy of biochar in immobilizing metals in soils is considerably dependent on soil type.

15.4.2 Bioavailability

The presence of sediment and heavy metals in soil could be a danger for plant adsorption. A pot experiment was conducted to investigate the bioaccumulation of heavy metals in polluted soil (Lai et al. 2019). Furthermore, bioavailability and environmental threats are determined by their presence in soil. Currently, the Community Bureau of Reference sequential extraction process and the Tessier sequential extraction method are successfully employed to analyze the soil's heavy metal forms. In 1979, Tessier et al. (Tessier, Campbell, and Bisson 1979) displayed a method for dividing particulate trace metals such as Pb^{2+}, Cu^{2+}, Fe^{3+}, zinc, Ni^{2+}, and Mn^{2+} into five segments: bound to carbonates, bound to organic matter, exchangeable, bound to iron-manganese oxides, and residual. The organic matter-bound fractions and the iron-manganese-oxide-bound segments are usually less bioavailable. The crystal lattice contained the residual fraction of primary and secondary minerals resulting in the most stable fraction with the lowest bioavailability. However, the bioavailability and environmental effects of heavy metals are represented by their fate and fractions. Heavy metal bioavailability in soil may be diminished using a biochar mixture and other substances. Zimmerman and Gao (2013) established a comparison study on the impacts of biochar, lime, and biochar-lime mixed as a healing agent on pH and meadow soil loaded with cadmium (Cd). The results from the investigation revealed that after employing these curing agents, the pH of the soil was considerably enhanced, and the amount of exchangeable Cd in the contaminated soil was diminished to some extent; the reduction is in the order of biochar-lime mix > lime > biochar. Other types of cadmium became more abundant, resulting in a reduction in cadmium bioavailability. In the meantime, the regression experiment revealed the subsequent detrimental interrelationship between the cadmium concentration in soil and the amount of pH. The impact of fruit-tree-based biochar on the rehabilitation of naturally contaminated ore was demonstrated by Beesley and Marmiroli (2011). The diffused amount of heavy metals in soil extract and pore water was evaluated by a pot experiment. Moreover, the results also revealed that the amount of free heavy toxic metals was diminished by applying biochar. The amount of liquefied organic carbon (OC) in manure has a big impact on heavy metals migration. Moreover, heavy metal bioavailability may be efficiently reduced by combined repair. These studies summarized that heavy metal's bioavailability and adsorption are reduced by biochar and plant, respectively; hence the heavy metal toxicity is significantly reduced.

15.4.3 Strength

The mechanical characteristics of processed soil are significantly impacted by the pore and density of biocharcoal. As indicated by Beesley et al., the mechanical properties of red soil were investigated using three distinct kinds of biochar to assess their influence on shear strength and unconfined compressive strength (UCS) (Beesley and Marmiroli 2011). The result revealed that the applied biochar considerably diminished the soil's shear strength and UCS. The study also invested a high quantity of biochar used in the soil, resulting in raised compressive strength; however, there was no significant impact on friction angle. This is because biochar raises the number of macropores in the soil and alters the biochar–soil interaction, resulting in the reduction of soil compressive strength. The findings of the pot experiment revealed an increase in the rate of biochar application, decreasing in tensile strength (Ventura et al. 2015). Also, biochar-added soil lost its strength, which may influence crop yield in agriculture; a reduction in soil strength was not predicted in the context of soil solidification. Usually, in solidification/stabilization, the biochar is applied with another solidifying material that raises the soil's mechanical characteristics. Ground granulated blast-furnace slag (GGBS) and lime-treated-biochar blended with clays was examined by laboratory UCS tests equipped with scanning electron microscopy (SEM) and powder XRD (Haque et al. 2014). It was discovered that increasing the GGBS concentration resulted in a considerable increase in soil strength and that extending the curing period hence helped to improve compressive strength. The correlation among soil materials and biochar, specifically surface deposition, interface cementation, and pore space filling by cementitious minerals, contributes to the improvement of UCS; their microstructural properties were also recognized by XRD and SEM analysis. The soil, blended with biochar and GGBS, resulted in the reduction of lime content that is more environmentally acceptable and endurable. In pragmatic projects, environmental situations, the kinds of heavy-metal-contaminated soil, curing time, and the physical and chemical characteristics of biochar should all be taken into account for determining the remediation approach. Biochar, a high-carbon-containing material synthesized from three varieties of waste sources (rice husk, pulp and paper mill sludge, and poultry litter), was used to replace the cement quantity up to 1% of the entire volume, as observed in Akhtar and Sarmah's (2018) study, and the impact of individual biochar combined with cement on the mechanical properties of concrete was investigated using different characterization methods such as XRD, TGA, and others. In typical concrete applications, biochar has been shown to improve the characteristics of concrete while substituting cement in small amounts.

15.4.4 Influencing Factors

Biomass feedstocks and pyrolysis conditions influence the characteristics of biochar. So many things, such as hazardous soil type, alkalinity, acidity, and aging, should be taken into account during the mitigation of heavy metal using biochar from contaminated sites of soil. Numerous studies have demonstrated that mineral composition, CEC, pH, and amount of nutrient of the hazardous soil have a major impact on biochar mitigation (Xie et al. 2014). The study also revealed that increasing biochar utilization raises the soil's pH, leading to heavy metal immobilization through precipitation. The findings also demonstrate that the motility of transition metal concentration in soil is reduced by employing a small amount of biochar. Moreover, certain biomass feedstocks have a high concentration of heavy metals. As an example, sludge-based biochar often includes a significant number of heavy metals like lead, nickel, chromium, and cadmium. But pyrolysis of sludge resulted in the reduction of the mobility and leaching threats of heavy metals, e.g., zinc, lead, nickel, chromium, and cadmium, as stated by Méndez et al. (2012). Biochar has lengthy durability, but long-range environmental exposure can affect its physical and chemical characteristics. The biochar surface oxidizes during the aging treatment, thus producing oxygen-containing functional groups like hydroxyl and carboxyl groups, forming higher negative charges and greater CEC of biochar, which may assist in heavy metal immobilization. Examining the variables that influence heavy-metal-contaminated soil remediation can help to improve the practical application of biochar.

15.5 POTENTIAL RISKS WITH BIOCHAR APPLICATIONS

When using biochar as part of in situ soil remediation, it's important to examine not just the immobilization effects of the HMs but also the availability of the biochars for a certain duration of time, along with the possible ecological threats, which vary with the type of biochar used and its performance. The biochar has been reported as a carrier of volatile organic compounds (VOCs) (Yao et al. 2011; Wang et al. 2015; Cao and Harris 2010; Shen et al. 2015; Lu et al. 2012), polycyclic aromatic hydrocarbons (PAHs), HMs and dioxin (PCDD/Fs), and other harmful pollutants, with remediation demands ranging from 1.5 to 72 tons/ha or even higher. Henceforth, the harmful substances could become mixed with the air, water, or soil in the environment due to the implementation of the double-edged biochar, demonstrating a secondary pollution and causing a major ecological concern. Until now, biochar has played an essential role in utilization of carbon along with the reduction in emission of the non-CO_2 greenhouse gases from the standpoint of the soil. Further investigations on enhancing the emissions of the greenhouse gases using biochar have demonstrated that, under some conditions, the use of biochar can boost N_2O, CO_2, and CH_4 (Cheng et al. 2020) emissions to a certain limit. Biochar has the ability to diminish CH_4 emissions, but in the process it may even raise the emissions of N_2O, which relies purely on the type of biochar application settings. As a result, the reduction effect on emissions by biochar cannot be assumed to be the same for all the greenhouse gases. Further, investigations have demonstrated that biochar may possibly decline the efficacy of the pesticides of the soil and their biodegradation effects (Cheng et al. 2020), thereby providing hindrance to the agricultural weeding and insecticide process. Pesticide residues could be linked to biochar's high adsorption and binding capabilities. Furthermore, while biochar may promote biological bacterial activity, it can also demonstrate ill effects on the growth and survival of a range of the biological communities such as acidophilic earthworms and fungi (Cheng et al. 2020).

15.5.1 POTENTIAL REMEDIATION TECHNIQUES AND MECHANISMS

Biochar has high pH value, aromatic substances, a porous structure, and active functional groups at the surface. Electrostatic interaction, physical adsorption, complexation, ion exchange, and precipitation are some of the heavy metal remediation mechanisms in soil that these properties play a role in.

15.5.1.1 Physical Adsorption

The mixing of the adsorbate and the adsorbent molecules leads to the physical adsorption process, also known as van der Waals adsorption. The process of adsorption is often reversible since physical adsorption is mostly generated by the force between the molecules. The surface energy, SSA, and volume of the pore of the biochar all influence the physical adsorption of heavy metals on the material (Zhao et al. 2018). The SSA of the biochar and the volume of the pores increase with the inclination in the pyrolysis temperature, increasing the area of contact between the biochar and the HM ions and improving the physical adsorption of the biochar. Copper and uranium (U) can be successfully immobilized by physical adsorption in biochar made from the switchgrass (300°C) and pine wood (700°C) (Z. Liu, Zhang, and Wu 2010). HM ions like the zinc, cadmium, and arsenic are immobilized at the biochar's surface via physical adsorption process (Beesley et al. 2014).

15.5.1.2 Ion Exchange

Heavy metal ions selectively exchange exchangeable metal ions/ionizable protons on the biochar surface. The efficacy of ion exchange is largely determined by the biochar's surface chemical characteristics. The ion exchange capacity among the particles of the biochar and the metal cations can be improved by biochar with a high CEC, as a greater pyrolysis temperature causes reduction in the quantity of the oxygen/carbon and the acidic oxygen functional groups, affecting the biochar's CEC. The best recorded biochar's CEC was reported at pyrolysis temperatures of 250–300°C (Lee et al. 2010). In a different study conducted by El-Shafey (2010), mercury (Hg) and zinc remediation methods of husk biochar were generated at 175–180°C (El-Shafey 2010). As shown in the following equations, the functional groups of the acidic oxygen at the surface of the biochar, like -OH and -COOH, could be replaced with mercury (II) (Hg^+) and zinc (Zn^{2+}) ions to release ionizable protons.

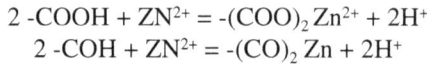

$$2 \text{ -COOH} + ZN^{2+} = \text{-(COO)}_2\, Zn^{2+} + 2H^+$$
$$2 \text{ -COH} + ZN^{2+} = \text{-(CO)}_2\, Zn + 2H^+$$

Biochar's ion exchange capacity is proportional to the pH value of the soil solution. With the reduction in the pH value of the soil solution lower than the biochar's pH at point of zero charge (PZC), the ion exchange mechanism attracts more HM ions at the surface of the biochar (Sánchez-Polo and Rivera-Utrilla 2002). As per the investigation conducted by Liu et al. (2010), an abundant amount of functional groups of oxygen are available in hydrothermal biochar; however, pyrolysis biochar has a lot of SSA, which helps with ion exchange and physical adsorption of copper (II) ions (Cu2+) (Z. Liu, Zhang, and Wu 2010). Furthermore, hydrothermal biochar has a higher adsorption capacity than that of the pyrolysis biochar, showing better ion exchange than that of physical adsorption.

15.5.1.3 Electrostatic Interactions

Biochar's large negative charge could improve the electrostatic interactions among the particles of the soil and the metal cations, allowing heavy metals to be immobilized via electrostatic attraction. The soil solution's pH, the radius of the ions, the valence of the HMs, and the biochar's PZC all play a role in the biochar–metal electrostatic interaction. Biochar application to soils raises the CEC and pH of the soil, increasing electrostatic interactions between heavy metal ions and soil particles (Mukherjee, Zimmerman, and Harris 2011). Because of the electrostatic interactions among the (Pb2+) ions, lead (II) ions, and the biochar's negative charge, Qiu et al. (2008) discovered that the biochar from the wheat and the rice straw has a better remediation effect than that of the activated carbon (Qiu et al. 2008). As a result, biochar uses electrostatic interaction to immobilize the HMs in the soil.

15.5.1.4 Complexation

Biomass carbon's surface contains many functional groups of oxygen like -COH, -COOH, and -OH, which form stable complexes with HM ions (Park et al. 2011). Low-temperature biochar includes a huge amount of functional groups of oxygen that actively immobilize the HMs via metal complexation. The amount of functional groups of oxygen present in the biochar rises with duration, which may be attributed to the oxidation of the surface and carboxyl production. The surface complexation of lead (II) ions, along with the available functional groups of the carboxyl and hydroxyl, and the inner-sphere complexation of the lead (II) ions, along with the available functional groups of the hydroxyl of mineral oxides, are examples of complexation between positively charged metal cations and the C=O ligand of the oxygen functional group (Lu et al. 2012). As per the investigation conducted by Chen, Chen, and Lv (2011), the reaction mechanisms of the functional groups along with the HM ions (M^{n+}) is as follows:

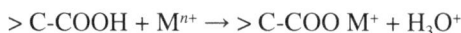

$$> \text{C-COOH} + M^{n+} \rightarrow\, > \text{C-COO } M^+ + H_3O^+$$

The surface complexation of the available carboxyl and hydroxyl functional groups with the lead (II) ions, along with the inner-sphere complexation of lead (II) ions with the abundantly available hydroxyl functional groups of mineral oxides, is certain evidence of the complexation among the positively charged metal cations and the oxygen functional groups' C=O ligand (Lu et al. 2012).

15.5.1.5 Precipitation

Biochar can disable the available HMs in the soil by coprecipitating with the available metal cations to create insoluble carbonates and phosphates (Shen et al. 2015). Pyrolysis of cellulose and hemicellulose in plant biomass at a temperature greater than 300°C produces alkaline biochar, which forms metal precipitates in soil (Cao and Harris 2010). Animal biochar often contains more ash, having the ability to combine with HMs to generate insoluble minerals like sodium, calcium, phosphorous, potassium, magnesium, silicon (Si), and sulfur (S). Furthermore, investigations have demonstrated that biochar made from dairy dung adsorbs lead extracted via aqueous solution using precipitation and surface sorption ranging from 13 to 16%. The findings imply that heavy metals can be successfully immobilized by precipitation.

15.6 FUTURE SCOPE AND PERSPECTIVES

Scholars are currently putting forth a lot of effort to look into the use of biochar in soil remediation. Following are a few tasks required to facilitate the soil remediation process using biochar.

- *Create a unified biochar classification standard*: Biochar formed from various biomass feedstock and pyrolysis settings has variable physical and chemical properties and thus has different remediation performance. It is critical to establish a consistent classification standard for biochar, since this will aid in the large-scale production and use of biochar.
- *Examine biochar's performance in soil with many pollutants*: The majority of biochar research now focuses on soil that has been polluted by a single heavy metal. Real polluted areas, on the other hand, usually contain a variety of chemicals. Heavy metals have a competitive adsorption difficulty in biochar remediation. As a result, research into the adsorption process of the HMs and the treatment efficacy of biochar on soil polluted with various HMs is required.
- *Determine how biochar works to immobilize HMs in the soil*: The role of multiple processes in biochar's HM inactivation in the soil is unknown. The majority of research is focused on the mechanics of adsorption in solution. However, soil is a complex system, with a variety of elements in the soil environment capable of influencing the biochar's ability to immobilize the available HMs in the soil. It is further essential to explore the mixing process of the biochar and the HMs in the complex soil environment.
- *Expand the biochar research scale for the treatment of soil containing HMs*: Future study should focus on long duration and bigger in situ trials. Low-cost biochar production systems are required for large-scale applications.
- *Conduct a solidification test using biochar*: Most studies involving biochar for the treatment of the HMs-containing soils currently focus on environmental sciences, agriculture, and soil, with an emphasis on remediation effects and the mechanisms involved. Biochar can be used to partially replace traditional cleanup materials like lime and cement, lowering project costs and ensuring long-term heavy metal stability. Mechanical properties research, on the other hand, is urgently required to set a direction for engineering applications.

15.7 CONCLUSION

Biochar is a potential substance for remediation of HM-contaminated soil. The production, characteristics, remediation impact, and process mechanism of biochar were discussed and summarized based on recent research advances. As illustrated, heavy metals present a significant threat, have a lengthy residence time, and are difficult to degrade. Heavy metals' solubility, mobility, and toxicity in soil can all be reduced by solidification/stabilization. In recent years, biochar has emerged as a viable remediation material. It has minimal costs and high remediation efficiency, is environmentally friendly, and improves soil. The biomass ingredients and pyrolysis temperature influence the physicochemical properties of biochar, which, combined, dictate the treatment effect on the soil contaminated with HMs. Biochar can be modified using techniques including activation, magnetization, oxidation, and digestion to boost its heavy metal adsorption considerably. The HMs-contaminated soil can be adsorbed by biochar, limiting their mobility and bioavailability. Biochar, on the other hand, can sometimes cause heavy metals to be mobilized in the soil. Biochar could further be applied with other solidified materials to improve soil and sediment strength. The best cleanup plan should be selected depending on site characteristics and other criteria in practical applications. The biochar's process mechanisms for remediation of the HMs-contaminated soil are determined by its physicochemical characteristics.

REFERENCES

Abdelhafez, Ahmed A., Jianhua Li, and Mohamed H.H. Abbas. 2014. "Feasibility of Biochar Manufactured from Organic Wastes on the Stabilization of Heavy Metals in a Metal Smelter Contaminated Soil." *Chemosphere* 117 (1). Elsevier Ltd: 66–71. doi:10.1016/j.chemosphere.2014.05.086

Ahmad, Mahtab, Sang Soo Lee, Jung Eun Lim, Sung Eun Lee, Ju Sik Cho, Deok Hyun Moon, Yohey Hashimoto, and Yong Sik Ok. 2014. "Speciation and Phytoavailability of Lead and Antimony in a Small Arms Range Soil Amended with Mussel Shell, Cow Bone and Biochar: EXAFS Spectroscopy and Chemical Extractions." *Chemosphere* 95. Elsevier Ltd: 433–441. doi:10.1016/j.chemosphere.2013.09.077

Ahmad, Zahoor, Bin Gao, Ahmed Mosa, Haowei Yu, Xianqiang Yin, Asaad Bashir, Hossein Ghoveisi, and Shengsen Wang. 2018. "Removal of Cu(II), Cd(II) and Pb(II) Ions from Aqueous Solutions by Biochars Derived from Potassium-Rich Biomass." *Journal of Cleaner Production* 180. Elsevier Ltd: 437–449. doi:10.1016/j.jclepro.2018.01.133

Akhtar, Ali, and Ajit K. Sarmah. 2018. "Novel Biochar-Concrete Composites: Manufacturing, Characterization and Evaluation of the Mechanical Properties." *Science of the Total Environment* 616–617. Elsevier B.V.: 408–416. doi:10.1016/j.scitotenv.2017.10.319

Bashir, Saqib, Muhammad Shaaban, Sajid Mehmood, Jun Zhu, Qingling Fu, and Hongqing Hu. 2018. "Efficiency of C3 and C4 Plant Derived-Biochar for Cd Mobility, Nutrient Cycling and Microbial Biomass in Contaminated Soil." *Bulletin of Environmental Contamination and Toxicology* 100 (6). Springer US: 834–838. doi:10.1007/s00128-018-2332-6

Beesley, Luke, Eduardo Moreno-Jiménez, and Jose L. Gomez-Eyles. 2010. "Effects of Biochar and Greenwaste Compost Amendments on Mobility, Bioavailability and Toxicity of Inorganic and Organic Contaminants in a Multi-Element Polluted Soil." *Environmental Pollution* 158 (6): 2282–2287. doi:10.1016/j.envpol.2010.02.003

Beesley, Luke, and Marta Marmiroli. 2011. "The Immobilisation and Retention of Soluble Arsenic, Cadmium and Zinc by Biochar." *Environmental Pollution* 159 (2). Elsevier Ltd: 474–480. doi:10.1016/j.envpol.2010.10.016

Beesley, Luke, Onyeka S Inneh, Gareth J Norton, Eduardo Moreno-Jimenez, Tania Pardo, Rafael Clemente, and Julian J. C Dawson. 2014. "Assessing the Influence of Compost and Biochar Amendments on the Mobility and Toxicity of Metals and Arsenic in a Naturally Contaminated Mine Soil." *Environmental Pollution* 186: 195–202. doi:https://doi.org/10.1016/j.envpol.2013.11.026

Bian, Rongjun, Stephen Joseph, Liqiang Cui, Genxing Pan, Lianqing Li, Xiaoyu Liu, Afeng Zhang, et al. 2014. "A Three-Year Experiment Confirms Continuous Immobilization of Cadmium and Lead in Contaminated Paddy Field with Biochar Amendment." *Journal of Hazardous Materials* 272. Elsevier B.V.: 121–128. doi:10.1016/j.jhazmat.2014.03.017

Bulasara, Vijaya Kumar, Harjyoti Thakuria, Ramgopal Uppaluri, and Mihir Kumar Purkait. 2011. "Effect of Process Parameters on Electroless Plating and Nickel-Ceramic Composite Membrane Characteristics." *Desalination* 268 (1): 195–203. doi:https://doi.org/10.1016/j.desal.2010.10.025

Cao, Xinde, Lena Ma, Bin Gao, and Willie Harris. 2009. "Dairy-Manure Derived Biochar Effectively Sorbs Lead and Atrazine." *Environmental Science and Technology* 43 (9): 3285–3291. doi:10.1021/es803092k

Cao, Xinde, and Willie Harris. 2010. "Properties of Dairy-Manure-Derived Biochar Pertinent to Its Potential Use in Remediation." *Bioresource Technology* 101 (14): 5222–5228. doi:https://doi.org/10.1016/j.biortech.2010.02.052

Chen, Baoliang, Zaiming Chen, and Shaofang Lv. 2011. "A Novel Magnetic Biochar Efficiently Sorbs Organic Pollutants and Phosphate." *Bioresource Technology* 102 (2): 716–723. doi:https://doi.org/10.1016/j.biortech.2010.08.067

Cheng, Sheng, Tao Chen, Wenbin Xu, Jian Huang, Shaojun Jiang, and Bo Yan. 2020. "Application Research of Biochar for the Remediation of Soil Heavy Metals Contamination: A Review." *Molecules* 25 (14): 3167. doi:10.3390/molecules25143167

Choppala, G. K., N.S. Bolan, M. Megharaj, Z. Chen, and R. Naidu. 2012. "The Influence of Biochar and Black Carbon on Reduction and Bioavailability of Chromate in Soils." *Journal of Environmental Quality* 41 (4): 1175–1184. doi:10.2134/jeq2011.0145

Chou, Jing-Dong, Ming-Yen Wey, and Shih-Hsien Chang. 2010. "Study on Pb and PAHs Emission Levels of Heavy Metals- and PAHs-Contaminated Soil during Thermal Treatment Process." *Journal of Environmental Engineering* 136 (1): 112–118. doi:10.1061/(asce)ee.1943–7870.0000133

Dong, Deming, Xingmin Zhao, Xiuyi Hua, Jinfu Liu, and Ming Gao. 2009. "Investigation of the Potential Mobility of Pb, Cd and Cr(VI) from Moderately Contaminated Farmland Soil to Groundwater in Northeast, China." *Journal of Hazardous Materials* 162 (2–3): 1261–1268. doi:10.1016/j.jhazmat.2008.06.032

El-Shafey, E I. 2010. "Removal of Zn(II) and Hg(II) from Aqueous Solution on a Carbonaceous Sorbent Chemically Prepared from Rice Husk." *Journal of Hazardous Materials* 175 (1): 319–327. doi:https://doi.org/10.1016/j.jhazmat.2009.10.006

Emani, Sriharsha, Ramgopal Uppaluri, and Mihir Kumar Purkait. 2013. "Preparation and Characterization of Low Cost Ceramic Membranes for Mosambi Juice Clarification." *Desalination* 317: 32–340. doi:https://doi.org/10.1016/j.desal.2013.02.024

Haque, Asadul, Chiak Kai Tang, Shahidul Islam, P. G. Ranjith, and Ha H. Bui. 2014. "Biochar Sequestration in Lime-Slag Treated Synthetic Soils: A Green Approach to Ground Improvement." *Journal of Materials in Civil Engineering* 26 (12): 06014024. doi:10.1061/(asce)mt.1943–5533.0001113

Hartley, William, Nicholas M. Dickinson, Philip Riby, and Nicholas W. Lepp. 2009. "Arsenic Mobility in Brownfield Soils Amended with Green Waste Compost or Biochar and Planted with Miscanthus." *Environmental Pollution* 157 (10). Elsevier Ltd: 2654–2662. doi:10.1016/j.envpol.2009.05.011

Hass, Amir, Javier M. Gonzalez, Isabel M. Lima, Harry W. Godwin, Jonathan J. Halvorson, and Douglas G. Boyer. 2012. "Chicken Manure Biochar as Liming and Nutrient Source for Acid Appalachian Soil." *Journal of Environmental Quality* 41 (4). John Wiley & Sons, Ltd: 1096–1106. doi:10.2134/jeq2011.0124

He, Lizhi, Huan Zhong, Guangxia Liu, Zhongmin Dai, Philip C. Brookes, and Jianming Xu. 2019. "Remediation of Heavy Metal Contaminated Soils by Biochar: Mechanisms, Potential Risks and Applications in China." *Environmental Pollution* 252: 846–855. doi:10.1016/j.envpol.2019.05.151

Ho, Shih Hsin, Shishu Zhu, and Jo Shu Chang. 2017. "Recent Advances in Nanoscale-Metal Assisted Biochar Derived from Waste Biomass Used for Heavy Metals Removal." *Bioresource Technology* 246 (July). Elsevier: 123–134. doi:10.1016/j.biortech.2017.08.061

Ibrahim, M., S. Khan, X. Hao, and G. Li. 2016. "Biochar Effects on Metal Bioaccumulation and Arsenic Speciation in Alfalfa (Medicago Sativa L.) Grown in Contaminated Soil." *International Journal of Environmental Science and Technology* 13 (10): 2467–2474. doi:10.1007/s13762-016-1081-5

Inyang, Mandu, Bin Gao, Pratap Pullammanappallil, Wenchuan Ding, and Andrew R. Zimmerman. 2010. "Biochar from Anaerobically Digested Sugarcane Bagasse." *Bioresource Technology* 101 (22): 8868–8872. doi:https://doi.org/10.1016/j.biortech.2010.06.088

Jiang, Tian Yu, Jun Jiang, Ren Kou Xu, and Zhuo Li. 2012. "Adsorption of Pb(II) on Variable Charge Soils Amended with Rice-Straw Derived Biochar." *Chemosphere* 89 (3). Elsevier Ltd: 249–256. doi:10.1016/j.chemosphere.2012.04.028

Kambo, Harpreet Singh, and Animesh Dutta. 2015. "A Comparative Review of Biochar and Hydrochar in Terms of Production, Physico-Chemical Properties and Applications." *Renewable and Sustainable Energy Reviews* 45. Elsevier: 359–378. doi:10.1016/j.rser.2015.01.050

Kambo, H.S., and A. Dutta. 2015. "A Comparative Review of Biochar and Hydrochar in Terms of Production, Physico-Chemical Properties and Applications." *Renew. Sustain. Energy Rev*, 45: 359–378. https://doi.org/10.1016/j.rser.2015.01.050

Lai, Cui, Fanglong Huang, Guangming Zeng, Danlian Huang, Lei Qin, Min Cheng, Chen Zhang, et al. 2019. "Fabrication of Novel Magnetic MnFe2O4/Bio-Char Composite and Heterogeneous Photo-Fenton Degradation of Tetracycline in near Neutral PH." *Chemosphere* 224. Elsevier Ltd: 910–921. doi:10.1016/j.chemosphere.2019.02.193

Lee, James W., Michelle Kidder, Barbara R. Evans, Sokwon Paik, A. C Buchanan III, Charles T. Garten, and Robert C. Brown. 2010. "Characterization of Biochars Produced from Cornstovers for Soil Amendment." *Environmental Science & Technology* 44 (20). American Chemical Society: 7970–7974. doi:10.1021/es101337x

Liang, Yuan, Xinde Cao, Ling Zhao, and Eduardo Arellano. 2014. "Biochar- and Phosphate-Induced Immobilization of Heavy Metals in Contaminated Soil and Water: Implication on Simultaneous Remediation of Contaminated Soil and Groundwater." *Environmental Science and Pollution Research* 21 (6): 4665–4674. doi:10.1007/s11356-013-2423-1

Liu, Wu Jun, Hong Jiang, and Han Qing Yu. 2015. "Development of Biochar-Based Functional Materials: Toward a Sustainable Platform Carbon Material." *Chemical Reviews* 115 (22): 12251–12285. doi:10.1021/acs.chemrev.5b00195

Liu, Zhengang, Fu-Shen Zhang, and Jianzhi Wu. 2010. "Characterization and Application of Chars Produced from Pinewood Pyrolysis and Hydrothermal Treatment." *Fuel* 89 (2): 510–514. doi:10.1016/j.fuel.2009.08.042

Li, Zhiyuan, Zongwei Ma, Tsering Jan van der Kuijp, Zengwei Yuan, and Lei Huang. 2014. "A Review of Soil Heavy Metal Pollution from Mines in China: Pollution and Health Risk Assessment." *Science of the Total Environment*: 468–469. Elsevier B.V.: 843–853. doi:10.1016/j.scitotenv.2013.08.090

Lu, Huanliang, Weihua Zhang, Yuxi Yang, Xiongfei Huang, Shizhong Wang, and Rongliang Qiu. 2012. "Relative Distribution of Pb2+ Sorption Mechanisms by Sludge-Derived Biochar." *Water Research* 46 (3): 854–862. doi:10.1016/j.watres.2011.11.058

Maguyon-Detras, Monet Concepcion, Maria Victoria P. Migo, Nguyen Van Hung, and Martin Gummert. 2020. *Sustainable Rice Straw Management*. Edited by Martin Gummert, Nguyen Van Hung, Pauline Chivenge, and Boru Douthwaite. Cham: Springer International Publishing. doi:10.1007/978-3-030-32373-8

Méndez, A., A. Gómez, J. Paz-Ferreiro, and G. Gascó. 2012. "Effects of Sewage Sludge Biochar on Plant Metal Availability after Application to a Mediterranean Soil." *Chemosphere* 89 (11): 1354–1359. doi:10.1016/j.chemosphere.2012.05.092

Meng, Jun, Mengming Tao, Lili Wang, Xingmei Liu, and Jianming Xu. 2018. "Changes in Heavy Metal Bioavailability and Speciation from a Pb-Zn Mining Soil Amended with Biochars from Co-Pyrolysis of Rice Straw and Swine Manure." *Science of the Total Environment* 633. Elsevier B.V.: 300–307. doi:10.1016/j.scitotenv.2018.03.199

Meng, Jun, Xiaoli Feng, Zhongmin Dai, Xingmei Liu, Jianjun Wu, and Jianming Xu. 2014. "Adsorption Characteristics of Cu(II) from Aqueous Solution onto Biochar Derived from Swine Manure." *Environmental Science and Pollution Research* 21 (11): 7035–7046. doi:10.1007/s11356–014–2627-z

Mukherjee, A., A. R. Zimmerman, and W. Harris. 2011. "Surface Chemistry Variations among a Series of Laboratory-Produced Biochars." *Geoderma* 163 (3): 247–255. doi:10.1016/j.geoderma.2011.04.021

Nandi, B. K., B. Das, R. Uppaluri, and M. K. Purkait. 2009. "Microfiltration of Mosambi Juice Using Low Cost Ceramic Membrane." *Journal of Food Engineering* 95 (4): 597–605. doi:10.1016/j.jfoodeng.2009.06.024

Nandi, B. K., R. Uppaluri, and M. K. Purkait. 2009. "Treatment of Oily Waste Water Using Low-Cost Ceramic Membrane: Flux Decline Mechanism and Economic Feasibility." *Separation Science and Technology* 44 (12): 2840–2869. doi:10.1080/01496390903136004

Park, Jin Hee, Girish Kumar Choppala, Nanthi Sirangie Bolan, Jae Woo Chung, and Thammared Chuasavathi. 2011. "Biochar Reduces the Bioavailability and Phytotoxicity of Heavy Metals." *Plant and Soil* 348 (1): 439. doi:10.1007/s11104–011–0948-y

Purkait, M. K., P. K. Bhattacharya, and S. De. 2005. "Membrane Filtration of Leather Plant Effluent: Flux Decline Mechanism." *Journal of Membrane Science* 258 (1–2): 85–96. doi:10.1016/j.memsci.2005.02.029

Qi, Fangjie, Saranya Kuppusamy, Ravi Naidu, Nanthi S. Bolan, Yong Sik Ok, Dane Lamb, Yubiao Li, Linbo Yu, Kirk T. Semple, and Hailong Wang. 2017. "Pyrogenic Carbon and Its Role in Contaminant Immobilization in Soils." *Critical Reviews in Environmental Science and Technology* 47 (10): 795–876. doi:10.1080/10643389.2017.1328918

Qin, Peng, Hailong Wang, Xing Yang, Lizhi He, Karin Müller, Sabry M. Shaheen, Song Xu, et al. 2018. "Bamboo- and Pig-Derived Biochars Reduce Leaching Losses of Dibutyl Phthalate, Cadmium, and Lead from Co-Contaminated Soils." *Chemosphere* 198: 450–459. doi:10.1016/j.chemosphere.2018.01.162

Qiu, Yuping, Haiyan Cheng, Chao Xu, and G. Daniel Sheng. 2008. "Surface Characteristics of Crop-Residue-Derived Black Carbon and Lead(II) Adsorption." *Water Research* 42 (3): 567–574. doi:10.1016/j.watres.2007.07.051

Saghir Abbas, Muhammad Tariq Javed, Qasim Ali, Hassan Javed Chaudhary, Muhammad Rizwan, Chapter 30 - Alteration of plant physiology by the application of biochar for remediation of organic pollutants, Editor(s): Mirza Hasanuzzaman, Majeti Narasimha Vara Prasad, *Handbook of Bioremediation*, Academic Press, 2021, 475-492, ISBN 9780128193822, https://doi.org/10.1016/B978-0-12-819382-2.00030-2.

Sánchez-Polo, M., and J. Rivera-Utrilla. 2002. "Adsorbent–Adsorbate Interactions in the Adsorption of Cd(II) and Hg(II) on Ozonized Activated Carbons." *Environmental Science & Technology* 36 (17). American Chemical Society: 3850–3854. doi:10.1021/es0255610

Shaheen, S.M., N.K. Niazi, N.E.E. Hassan, I. Bibi, H. Wang, D.C.W. Tsang, Y.S. Ok, N. Bolan, J. Rinklebe. 2019. "Wood-Based Biochar for the Removal of Potentially Toxic Elements in Water and Wastewater: A Critical Review." *International Materials Reviews* 64 (4): 216–247. doi:10.1080/09506608.2018.1473096

Sharma, Mukesh, P. Mondal, Ankush Sontakke, Arun Chakraborty, and M. K. Purkait. 2021. "High Performance Graphene-Oxide Doped Cellulose Acetate Based Ion Exchange Membrane for Environmental Remediation Applications." *International Journal of Environmental Analytical Chemistry*. Taylor & Francis. doi:10.1080/03067319.2021.1975276

Shen, Zhengtao, Fei Jin, Fei Wang, Oliver McMillan, and Abir Al-Tabbaa. 2015. "Sorption of Lead by Salisbury Biochar Produced from British Broadleaf Hardwood." *Bioresource Technology* 193: 553–556. doi:10.1016/j.biortech.2015.06.111

Shen, Zhengtao, Jingzhuo Zhang, Deyi Hou, Daniel C.W. Tsang, Yong Sik Ok, and Daniel S. Alessi. 2019. "Synthesis of MgO-Coated Corncob Biochar and Its Application in Lead Stabilization in a Soil Washing Residue." *Environment International* 122 (November 2018). Elsevier: 357–362. doi:10.1016/j.envint.2018.11.045

Shu, Rui, Yongjie Wang, and Huan Zhong. 2016. "Biochar Amendment Reduced Methylmercury Accumulation in Rice Plants." *Journal of Hazardous Materials* 313. Elsevier B.V.: 1–8. doi:10.1016/j.jhazmat.2016.03.080

Sinha, M. K., and M. K. Purkait. 2014. "Preparation and Characterization of Novel Pegylated Hydrophilic PH Responsive Polysulfone Ultrafiltration Membrane." *Journal of Membrane Science* 464: 20–32. doi:10.1016/j.memsci.2014.03.067

Sinha, M. K., and M. K. Purkait. 2015. "Preparation of Fouling Resistant PSF Flat Sheet UF Membrane Using Amphiphilic Polyurethane Macromolecules." *Desalination* 355: 155–168. doi:10.1016/j.desal.2014.10.017

Tessier, A., P. G.C. Campbell, and M. Bisson. 1979. "Sequential Extraction Procedure for the Speciation of Particulate Trace Metals." *Analytical Chemistry* 51 (7): 844–851. doi:10.1021/ac50043a017

Tong, Xue Jiao, Jiu Yu Li, Jin Hua Yuan, and Ren Kou Xu. 2011. "Adsorption of Cu(II) by Biochars Generated from Three Crop Straws." *Chemical Engineering Journal* 172 (2–3). Elsevier B.V.: 828–834. doi:10.1016/j.cej.2011.06.069

Uchimiya, Minori, Se Chin Chang, and K. Thomas Klasson. 2011. "Screening Biochars for Heavy Metal Retention in Soil: Role of Oxygen Functional Groups." *Journal of Hazardous Materials* 190 (1–3). Elsevier B.V.: 432–441. doi:10.1016/j.jhazmat.2011.03.063

Venegas, A, A. Rigol, and M. Vidal. 2016. "Changes in Heavy Metal Extractability from Contaminated Soils Remediated with Organic Waste or Biochar." *Geoderma* 279: 132–140.

Ventura, Maurizio, Giorgio Alberti, Maud Viger, Joseph R. Jenkins, Cyril Girardin, Silvia Baronti, Alessandro Zaldei, et al. 2015. "Biochar Mineralization and Priming Effect on SOM Decomposition in Two European Short Rotation Coppices." *GCB Bioenergy* 7 (5): 1150–1160. doi:10.1111/gcbb.12219

Wang, Shengsen, Bin Gao, Andrew R. Zimmerman, Yuncong Li, Lena Ma, Willie G. Harris, and Kati W. Migliaccio. 2015. "Removal of Arsenic by Magnetic Biochar Prepared from Pinewood and Natural Hematite." *Bioresource Technology* 175: 391–395. doi:10.1016/j.biortech.2014.10.104

Warren, Janet M., C.J.K. Henry, H. J. Lightowler, S. M. Bradshaw, and S. Perwaiz. 2003. "Evaluation of a Pilot School Programme Aimed at the Prevention of Obesity in Children." *Health Promotion International* 18 (4): 287–296. doi:10.1093/heapro/dag402

Xie, Mengxing, Wei Chen, Zhaoyi Xu, Shourong Zheng, and Dongqiang Zhu. 2014. "Adsorption of Sulfonamides to Demineralized Pine Wood Biochars Prepared under Different Thermochemical Conditions." *Environmental Pollution* 186. Elsevier Ltd: 187–194. doi:10.1016/j.envpol.2013.11.022

Yao, Ying, Bin Gao, Mandu Inyang, Andrew R Zimmerman, Xinde Cao, Pratap Pullammanappallil, and Liuyan Yang. 2011. "Biochar Derived from Anaerobically Digested Sugar Beet Tailings: Characterization and Phosphate Removal Potential." *Bioresource Technology* 102 (10): 6273–6278. doi:https://doi.org/10.1016/j.biortech.2011.03.006

Zhao, Bin, David O'Connor, Junli Zhang, Tianyue Peng, Zhengtao Shen, Daniel C. W Tsang, and Deyi Hou. 2018. "Effect of Pyrolysis Temperature, Heating Rate, and Residence Time on Rapeseed Stem Derived Biochar." *Journal of Cleaner Production* 174: 977–987. doi:10.1016/j.jclepro.2017.11.013

Zimmerman, Andrew, and Bin Gao. 2013. "The Stability of Biochar in the Environment." *Biochar and Soil Biota*, no. February: 1–40. doi:10.1201/b14585-2

16 Treatment of Industrial Wastewater by Utilization of Biochar as a Green Technology

Mukesh Sharma, Pranjal P. Das, Rahul Raghudhas, Arun Chakraborty, and Mihir K. Purkait

CONTENTS

16.1 INTRODUCTION

Soil has been the basis of food, fuel, fiber, and animal feed for millennia. It supplies raw materials as well as groundwater. Humans and other billions of living organisms use it as a water purifier and habitat (Cristaldi et al. 2017). Food security, purified water, improved sanitation, and excellent health and well-being are all important goals acknowledged by the United Nations Sustainable Development Program (UNSDP), and they are all achieved through critical resources such as soil. In recent decades, strong linearity exists between human population

DOI: 10.1201/9781003203438-16

and industrialization and urbanization, resulting in contamination of nearly all the ecosystems on the Earth (Ali, Khan, and Ilahi 2019). Polycyclic aromatic hydrocarbons (PAHs), crude oil and its petrochemical derivatives, heavy metals/metalloids (HMs), insecticides, herbicides, and chlorophenols are the major sources of common soil contaminants (M. Chen et al. 2015), and HMs PAHs causes major damage (Pérez et al. 2010), which is attributed to their toxic effects, endurance, and bioaccumulation potential, as well as their nonbiodegradability qualities in the case of HMs (Pérez et al. 2010).

Macro- and micronutrients belong to HMs that can be detrimental when present in excessive amounts even though they are useful to plant growth, development, and reproduction. Highly toxic elements, on the other hand, can be hazardous at low concentrations, depending on the HM ion species and the individual organisms impacted (Sall et al. 2020). In the last 50 years, environmental and analytical research has widely used a standard collection of compounds approved by the United States Environmental Protection Agency (USEPA) concerning polycyclic aromatic hydrocarbons (PAHs) (Andersson and Achten 2015). The newly included PAHs compounds are even more dangerous than the compounds previously included by the USEPA (Andersson and Achten 2015). Among the various kinds of PAHs, the following are some of the most frequently used and investigated: pyrene (Pyr), naphthalene (Nap), benzo[a]anthracene (B[a]A), chrysene (Chr), benzo[a]pyrene (B[a]P), anthracene (Ant), phenanthrene (Phe), fluoranthene (Flu), and benzo[b]fluoranthene (B[b]F) (Andersson and Achten 2015; Cipullo et al. 2019). Individual PAH compounds are categorized based on their genotoxic and carcinogenic properties. The volatile property of gas is also detected among two-, three-, and four-ringed PAHs in the presence of temperature fluctuation (Atkinson and Arey 1994; Srogi 2007). High-molecular-weight PAHs (HMWPAHs) are molecules with more than four rings that are mostly in solid form and attached to the soil and organic materials (M. Chen et al. 2015; Hassan et al. 2020; Li et al. 2018; J. Wang and Wang 2019). Apart from HMWPAHs, PAHs, including two- and three-ringed PAHs, are noncarcinogenic and less hazardous to animals and humans (J. Wang and Wang 2019). The PAH metabolites directly attach to DNA, resulting in cancer-like diseases (X. Zhang et al. 2013a; Barnes et al. 2018).

The primary effects and diseases are defined by HMs and PAHs, which are influenced by the absorbed dosage, mode of exposure, and life span of exposure (i.e., chronic or acute) (Barnes et al. 2018). The harmful effects of different pollutants on plants, soil microbes, and people result from various anthropogenic elements that lead to the discharge of HMs/PAHs into the environment. Chemical industries, mining, agriculture, and petroleum products are the HMs and PAHs, but when we consider the soil environment, natural geological processes become the sources (Barnes et al. 2018; Aydin and Aksoy 2009). Carcinogenic or mutagenic products can be produced from the combination of HMs and PAHs with natural inorganic compounds in soil (Singh et al. 2017). As a result, its presence in the soil poses a concern that necessitates effective contamination procedures. Eco-friendly techniques have been proposed as a low-cost and more effective approach to minimize these pollutants that nowadays are considered as the better solution. The use of biochar, especially in environmental engineering, is one such technique that has sparked new attention. To our understanding, this is the beginning chapter that summarizes the potential of combining biochar with hydrogel, bioaugmentation, and digestate to improve the decontamination of contaminated (HMs and PAHs) soil. The effectiveness of biochar and modified biochar for environmental remediation is examined in this paper. In addition, we incorporated remediation methods for a few potential extraction approaches along with analyzing their benefits and drawbacks.

The biochar has tremendous potential for the rehabilitation of soil polluted with HMs and/or PAHs. Numerous key features, ideas, and knowledge gaps must be considered for it to be successful in the future:

1. A better knowledge of biochar's sorption mechanism and the variables that influence its adsorption characteristics, notably physicochemical aspects

2. Using efficiently integrated (bio)engineered remediation techniques to improve the efficacy of soil remediation
3. Investigating the variables that impact the remediation features of a certain technique and to understand the pollutant removal processes of digestate, hydrogel, and bioaugmentation
4. Promoting eco-friendly and long-term solutions based on the circular bioeconomy idea
5. Investigating the regulatory and industry implications of (bio)engineered remediation techniques

16.2 SIGNIFICANT ROLE AND REQUIREMENT OF INTEGRATED ENGINEERED REMEDIATION APPROACHES

16.2.1 DISADVANTAGES OF TRADITIONAL TECHNOLOGY

Current physical and chemical treatments, as well as physicochemical technologies, such as soil washing, vitrification, and electrokinetics, can be used to decontaminate soils by removing HMs and PAHs (Sharma et al. 2018). Even while inefficiency, lack of practicality, and increasing cost pose a threat, there might be minor or major flaws in these techniques (Khalid et al. 2017). Several current methods can potentially harm soil quality, lead to the loss of other precious metals, or work only in particular situations, including limited soil permeability and low carbonate concentrations (Khalid et al. 2017; Rascio and Navari-Izzo 2011). The primary alternative approaches are encapsulation, solidification (S/S), and stabilization, which cannot remove, convert to nonhazardous compounds, or eliminate pollutants, which is considered to be the major remediation but which can only stabilize and entrap them in a solid state (Khalid et al. 2017; Rascio and Navari-Izzo 2011).

Bioremediation is an environmentally beneficial and cost-effective technology that is becoming more popular in the present climate, but it has certain limitations, such as the requirement for slow-growing hyperaccumulators, which decreases the pollutant accumulation rate. Plants that can collect huge levels of pollutants such as HMs and PAHs without incurring phytotoxic effects are known as hyperaccumulators, and they thrive in polluted soil (Rascio and Navari-Izzo 2011). Using invasive hyperaccumulator species can put natural habitats at risk (Chibuike and Obiora 2014).

16.2.2 BIOCHAR: A POTENTIAL GREEN TECHNOLOGY

Even though a modest amount of research has been done in this study area, which comprises the use of biochar for eradicating HMs and PAHs but very little in this research field, numerous industrial applications are necessary to determine the practicality and viability of this approach.

A different approach in this field of research found that increasing the rate of sugarcane bagasse biochar diminished interchangeable Cd by 8.5% and lowered the bioavailability of HMs (Pb, Cd, and Cu) to plant shoots and roots (Nie et al. 2018). A detailed investigation comprising biochar treatments of 16.5 tons/ha of biochar application dosage in 2010 has been used to examine the PAH sorptive effectiveness of biochar. Following that, PAH amounts fell substantially over time, from 153 ± 38 ng/g at the time of the first testing to 80 ± 12 ng/g at the time of the second testing (Rombolà et al. 2019; Stefaniuk, Oleszczuk, and Różyło 2017) in an 18-month field trial; researchers eliminated total PAHs (Σ16USEPA PAH) from podzolic soil using a 2.5% biochar application rate. In virtually many situations, the surface area and, as a result, the absorption capacity of traditional biochars produced by pyrolysis would be reduced. Engineering applications can benefit from customizing their physicochemical characteristics (Rajapaksha et al. 2016). Biochar modification is another term for biochar tailoring.

Engineered biochars are biochars that have been mass produced and changed in some way (Duan et al. 2019). Bioengineered biochars are biochars that have been modified or engineered utilizing biological theories and engineering techniques. They are made using biological feedstock materials. As a result, a technique is to combine multiple methods to solve some of the limitations of each while improving their remediation efficacy. Biochar is successful in the laboratory, both theoretically and in short-term investigations, based on the kind of biomass feedstock material, carbonization process, pyrolysis temperatures, and biochar dose (Duan et al. 2019; B. Chen, Chen, and Lv 2011; Xueyan Chen et al. 2020; Xincai Chen et al. 2011; Tu et al. 2020; Bian et al. 2013; M. Chen et al. 2015; Z. Chen et al. 2012; Hassan et al. 2020). At the laboratory scale, it is economically feasible and sustainable from the perspective that:

1. Biomass is derived from renewable biowastes that are either free or inexpensive.
2. Because of the modest scale of the lab projects, no transportation costs are involved.
3. Biochar's potential can be regenerated and reused.

Biochar can include HMs and PAHs and act as a transmitter for these pollutants, resulting in secondary contamination after treatment (Duan et al. 2019; LAHORI et al. 2017). Previous research has revealed that remediation utilizing hydrogel, digestate, and bioaugmentation is efficient in eliminating HMs and PAHs when employed sequentially. Further research needs to be done, nevertheless, to get a greater knowledge of the remediation effectiveness and promise of combining biochar with each of the preceding remediation products, as well as to examine their synergistic implications on pollutant activity and bioavailability. As of in 2022, the minimal adsorption experiments that have been done have yielded encouraging findings. When a biochar-digestate composite was administered at a dose of 1 g L^{-1}, the removal efficiency of Pb increased by approximately 100%. The eradication percentage for Cd rose from 13% to almost 100% in this dose range, suggesting a similar pattern (Tu et al. 2020; Pandey, Daverey, and Arunachalam 2020). The biochar was synthesized by pyrolysis of digestate granules at 700°C for 15 min, and the digestate was formed by anaerobic decomposition of waste. The previously reported investigations measured total pollution levels rather than the bioavailable fraction. The percentage of a freely accessible pollutant in the ecosystem that is transportable and hence most likely to cause humans to be exposed is known as bioavailability (Dean and Scott 2004). Due to their bio-based origins and composition, such remediation technologies are considered environmentally benign, cost-effective, and long-term solutions for bioremediation.

16.3   BIOCHAR PRODUCTION, PROPERTIES, AND APPLICATIONS

The common processes for producing biochar are microwave hydrothermal carbonization, carbonization, pyrolysis, torrefaction, flash carbonization, gasification, laser, and plasma cracking (Z. Liu et al. 2013; Tu et al. 2020; Xie et al. 2015). The most popular processes for making biochar are pyrolysis, gasification, and hydrothermal carbonization (Kwak et al. 2019). Biochars are made using the pyrolysis process at extreme temperatures (usually between 350 and 800°C) and in an oxygen-free environment (Pandey, Daverey, and Arunachalam 2020; Varma, Shankar, and Mondal 2018). Pyrolysis research has recently shifted toward the creation of novel pyrolysis techniques as well as the alteration of existing ones. An ultimate purpose is to create designed biochars that function well, or that improve specific physicochemical features of biochar, or that decrease the potential hazards by using polluted renewable resources, which might bring pollutants into the environment after biochar application. Microwave-assisted pyrolysis can produce biochars with a greater degree of carbonization, better physicochemical characteristics, and fewer oxygen-containing functional groups (Crombie et al. 2013; Zhao et al. 2013; Mašek et al. 2013; Paunovic et al. 2019). Low-oxygen-containing functional groups are also produced using steam-assisted pyrolysis (Braghiroli, Bouafif, and Koubaa 2019; L. Wang et al. 2020). The efficiency of a biochar adsorbent is determined by the kind of biomass feedstock used and the physicochemical characteristics of individual biochars, such as yield, pore structure, ash, functional group type, specific surface area and number, and cation exchange capacity (Xie et al. 2015; Pandey, Daverey, and Arunachalam 2020). Numerous pyrolysis operational factors, such as the kind of feedstock material, residence duration, pyrolysis

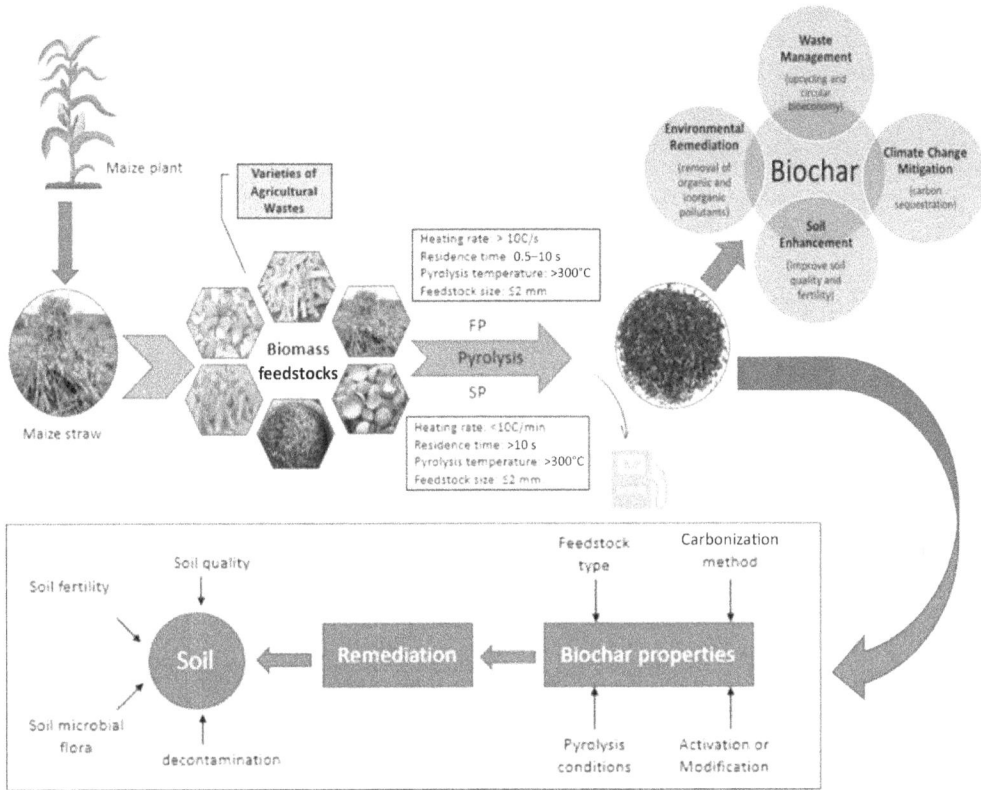

FIGURE 16.1 Preparation of biochar using agriculture waste and their applications.

temperature, heating rate, and reaction environment, impact the physicochemical characteristics of biochar (Xincai Chen et al. 2011; Pandey, Daverey, and Arunachalam 2020; Chibuike and Obiora 2014; X. Zhang et al. 2013b). Nevertheless, as the notion of upcycling and the cyclical bioeconomy has gained traction in recent years, more focus has shifted to the use of agricultural waste materials, especially agricultural food wastes. Several common agricultural biomass wastes have been utilized to make biochar, as well as their remediation impacts on HMs and PAHs.

Agriculture-related biomass wastes are being utilized to make biochars with different uses, as shown in Figure 16.1. Figure 16.1 demonstrates several steps involving the selection of proper material like maize straw, pyrolysis process, and obtaining the final product. Further, the figure also demonstrates the various advantages of using biochar, which includes waste management, environmental remediation, climate change mitigation, and soil enhancement. The carbon concentration, carbon sequestration capability, and ash content of biochar are all affected by the feedstock type (Zhao et al. 2013). A significant characteristic that affects the physicochemical characteristics of biochar is the reaction temperature. The pore diameter, ash content, specific surface area, and pH increase as the pyrolysis temperature rises. Negative charges are reduced when the number of acidic functional groups decreases. Low pyrolysis temperatures, on the other hand, can keep the structure of biomass materials while dehydrating them (W. J. Liu, Jiang, and Yu 2015).

16.4 BIOCHAR AMENDMENT FOR ENHANCED REMEDIATION APPROACHES

Increased binding sites and the introduction of HM/PAH tolerant microorganisms' ways of destroying HMs and decomposing PAHs are two ways to lower the mobility and bioavailability of HMs and

PAHs. A derivative of pristine biochar is called engineered biochar, and, when compared to pristine biochar, it has been changed to improve its chemical, physical, and biological characteristics (e.g., surface functional group, specific surface area, pH, porosity, cation exchange capacity, etc.) and adsorption capacity (Mohamed et al. 2017; Rajapaksha et al. 2016). Physicochemical–biological modifications and other modification techniques like UV and magnetic exposures are the four most frequently designed techniques for modifying biochars (Xueyan Chen et al. 2020; Pandey, Daverey, and Arunachalam 2020). Physical alterations primarily influence the physical structure of biochar, whereas chemical modifications have a significant impact on the chemical structure of biochar, including bonds and functional groups, and biological modifications encourage microbial communities and activities (Bianco et al. 2020; Pandey, Daverey, and Arunachalam 2020; L. Wang et al. 2020).

16.4.1 Remediation Using Hydrogel

16.4.1.1 Properties and Applications

A hydrogel is a cross-linked, extensible three-dimensional web of hydrophilic polymers that can expand in water or biological fluid and hold a significant amount of water while keeping its structure (Paunovic et al. 2019; J. Wang and Wang 2019; Sophia A. and Lima 2018). Hydrogels have a wide range of applications in agriculture, bioengineering, sensors, the food industry, biomedicine, water purification, and separation processes (Hassan et al. 2020). The hydrophilic characteristics of hydrogels are ascribed to their use in removing contaminants (Ali, Khan, and Ilahi 2019; Singh et al. 2017).

16.4.1.2 Adsorption Process and Mechanism

Chemical and physical sorption are the two categories that make up the adsorption mechanism of hydrogels. Physical sorption is described by reversible feeble physical interactions between adsorbate and the adsorbent, such as van der Waals bonds, electrostatic interactions, coordination bonds, hydrogen bonding, and hydrophobic interactions. The development of the chemical bond seen between adsorbate and the adsorbent is an irreversible characteristic of contacts and chemical sorption (K. Lu et al. 2014).

16.4.1.3 Synthesis: Method and Key Ingredients

Freeze-drying, microemulsion creation, pyrogenation, and segment separation are some of the processes used to make hydrogels (Kwak et al. 2019). Currently, the much more frequently used technique for creating hydrogels and certain other polymers with repeated additions of free radical building blocks is free radical polymerization (FRP) (J. Wang et al. 2008). FRP is extensively utilized and may be employed as a synthesis method to create a wide range of polymers and material composites. FRP is also one of the greatest flexible ways for producing polymers due to the nonspecific nature of free radical chemical interactions (Xueyan Chen et al. 2020; Pandey, Daverey, and Arunachalam 2020; Hassan et al. 2020; L. Wang et al. 2020; J. Wang et al. 2008). The characteristics of the hydrogel construct are tethered by weak and strong interactions, resulting in the development of a gel. Different studies have stated that using synthetic chemical backbone polymers may cause pollution problems and have instead advised using biopolymers (biosorbents). The modern study looked at the potential of synthetic biopolymers like proteins, polysaccharides, and lipids.

Numerous hydrogels, such as the PVA-hydrogel biomass of *Penicillium cyclopium*, cellulose graft acrylic acid (C-g-AA), chitosan hydrogel, and starch graft acrylic acid/montmorillonite (S-g-AA/MMT), were already utilized as biosorbents (Xueyan Chen et al. 2020; L. Wang et al. 2020; Dai et al. 2019). When biosorbent polymers are used to remediate polluted soils, soil microorganisms can degrade them, limiting their remedial life span and preventing them from performing their intended purpose.

16.4.1.4 Factors Altering Adsorption Capacity

Variables such as the pH of the polluted medium, ionic energy, the constitution of adsorbate and adsorbent, temperature, contact duration, light chemical species, starting concentration, and electric and magnetic fields all impact the swelling and sorptive activity of a hydrogel (Bian et al. 2013; W. Zhang et al. 2013). Furthermore, pH can be said to have a considerable impact, mainly in terms of metal species selective adsorption relative to most other variables. According to researchers, the best adsorption effectiveness is attained when the pH is between 4 and 6 (Rascio and Navari-Izzo 2011).

16.4.2 SOURCE AND CLASSIFICATION

Very little is known about the role of microorganisms from digestate to total contaminant removal effectiveness and the function of digestate in the remediation of HMs and PAHs. The substantial nutritional composition of digestate contributes to a rise in microbial population and activity, and now, in turn, the microorganisms eliminate contaminants (Bianco et al. 2020). By employing OFMSW and AFW to increase pollution remediation, digestate, like biochar, uses and supports the notion of upcycling and circular bioeconomy, with ecological, agricultural, and financial gains (M. Chen et al. 2015; Chibuike and Obiora 2014; H. Lu et al. 2012). Except for AFW, which is made completely of organic wastes, MSW is made up of a variety of wastes such as textiles, paper, metal, glass, plastic, and other organic wastes. MSW differs in composition and proportion (approximately equal percentages of a certain form of waste) throughout the world. MSW in underdeveloped nations has a high organic percentage, but in industrialized countries, plastic and paper make up the majority of MSW. Organic wastes, as an illustration, account for more than half of Thailand's MSW (Dai et al. 2019). Inadequate disposal and management of OFMSW and AFW in landfills may end in unregulated anaerobic decomposition, causing soil and water contamination, unpleasant odors, and the release of methane and certain other pollutants into the air. Alternatively, the OFMSW and AFW may be upcycled to create nutrient-dense digestate (organic fertilizers) that would be less expensive than inorganic fertilizers. It can transport microorganisms that can remove HMs and degrade PAHs. Additionally, it promotes carbon storage in soil organic matter (SOM), lowering emissions of CO_2 (Duan et al. 2019; Johnsen, Wick, and Harms 2005; Tomczyk, Sokołowska, and Boguta 2020; M. Chen et al. 2015; Srogi 2007).

16.4.3 DIGESTATE-ASSISTED REMEDIATION WITH BIOCHAR

An experiment using Pb was set up on a French technosol for 60 days. When compared with a control, cow manure biochar decreased Cd and Zn by 90% and 80%, accordingly, and digestate by 63% and 73%, respectively, after 30 days of incubation (Cipullo et al. 2019). Following 120 days, polluted soil treated with digestate decreased PAHs by 43%, according to Bianco et al. (2020; Bianco et al. 2020). Digestate was investigated as a possible nutrient and microbial seeding for bioremediation of weathered (aged) petroleum hydrocarbon polluted (TPH) soils in related studies. Numerous comparable research has revealed digestate's excellent potential as a source of nutrients and microorganisms for soil bioremediation. As a result, biochar-digestate composites may be bioengineered to harbor certain microbial consortia that form a biochemical system that will improve polluted soil reclamation surpassing current approaches. It is also considered cost-effective, ecologically friendly, and stable, as well as requiring less work and equipment.

16.5 REMEDIATION PROCESS AND MECHANISM

Microorganisms can interact with biochar to improve their remediation effectiveness, just like hydrogels and digestate. To break down organic pollutants like PAHs, microorganisms might release contaminant-degradative enzymes into the environment. PAHs are destroyed during microbial degradation, and if the process is completed, they will be mineralized to CO_2 and H_2O (J. Wang and Wang 2019). Microorganisms can take up and store HMs by an active process termed bioaccumulation or by a passive mechanism termed adsorption. Microorganisms that can accumulate HMs in their bodies can transform harmful forms of HMs into nontoxic or less toxic forms through a variety of processes, therefore avoiding the negative health consequences of metal poisoning. Although a metal-tolerant microbe that can go through all bioaccumulation pathways would be perfect for remediating soil contaminated with HMs and/or PAHs, such an organism has yet to be found or bioengineered. As a result, microorganisms with different mechanisms are brought together to create a mixed culture with enhanced processes and characteristics (J. Wang and Wang 2019).

16.5.1 Bactoremediation, Mycoremediation, and Phycoremediation

Because of their demonstrated flexibility and availability in the ecosystem, bactoremediation (using bacteria for detoxification) utilizing actinobacteria has gained significant interest as a possibility for bioremediation in current history. The biggest bacterial phylum is *Actinobacteria*, a group of Gram-positive bacteria having high guanine and cytosine concentration in their DNA (Sall et al. 2020). For instance, it was explored that actinobacteria may remove both HMs and pesticides at the same time (J. Wang and Wang 2019). Gram-positive bacteria, which include glycoproteins, are thought to be more successful at removing HMs than Gram-negative bacteria. Additionally, Gram-positive bacteria are efficient PAH degraders and are frequently identified as the major bacterial species in PAH-contaminated soil (Bianco et al. 2020). Fungi and algae have been examined for their ability to eliminate HMs and PAHs from polluted soils, in addition to frequently employed bacteria for microbial remediation.

16.5.2 Microbe-assisted Remediation Studies

Biochar was employed as a carrier for immobilizing Cd-Cu tolerant strain NT-2 in a 75-day pot experiment (*Pseudomonas* sp.). The enhanced biochar significantly strengthened the soil microbial community profile and boosted soil enzyme activity (Tu et al. 2020). Phosphate-solubilizing–bacteria- (PSB-) altered biochar was used to improve Pb^{2+} immobilization by rice biochar (RB) and sludge biochar (SB) (Pandey, Daverey, and Arunachalam 2020; Pandey, Daverey, and Arunachalam 2020). The RB and SB were shown in the research to efficiently remove Pb^{2+} (18.61 and 53.89%, respectively).

Three distinct endophytic bacteria strains were used: FD-17, KS-54, and PsJN. Effective soil augmentation requires not only a thorough understanding of the types and levels of pollutants but also the right microorganism strains and their consortia. Mrozik and Piotrowska-Seget (2010) stated that certain characteristics of microorganisms should be examined while selecting appropriate cultures. This comprises the capacity to develop quickly, to be easily cultivated, to tolerate high levels of pollutants, and to live in a variety of environments (M. Chen et al. 2015).

16.6 REMEDIATION MECHANISM OF BIOCHAR

Similar studies have established the remediation or sorption processes of biochar (Oliveira et al. 2017; Cristaldi et al. 2017). Nevertheless, limited research has been done on the remediation process of (bio)engineered biochar and how the incorporated material may increase biochar's sorption capacities, therefore leading to improved pollution extraction efficiency. In summary, remediation processes vary depending on the kind of pollutant and biochar input material. The biochar–pollutants interactions are depicted in Figure 16.2, and the involved properties are those

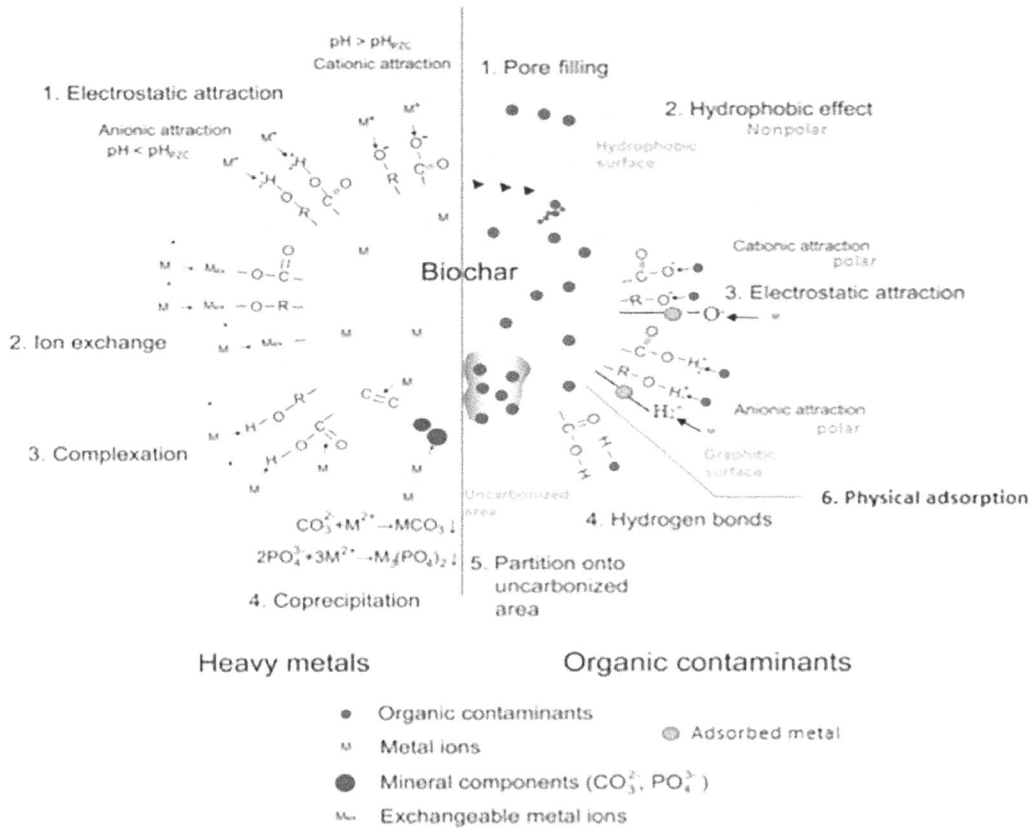

FIGURE 16.2 Sorption interactions between pollutants and biochar.

Source: Ambaye et al. (2020).

of the frequently employed removal methods. As illustrated, the several sorption mechanisms for heavy metals removal include electrostatic attraction, ion exchange, complexation, and coprecipitation, whereas for the removal of organic contaminants, the mechanisms are pore filling, hydrophobic effect, electrostatic attraction, physical adsorption, hydrogen bonds, and partition onto an uncarbonized area.

Biochar's remediation processes for inorganic pollutants like HMs are highly dependent on the valence state of the target metal ion(s) of interest at various pH levels (Li et al. 2018). The addition of biochar to the soil can raise the pH, and HM ions react with -OH, PO_4^{3}, CO_3^{2} to efficiently solidify HM contaminants. Pyrolysis temperature, feedstock type, soil and biochar pH, physical and chemical characteristics of HM ions, and rate of application all impact the sorption association of HM ions with biochar (Li et al. 2018). Biochar, on the other hand, mediates five major PAH-remediation pathways (H. Y. Xu et al. 2020; L. Wang et al. 2020). Table 16.1 presents several research works illustrating favorable outcomes of biochar application in eliminating HM and PAH contaminants, owing to the remediation processes of biochar as discussed. Furthermore, relative to (bio)engineered biochar, remediation operations employing biochar may simply need more time and/or greater quantities of biochar or may only be successful in eliminating particular contaminants.

TABLE 16.1
Adsorption Capacities of Different Biochars toward Organic Contaminants from Industrial Wastewater (Ambaye et al. 2020)

Type of biochar	pH	Temp. (°C)	Targeted species	Concentration range (mg/L)	Langmuir adsorption capacity (mg/g)	References
P-(acrylamide)-wood biochar	–	25	Phenol	5–50	23.14	Ambaye et al. (2020)
Rice husk biochar	–	40	Phenol	100–15,000	409.8	Y. Liu et al. (2011)
Hardwood (laurel oak, *Quercus*)	–	–	Humic acid	0–13	–	Mukherjee, Zimmerman, and Harris (2011)
Orange peel biochar	–	25	Naphthalene	18	–	Ambaye et al. (2020)
Wood char	7	25	Pyrene	0.002–0.12	–	H. Wang et al. (2014)
Pine needle biochar	–	–	m-Dinitrobenzene	0.02–1.0	–	B. Chen, Chen, and Lv (2011)
Peanut straw char	5–9	25	Methyl violet	40–816	104.61	H. Y. Xu et al. (2020)
Kenaf fiber char	8.5	30	Methylene blue	50–200	18.18	H. Y. Xu et al. (2020)
Rice-straw-derived char	5	30	Malachite green	25–300	128.14	W. J. Liu, Jiang, and Yu (2015)
Bamboo biochar	1	40	Acid black 172	50–500	358.14	W. J. Liu, Jiang, and Yu (2015)
Sewage sludge/			2,4-Dichlorophenol	100% removal	–	Kalderis et al. (2017)
Corn straw	5	22	Bisphenol	500	–	Kalderis et al. (2017)
Dissolved black carbon biochar (DBC)	5	25	Naphthalene	91	–	Fu et al. (2018)
Miscanthus-derived biochar	5	30	Naphthalene	4–35	–	Kim and Hyun (2018)
Bamboo hydrochars	1	35	Congo red	100	–	Li et al. (2018)
Sugarcane bagasse biochar	7	25	Malachite green	3000	–	Vyavahare et al. (2018)
Corn straw	5	22	Atrazine	70	–	Vyavahare et al. (2018)
Macadamia nutshells	7	30	Phenanthrene	60	–	Vyavahare et al. (2018)

16.7 TREATMENT OF TYPICAL POLLUTANTS

Biochar has been widely investigated for water/wastewater treatment owing to its capacity to adsorb contaminants in the liquid phase (B. Chen, Chen, and Lv 2011; Xincai Chen et al. 2011; M. Chen et al. 2015). It is primarily utilized for the adsorption of potentially hazardous metals (46%) and organic contaminants (39%). It can also be utilized for phosphorus and nitrogen adsorption (13%). Because of its huge specific surface area, porous structure, and surface functional groups, the remaining 2% is used for the adsorption of various contaminants (Xincai Chen et al. 2011; B. Chen, Chen, and Lv 2011).

16.7.1 ADSORPTION ISOTHERM

The release of organic and inorganic contaminants into industrial effluents has become a worldwide issue. Biochar was used by researchers to eliminate these harmful contaminants from an aqueous solution in the application of biochar for the adsorption of contaminants in the water. As a result, using adsorption isotherms at a steady temperature range, it is important to measure and optimize the relationship between the biochar's adsorption capacity and the adsorbates (Fan et al. 2017). To explain the equilibrium of organic and inorganic contaminant adsorption to biochar, several empirical equilibrium models were used. The findings revealed that the adsorption equilibrium isotherms are influenced by the contaminants' binding affinity on the biochar surface. During the measurement of inorganic pollutant adsorption on biochar, the Freundlich and Langmuir isotherm models performed better than other models. During the binding, these models only include one layer of adsorption of the adsorbate on the homogeneous surface of the adsorbent (Aydin and Aksoy 2009; Reddy and Lee 2013). Biochar has a strong aversion to organic contaminants and generates great efficacy from biomass in eliminating organic contaminants such as pesticides, antibiotics, herbicides, and dyes. Organic pollutants' adsorption was consistent with the Langmuir and Freundlich's models (Sophia and Lima 2018). Trichloroethylene was removed from water using biochar made from agricultural wastes (Ahmad et al. 2012). The researchers demonstrated that the experimental trichloroethylene adsorption suited the Langmuir model. Biochar as an adsorbent for the extraction of various contaminants from water performed well overall. Furthermore, the physicochemical characteristics of the biochar influenced the biochar's ability to bind pollutants.

16.7.2 ADSORPTION KINETICS

The adsorption behavior of biochar toward organic/inorganic pollutants is drastically influenced by its physical and chemical characteristics. The chemical and physical properties of the biochar may have a significant impact on adsorption of the organic and inorganic pollutants. Such characteristics have an impact on the kinetics and mechanism of adsorption, which include chemical binding as well as mass transfer (Boutsika, Karapanagioti, and Manariotis 2014). According to Reddy and Lee (2013), using the intraparticle diffusion model, it is important to understand the diffusion mechanisms, adsorbent surface area, and chemical reactions engaged in the adsorption process to implement biochar on a wide scale for the degradation of different pollutants from wastewater (Mohan et al. 2011). Lu et al. (2012) utilized biochar made from sewage sludge to adsorb Pb^{2+} from water with a pH range of 2–5 (H. Lu et al. 2012).

16.7.3 HEAVY METAL POLLUTANTS

The industrial contaminants include heavy metals that provide the major source of significant contaminants with even very tiny quantities of heavy metals such as Cd, Cu, Zn, and Pb in wastewater having a major environmental impact. Despite being made at the same pyrolysis temperature, biochars made from various feedstocks have variable heavy metal adsorption capacities. It should be linked to the presence or lack of minerals in the feedstock, such as $CO_2\_3CO_3^{2-}$ and $PO_3\_4PO_4^{3-}$, which influence biochar adsorption ability. Such minerals may aid in the development of active sites that improve heavy metal adsorption from wastewater (Cao et al. 2009; X. Xu et al. 2013).

16.7.4 TREATMENT OF NITROGEN AND PHOSPHORUS CONTAMINATION

Biochar has also been studied to eliminate nitrogen and phosphorus from industrial wastewater to diminish the effect of eutrophication on the environment, in addition to the aforementioned organic and heavy metal contaminants. After that process, the digestate can even be utilized as a high-quality fertilizer (Hale et al. 2012) Biochar made from cow dung is being used to reduce PO_4-P and NH_4-N, from wastewater. Adsorption capability was shown to be high by Wang et al. (2014), Xie et al. (2015), and H. Wang et al. (2014). Researchers also discovered that the adsorption capacity for nitrogen and phosphorus was determined by the charcoal feedstock. As a result, biochar has a high capacity for removing PO_4-P and NH_4-N, allowing nutrients to be released and boosting soil fertility, which helps to enhance crop output.

16.8   ECONOMIC AND ENVIRONMENTAL BENEFIT ANALYSES

Biochar has tremendous application potential in the elimination of industrial pollutants, as evidenced by the research just discussed. Moreover, several factors, such as the kind of biochar, the relative amount of the source, the pyrolysis temperature, the availability of industrial-scale reactors, the regeneration technique, and the life duration, should be adjusted before large-scale use to increase the process economic feasibility.

Usually, the use of biochars to remove contaminants has been extensively addressed in recent years. Moreover, because the study on the biochar price is still in its initial phases, additional studies need to be done to examine the considerable heterogeneity in biochar pricing and associated advantages in order to use biochar at a wide scale for pollution reduction. Biochar is a novel strategy to mitigate the effects of climate change and enhance sustainable environmental and agricultural growth (Oliveira et al. 2017). Nowadays, the globe is still dealing with the effects of greenhouse gases and global warming on the ecosystem (P. Wang et al. 2018; Nie et al. 2018). Biochar can cut CO_2 emissions by 0.10–0.30 billion tonnes per year, according to (Liu, Jiang, and Yu 2015). This demonstrates that biochar can help to lower CO_2 levels in the atmosphere, resulting in lower GHG emissions. In addition, Dai et al. (2019) claimed that biochar-modified soil might minimize N_2O emissions. Furthermore, N_2O has a 310-fold greater greenhouse impact than CO_2 (Dai et al. 2019).

The four complementary and quite often synergistic societal and economical uses of the biochar technique are:

- Soil enhancement (to boost productivity while reducing pollutant toxicity),
- Climate change mitigation,
- Energy generation, and
- Waste management.

The kind of biochar used to eliminate organic or inorganic pollutants and other environmental activities, on the other hand, is dependent on the type of biochar employed. Biochar must be able to stay in the soil for an extended period to accomplish carbon sequestration. The mineralization of the biochar over several years is primarily responsible for its durability. Infrared spectroscopy, biomarkers, and nuclear magnetic resonance are just a few of the methods that may be used to track biochar mineralization over time.

16.9   FUTURE PERSPECTIVES

Biochar is a potential material for water treatment as well as agricultural soil fertility preservation and restoration. Meanwhile, research on the role of biochar in the purification of water from wastewaters, the impact of various feedstock sources on biochar quality, and physicochemical variables influencing the effectiveness of pollutants removal are still in their early stages. As a result, shortly, the following literature review fields should be considered:

1. *Developing a model for a biochar adsorption process that has been changed.* Surface functionality and porosity must be adjusted to meet requirements. Surface oxidation, amination, sulfonation, and pore structure change are some of the techniques that may be used to improve the production of selective biochar materials.
2. *Improving adsorption efficiency using improved techniques.* Several studies need to be conducted to investigate additional regulators or activators that affect biochar's ability to remove organic and inorganic contaminants.
3. *Determining the best dose and technique for producing biochar.* The aim is to enable biochar to swiftly remove various pollutants from wastewater.

4. *The economic visibility of biochar and its regeneration.* Also needed are standard operating procedures for eliminating hazardous contaminants from industrial effluent.
5. *Biochar's pretreatment for hazardous chemicals elimination and subsequent biological treatment.* Additional research is required before biochar may be used on a wide scale.
6. *More research into absorption processes.* Chemically activated biochar has a higher adsorption capability for contaminants than nonactivated biochar, but more research is needed.

Only a few more studies have been done in the biochar field; therefore further study is needed on subjects such as the link between raw materials and processing parameters, as well as biochar regeneration to minimize ecological consequences.

16.10   CONCLUSION

According to the study, biochar has a lot of potential for removing common organic and inorganic contaminants found in industrial effluent. It's a fascinating adsorbent with high efficiency. Due to its high surface area, charged surface, and functional groups, it is an environmentally beneficial sorbing material. Biochar has been shown to have a high potential for adsorbing inorganic and organic pollutants through a variety of mechanisms, including pore filling, electrostatic interactions, ion exchange, precipitation, and surface sorption, all of which are influenced by the physico-chemical properties of biochar, such as biochar dosage, pyrolysis temperature, and the pH of the treated matrix. Biochar is also useful for the elimination of pollutants at large-scale levels because it reduces the bioavailability, toxicity, and mobility of organic and inorganic pollutants. Biochar may be utilized as an effective, low-cost, and environmentally acceptable adsorbent for the removal of organic and inorganic contaminants in industrial wastewater, according to this review.

REFERENCES

Ahmad, Mahtab, Sang Soo Lee, Xiaomin Dou, Dinesh Mohan, Jwa Kyung Sung, Jae E. Yang, and Yong Sik Ok. 2012. "Effects of Pyrolysis Temperature on Soybean Stover- and Peanut Shell-Derived Biochar Properties and TCE Adsorption in Water." *Bioresource Technology* 118. Elsevier Ltd: 536–544. doi:10.1016/j.biortech.2012.05.042

Ali, Hazrat, Ezzat Khan, and Ikram Ilahi. 2019. "Environmental Chemistry and Ecotoxicology of Hazardous Heavy Metals: Environmental Persistence, Toxicity, and Bioaccumulation." *Journal of Chemistry* 2019 (Cd). doi:10.1155/2019/6730305

Ambaye, T. G., M. Vaccari, E. D. van Hullebusch, A. Amrane, and S. Rtimi. 2020. "Mechanisms and Adsorption Capacities of Biochar for the Removal of Organic and Inorganic Pollutants from Industrial Wastewater." *International Journal of Environmental Science and Technology.* doi:10.1007/s13762–020–03060-w

Andersson, Jan T., and Christine Achten. 2015. "Time to Say Goodbye to the 16 EPA PAHs? Toward an Up-to-Date Use of PACs for Environmental Purposes." *Polycyclic Aromatic Compounds* 35 (2–4): 330–354. doi:10.1080/10406638.2014.991042

Atkinson, R., and J. Arey. 1994. "Atmospheric Chemistry of Gas-Phase Polycyclic Aromatic Hydrocarbons: Formation of Atmospheric Mutagens." *Environmental Health Perspectives* 102 (SUPPL. 4): 117–126. doi:10.1289/ehp.94102s4117

Aydin, Yaşar Andelib, and Nuran Deveci Aksoy. 2009. "Adsorption of Chromium on Chitosan: Optimization, Kinetics and Thermodynamics." *Chemical Engineering Journal* 151 (1–3): 188–194. doi:10.1016/j.cej.2009.02.010

Barnes, Jessica L., Maria Zubair, Kaarthik John, Miriam C. Poirier, and Francis L. Martin. 2018. "Carcinogens and DNA Damage." *Biochemical Society Transactions* 46 (5): 1213–1224. doi:10.1042/BST20180519

Bianco, F., M. Race, S. Papirio, and G. Esposito. 2020. "Removal of Polycyclic Aromatic Hydrocarbons during Anaerobic Biostimulation of Marine Sediments." *Science of the Total Environment* 709. Elsevier B.V.: 136141. doi:10.1016/j.scitotenv.2019.136141

Bian, Rongjun, De Chen, Xiaoyu Liu, Liqiang Cui, Lianqing Li, Genxing Pan, Dan Xie, et al. 2013. "Biochar Soil Amendment as a Solution to Prevent Cd-Tainted Rice from China: Results from a Cross-Site Field Experiment." *Ecological Engineering* 58. Elsevier B.V.: 378–383. doi:10.1016/j.ecoleng.2013.07.031

Boutsika, Lamprini G., Hrissi K. Karapanagioti, and Ioannis D. Manariotis. 2014. "Aqueous Mercury Sorption by Biochar from Malt Spent Rootlets." *Water, Air, and Soil Pollution* 225 (1). doi:10.1007/s11270-013-1805-9

Braghiroli, Flavia Lega, Hassine Bouafif, and Ahmed Koubaa. 2019. "Enhanced SO2 Adsorption and Desorption on Chemically and Physically Activated Biochar Made from Wood Residues." *Industrial Crops and Products* 138 (December 2018). doi:10.1016/j.indcrop.2019.06.019

Cao, Xinde, Lena Ma, Bin Gao, and Willie Harris. 2009. "Dairy-Manure Derived Biochar Effectively Sorbs Lead and Atrazine." *Environmental Science and Technology* 43 (9): 3285–3291. doi:10.1021/es803092k

Chen, Baoliang, Zaiming Chen, and Shaofang Lv. 2011. "A Novel Magnetic Biochar Efficiently Sorbs Organic Pollutants and Phosphate." *Bioresource Technology* 102 (2). Elsevier Ltd: 716–723. doi:10.1016/j.biortech.2010.08.067

Chen, Ming, Piao Xu, Guangming Zeng, Chunping Yang, Danlian Huang, and Jiachao Zhang. 2015. "Bioremediation of Soils Contaminated with Polycyclic Aromatic Hydrocarbons, Petroleum, Pesticides, Chlorophenols and Heavy Metals by Composting: Applications, Microbes and Future Research Needs." *Biotechnology Advances* 33 (6). Elsevier Inc.: 745–755. doi:10.1016/j.biotechadv.2015.05.003

Chen, Xincai, Guangcun Chen, Linggui Chen, Yingxu Chen, Johannes Lehmann, Murray B. McBride, and Anthony G. Hay. 2011. "Adsorption of Copper and Zinc by Biochars Produced from Pyrolysis of Hardwood and Corn Straw in Aqueous Solution." *Bioresource Technology* 102 (19). Elsevier Ltd: 8877–8884. doi:10.1016/j.biortech.2011.06.078

Chen, Xueyan, Deepika Kumari, C. J. Cao, Grażyna Plaza, and Varenyam Achal. 2020. "A Review on Remediation Technologies for Nickel-Contaminated Soil." *Human and Ecological Risk Assessment* 26 (3): 571–585. doi:10.1080/10807039.2018.1539639

Chen, Zaiming, Baoliang Chen, Dandan Zhou, and Wenyuan Chen. 2012. "Bisolute Sorption and Thermodynamic Behavior of Organic Pollutants to Biomass-Derived Biochars at Two Pyrolytic Temperatures." *Environmental Science and Technology* 46 (22): 12476–12483. doi:10.1021/es303351e

Chibuike, G. U., and S. C. Obiora. 2014. "Heavy Metal Polluted Soils: Effect on Plants and Bioremediation Methods." *Applied and Environmental Soil Science* 2014: 752708. doi:10.1155/2014/752708

Cipullo, S., I. Negrin, L. Claveau, B. Snapir, S. Tardif, C. Pulleyblank, G. Prpich, P. Campo, and F. Coulon. 2019. "Linking Bioavailability and Toxicity Changes of Complex Chemicals Mixture to Support Decision Making for Remediation Endpoint of Contaminated Soils." *Science of the Total Environment* 650. Elsevier B.V.: 2150–2163. doi:10.1016/j.scitotenv.2018.09.339

Cristaldi, Antonio, Gea Oliveri Conti, Eun Hea Jho, Pietro Zuccarello, Alfina Grasso, Chiara Copat, and Margherita Ferrante. 2017. "Phytoremediation of Contaminated Soils by Heavy Metals and PAHs. A Brief Review." *Environmental Technology and Innovation* 8. Elsevier B.V.: 309–326. doi:10.1016/j.eti.2017.08.002

Crombie, Kyle, Ondřej Mašek, Saran P. Sohi, Peter Brownsort, and Andrew Cross. 2013. "The Effect of Pyrolysis Conditions on Biochar Stability as Determined by Three Methods." *GCB Bioenergy* 5 (2): 122–131. doi:10.1111/gcbb.12030

Dai, Yingjie, Naixin Zhang, Chuanming Xing, Qingxia Cui, and Qiya Sun. 2019. "The Adsorption, Regeneration and Engineering Applications of Biochar for Removal Organic Pollutants: A Review." *Chemosphere* 223. Elsevier Ltd: 12–27. doi:10.1016/j.chemosphere.2019.01.161

Dean, J. R., and W. C. Scott. 2004. "Recent Developments in Assessing the Bioavailability of Persistent Organic Pollutants in the Environment." *TrAC—Trends in Analytical Chemistry* 23 (9): 609–618. doi:10.1016/j.trac.2004.06.008

Duan, Wenyan, Patryk Oleszczuk, Bo Pan, and Baoshan Xing. 2019. "Environmental Behavior of Engineered Biochars and Their Aging Processes in Soil." *Biochar* 1 (4). Springer Singapore: 339–351. doi:10.1007/s42773-019-00030-5

Fan, Shisuo, Yi Wang, Zhen Wang, Jie Tang, Jun Tang, and Xuede Li. 2017. "Removal of Methylene Blue from Aqueous Solution by Sewage Sludge-Derived Biochar: Adsorption Kinetics, Equilibrium, Thermodynamics and Mechanism." *Journal of Environmental Chemical Engineering* 5 (1). Elsevier B.V.: 601–611. doi:10.1016/j.jece.2016.12.019

Fu, Heyun, Chenhui Wei, Xiaolei Qu, Hui Li, and Dongqiang Zhu. 2018. "Strong Binding of Apolar Hydrophobic Organic Contaminants by Dissolved Black Carbon Released from Biochar: A Mechanism of Pseudomicelle Partition and Environmental Implications." *Environmental Pollution* 232: 402–410. doi:https://doi.org/10.1016/j.envpol.2017.09.053

Hale, Sarah E., Johannes Lehmann, David Rutherford, Andrew R. Zimmerman, Robert T. Bachmann, Victor Shitumbanuma, Adam O'Toole, Kristina L. Sundqvist, Hans Peter H. Arp, and Gerard Cornelissen. 2012. "Quantifying the Total and Bioavailable Polycyclic Aromatic Hydrocarbons and Dioxins in Biochars." *Environmental Science and Technology* 46 (5): 2830–2838. doi:10.1021/es203984k

Hassan, Masud, Yanju Liu, Ravi Naidu, Sanjai J. Parikh, Jianhua Du, Fangjie Qi, and Ian R. Willett. 2020. "Influences of Feedstock Sources and Pyrolysis Temperature on the Properties of Biochar and Functionality as Adsorbents: A Meta-Analysis." *Science of the Total Environment* 744. Elsevier B.V.: 140714. doi:10.1016/j.scitotenv.2020.140714

Johnsen, Anders R., Lukas Y. Wick, and Hauke Harms. 2005. "Principles of Microbial PAH-Degradation in Soil." *Environmental Pollution* 133 (1): 71–84. doi:10.1016/j.envpol.2004.04.015

Kalderis, Dimitrios, Berkant Kayan, Sema Akay, Esra Kulaksız, and Belgin Gözmen. 2017. "Adsorption of 2,4-Dichlorophenol on Paper Sludge/Wheat Husk Biochar: Process Optimization and Comparison with Biochars Prepared from Wood Chips, Sewage Sludge and Hog Fuel/Demolition Waste." *Journal of Environmental Chemical Engineering* 5 (3): 2222–2231. doi:https://doi.org/10.1016/j.jece.2017.04.039

Khalid, Sana, Muhammad Shahid, Nabeel Khan Niazi, Behzad Murtaza, Irshad Bibi, and Camille Dumat. 2017. "A Comparison of Technologies for Remediation of Heavy Metal Contaminated Soils." *Journal of Geochemical Exploration* 182. Elsevier B.V.: 247–268. doi:10.1016/j.gexplo.2016.11.021

Kim, Juhee, and Seunghun Hyun. 2018. "Sorption of Ionic and Nonionic Organic Solutes onto Giant Miscanthus-Derived Biochar from Methanol-Water Mixtures." *Science of the Total Environment* 615: 805–813. doi:https://doi.org/10.1016/j.scitotenv.2017.09.296

Kwak, Jin Hyeob, Md Shahinoor Islam, Siyuan Wang, Selamawit Ashagre Messele, M. Anne Naeth, Mohamed Gamal El-Din, and Scott X. Chang. 2019. "Biochar Properties and Lead(II) Adsorption Capacity Depend on Feedstock Type, Pyrolysis Temperature, and Steam Activation." *Chemosphere* 231. Elsevier Ltd: 393–404. doi:10.1016/j.chemosphere.2019.05.128

Lahori, Altaf Hussain, Zhanyu Guo, Zengqiang Zhang, Ronghua Li, Amanullah Mahar, Mukesh Kumar Awasthi, Feng Shen, et al. 2017. "Use of Biochar as an Amendment for Remediation of Heavy Metal-Contaminated Soils: Prospects and Challenges." *Pedosphere* 27 (6). Soil Science Society of China: 991–1014. doi:10.1016/S1002-0160(17)60490-9

Li, Hao, Yanbei Cao, Di Zhang, and Bo Pan. 2018. "PH-Dependent KOW Provides New Insights in Understanding the Adsorption Mechanism of Ionizable Organic Chemicals on Carbonaceous Materials." *Science of the Total Environment* 618. Elsevier B.V.: 269–275. doi:10.1016/j.scitotenv.2017.11.065

Liu, Wu Jun, Hong Jiang, and Han Qing Yu. 2015. "Development of Biochar-Based Functional Materials: Toward a Sustainable Platform Carbon Material." *Chemical Reviews* 115 (22): 12251–12285. doi:10.1021/acs.chemrev.5b00195

Liu, Yuxue, Min Yang, Yimin Wu, Hailong Wang, Yingxu Chen, and Weixiang Wu. 2011. "Reducing CH4 and CO2 Emissions from Waterlogged Paddy Soil with Biochar." *Journal of Soils and Sediments* 11 (6): 930–939. doi:10.1007/s11368-011-0376-x

Liu, Zhengang, Augustine Quek, S. Kent Hoekman, and R. Balasubramanian. 2013. "Production of Solid Biochar Fuel from Waste Biomass by Hydrothermal Carbonization." *Fuel* 103. Elsevier Ltd: 943–949. doi:10.1016/j.fuel.2012.07.069

Lu, Huanliang, Weihua Zhang, Yuxi Yang, Xiongfei Huang, Shizhong Wang, and Rongliang Qiu. 2012. "Relative Distribution of Pb2+ Sorption Mechanisms by Sludge-Derived Biochar." *Water Research* 46 (3). Elsevier Ltd: 854–862. doi:10.1016/j.watres.2011.11.058

Lu, Kouping, Xing Yang, Jiajia Shen, Brett Robinson, Huagang Huang, Dan Liu, Nanthi Bolan, Jianchuan Pei, and Hailong Wang. 2014. "Effect of Bamboo and Rice Straw Biochars on the Bioavailability of Cd, Cu, Pb and Zn to Sedum Plumbizincicola." *Agriculture, Ecosystems and Environment* 191. Elsevier B.V.: 124–132. doi:10.1016/j.agee.2014.04.010

Mašek, Ondřej, Peter Brownsort, Andrew Cross, and Saran Sohi. 2013. "Influence of Production Conditions on the Yield and Environmental Stability of Biochar." *Fuel* 103: 151–155. doi:10.1016/j.fuel.2011.08.044

Mohamed, Badr A., Naoko Ellis, Chang Soo Kim, and Xiaotao Bi. 2017. "The Role of Tailored Biochar in Increasing Plant Growth, and Reducing Bioavailability, Phytotoxicity, and Uptake of Heavy Metals in Contaminated Soil." *Environmental Pollution* 230. Elsevier Ltd: 329–338. doi:10.1016/j.envpol.2017.06.075

Mohan, Dinesh, Shalini Rajput, Vinod K. Singh, Philip H. Steele, and Charles U. Pittman. 2011. "Modeling and Evaluation of Chromium Remediation from Water Using Low Cost Bio-Char, a Green Adsorbent." *Journal of Hazardous Materials* 188 (1–3). Elsevier B.V.: 319–333. doi:10.1016/j.jhazmat.2011.01.127

Mrozik, A. and   Piotrowska-Seget Z 2010. Bioaugmentation as a strategy for cleaning up of soils contaminated with aromatic compounds. *Microbiological Research* 165(5): 363–375. https://doi.org/10.1016/j.micres.2009.08.001.

Mukherjee, A., A. R. Zimmerman, and W. Harris. 2011. "Surface Chemistry Variations among a Series of Laboratory-Produced Biochars." *Geoderma* 163 (3): 247–355. doi:https://doi.org/10.1016/j.geoderma.2011.04.021

Nie, Chengrong, Xing Yang, Nabeel Khan Niazi, Xiaoya Xu, Yuhui Wen, Jörg Rinklebe, Yong Sik Ok, Song Xu, and Hailong Wang. 2018. "Impact of Sugarcane Bagasse-Derived Biochar on Heavy Metal Availability and Microbial Activity: A Field Study." *Chemosphere* 200: 274–382. doi:10.1016/j.chemosphere.2018.02.134

Oliveira, Fernanda R., Anil K. Patel, Deb P. Jaisi, Sushil Adhikari, Hui Lu, and Samir Kumar Khanal. 2017. "Environmental Application of Biochar: Current Status and Perspectives." *Bioresource Technology* 246 (August). Elsevier: 110–122. doi:10.1016/j.biortech.2017.08.122

Pandey, Deepshikha, Achlesh Daverey, and Kusum Arunachalam. 2020. "Biochar: Production, Properties and Emerging Role as a Support for Enzyme Immobilization." *Journal of Cleaner Production* 255. Elsevier Ltd: 120267. doi:10.1016/j.jclepro.2020.120267

Paunovic, Olivera, Sabolc Pap, Snezana Maletic, Mark A. Taggart, Nikola Boskovic, and Maja Turk Sekulic. 2019. "Ionisable Emerging Pharmaceutical Adsorption onto Microwave Functionalised Biochar Derived from Novel Lignocellulosic Waste Biomass." *Journal of Colloid and Interface Science* 547. Elsevier Inc.: 350–360. doi:10.1016/j.jcis.2019.04.011

Pérez, R. M., G. Cabrera, J. M. Gómez, A. Ábalos, and D. Cantero. 2010. "Combined Strategy for the Precipitation of Heavy Metals and Biodegradation of Petroleum in Industrial Wastewaters." *Journal of Hazardous Materials* 182 (1–3): 896–902. doi:10.1016/j.jhazmat.2010.07.003

Rajapaksha, Anushka Upamali, Season S. Chen, Daniel C.W. Tsang, Ming Zhang, Meththika Vithanage, Sanchita Mandal, Bin Gao, Nanthi S. Bolan, and Yong Sik Ok. 2016. "Engineered/Designer Biochar for Contaminant Removal/Immobilization from Soil and Water: Potential and Implication of Biochar Modification." *Chemosphere* 148. Elsevier Ltd: 276–291. doi:10.1016/j.chemosphere.2016.01.043

Rascio, Nicoletta, and Flavia Navari-Izzo. 2011. "Heavy Metal Hyperaccumulating Plants: How and Why Do They Do It? And What Makes Them so Interesting?" *Plant Science* 180 (2). Elsevier Ireland Ltd: 169–181. doi:10.1016/j.plantsci.2010.08.016

Reddy, D. Harikishore Kumar, and Seung Mok Lee. 2013. "Application of Magnetic Chitosan Composites for the Removal of Toxic Metal and Dyes from Aqueous Solutions." *Advances in Colloid and Interface Science* 201–202. Elsevier B.V.: 68–93. doi:10.1016/j.cis.2013.10.002

Rombolà, Alessandro G., Daniele Fabbri, Silvia Baronti, Francesco Primo Vaccari, Lorenzo Genesio, and Franco Miglietta. 2019. "Changes in the Pattern of Polycyclic Aromatic Hydrocarbons in Soil Treated with Biochar from a Multiyear Field Experiment." *Chemosphere* 219: 662–670. doi:10.1016/j.chemosphere.2018.11.178

Sall, Mohamed Lamine, Abdou Karim Diagne Diaw, Diariatou Gningue-Sall, Snezana Efremova Aaron, and Jean Jacques Aaron. 2020. "Toxic Heavy Metals: Impact on the Environment and Human Health, and Treatment with Conducting Organic Polymers, a Review." *Environmental Science and Pollution Research* 27 (24). Environmental Science and Pollution Research: 29927–29942. doi:10.1007/s11356-020-09354-3

Sharma, Swati, Sakshi Tiwari, Abshar Hasan, Varun Saxena, and Lalit M. Pandey. 2018. "Recent Advances in Conventional and Contemporary Methods for Remediation of Heavy Metal-Contaminated Soils." *3 Biotech* 8 (4). Springer Berlin Heidelberg: 1–18. doi:10.1007/s13205-018-1237-8

Singh, Nitika, Vivek Kumar Gupta, Abhishek Kumar, and Bechan Sharma. 2017. "Synergistic Effects of Heavy Metals and Pesticides in Living Systems." *Frontiers in Chemistry* 5 (October): 1–9. doi:10.3389/fchem.2017.00070

Sophia, A., Carmalin, and Eder C. Lima. 2018. "Removal of Emerging Contaminants from the Environment by Adsorption." *Ecotoxicology and Environmental Safety* 150 (June 2017). Elsevier Inc.: 1–17. doi:10.1016/j.ecoenv.2017.12.026

Srogi, Krystyna. 2007. "Monitoring of Environmental Exposure to Polycyclic Aromatic Hydrocarbons: A Review." *Environmental Chemistry Letters* 5 (4): 169–195. doi:10.1007/s10311-007-0095-0

Stefaniuk, Magdalena, Patryk Oleszczuk, and Krzysztof Różyło. 2017. "Co-Application of Sewage Sludge with Biochar Increases Disappearance of Polycyclic Aromatic Hydrocarbons from Fertilized Soil in Long Term Field Experiment." *Science of the Total Environment* 599–600: 854–862. doi:10.1016/j.scitotenv.2017.05.024

Tomczyk, Agnieszka, Zofia Sokołowska, and Patrycja Boguta. 2020. "Biochar Physicochemical Properties: Pyrolysis Temperature and Feedstock Kind Effects." *Reviews in Environmental Science and Biotechnology* 19 (1). Springer Netherlands: 191–215. doi:10.1007/s11157-020-09523-3

Tu, Chen, Jing Wei, Feng Guan, Ying Liu, Yuhuan Sun, and Yongming Luo. 2020. "Biochar and Bacteria Inoculated Biochar Enhanced Cd and Cu Immobilization and Enzymatic Activity in a Polluted Soil." *Environment International* 137 (February). Elsevier: 105576. doi:10.1016/j.envint.2020.105576

Varma, Anil Kumar, Ravi Shankar, and Prasenjit Mondal. 2018. "A Review on Pyrolysis of Biomass and the Impacts of Operating Conditions on Product Yield, Quality, and Upgradation." *Recent Advancements in Biofuels and Bioenergy Utilization*, 227–259. doi:10.1007/978-981-13-1307-3_10

Vyavahare, Govind D., Ranjit G. Gurav, Pooja P. Jadhav, Ravishankar R. Patil, Chetan B. Aware, and Jyoti P. Jadhav. 2018. "Response Surface Methodology Optimization for Sorption of Malachite Green Dye on Sugarcane Bagasse Biochar and Evaluating the Residual Dye for Phyto and Cytogenotoxicity." *Chemosphere* 194: 306–315. doi:https://doi.org/10.1016/j.chemosphere.2017.11.180

Wang, Hou, Xingzhong Yuan, Guangming Zeng, Lijian Leng, Xin Peng, Kailingli Liao, Lijuan Peng, and Zhihua Xiao. 2014. "Removal of Malachite Green Dye from Wastewater by Different Organic Acid-Modified Natural Adsorbent: Kinetics, Equilibriums, Mechanisms, Practical Application, and Disposal of Dye-Loaded Adsorbent." *Environmental Science and Pollution Research* 21 (19): 11552–11564. doi:10.1007/s11356-014-3025-2

Wang, Jianlong, and Shizong Wang. 2019. "Preparation, Modification and Environmental Application of Biochar: A Review." *Journal of Cleaner Production* 227. Elsevier Ltd: 1002–1022. doi:10.1016/j.jclepro.2019.04.282

Wang, Jianlong, Shizong Wang, Kok Hui Goh, Teik Thye Lim, Zhili Dong, Deepshikha Pandey, Achlesh Daverey, and Kusum Arunachalam. 2008. "Preparation, Modification and Environmental Application of Biochar: A Review." *Journal of Cleaner Production* 227 (6–7). Elsevier Ltd: 1343–1368. doi:10.1016/j.jclepro.2020.120267

Wang, Liuwei, Yong Sik Ok, Daniel C.W. Tsang, Daniel S. Alessi, Jörg Rinklebe, Hailong Wang, Ondřej Mašek, Renjie Hou, David O'Connor, and Deyi Hou. 2020. "New Trends in Biochar Pyrolysis and Modification Strategies: Feedstock, Pyrolysis Conditions, Sustainability Concerns and Implications for Soil Amendment." *Soil Use and Management* 36 (3): 358–386. doi:10.1111/sum.12592

Wang, Pingping, Xingang Liu, Xiaohu Wu, Jun Xu, Fengshou Dong, and Yongquan Zheng. 2018. "Evaluation of Biochars in Reducing the Bioavailability of Flubendiamide in Water/Sediment Using Passive Sampling with Polyoxymethylene." *Journal of Hazardous Materials* 344. Elsevier B.V.: 1000–1006. doi:10.1016/j.jhazmat.2017.12.003

Xie, Tao, Krishna R. Reddy, Chengwen Wang, Erin Yargicoglu, and Kurt Spokas. 2015. "Characteristics and Applications of Biochar for Environmental Remediation: A Review." *Critical Reviews in Environmental Science and Technology* 45 (9): 939–969. doi:10.1080/10643389.2014.924180

Xu, Hong Ying, Jing Li Ge, Ling Li Zhang, Chan Zhang, Ru Jin, and Xiao Hui Wang. 2020. "A Dibenz[a,h] Anthracene-Degrading Strain Amycolatopsis Sp. Y1–2 from Soils in the Coal Mining Areas." *Polycyclic Aromatic Compounds* 40 (1). Taylor & Francis: 166–178. doi:10.1080/10406638.2018.1539019

Xu, Xiaoyun, Xinde Cao, Ling Zhao, Hailong Wang, Hongran Yu, and Bin Gao. 2013. "Removal of Cu, Zn, and Cd from Aqueous Solutions by the Dairy Manure-Derived Biochar." *Environmental Science and Pollution Research* 20 (1): 358–368. doi:10.1007/s11356-012-0873-5

Zhang, Weihua, Shengyao Mao, Hao Chen, Long Huang, and Rongliang Qiu. 2013. "Pb(II) and Cr(VI) Sorption by Biochars Pyrolyzed from the Municipal Wastewater Sludge under Different Heating Conditions." *Bioresource Technology* 147. Elsevier Ltd: 545–552. doi:10.1016/j.biortech.2013.08.082

Zhang, Xiaokai, Hailong Wang, Lizhi He, Kouping Lu, Ajit Sarmah, Jianwu Li, Nanthi S. Bolan, Jianchuan Pei, Huagang Huang, et al. 2013a. "Transport and Remediation of Subsurface Contaminants." *Pedosphere* 20 (6). Soil Science Society of China: 991–1014. doi:10.1007/s11356-013-1659-0

Zhang, Xiaokai, Hailong Wang, Lizhi He, Kouping Lu, Ajit Sarmah, Jianwu Li, Nanthi S. Bolan, Jianchuan Pei, Huagang Huang, et al. 2013b. "Using Biochar for Remediation of Soils Contaminated with Heavy Metals and Organic Pollutants." *Environmental Science and Pollution Research* 20 (12): 8472–8483. doi:10.1007/s11356-013-1659-0

Zhao, Ling, Xinde Cao, Ondřej Mašek, and Andrew Zimmerman. 2013. "Heterogeneity of Biochar Properties as a Function of Feedstock Sources and Production Temperatures." *Journal of Hazardous Materials* 256–257: 1–9. doi:10.1016/j.jhazmat.2013.04.015

17 Biochar Application for Treatment of Industrial Effluent as a Green Technology
Opportunities, Challenges, and Future Perspectives

Swati Sambita Mohanty

CONTENTS

17.1 INTRODUCTION

Every day, a substantial volume of wastewater is produced by several industrial sectors, which has a significant impact on the ecosystem (Inyang et al., 2012). Since the green revolution, technological advancement and increased usage of agrochemical-based crop yield strategies have

DOI: 10.1201/9781003203438-17

significantly enhanced the perseverance of organic and heavy metal pollution in the food system and the ecosystem. This has sparked widespread awareness about the safety of the atmosphere and human health (Spokas et al., 2009). Ion exchange, adsorption (with activated carbon), chemical precipitation, and membrane separation procedures are the most common techniques for extracting persistent contaminants from aqueous and gaseous systems. These procedures are expensive, and they frequently produce large amounts of chemical by-products that have no commercial importance.

Biochar, a low-cost carbonaceous substance, is gaining popularity as a cost-effective substitute to activated carbon for removing a wide range of organic contaminants, including agrochemicals, aromatic dyes, polycyclic aromatic hydrocarbons (PAHs), polychlorinated biphenyls (PCBs), antibiotics, and volatile organic compounds (VOCs) (Beesley et al., 2010; T. Xu et al., 2012), as well as a variety of inorganic pollutants (e.g., phosphate, heavy metals, nitrate, ammonia, sulfide, etc.) from solid, liquid, and gaseous phases (Ahmad et al., 2014; Jung et al., 2015). Recent biochar studies have primarily targeted optimizing its properties to increase organic and inorganic contaminants removal efficacy (Ahmed et al., 2016). Biochar is a by-product formed by thermochemical transformation of carbonaceous organic substances (agricultural and forest residues, activated sludge, algal biomass, manures, digestate, energy crops, and others) at high temperatures (300–900°C) and O2-limiting environments (including pyrolysis, torrefaction, hydrothermal carbonization, and gasification) (Ahmad et al., 2012).

Biochar offers a broad range of global advantages due to its unique characteristics, including higher specific surface area, ion exchange capability, high adsorption ability, and microporosity (Ahmad et al., 2014; Mohanty et al., 2013). Various biochar physicochemical features, which are related to substrate type and pyrolysis conditions utilized during its manufacture, determine the diversity and superiority of a specific reaction. These two parameters have a major influence on the physicochemical features of biochar, including pH, polarity, surface area, element content, and atomic ratio, as well as the biochar's increasing surface property (Ahmad et al., 2014; Ronsse et al., 2013; Uchimiya et al., 2013). These differences in biochar properties have a major impact on its acceptability and efficiency for removing specific contaminants.

Biochar adsorption methodology is now being implemented in the wastewater treatment industry. It has been proven to have a good adsorption impact on organic contaminants (Chen et al., 2011), heavy metals (X. Xu et al., 2016), nitrogen, and phosphorus (Fang et al., 2014) found in livestock wastewater. Biochar with high magnetic properties may be easily isolated from liquid after magnetization (Han et al., 2016), making it more suited for livestock effluent than conventional activated carbon compounds. It can also be utilized as a slow-moving fertilizer and has agronomic properties (Y. Yao et al., 2011) due to its high nitrogen and phosphorus adsorption ability.

Though biochar has a promising future in treating wastewater, the potential adverse consequences of its use must be considered as well. Biochar may involve heavy metal ions and other pollutants that could have been discharged upon its application in aqueous systems, based on the quality of the feedstock and the conversion procedure used for its manufacture (Jin et al., 2016; Kim et al., 2015). However, increased analysis into the sustainability of biochar and its relationship to the experimental settings employed during biochar formation is required. The latest research on the manufacture, transformation, and application of biochar pyrolysis and chars from various thermal transformation methods for the remediation of organic and inorganic wastewater contaminants is summarized in this chapter. The key mechanisms associated with the adsorption phase, as well as current advancements in utilizing biochar as a catalytic support, in filtration media, and in the anaerobic digestion of wastewater, are highlighted.

17.2 BIOCHAR SYNTHESIS AND PROPERTIES

17.2.1 Preparation Method

Biochar is produced using a variety of techniques. Table 17.1 illustrates some of the biochar synthesis methods, circumstances, and absorption yields that have recently become accessible. Pyrolysis

TABLE 17.1

Conventional Methodology Utilized for Biochar Synthesis (Y. Sun et al., 2014; X. Xu et al., 2016)

Synthesis methodology	Temperatures (°C)	Heating speed	Reaction speed (time)	Yield percentage (%)		
				Solid	Liquid	Gas
Slow pyrolysis	350–980° C	Slower	h	35	30	35
Fast pyrolysis	400–1000°C	Faster	s	10	70	20
Flash pyrolysis	775–1025°C	Faster	s	10–15	70–80	5–20
Gasification	700 to 1500°C	Faster	s to min	10	5	85
Hydro-carbonization	< 350°C	Slower	min to h	50 to 80	–	–

is anticipated to be a common method of biochar production. Pyrolysis is a type of decomposition that can occur without oxygen and at higher temperatures. Dependent on the temperature and reaction speed, it could be classified as rapid, slow, or "flash" pyrolysis (Y. Sun et al., 2014). Reduced heating and lower pyrolysis temperatures enable the formation of solid by-products. The production of solid by-products from slow pyrolysis is typically 35%, implying that it is the most common means of biochar synthesis among the three pyrolysis procedures stated in Table 17.1 (Park et al., 2013). The biomass reaction in an underwater stagnant environment for 5 min to 16 h at a pressure of 2–6 MPa and a considerably moderate temperature (350°C) is known as hydrothermal carbonization (HTC) (Berge et al., 2015; Kambo & Dutta, 2015; Tan et al., 2015). It is difficult to develop hazardous compounds in the hydrothermal carbonization process since it utilizes water as the basic medium under high temperature and pressure conditions. As a result, biochar synthesized using this approach is better for the adsorption of water contaminants (Kumar et al., 2011; Regmi et al., 2012; Z. Bin Zhang et al., 2013). However, the preparation conditions, as well as the expensive reactor's requirement for high temperature and pressure, restrict this process. The operational application is harder to popularize due to its high preparatory cost (Regmi et al., 2012). Other processes, including gasification, rapid pyrolysis, drying, and flash pyrolysis, are mostly employed to manufacture bio-oil or gaseous products (Tan et al., 2015) due to the reduction in the yield of solid products acquired, like that of the gas product composition of gasification, which is roughly 90% and increases with temperature.

17.2.2 CHARACTERISTICS

Biochar has a lot of permeabilized oxygen-rich functional groups because of its higher carbon content and void systems. The physicochemical parameters of biochar differ based on the kind of raw material, substrate particle size, pyrolysis method, temperature (such as rate of increase in temperature), pyrolysis time, and modification conditions (H. Sun et al., 2012; Tan et al., 2015). Though several variables influence the composition of biochar, it has immense surface functional groups (-COOH, -OH, -CH3, and -C=O) (Beesley et al., 2011), a defined pore structure, a large surface area, and a stabilized molecular structure (H. Yao et al., 2013), all of which favor adsorption of contaminants in livestock wastewater.

17.3 BIOCHAR AS AN EFFECTIVE ORGANIC AND INORGANIC CONTAMINANTS ADSORBENT

Organic matters, N, P, heavy metals, and several common contaminants exist in livestock effluent, posing a severe environmental threat. In the liquid phase, biochar has a higher adsorption efficiency for pollutants. Tan et al. summarized the applicability of biochar in water pollution with adsorption efficiency for organic contaminants (39%), N and P (13%), heavy metals (46%), whereas only 2% for other investigations (Tan et al., 2015).

17.3.1 Organic Contaminants

Biochar has a higher adsorption impact on organic contaminants like antibiotics, phenols, herbicides, and other chemicals (H. Yao et al., 2013). It has acquired an interest in agricultural sources and the ecosystem attributed to the closeness among both the types of contaminants absorbed as well as the nature of organic contamination in livestock effluent. Biocarbon can absorb antibiotics involving fluoroquinolone, sulfamethoxazole, etc. in the aqueous phase, with absorption mediated mostly by hydrogen bonding, π–π electron donor/receptor, and cationic bridge. H. Yao et al. (2013) synthesized biochar by pyrolyzing sludge for 60 min at 500°C, and it had the highest absorbance of 19.80 ± 0.40 mg/g for fluoroquinolone (an antibiotic used to treat infections of the urinary system, respiratory, skin soft tissue, abdominal cavity, intestinal system, and joints). Furthermore, the quantity of volatile matter in the originating effluent was found to be significantly linked with the quantity of fluoroquinolones absorption by biochar. Inorganic material in the feedstock enhanced sulfamethoxazole adsorption capability in the lower-temperature pyrolysis of biochar and reduced sulfamethoxazole adsorption ability in higher-temperature pyrolysis, according to Zheng et al. (2013).

In the aqueous environment, biochar has a major adsorption impact on high-chroma organic contaminants (L. Sun et al., 2013; R. kou Xu et al., 2011), phenols, herbicides, and other compounds, and its adsorption system includes a wide range of physicochemical properties, primarily based on the polarization of organic contaminants and biochar, aromatic properties of specific functional groups. The electrostatic force and intermolecular gravity among biochar and organic contaminants play a major role in its physical adsorption. Chemical adsorption occurs primarily through the formation of hydrogen bonds, coordination bonds, and π–π interactions between charcoal and organic contaminants.

Xu et al. utilized peanut, cole, and rapeseed straw as feedstock to make biochar that had a methyl violet adsorption capability of 123.5–195.4 mg/g after being pyrolyzed at 350°C (R. kou Xu et al., 2011). At ambient temperature, the adsorption of methyl violet on rapeseed straw biochar was the maximum. The electrostatic attraction was established between methyl violet and charcoal using zeta potential and FTIR studies. The methyl violet adsorption on -COO- and hydrophilic was dominating. Y. Sun et al. (2014) performed pyrolysis at 400°C for 30 min to produce biochar from anaerobic digestion by-products, palm bark, tree, etc. The elimination efficacies were 99.5, 99.3, and 86.1%, respectively, at pH 7, the temperature of 40°C, and 4 mg/L of methyl blue. The findings indicate that the pyrolysis temperature had a significant impact on the methyl blue elimination efficacy.

Yang et al. (2014) produced biochar out of peanut shells and wheat straw at temperatures of 300, 400, and 600°C, respectively. The highest adsorption efficiency of wheat straw and peanut shell biochar was found to be 20.61 mg/g and 58.82 mg/g, respectively. Zheng et al. (2013) generated biochar from combined wood debris that had been pyrolyzed for 1 h at 450°C. Atrazine (herbicide) and simazine (herbicide) had maximal adsorption ability of 1158 mg/g and 1066 mg/g, respectively, and adsorption efficiency was greater in acidic environments.

17.3.2 Heavy Metal Contaminants

Heavy metals are hazardous and nonbiodegradable. Though the concentrations are negligible, they pose a health risk to humans. Increased heavy metal discharges, including Pb, Cu, Cd, and Zn, are observed in livestock effluent and could pollute the environment (Devi & Saroha, 2015). Heavy metal ion adsorption on biochar is influenced by ion exchange, functional groups, chemical cross-linking among heavy metal ions, as well as surface accumulation between ashes.

Inyang et al. (2012) reported that the elimination efficacy of Ni(II), Pb(II), Cu(II), and Cd(II) in biochar obtained from sugar beetroot through anaerobic digestion and pyrolyzing at 600°C for 2 h was found to be 97%. Furthermore, the adsorption efficiency of the four types of ions was reduced, with Cd(II) adsorption being the highest and Cu(II) adsorption being the least. Z. Bin Zhang et al. (2013) observed that increasing the pyrolysis temperature lead to reduced adsorption capacities for Cr(VI) on wheat straw biochar, and the highest adsorption of 35.78 mg/g was attained at 200°C for Cr(VI).

17.3.3 NITROGEN AND PHOSPHORUS CONTAMINANTS

Livestock waste is high in nutrients, particularly N and P. Biochar for adsorption and fixation reduces eutrophication, and it can be reused, enhancing soil fertility, and reapplied to the soil, and regenerating nutrient resources, which is the subject of the current study. F. Zhang et al. (2015) reported that biochar from corn cob pyrolyzed at 600°C for 2 h had the highest ammonia nitrogen adsorption efficiency of 9.67 mg/g. At 25°C, Deng et al. (2017) proved that cow dung biochar had a maximal ammonia nitrogen adsorption efficacy of approximately 25.84 mg/g. Gabhane et al. (2020) generated biochar by pyrolyzing anaerobic digestion waste from municipal effluent. The findings indicate that effluent digestion assisted in the creation of biochar pores and increased nitrogen and phosphorus adsorption, as predicted by the Langmuir model.

Biochar was made by a pyrolyzing corncob and infusing calcium and magnesium, according to Fang et al (2015). The modified biochar had a high phosphate adsorption capability, with a maximum of 319.63 mg/g, according to the findings. The results suggest that adding Ca and Mg cations to corncob biochar can improve its anion exchange capability, hence increasing its phosphate adsorption efficacy. As a result of the findings, it was determined that the biochar absorbs P efficiently and could be used as fertilizer.

17.4 ADSORPTION MECHANISM OF BIOCHAR

The variability of the biochar surface allows for a variety of adsorption mechanisms to take place. The various mechanisms of biochar's interaction with inorganic and organic contaminants are illustrated in Figure 17.1. The type of pollutants and the chemical features of the adsorbent surface

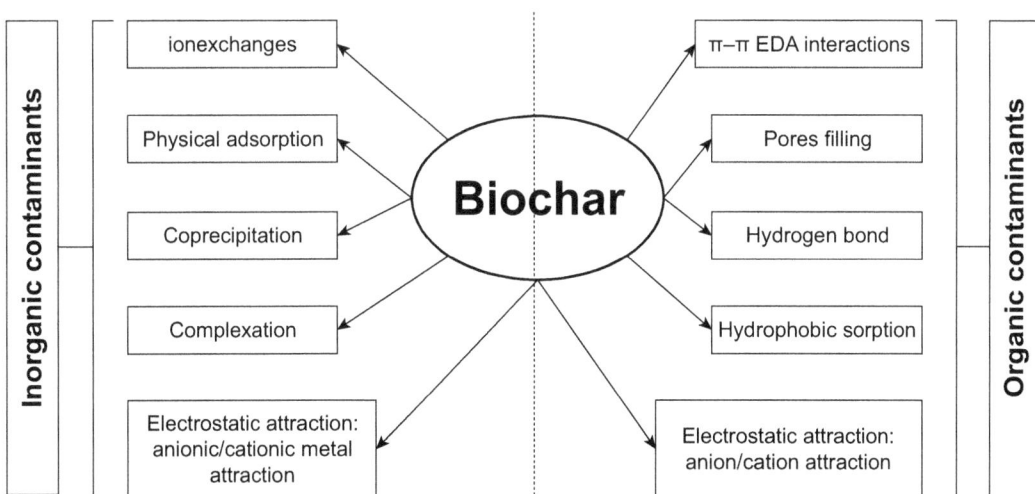

FIGURE 17.1 The interactions of biochar with inorganic and organic pollutants are hypothesised to entail a number of mechanisms in two categories. The adsorption process is facilitated by different interactions for elimination of organic and inorganic pollutants.

influence the adsorption mechanism (Rosales et al., 2017). According to (Pignatello et al., 2011), the main adsorption paths are the physical route (the sorbent accumulates on the adsorbent surface), the precipitation route (the adsorbent forms layers on the adsorbent surface), and the pore filling route (adsorbate condenses into the adsorbent pores). Electrostatic attraction, the interaction between π–π electron-donor/acceptor, pore filling, complex adsorption, hydrogen bonding, as well as hydrophobic interactions, aid the adsorption of organic contaminants. The adsorption of organic contaminants onto biochar by pore filling, for example, is directly proportional to the total micropore and mesopore volumes; the pollutant's absorption on the biochar structure is facilitated because of its smaller ionic radius, resulting in the biochar's higher adsorption capacity (Ahmad et al., 2014; Pignatello et al., 2011; Rosales et al., 2017).

When soluble contaminants have a hydrophobic functional group, they can be linked to hydrophobic biochar or precipitate on alkaline biochar surfaces. Due to the release of oxygen-containing functional groups on the surface of biochar, it is generally negatively charged, causing electrostatic interaction between the biochar and positively charged molecules (Ahmad et al., 2014; Qambrani et al., 2017). The reduction of O_2- and H-rich functional groups in biochar synthesized at high temperatures makes it less polar and aromatic, making it less suitable for removing polar organic pollutants. However, H-bonding could occur as a consequence of the electrostatic repulsion between negatively charged anionic organic molecules and charcoal, resulting in adsorption. The loss of hydrogen bonds among water and O_2-functional groups makes nonpolar pollutant absorption to hydrophobic regions more prominent (Ahmad et al., 2014). A variety of mechanisms could hamper the detoxification of inorganic contaminants like heavy metals, involving surface precipitation under an alkaline environment, ion exchange complexation, as well as cationic and anionic electrostatic attraction.

The actual contribution of several pathways to Pb adsorption on sludge-derived biochar was investigated by Lu et al. (2012) who postulated the underlying mechanisms: (1) coprecipitation and complex formation with organic matter and mineral oxides of the biochar owing to metal exchange with cations (K and Na) accessible in the biochar; (2) electrostatic complexation due to metal exchange with cations (K and Na) present in the biochar; (3) surface complexation with the biochar's free -COOH and -OH functional groups; (4) surface precipitation as lead-phosphate-silicate ($5PbO.P_2O_5.SiO_2$). It is critical to understand the contaminant adsorption mechanism of biochar in order to enhance its adsorption efficiency on contaminants, particularly for the usual contaminants found in livestock wastewater. Because biochar's adsorption capability varies depending on the nature and properties of pollutants, the adsorption mechanism of biochar can be examined from four perspectives: organic structure, surface functional group, surface electrical property, and mineral composition.

17.4.1 ORGANIC STRUCTURE

The biochar's organic structure is made up of two layers: stacked graphene layers and aromatic arrangements interleaved with the graphene coating, armoring the biochar with higher surface areas and dense pore arrangements (Verheijen et al., 2010). Biochar's physical adsorption capacity is enhanced by its higher surface areas, and its rich pore structure aids in the adsorption of organic materials of almost the same molecular weight (Tan et al., 2015). According to Y. Wang et al. (2015), bamboo biochar has a large number of pore structures, with mesoporous structures accounting for almost 90% of total pore structure and the adsorption of quinolone (antibiotics) occurring primarily in the mesoporous assemblies.

17.4.2 SURFACE FUNCTIONAL GROUPS

Electrostatic attraction, surface precipitation, and complexation are used by the functional groups (-OH, -COOH, etc.) on biochar's surface for metals fixation. F. Zhang et al. (2015) established that by raising the pyrolysis temperature to 500°C and above, the surface functional groups (-OH and -COOH) on the surface of effluent biochar were destructed, though the adsorption rate of Pb(II) was lower in the surface areas prepared under higher temperature settings, signifying the presence of the functional groups (-OH and -CH) that help in heavy metals adsorption. Nguyen and Lee (2015) observed that adding HNO_3/NH_3 to the surface of chicken dung biochar might result in the formation of novel amino functional groups that can promote dimethyl sulfide adsorption.

17.4.3 Surface Electrical Characteristics

The capability of electrostatic attraction on the biochar surface to adsorb contaminants is highly essential. Biochar has high adsorption characteristics for positive particles like ammonia and heavy metals due to its negative surface voltage. If the surface electricity of the biochar is modified to be positively charged, anions including phosphate can be absorbed. Z. Bin Zhang et al. (2013) showed that biochar had a higher adsorption efficacy on ammonia nitrogen but only a limited adsorption effect on phosphate. Fang et al. (2014) added magnesium salt to corncob biochar to increase its phosphate adsorption efficiency by making the surface electricity of the biochar positive.

17.4.4 Mineral Components

Mineral constituents in biochar, including (CO3)2, (PO3)2, and others, can improve adsorption characteristics. The major function of Pb(II) adsorption by digested cow dung biochar, according to Inyang et al. (2012) was surface precipitation. On the biochar surface, Pb(II) reacted with (CO3)2, HCO3, as well as H2PO4 ions to generate Pb(CO3)2(OH)2, PbCO3, and Pb5(PO4)3XS (X might be -F, -OH, -Cl, or -Br). In addition, specific contaminants are adsorbable due to different synergistic actions. In general, biochar adsorbs organic contaminants primarily by the conjunction of pore immobilization and organic functional group electrostatic attraction. Heavy metals adsorption occurs primarily via ion exchange, electrostatic attraction, functional group complexation reactions, and mineral component precipitation.

17.5 PARAMETERS INFLUENCING POLLUTANTS ADSORPTION ON BIOCHAR

17.5.1 Biochar Characteristics

The available volume of micropores controls the sorption of an adsorbate by an adsorbent (Zabaniotou et al., 2008). Pores of varying sizes can be found in adsorbent materials, which are classified as macropores, micropores, or mesopores depending on the thickness of the pore (Zhou et al., 2010). The size and distribution of pores are heavily influenced by the experimental parameters used during biochar formation, with temperature having the greatest influence (Zhou et al., 2010). The most widespread micropores inside the biochar structure are said to be accountable for their high surface area and adsorptive capability.

According to (Zabaniotou et al., 2008), biochar formed by high-temperature pyrolysis has a large micropore volume, ranging from 50 to 78% of the total pores. The adsorbate size is another crucial factor that influences the biochar's sorption rate. Small particle sizes decrease the mass transfer limitations and enhance the van der Waal force of diffusion into the adsorbent, whereas bigger adsorbate sizes can result in exclusions of or blocks on active sites (Daifullah & Girgis, 1998). The amounts and types of surface functional groups affect the adsorption ability of biochar (Qambrani et al., 2017). The chemical properties of the feedstock, the carbonization process, and the carbonization temperature all affect the arrangement of the surface functional groups (Ahmad et al., 2012).

Biochar and hydrochar produced by pyrolysis and HTC of pig manure were investigated by Gascó et al. (2018). The significant peak at 3400 cm^{-1}, attributable to -OH stretching vibration in -COOH and -OH groups, is much less apparent in biochars as the pyrolysis temperature rises, compared to the raw material. HTC hydrochars also revealed a prominent peak at 3400 cm^{-1} with lower intensity than that of the raw material because of the dehydration and decarboxylation events that happened during the HTC procedure. The authors stated that the elevated pyrolysis temperature (600°C) synthesizes biochars with the strong aromatic arrangement, but the HTC procedure at 200–240°C for 2 h favors biochar with more aliphatic structures.

According to (Qambrani et al., 2017), the -OH, -CH2, C=O, C=C, and -CH3 functional groups in biochar were disturbed due to the pyrolytic temperatures, thus increasing biochar hydrophobic associations. The hydrophobic character of biochars is determined by the quantity of O2- and N2-rich functional groups; the fewer there are of the O2-and N2-rich functional groups in the biochar, the more hydrophobic the biochar will be (Moreno-Castilla, 2004). Biochars with a hydrophobic characteristic are thought to aid in the adsorption of insoluble adsorbents, whereas biochars with a hydrophilic character are known to be less efficient due to water sorption. The availability of O2-rich functional groups on the hydrophilic biochar surface enables water to penetrate via hydrogen bonding, resulting in competition for accessible sites on the biochar surface involving water and adsorbate (K. Wang et al., 2006). The insoluble adsorbates seem to be most likely adsorbed in the pores of the biochar in aqueous systems (K. Wang et al., 2006).

17.5.2 Solution pH

The adsorption process is controlled by pH, which affects the charge on the adsorbent surface as well as the speciation and ionization frequency of the adsorbent (Nguyen & Lee, 2015). When the pH of the medium exceeds the threshold limit, a negative charge forms on the adsorbent's surface due to the deprotonation of -C6H5 and -COOH groups, and the competition between protons and cation pollutants is reduced. Basic functional groups, including -NH3, become protonated and have a positive electric charge at low pH values, facilitating the adsorption of anions (Kumar et al., 2011; Moreno-Castilla, 2004). This means that biochar adsorption behavior is influenced by the medium's pH and the deprotonation of functional groups.

Kizito et al. (2015) and Hu et al. (2020) revealed the effect of pH on biochar adsorption capacity toward ammonium (NH^{4+}-N). The authors discovered that when the pH was reduced (pH = 3/4), the adsorption of NH^{4+}-N on biochar was decreased. The adsorption capacity of NH^{4+}-N was enhanced when the initial pH solution was raised in the range of 4–8, then declined as the pH was raised to the value of 9.

17.5.3 Adsorbent Dosage

The sorbent–sorbate equilibrium of an adsorption process is also influenced by the adsorbent dosage. The presence of more active sites enhanced the elimination efficacy of organic and inorganic contaminants when the adsorbent dosage was improved (Hu et al., 2020; Kizito et al., 2015). When the dose rate is too high, the biochar adsorption capacity is reduced, and the adsorption layers overlap, shielding the accessible active sites on the absorbent surface (Qambrani et al., 2017). To maximize adsorption capacity while keeping the process cost- effective, the adsorbent dosage must be optimized.

17.5.4 Temperature

The temperature of the solution in which biochar is administered has been found to alter its adsorption capability. The majority of investigations found that when the temperature increases, the adsorption efficiency was improved, indicating that the adsorption mechanism is endothermic. Enaime et al. (2017) found that the adsorption of indigo carmine over KOH-activated biochar rises as the temperature is raised because of the endothermicity of the adsorption processes. The movement of dye molecules increases as the temperature rises, and the porosity of the adsorbent may be enhanced as well. This could be attributed to the expanding impact of the adsorbent's inner structure, as the temperature rises, allowing the excess dye to permeate further (Jin et al., 2016). Kizito et al. (2015) demonstrated that increasing the experimental temperatures for NH_4^+-N adsorption from 15 to 45°C improved adsorption effectiveness. They reported that raising the temperature beyond 30°C to a maximum of 45°C is excellent for optimal elimination efficacy.

17.6 BIOCHAR FOR WASTEWATER TREATMENT: AN ENVIRONMENTAL APPLICATION

17.6.1 As a Support Substrate under Anaerobic Digestion

Anaerobic digestion is a highly attractive remediation strategy for wastewater treatment due to its potential to integrate bioremediation and electricity generation that is regarded as ecologically acceptable. Scientists are concerned that some components of the raw material or their metabolic intermediates can restrict microbes and adversely affect their action, making it difficult to accomplish the anaerobic digestion operation under steady conditions with increased energy conversion and effective contaminants elimination. Acclimation of microorganisms, incorporation of thermophilic system parameters, and lowering of inhibitor concentrations via dilution or codigestion with several other substrates have all been reported as ways to combat inhibition in anaerobic reactors. In anaerobic digestion, packing materials are an effective way to minimize inhibitors movement and bioavailability, thus causing the formation of high microbial biomass inside the digester.

Biochar is gaining popularity as a porous, biostable, and readily affordable substitute to adsorbents such as activated carbon and zeolite, and its use in anaerobic digestion is rising steadily (Rosales et al., 2017). Furthermore, the leftover biochar in digestate can be employed as a supplement to enhance soil qualities without posing damage to the environment (Daifullah & Girgis, 1998). Although biochar's sorption capability for various organic and inorganic pollutants has been widely described in the literature, there is still a paucity of data on its behavior toward inhibiting chemicals during anaerobic digestion. Biochar can absorb NH_4^+ and remain unchanged in ambient air, according to (Devi & Saroha, 2015). Biochar also mitigates NH_4^+ inhibition during anaerobic digestion of glucose solution at an NH_4^+ concentration of 7 g/L, according to (Devi & Saroha, 2015). Torri and Fabbri (2014) investigated the anaerobic digestion of wet pyrolysis solution in batch testing. The results revealed that the anaerobic procedure performed poorly and that the biological activity was inhibited even with nutritional supplementation, whereas charcoal addition enhanced the output of CH4 and elevated the reaction rate.

Sunyoto et al. (2016) examined the impact of biochar accumulation on H2 and CH4 generation in a two-phase batch anaerobic digestion system. Biochar addition improved H2 and CH4 yields, as well as a volatile fatty acid formation, through H2 production and VFA decomposition towards CH4 production, according to their findings. Inyang et al. (2012) found that biochar is less effective in process-stabilizing capability than zeolite. The utilization of chars in anaerobic digestion might be significantly improved, as per the same research. Another possibility for using biochar in the anaerobic digestion system is to use it to enhance the buffering capacity of digesters. The inclusion of alkaline substances such as lime in digesters can typically prevent acidification. Moreover, several investigations have shown that adding alkaline biochar to the anaerobic medium regularly can improve its buffering ability (Kambo & Dutta, 2015; Uchimiya et al., 2013). L. Sun et al. (2013) examined biochar and nonbiochar incubation with glucose as a feedstock and found that the CH4 output improved by around 90% while acidification was minimized in biochar testing.

Sunyoto et al. (2016) found that using biochar can support bacterial growth and metabolism as well as buffering pH throughout biohydrogen generation. Biochar alkalinity is affected by the biomass supply and the carbonization temperature, with increased alkalinity as the pyrolysis temperature rises (Beesley et al., 2011; Spokas et al., 2009). Immobilizing biomass on a packing medium with a wide surface area for bacterial growth has proven to be a realistic approach, as it enables the development of enormous biomass that could be established inside the reactors for an extended period. It also reduces the distance among syntrophic bacteria and methanogens by facilitating electron transport within species (Ahmed et al., 2016; Beesley et al., 2010; Inyang et al., 2012; Spokas et al., 2009). This can help to speed up the beginning of the operation, ensure improved stability and resistance for maximum loading rates, and assist microorganisms to swiftly regain their function after a period of starvation (Spokas et al., 2009).

Microbial adhesion and growth have been supported by packing materials like clay, activated carbon, zeolite, and other polymeric components (Ahmed et al., 2016; Beesley et al., 2011; Berge et al., 2015; Mohanty et al., 2013). Biochar's use for cell immobilization is, however, less widespread than that of most other adsorbents. Sunyoto et al. (2016) found that biochar stimulated the establishment of methanogenic biofilms, which increased microbial growth, volatile fatty acid breakdown, and CH_4 generation. During the anaerobic digestion of glucose, L. Sun et al. (2013) detected *Methanosarina* colonization on charcoal material. When compared to the nonbiochar trial, the authors found that using biochar increases biogas generation by roughly 86% while also improving the decomposition of intermediate acids. Biochar enhances bacterial metabolism and growth, as well as providing ideal conditions for microbes, according to many researchers (Jin et al., 2016; Ronsse et al., 2013). However, there is no detailed analysis of the microorganism arrangement on the biochar surface or the link among its structure and the number of microbial immobilization.

17.6.2 As a Filtration Support Media

Biochar is a material that can be utilized in a variety of water treatment systems. It can, for example, be used in a biochar layer/column in sand filters or bio-sand filter system for treating wastewater. Biochar filtration methods have earned more consideration due to their ability to eliminate particulates, heavy metals, and microbial loads. The high elimination effectiveness of a *Miscanthus* biochar filter was discovered in research shown by Kim et al. (2015), which was greater than or equivalent to a sand filter. The biochar filter's mean reduction of chemical oxygen demand (74 to 18%) was much greater than the sand filter's (61 to 12%). *E. coli* was also removed more effectively by the biochar filter. Bacteria had a mean reduction of 1.35 ± 0.27 log units, while sand filters had a mean decrease of 1.18 ± 0.31 log units.

The practical utilization of biochar as a filtration system for the elimination of total suspended particles, heavy metals, and nutrients was shown by Ahmed et al. (2016). *E. coli* was isolated from synthetic stormwater. The stormwater effluent was filtered using biochar, which removed 86% suspended particles, 86% nitrate, and 47% phosphate. Pb, Cd, Zn, Cu, Ni, and Cr (heavy metals) concentrations were reduced by 17, 19, 18, 24, 65, and 75%, respectively, after filtration, whereas *E.coli* shows the removal effectiveness of about 27%. This removal percentage was significantly less than that obtained for zeolite (Beesley et al., 2011) and activated carbon (Ahmad et al., 2014; Ronsse et al., 2013), indicating that *E.coli* removal efficiency from stormwater was significantly lower at roughly 53 and 98%. This variation might be attributable to the inflow concentration and antecedent bacterial levels, as Ronsse et al. (2013) also found.

In an anaerobic biochar filter, the removal of suspended particles and organic materials could be accomplished in a sequence of phases, as reported by Kim et al. (2015). It began with the dispersion of suspended solids in the supernatant liquid phase, followed by centrifugation of large particles on the filter medium, concentrating and adsorption of smaller particle size in inner filter zones, and organic matter hydrolysis under anaerobic environments. The surface area and active sites on the charcoal surface enhance the performance of the biochar filter, which further promotes biofilm development and bacterial accumulation within the filter column (Jin et al., 2016; Kizito et al., 2015).

The filtration of macroscopic pathogens such as amoeba and protozoa, as well as the absorption of negatively charged bacterial and viral cells, are two essential strategies that potentially control pathogen reductions. Other mechanisms may interfere, including the electrostatic interactions of microbes to the biological film formed on the biochar filters (Kizito et al., 2015), as well as the adsorption of *E. coli* to the biochar filter in combination with an improvement in water holding capability. Following filtration, the biochar becomes nutrient-rich and can be utilized as a fertilizer or soil amendment. The biochar filter removed the same pathogens as a typical sand filter (i.e., 1.4 log units on average) in a finding established by Fang et al. (2015). During filtration, the concentrations of P, Mg, and K were lowered, while the N content remained unaltered. On an acidic fine sand soil, the agronomic impacts of the biochar filtration system on spring wheat biomass generation were assessed. When compared to the unamended standard, the biochar filter (37%) treatment (20 t/ha) produced more wheat biomass (Spokas et al., 2009).

17.6.3 As a Catalyst during Heterogeneous Oxidation

Catalytic ozonation procedures, Fenton-like activities, and photocatalytic techniques have all used biochar-based catalysts for the destruction of recalcitrant chemicals and the remediation of pollutants in effluents. Devi and Saroha (2015) investigated the catalytic ability of biochar made from pistachio hull biomass, which had a macroporous shape, and -OH and -C6H5OH surface functional groups for ozonation of a recalcitrant pollutant in water (reactive red 198 dye). The produced biochar had a higher catalytic activity at 10.0 pH, with a catalytic potential of roughly 58% and dye solution remediation of 71% after 1 h of reaction. Carbon materials' unique ability to trigger various oxidants, including O2 and H2O2, and produce reactive O2 sp. for the destruction of refractory organic pollutants was well documented in the previous findings (Mohanty et al., 2013; Moreno-Castilla, 2004).

In aerobic conditions, biochars made from wheat, pine needles, and maize straw can stimulate -OH production, according to (Fang et al., 2014). The availability of free radicals in biochar transport electrons to O2 to make a superoxide radical anion and H2O2, which then combines by free radicals to generate -OH. The -OH formed from biochar suspensions was shown to be particularly effective at degrading organic pollutants, according to the authors. According to Chen et al. (2011), HTC hydrochar improved alachlor decomposition in the Fe(III)/hydrogen peroxide Fenton-like mechanism by enhancing Fe(III)/Fe(II) cycling through electron transfer from biochar to Fe(II)/(III). Hydrochars were generated by HTC from two-phase olive mill debris coated with ferric chloride and utilized as catalysts for the breakdown of methylene blue in another investigation by (Zabaniotou et al., 2008). The findings demonstrate that hydrochar generated at 250°C HTC temperatures for 4 h and using a ferric-chloride-to-biomass ratio of 1.5 was efficient as a precursor in heterogeneous Fenton-like oxidation, enabling a methylene blue removal percentage of 91%.

17.6.4 Indirect Water and Wastewater Remediation

Constructed wetlands (CWs) were frequently utilized in treating industrial effluent in recent years, including the elimination of nitrogen, phosphorus, and certain organic pollutants (Berge et al., 2015). However, the percentage removal for N and P is significantly hampered due to inadequate oxygen availability and transportability, poor sorption capability of the substrate, and suppression of microorganisms and plant metabolism at relatively reduced temperatures. Researchers have worked to investigate certain substrates to enhance the activities of CWs with high pollutants concentrations, with biochar being a popular choice (Spokas et al., 2009). With a varied range of low C/N ratio influent levels, Zhou et al. (2010) utilized biochar as a substrate/medium in vertical flow-built wetlands (VFCWs) to improve elimination capacity. They looked at eliminating N and organic pollutants in VFCWs with and without biochar. The maximum percentage removal of organic pollutants (85%), NH_4^+-N (39%), and TN (39%) were higher than that of traditional VFCWs, particularly for higher-strength effluent.

Enaime et al. (2017) conducted a seven-month investigation and found that enriched biochar was an effective substrate for removing PO4-P. Waste biochar can regenerate and could be used as a soil fertilizer to enhance soil fertility, though further research is needed on this application. Deng et al. (2017) developed four subsurface flow constructed wetlands (SFCWs) using biochar modified in conventional gravel at various volume percentages (0–30%). The elimination percentage of NH^{4+}-N and TN by SFCWs containing biochar was stronger than that with pure gravel-filled SFCWs, according to the findings. By modifying the composition of microbial populations and raising the abundance of dominating species, biochar enhances N removal. Biochar also increases the metabolism of high-molecular-weight molecules by converting them to low-molecular-weight ones. These findings add to our understanding of how biochar can help improve N removal via microbial metabolism.

Surface water runoff and soil erosion in the river system, particularly in some degraded farms through excessive precipitation, may pollute the aquatic environment. Biochar has been shown in various studies to have the ability to minimize surface runoff and soil erosion (Gabhane et al., 2020; Mohanty et al., 2013; Ronsse et al., 2013). Biochar particles have -C6H5OH and -COOH functional groups that can interact with soil mineral surfaces, improving soil accumulation and structure durability (Mohanty et al., 2013). Furthermore, interchangeable divalent cations by increased charge density (Ca^{2+}, Mg^{2+}) on biochar surfaces can substitute monovalent cations (Na^+, K^+) on clay particle exchangeable sites, enhancing clay flocculation and, as a result, improving pore size distribution size and connections in the soil (Fang et al., 2015), increasing infiltration capacity. As a result, the biochar addition has been found to improve soil physical qualities, which minimizes erosion, runoff, and waterlogging (Y. Yao et al., 2011). Furthermore, biochar with a high water holding capacity that is put on the soil surface can adsorb the force of rain, lengthening the runoff duration (Fang et al., 2014; Y. Yao et al., 2011).

17.7 BIOCHAR INFLUENCE IN MICROBIAL GROWTH AND CONTAMINANT REMEDIATION

In Section 3, the removal methods of biochar for organic and inorganic contaminants are comprehensively explored based on physicochemical properties. However, biochar's biological properties related to pollution removal emerge later in the soil application process (H. Yao et al., 2013). Biochar pores function as a residence for microbes including mycorrhizae and bacteria, which rely on these microhabitats for their metabolic demands (Zabaniotou et al., 2008). The labile soil organic matter (SOM) on the surface of biochar promotes bacterial growth, resulting in increased microbial population, activity, and degradation (Ahmad et al., 2014; Beesley et al., 2011; Tan et al., 2015; H. Yao et al., 2013). Labile SOMs are remnants of volatile and semivolatile organic materials (e.g., benzoic acids, n-alkanoic acids, acetoxy acids and hydroxyl, diols, triols, phenols) precipitated on the surface of biochar all through the freezing procedure, which distinguishes biochar from supplementary chars (Ronsse et al., 2013).

Biochar similarly aids allelochemical decontamination (Gabhane et al., 2020), increasing the growth of rhizobacteria (*Rhizobium* sp., *Paenibacillus* sp., *Pseudomonas* sp., *Bradyrhizobium* sp., and many more) and mycorrhizae (primarily ectomycorrhizal, arbuscular, and ericoid). Furthermore, the biochar surface's abundant O2-rich functional groups allow the adsorption of NH^{4+} ions, dissolved organic molecules, and simple organic materials resulting in a desirable microbial habitat (Hu et al., 2020; Kizito et al., 2015). However, the biochar surface experiences desirable alterations in the soil, affecting microbial population and/or activity. Mineralization is the term used to describe the monomeric products that come from the microbial breakdown of complex substances.

Mineralization is extremely important in the environmental rehabilitation process. It aids not only in the decomposition of the organic matter (adsorbed from the soil) for efficient waste management but also in the nutrient cycle to maintain soil equilibrium (Ronsse et al., 2013). Although the significance of biochars in bacterial growth is well established, their interactions with distinct microbes are not. This is owing to the diverse groups of bacteria involved. The decline in ectomycorrhizal quantity could be linked to the addition of nutrients and water, resulting in unfavorable symbiotic relationships as a result of pH changes. Biochar appears to have a favorable effect on mycorrhizal fungi, with colonization rates ranging between 19 to 157% based on the biochar substrate and its physicochemical parameters (Ahmed et al., 2016; Ronsse et al., 2013).

The relationship of biochar with soil biota must be studied to fully understand the processes affecting soil fertility. Z. Bin Zhang et al. (2013) investigated the influence of biochar on bacterial development in heavy-metal-polluted soil in a recent survey. The investigation of soil microbial metabolism led to the discovery of bacterial carbon utilization in biochar-enriched soil. After applying 2.5% (w/w) biochar to the Cd-polluted soil, the McIntosh index demonstrated a considerable increase in soil microbial populations from 32.68 to 135.52%. The effect of free and immobilized bacterial cells (on charcoal and alginate) on the phytoremediation of pollutants in the soil was also investigated. Bioaugmentation of PCB phytoremediation by charcoal-immobilized microbes was more efficient (30.3%) than that of alginate immobilized cells (6.8%) or unbound cells (5%) (Pignatello et al., 2011).

Another study looked at the impact of rape-straw-derived biochar application on microbial populations and pentachlorophenol (PCP) conversion in paddy soil under anaerobic conditions. In biochar-treated soils, PCP adsorption and conversion, Fe(II) production, and microorganism growth and activity all improved dramatically. The PCP conversion was driven by an increase in the transferable extracellular electron in biochar-enhanced soils. In addition, biochar aided the growth of Fe(III)-reducing and dechlorinating microbes, which improved PCP biotransformation (F. Zhang et al., 2015). Biochar supplementation has strong favorable impacts on bacterial growth and metabolism and on the mineralization of pollutants in polluted soils, according to these researches.

17.8 ENVIRONMENTAL ISSUES AND THE FUTURE PERSPECTIVES

Biochar, on the other hand, is not yet commonly used and is still at the experimental stage of study. Because of various environmental problems that cannot be overlooked in actual application, the manufacture and application of biochar are not widespread at the moment, especially in some developing countries wherever comprehensive industrial chains are absent. In this scenario, extensive study is required to address potential environmental issues and to give developing countries practicable research findings for expanding biochar utilization. Although biochar substrates are plentiful and easy to obtain, they must first be organized (drying, cleaning, and grinding) before being pyrolyzed for the existing biochar. For the best sorption effect, more modifications are required. These biochar treatments will invariably raise the expense of manufacturing as compared to regular activated carbon. As a result, future research needs to strike a balance between optimizing the production method and enhancing biochar's applications to reduce costs (Mohanty et al., 2013). Meanwhile, selecting high-quality feedstocks, manufacturing conditions, and modification processes are crucial for producing high-performance biochars.

The aggregation of a large number of previous study findings can aid in the search for the best alternatives. Because of the resistance of lignin, the mesopore region of cellulose biochar (280 m^2/g) was greater than that of lignin biochar (200 m^2/g) once carbonized at a similar temperature, indicating that cellulose biomass is preferred to lignin biomass for biochar synthesis. The surface area and micropore volume of pinewood biochar pyrolyzed at extremely high temperatures have been significantly greater owing to the more effective carbonization of lignin, demonstrating that rising temperatures promote pore size very well (Beesley et al., 2011). The long-term viability of biochar and biochar-based composites must be considered in biochar applications. Lu et al. (2012) identified the probable dissociation of organic substances from biochar, mostly after complex formation with heavy metals, that can boost the carbon content of water, due to the strong aromaticity and persistence of organic materials. Furthermore, biochars, particularly those made from sewage effluent, might contain major levels of heavy metals, which may leach out after application, resulting in increased heavy metal pollution (J. Wang & Wang, 2019). If the embedded components in biochar-based composites are not well fixed, they may leak out from the biochar matrix.

Because biochar sustainability is defined in terms of the stability of its carbon network (J. Wang & Wang, 2019), research on the effects of carbonization settings on carbon composition is required. Biochar made through hydrothermal carbonization, for example, has more carbon than biochar made through gasification or pyrolysis (Axelsson et al., 2012). Furthermore, continuous water quality checking is intensely advised throughout the sorbents' life-cycle procedure. To detect whether harmful compounds are dissolved from the biochar, leaching or toxicity experiments utilizing water fleas, algae, fish, or luminous bacteria have been proposed (J. Wang & Wang, 2019). The majority of studies to date have concentrated on the adsorption of single contaminants in aqueous systems. Moreover, in real-world applications, the coexistence of a range of pollutants is common, with both synergistic and antagonistic adsorption effects seen.

Multiple pollutants may cause ionic intervention and competition for sorption sites, lowering removal effectiveness. Empirical evidence on cocontaminant sorption is currently scarce, necessitating the development of real-time adsorption models that might disclose the associated synergistic or antagonistic adsorption mechanisms. To make such investigations easier, publications on biochar sorption must include as much detail as possible about both the sorbent characteristics and sorption settings in case further directions are needed. Several attempts have been published, like simulated molecular calculations for analyzing competitive adsorption of cocontaminants (Bahamon et al., 2017), new meta-analysis approaches (K. Wang et al., 2006), and in-depth investigation (Tan et al., 2015; F. Zhang et al., 2015). Although there is agreement that biochar is less expensive, renewable, and sustainable than activated carbon (Moreno-Castilla, 2004), it is necessary to find methods of waste biochar recovery and desorption, such as biochar magnetization, which allows for the separation of contaminant-laden biochar from water using an external magnetic field. Otherwise, desorption of waste biochar could be expensive.

On either hand, if pollutants sorbed on biochar cannot be properly desorbed and retrieved, waste biochar can be used as a resource, allowing waste biochar to be recycled in another way. Biochar containing N and P, for example, could be used as a slow-release fertilizer in agriculture or for ecological rehabilitation (Roy, 2017). As a result, charcoal containing Cu or Zn can also be utilized as a micronutrient fertilizer. However, it is important to consider whether any hazardous components released from the biochar could be absorbed by crops and hence enter the food chain. As a consequence, the protection of incorporating waste biochar into soil must be investigated.

Biochar is a distinct renewable resource that can address several environmental challenges that have arisen in recent years, such as pollutant remediation in soil, water, and gaseous environments. This could have a synergistic effect on soil, water, and air quality, as well as carbon sequestration and greenhouse gas reduction. Because the productivity and efficiency of biochar differ greatly based on the raw materials and pyrolysis parameters, future biochar advancement will likely focus on "tuning" the features for specific purposes. The International Biochar Program has made tremendous progress in creating stakeholder collaborations and industrial practices and in establishing ecological and ethical values to enable biochar systems that are both safe and commercially viable. Biochar production companies, users, and policy experts will collaborate to explore economical usage of biochar for efficient mitigation and agronomic implementations by unifying methodologies and developing principles for manufacturing procedures, characterization, and life-cycle assessment using both experimental and model strategies.

Another key area to adapt the applications of biochar for the elimination of certain pollutants is biochar activation. Tannery waste stimulation of pine-wood-derived biochar, for example, was found to improve the adsorption of NH^{4+}-N and other inorganic contaminants from wastewater effluent; chemical activation of biochar resulted in higher adsorption of contaminants with minimal desorption. More investigation is necessary to discover various activation strategies as well as pollutant adsorption and desorption methods. Microbial populations and dispersion in biochar-amended soil have not yet been thoroughly studied, particularly with biochar characteristics (e.g., pH, microporosity, ion exchange, particle size, nutrient content). With the rising awareness about soil contamination and infertility, the implementation of biochar can provide a new avenue of macro- and micronutrients (acquired from waste products) in nutrient-deficient soils, not only for their decontamination but as a source of macro- and micronutrients.

To thoroughly evaluate the application of bacteria in soil remediation and mineralization operations, more study is required. Another relevant research field might be agronomic welfare, namely biochar surface chemistry, and collaboration with different soil elements, particularly the binding of micronutrients on biochar and their interchange processes. Biochar uses in wastewater treatment, particularly the elimination of hazardous chemicals from diverse industries, are gaining popularity. As a result, biochar management can be used as a pretreatment to remove hazardous chemicals before biological treatment. To understand the elimination methods of dangerous substances, a thorough investigation is required.

17.9 CONCLUSIONS

Biochar's utilization for wastewater treatment is becoming more viable due to the low cost of the substrate and sample preparation technique, as well as its improved physicochemical properties. This chapter systematically gave an outline of several biochar synthesis procedures, adsorption mechanisms against organic and inorganic contaminants, and potential uses in wastewater treatment. Biochar's ability to eliminate contaminants from aqueous environments is directly related to its physicochemical properties, which are influenced by the type of the raw materials, the thermal transformation methodology, and the preparation parameters. It has been stated that areas to investigate are altering biochar's surface functional groups, increasing its surface area and pore structure, and increasing its surface O2-rich functional groups through physical and chemical activation methods. Biochar was successfully included in a wide range of uses aimed at the degradation of polluted effluents, such as the adsorption of lethal heavy metals and dyes from the water stream, as a support for catalysts, as an immobilization substrate for microbes, and as an adsorbent of inhibitive substances throughout anaerobic digestion, due to its distinctive and extremely versatile features. However, it is apparent that using biochar has multiple advantages, including possible environmental and economic benefits, and its efficacy in removing various pollutants in the laboratory is broadly known. Therefore, further in situ studies must be conducted to assess biochar efficacy utilizing real wastewater and to investigate the true environmental impact of biochar before it is used on a broad scale. Biochar's stability and regeneration after numerous cycles of use will have to be researched further.

REFERENCES

Ahmad, M., Lee, S. S., Dou, X., Mohan, D., Sung, J. K., Yang, J. E., & Ok, Y. S. (2012). Effects of pyrolysis temperature on soybean stover- and peanut-shell-derived biochar properties and TCE adsorption in water. *Bioresource Technology, 118*, 536–544. https://doi.org/10.1016/j.biortech.2012.05.042

Ahmad, M., Rajapaksha, A. U., Lim, J. E., Zhang, M., Bolan, N., Mohan, D., Vithanage, M., Lee, S. S., & Ok, Y. S. (2014). Biochar as a sorbent for contaminant management in soil and water: A review. *Chemosphere, 99*, 19–33. https://doi.org/10.1016/j.chemosphere.2013.10.071

Ahmed, M. B., Zhou, J. L., Ngo, H. H., Guo, W., & Chen, M. (2016). Progress in the preparation and application of modified biochar for improved contaminant removal from water and wastewater. *Bioresource Technology, 214*, 836–851. https://doi.org/10.1016/j.biortech.2016.05.057

Axelsson, L., Franzén, M., Ostwald, M., Berndes, G., Lakshmi, G., & Ravindranath, N. H. (2012). Perspective: Jatropha cultivation in southern India: Assessing farmers' experiences. *Biofuels, Bioproducts and Biorefining, 6*(3), 246–256. https://doi.org/10.1002/bbb

Bahamon, D., Carro, L., Guri, S., & Vega, L. F. (2017). Computational study of ibuprofen removal from water by adsorption in realistic activated carbons. *Journal of Colloid and Interface Science, 498*, 323–334. https://doi.org/10.1016/j.jcis.2017.03.068

Beesley, L., Moreno-Jiménez, E., & Gomez-Eyles, J. L. (2010). Effects of biochar and greenwaste compost amendments on mobility, bioavailability and toxicity of inorganic and organic contaminants in a multi-element polluted soil. *Environmental Pollution*, *158*(6), 2282–2287. https://doi.org/10.1016/j.envpol.2010.02.003

Beesley, L., Moreno-Jiménez, E., Gomez-Eyles, J. L., Harris, E., Robinson, B., & Sizmur, T. (2011). A review of biochars' potential role in the remediation, revegetation and restoration of contaminated soils. *Environmental Pollution*, *159*(12), 3269–3282. https://doi.org/10.1016/j.envpol.2011.07.023

Berge, N. D., Li, L., Flora, J. R. V., & Ro, K. S. (2015). Assessing the environmental impact of energy production from hydrochar generated via hydrothermal carbonization of food wastes. *Waste Management*, *43*, 203–217. https://doi.org/10.1016/j.wasman.2015.04.029

Chen, B., Chen, Z., & Lv, S. (2011). A novel magnetic biochar efficiently sorbs organic pollutants and phosphate. *Bioresource Technology*, *102*(2), 716–723. https://doi.org/10.1016/j.biortech.2010.08.067

Daifullah, A. A. M., & Girgis, B. S. (1998). Removal of some substituted phenols by activated carbon obtained from agricultural waste. *Water Research*, *32*(4), 1169–1177. https://doi.org/10.1016/S0043-1354(97)00310-2

Deng, Y., Zhang, T., & Wang, Q. (2017). Biochar adsorption treatment for typical pollutants removal in livestock wastewater : A review. *Engineering Applications of Biochar*, 71–82. www.intechopen.com/books/engineering-applications-of- biochar

Devi, P., & Saroha, A. K. (2015). Simultaneous adsorption and dechlorination of pentachlorophenol from effluent by Ni-ZVI magnetic biochar composites synthesized from paper mill sludge. *Chemical Engineering Journal*, *271*, 195–203. https://doi.org/10.1016/j.cej.2015.02.087

Enaime, G., Ennaciri, K., Ounas, A., Baçaoui, A., Seffen, M., Selmi, T., & Yaacoubi, A. (2017). Preparation and characterization of activated carbons from olive wastes by physical and chemical activation: Application to Indigo carmine adsorption. *Journal of Materials and Environmental Science*, *8*(11), 4125–4137.

Fang, C., Zhang, T., Li, P., Jiang, R. F., & Wang, Y. C. (2014). Application of magnesium modified corn biochar for phosphorus removal and recovery from swine wastewater. *International Journal of Environmental Research and Public Health*, *11*(9), 9217–9237. https://doi.org/10.3390/ijerph110909217

Fang, C., Zhang, T., Li, P., Jiang, R., Wu, S., Nie, H., & Wang, Y. (2015). Phosphorus recovery from biogas fermentation liquid by Ca-Mg loaded biochar. *Journal of Environmental Sciences (China)*, *29*, 106–114. https://doi.org/10.1016/j.jes.2014.08.019

Gabhane, J. W., Bhange, V. P., Patil, P. D., Bankar, S. T., & Kumar, S. (2020). Recent trends in biochar production methods and its application as a soil health conditioner: A review. *SN Applied Sciences*, *2*(7), 1–21. https://doi.org/10.1007/s42452-020-3121-5

Gascó, G., Paz-Ferreiro, J., Álvarez, M. L., Saa, A., & Méndez, A. (2018). Biochars and hydrochars prepared by pyrolysis and hydrothermal carbonisation of pig manure. *Waste Management*, *79*, 395–403. https://doi.org/10.1016/j.wasman.2018.08.015

Han, Y., Cao, X., Ouyang, X., Sohi, S. P., & Chen, J. (2016). Adsorption kinetics of magnetic biochar derived from peanut hull on removal of Cr (VI) from aqueous solution: Effects of production conditions and particle size. *Chemosphere*, *145*, 336–341. https://doi.org/10.1016/j.chemosphere.2015.11.050

Hu, X., Zhang, X., Ngo, H. H., Guo, W., Wen, H., Li, C., Zhang, Y., & Ma, C. (2020). Comparison study on the ammonium adsorption of the biochars derived from different kinds of fruit peel. *Science of the Total Environment*, *707*, 135544. https://doi.org/10.1016/j.scitotenv.2019.135544

Inyang, M., Gao, B., Yao, Y., Xue, Y., Zimmerman, A. R., Pullammanappallil, P., & Cao, X. (2012). Removal of heavy metals from aqueous solution by biochars derived from anaerobically digested biomass. *Bioresource Technology*, *110*, 50–56. https://doi.org/10.1016/j.biortech.2012.01.072

Jin, J., Li, Y., Zhang, J., Wu, S., Cao, Y., Liang, P., Zhang, J., Wong, M. H., Wang, M., Shan, S., & Christie, P. (2016). Influence of pyrolysis temperature on properties and environmental safety of heavy metals in biochars derived from municipal sewage sludge. *Journal of Hazardous Materials*, *320*, 417–426. https://doi.org/10.1016/j.jhazmat.2016.08.050

Jung, C., Oh, J., & Yoon, Y. (2015). Removal of acetaminophen and naproxen by combined coagulation and adsorption using biochar: influence of combined sewer overflow components. *Environmental Science and Pollution Research*, *22*(13), 10058–10069. https://doi.org/10.1007/s11356-015-4191-6

Kambo, H. S., & Dutta, A. (2015). A comparative review of biochar and hydrochar in terms of production, physico-chemical properties and applications. *Renewable and Sustainable Energy Reviews*, *45*, 359–378. https://doi.org/10.1016/j.rser.2015.01.050

Kim, J. H., Ok, Y. S., Choi, G. H., & Park, B. J. (2015). Residual perfluorochemicals in the biochar from sewage sludge. *Chemosphere*, *134*, 435–437. https://doi.org/10.1016/j.chemosphere.2015.05.012

Kizito, S., Wu, S., Kipkemoi Kirui, W., Lei, M., Lu, Q., Bah, H., & Dong, R. (2015). Evaluation of slow pyrolyzed wood and rice husks biochar for adsorption of ammonium nitrogen from piggery manure anaerobic digestate slurry. *Science of the Total Environment*, *505*, 102–112. https://doi.org/10.1016/j.scitotenv.2014.09.096

Kumar, S., Loganathan, V. A., Gupta, R. B., & Barnett, M. O. (2011). An Assessment of U(VI) removal from groundwater using biochar produced from hydrothermal carbonization. *Journal of Environmental Management*, *92*(10), 2504–2512. https://doi.org/10.1016/j.jenvman.2011.05.013

Lu, H., Zhang, W., Yang, Y., Huang, X., Wang, S., & Qiu, R. (2012). Relative distribution of Pb2+ sorption mechanisms by sludge-derived biochar. *Water Research*, *46*(3), 854–862. https://doi.org/10.1016/j.watres.2011.11.058

Mohanty, P., Nanda, S., Pant, K. K., Naik, S., Kozinski, J. A., & Dalai, A. K. (2013). Evaluation of the physiochemical development of biochars obtained from pyrolysis of wheat straw, timothy grass and pinewood: Effects of heating rate. *Journal of Analytical and Applied Pyrolysis*, *104*, 485–493. https://doi.org/10.1016/j.jaap.2013.05.022

Moreno-Castilla, C. (2004). Adsorption of organic molecules from aqueous solutions on carbon materials. *Carbon*, *42*(1), 83–94. https://doi.org/10.1016/j.carbon.2003.09.022

Nguyen, M. V., & Lee, B. K. (2015). Removal of dimethyl sulfide from aqueous solution using cost-effective modified chicken manure biochar produced from slow pyrolysis. *Sustainability (Switzerland)*, *7*(11), 15057–15072. https://doi.org/10.3390/su71115057

Park, J., Hung, I., Gan, Z., Rojas, O. J., Lim, K. H., & Park, S. (2013). Activated carbon from biochar: Influence of its physicochemical properties on the sorption characteristics of phenanthrene. *Bioresource Technology*, *149*, 383–389. https://doi.org/10.1016/j.biortech.2013.09.085

Pignatello, J. J., Boyd, S. A., Johnston, C. T., Laird, D. A., Teppen, B. J., Li, H., Simpson, M. J., Simpson, A. J., Chen, C., Wang, Z., Ma, W., Ji, H., & Zhao, J. (2011). Fundamental biophysico-chemical processes of anthropogenic organic compounds in the interactions of anthropogenic organic. In *Biophysico-Chemical Processes of Anthropogenic Organic Compounds in Environmental Systems*, Environmental Chemistry: Wiley online library (pp. 1–50).

Qambrani, N. A., Rahman, M. M., Won, S., Shim, S., & Ra, C. (2017). Biochar properties and eco-friendly applications for climate change mitigation, waste management, and wastewater treatment: A review. *Renewable and Sustainable Energy Reviews*, *79*(February), 255–273. https://doi.org/10.1016/j.rser.2017.05.057

Regmi, P., Garcia Moscoso, J. L., Kumar, S., Cao, X., Mao, J., & Schafran, G. (2012). Removal of copper and cadmium from aqueous solution using switchgrass biochar produced via hydrothermal carbonization process. *Journal of Environmental Management*, *109*, 61–69. https://doi.org/10.1016/j.jenvman.2012.04.047

Ronsse, F., van Hecke, S., Dickinson, D., & Prins, W. (2013). Production and characterization of slow pyrolysis biochar: Influence of feedstock type and pyrolysis conditions. *GCB Bioenergy*, *5*(2), 104–115. https://doi.org/10.1111/gcbb.12018

Rosales, E., Meijide, J., Pazos, M., & Sanromán, M. A. (2017). Challenges and recent advances in biochar as low-cost biosorbent: From batch assays to continuous-flow systems. *Bioresource Technology*, *246*, 176–192. https://doi.org/10.1016/j.biortech.2017.06.084

Roy, E. D. (2017). Phosphorus recovery and recycling with ecological engineering: A review. *Ecological Engineering*, *98*, 213–227. https://doi.org/10.1016/j.ecoleng.2016.10.076

Spokas, K. A., Koskinen, W. C., Baker, J. M., & Reicosky, D. C. (2009). Impacts of woodchip biochar additions on greenhouse gas production and sorption/degradation of two herbicides in a Minnesota soil. *Chemosphere*, *77*(4), 574–581. https://doi.org/10.1016/j.chemosphere.2009.06.053

Sun, H., Hockaday, W. C., Masiello, C. A., & Zygourakis, K. (2012). Multiple controls on the chemical and physical structure of biochars. *Industrial and Engineering Chemistry Research*, *51*(9), 1587–1597. https://doi.org/10.1021/ie201309r

Sun, L., Wan, S., & Luo, W. (2013). Biochars prepared from anaerobic digestion residue, palm bark, and eucalyptus for adsorption of cationic methylene blue dye: Characterization, equilibrium, and kinetic studies. *Bioresource Technology*, *140*, 406–413. https://doi.org/10.1016/j.biortech.2013.04.116

Sun, Y., Gao, B., Yao, Y., Fang, J., Zhang, M., Zhou, Y., Chen, H., & Yang, L. (2014). Effects of feedstock type, production method, and pyrolysis temperature on biochar and hydrochar properties. *Chemical Engineering Journal*, *240*, 574–578. https://doi.org/10.1016/j.cej.2013.10.081

Sunyoto, N. M. S., Zhu, M., Zhang, Z., & Zhang, D. (2016). Effect of biochar addition on hydrogen and methane production in two-phase anaerobic digestion of aqueous carbohydrates food waste. *Bioresource Technology*, *219*, 29–36. https://doi.org/10.1016/j.biortech.2016.07.089

Tan, X., Liu, Y., Zeng, G., Wang, X., Hu, X., Gu, Y., & Yang, Z. (2015). Application of biochar for the removal of pollutants from aqueous solutions. *Chemosphere*, *125*, 70–85. https://doi.org/10.1016/j.chemosphere.2014.12.058

Torri, C., & Fabbri, D. (2014). Biochar enables anaerobic digestion of aqueous phase from intermediate pyrolysis of biomass. *Bioresource Technology*, *172*, 335–341. https://doi.org/10.1016/j.biortech.2014.09.021

Uchimiya, M., Ohno, T., & He, Z. (2013). Pyrolysis temperature-dependent release of dissolved organic carbon from plant, manure, and biorefinery wastes. *Journal of Analytical and Applied Pyrolysis*, *104*, 84–94. https://doi.org/10.1016/j.jaap.2013.09.003

Verheijen, F., Jeffery, S., Bastos, A. C., Van Der Velde, M., & Diafas, I. (2010). Biochar application to soils: A critical scientific review of effects on soil properties, processes and functions. *Environment*, *8*(4). https://doi.org/10.2788/472

Wang, J., & Wang, S. (2019). Preparation, modification and environmental application of biochar: A review. *Journal of Cleaner Production*, *227*, 1002–1022. https://doi.org/10.1016/j.jclepro.2019.04.282

Wang, K., Zhang, Y., Qi, R., Yang, M., & Deng, R. (2006). Effects of activated carbon surface chemistry and pore structure on adsorption of HAAs from water. *Huagong Xuebao/Journal of Chemical Industry and Engineering (China)*, *57*(7), 1659–1663.

Wang, Y., Lu, J., Wu, J., Liu, Q., Zhang, H., & Jin, S. (2015). Adsorptive removal of fluoroquinolone antibiotics using bamboo biochar. *Sustainability (Switzerland)*, *7*(9), 12947–12957. https://doi.org/10.3390/su70912947

Xu, R. kou, Xiao, S. cheng, Yuan, J. hua, & Zhao, A. zhen. (2011). Adsorption of methyl violet from aqueous solutions by the biochars derived from crop residues. *Bioresource Technology*, *102*(22), 10293–10298. https://doi.org/10.1016/j.biortech.2011.08.089

Xu, T., Lou, L., Luo, L., Cao, R., Duan, D., & Chen, Y. (2012). Effect of bamboo biochar on pentachlorophenol leachability and bioavailability in agricultural soil. *Science of the Total Environment*, *414*, 727–731. https://doi.org/10.1016/j.scitotenv.2011.11.005

Xu, X., Schierz, A., Xu, N., & Cao, X. (2016). Comparison of the characteristics and mechanisms of Hg(II) sorption by biochars and activated carbon. *Journal of Colloid and Interface Science*, *463*, 55–60. https://doi.org/10.1016/j.jcis.2015.10.003

Yang, Y., Lin, X., Wei, B., Zhao, Y., & Wang, J. (2014). Evaluation of adsorption potential of bamboo biochar for metal-complex dye: Equilibrium, kinetics and artificial neural network modeling. *International Journal of Environmental Science and Technology*, *11*(4), 1093–1100. https://doi.org/10.1007/s13762-013-0306-0

Yao, H., Lu, J., Wu, J., Lu, Z., Wilson, P. C., & Shen, Y. (2013). Adsorption of fluoroquinolone antibiotics by wastewater sludge biochar: Role of the sludge source. *Water, Air, and Soil Pollution*, *224*(1). https://doi.org/10.1007/s11270-012-1370-7

Yao, Y., Gao, B., Inyang, M., Zimmerman, A. R., Cao, X., Pullammanappallil, P., & Yang, L. (2011). Removal of phosphate from aqueous solution by biochar derived from anaerobically digested sugar beet tailings. *Journal of Hazardous Materials*, *190*(1–3), 501–507. https://doi.org/10.1016/j.jhazmat.2011.03.083

Zabaniotou, A., Stavropoulos, G., & Skoulou, V. (2008). Activated carbon from olive kernels in a two-stage process: Industrial improvement. *Bioresource Technology*, *99*(2), 320–326. https://doi.org/10.1016/j.biortech.2006.12.020

Zhang, Z. Bin, Cao, X. H., Liang, P., & Liu, Y. H. (2013). Adsorption of uranium from aqueous solution using biochar produced by hydrothermal carbonization. *Journal of Radioanalytical and Nuclear Chemistry*, *295*(2), 1201–1208. https://doi.org/10.1007/s10967-012-2017-2

Zhang, F., Wang, X., Yin, D., Peng, B., Tan, C., Liu, Y., Tan, X., & Wu, S. (2015). Efficiency and mechanisms of Cd removal from aqueous solution by biochar derived from water hyacinth (Eichornia crassipes). *Journal of Environmental Management*, *153*, 68–73. https://doi.org/10.1016/j.jenvman.2015.01.043

Zheng, H., Wang, Z., Zhao, J., Herbert, S., & Xing, B. (2013). Sorption of antibiotic sulfamethoxazole varies with biochars produced at different temperatures. *Environmental Pollution*, *181*, 60–67. https://doi.org/10.1016/j.envpol.2013.05.056

Zhou, Z., Shi, D., Qiu, Y., & Sheng, G. D. (2010). Sorptive domains of pine chars as probed by benzene and nitrobenzene. *Environmental Pollution*, *158*(1), 201–206. https://doi.org/10.1016/j.envpol.2009.07.020

Index

For Product Safety Concerns and Information please contact our EU
representative GPSR@taylorandfrancis.com
Taylor & Francis Verlag GmbH, Kaufingerstraße 24, 80331 München, Germany

www.ingramcontent.com/pod-product-compliance
Lightning Source LLC
Chambersburg PA
CBHW080927220326
41598CB00034B/5703

9 781032 066967